Sergey Stepanov
Relativistic World

Also of interest

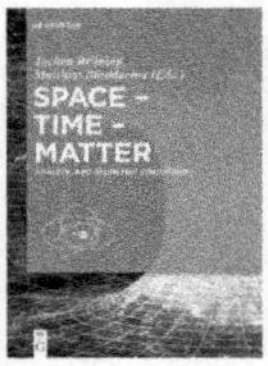

Space – Time – Matter
Analytic and Geometric Structures
Jochen Brüning, Matthias Staudacher, 2018
ISBN 978-3-11-045135-1, e-ISBN: 978-3-11-045215-0

Oscillatory Models in General Relativity
De Gruyter Studies in Mathematical Physics 41
Esra Russell, Oktay K.Pashaev, 2017
ISBN 978-3-11-051495-7, e-ISBN: 978-3-11-051536-7

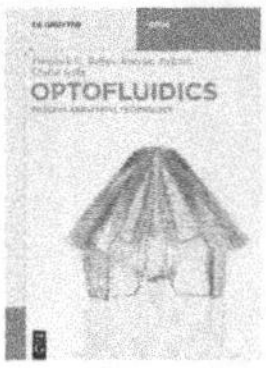

Optofluidics
Dominik G. Rabus, Cinzia Sada, Karsten Rebner, 2018
ISBN 978-3-11-054614-9, e-ISBN 978-3-11-054615-6

Computational Physics
With Worked Out Examples in FORTRAN and MATLAB
Michael Bestehorn, 2018
ISBN 978-3-11-051513-8, e-ISBN 978-3-11-051514-5

Zeitschrift für Naturforschung A
A Journal of Physical Sciences
Martin Holthaus (Editor-in-Chief)
ISSN 1865-7109

Sergey Stepanov

Relativistic World

Volume 1: Mechanics

DE GRUYTER

Mathematics Subject Classification 2010
35-02, 65-02, 65C30, 65C05, 65N35, 65N75, 65N80

Author
Dr. Sergey Stepanov
Dnipropetrovsk National University
Honchar, Dneropetrovsk
Ukraine
steps137@gmail.com

ISBN 978-3-11-051587-9
e-ISBN (E-BOOK) 978-3-11-051588-6
e-ISBN (EPUB) 978-3-11-051600-5

Library of Congress Control Number: 2018942831

Bibliografische Information der Deutschen Nationalbibliothek
The Deutsche Nationalbibliothek lists this publication in the Deutsche Nationalbibliografie; detailed bibliographic data are available on the Internet at http://dnb.dnb.de.

Printing and binding: CPI books GmbH, Leck
Coverimage: Maxiphoto / DigitalVision Vectors / getty images

www.degruyter.com

Contents

Introduction

The theory of relativity was first formulated more than a century ago. Since then, it has been confirmed several times by experiment and actually become an engineering science, where its implications affect calculations in the nuclear power industry, in the construction of particle accelerators, and even GPS-navigation. There is an extensive literature dedicated to its analysis and practical applications. Equally numerous are the counter-theories, whose authors are usually trying to find contradictions in the basic propositions or conclusions of the theory of relativity.

This naturally prompts the question: "Why do wee need yet another book on the theory of relativity?" We won't answer this directly, and suggest instead that the reader take a short test, expressing his or her agreement or disagreement with the following statements:

TEST	yes	no
1. The theory of relativity is a generalization of classical mechanics and thus requires more postulates.	□	□
2. The theory of relativity relies upon the analysis of light signals and the constancy of the speed of light.	□	□
3. Fast moving objects appear contracted in the direction of their motion.	□	□
4. The spokes of the wheels of a bicycle moving quickly past an observer are straight and equally spaced.	□	□
5. To explain the twin paradox one must introduce non-inertial reference frames.	□	□
6. The faster a body moves, the larger its mass becomes.	□	□
7. To explain the Mercury perihelion precession one must introduce curved space.	□	□
8. To describe non-inertial reference frames one must use the general theory of relativity.	□	□

For every "no" or "not necessarily" add one point to your score. In all other cases (that is, "yes" or "do not know") no points are awarded. If you scored more than 8 points, you have already done your homework, and this book may not teach you anything new. If the number of points you earned is *less* than 8, read on.

In the introduction we will give a brief explanation of the reasons for the negative answers to the first two questions of the test. In the process of further analysing the assumptions of the theory we will present more detailed arguments and consider a variety of other topics.

https://doi.org/10.1515/9783110515886-201

▷ Thought experiments involving light signals are often made use of in justifying the theory of relativity. The lack of invariance of Maxwell's equations with respect to Galilean transformations led to the discovery of the Lorentz transformations. There is no doubt that the equations of any *correctly* constructed theory of interactions (such as electromagnetism) reflect the general properties of space and time. Their invariance under certain transformations allows us to determine how the laws of physics appear in different frames of reference. However, within the deductive approach to physics it seems reasonable that one first establish general properties, on whose basis a particular theory can then be built. Thus, the Lorentz transformations are applicable to all types of interactions, and should not depend on the speed of electromagnetic waves, which is a parameter related to a specific interaction.

In our world, photons appear to have no mass. Nevertheless, none of the known physical principles prohibits the existence of a massive photon. This would not contradict the theory of relativity. It is possible to have a world where there are no massless particles in which the theory of relativity is still valid. Therefore, the speed of light "c_{em}" and the fundamental speed "c" from the formulas of the theory of relativity are two distinct constants. For photons with zero mass, they are numerically the same, but their physical meaning is different. The fundamental speed "c" is the maximum speed of motion possible for any material object, regardless of its nature. The speed of light "c_{em}" is an extremely important physical constant, but it is still specific to one particular interaction, and is related to the value of the photon mass.

In the derivation of the Lorentz transformations, a propagating light wave or light pulse is usually considered. Following Einstein, in numerous textbooks on the theory of relativity, these transformations are derived from the postulate of the constancy of the speed of light. However, only five years after the work of Einstein, it was shown by Ignatovsky and then by Frank and Rothe that the Lorentz transformations can be obtained on the basis of *a subset* of the postulates of classical mechanics. In other words, the theory of relativity does not require Einstein's second postulate for its justification and can be obtained from the principle of relativity, augmented by some natural assumptions. Despite the elegance of this result, it did not make its way into the general physics consciousness, although there are many methodological articles and other materials on the subject on the internet.

▷ You might ask: "Why does it matter how we derive the Lorentz transformations?" The answer to this question is twofold. First of all, a valid justification for the theory of relativity is essential simply because of human nature. No physical theory is exposed to so many attacks from would-be debunkers as the theory of relativity. This is because, on the one hand, of its mathematical simplicity, and on the other hand, of the psychological difficulty in accepting the initial postulate that "you cannot catch up with light." This runs counter to our everyday experience and thus casts a shadow of doubt on the predictions of the theory.

Most students just get used to this idea and do not see the problem, but some prospective physicists are not satisfied and join the dissenters. One of the aims of this book is to reduce their number, by at least a little.

Secondly, the lack of understanding of the logical foundations of physics impoverishes science. In the theory of relativity we have a great example of the fact that, to create a new theory, it is not always necessary to increase, and sometimes pays to *reduce*, the number of initial postulates. This deductive method is very inviting, and it can be used to build new, as yet unknown theories.

Physical and mathematical theories are based on initial postulates or axioms. A system of axioms is considered complete if any statement regarding the entities which the theory describes can be either proved or disproved. If one wants to build a more general theory about the same entities, one cannot simply add a new axiom. In a *complete* theory this axiom would either be a theorem (and the extended axiom set would not be independent), or it would contradict the existing axioms.

Generalization means that the old theory is included in the new one as a special case. In physics, this is reflected in *the correspondence principle*: in a more general theory there are fundamental constants which, when sent to certain specific values, lead to a more particular theory. So, most of the formulas of the theory of relativity turn into their classic counterparts as $c \to \infty$. This happens in quantum mechanics when $\hbar \to 0$. In this regard, the questions arise: What is the origin of these fundamental constants? Why do they not appear in classical mechanics, but only in relativity or quantum physics? The short answer is: the theory of relativity requires for its justification a smaller number of axioms than the classical theory. Reducing the number of incoming axioms results in the appearance of additional parameters (constants), whose values can be determined only by experiment. It's these parameters that become the fundamental constants.

Summary

Nature speaks to us in the language of mathematics. However, the mathematics should not obscure the physical significance of a theory. Therefore, in the first chapters of the book we assume of the reader only basic mathematical qualifications, and we will use only elementary tools.

Further on, the degree of abstraction gradually increases. Unfortunately, this is inevitable, and a deeper understanding of the structure of our world requires more and more sophisticated mathematical language.

Lest the more mathematically prepared readers be bored by the elementary mathematical explanations, we collect the basic information on vector and tensor analysis, special functions and coordinate systems, which may be required when reading the book, in the "*Mathematical appendix*". Despite its brevity, this annex is more than just a handbook, and for the attentive reader it may well serve as an introduction to a number of issues. It is also strongly recommended that one reads the book "*Vectors, tensors and forms. Instructions for use*" [1] and completes all of the exercises included therein. The reader is expected to have a working knowledge of vector calculus throughout this book.

The book "*Relativistic world*" consists of two parts. The first part is comprised of 12 chapters and covers issues at approximately the same level as general university courses in physics. The second part uses more advanced mathematical apparatus. It deals with field theory, quaternions, spinors, group theory, and other issues related to the theory of relativity.

A summary of the first part of the book, chapter by chapter, follows:

The first chapter discusses the logical foundations of the theory of relativity. We will discuss in detail the measurement procedures performed by different observers to agree upon units of length and time, and we give the "correct" derivation of the Lorentz transformations.

The second chapter is devoted to the methodological issues related to the axiomatic theory of relativity and its history. In particular, the principle of parametric incompleteness is formulated, allowing, at least in principle, for the creation of new physical theories. Also, the classical relativistic experiments are briefly outlined.

In *the third chapter* kinematic effects are studied in detail. The famous twin paradox is considered, along with the Doppler effect, relativistic aberration and the photographing of relativistic objects.

The fourth chapter is devoted to dynamics. Based only on the law of energy-momentum conservation, one can tell a lot about the character of an interaction between particles, even if the exact nature of the interaction is unknown.

The fifth chapter continues the investigation of dynamical problems. We will discuss the concepts of force and momentum, and analyze various examples of relativistic dynamic effects in detail.

The sixth chapter is devoted to non-inertial frames of reference. Contrary to popular belief, these systems can be analyzed without resorting to Einstein's theory of gravity. We consider the physical phenomena occurring in uniformly accelerated and rotating frames of reference, and discuss some of the related "paradoxes".

The seventh chapter is concerned with covariant formalism in the theory of relativity. This elegant mathematical toolset allows many expressions to be re-written in a very compact form.

In *the eighth chapter* this covariant formalism is applied to dynamical problems. We describe particle collisions, as well as the angular momentum and spin of a classical particle.

The ninth chapter continues with the theme of non-inertial reference frames, now from a more formal mathematical perspective. Curvilinear coordinates are introduced, as well as methods of defining physical quantities in these coordinates.

In *the tenth chapter* we examine the specific effects of the theory of relativity related to non-uniform motion and rotation. The equations of motion of an accelerated rod and a rotating gyroscope will be derived. Also discussed are the peculiarities of the conservation of angular momentum in the theory of relativity.

In *Chapter 11* we introduce a new mathematical object - the quaternion. Quaternions are a convenient tool for describing rotations and compositions of Lorentz transformations. The development of quaternion mathematics in the next volume will lead us to spinors, which are very general mathematical objects.

Chapter 12 is devoted to spaces of constant curvature. Extensive mathematical material is followed by a simple physical statement - the space of velocities in the theory of relativity has hyperbolic geometry.

▷ The following chapters will give the reader a basic understanding of the theory of relativity:

$$1 \mapsto 3 \mapsto 4 \mapsto 5 \mapsto 6.$$

The tenth chapter can be skipped on the first read, and the same can be said about Chapter 9, which is devoted to non-inertial reference frames.

Reading this book will be beneficial to readers with different levels of mathematical skill. In all cases where a result can be obtained by elementary methods, we try to do this. However, some issues require more abstract mathematical tools. Such advanced sections are marked with an asterisk (or two), and during the first reading, the reader can skip them, and return to them later on.

Throughout the text of the books are scattered many small tasks, marked with a stylized eye: ($\lessdot$ H_i), where i is the number of the solution in the appendix "*Help*". Also, there are references of the form ($\lessdot$ C_i), which the reader should follow if they have questions about the adjacent text. The answer could, perhaps, be found in the appendix "*Comments*" under the number i.

The author thanks the world around him, in which he has been fortunate to encounter scores of great people and wonderful books. All of them, in one way or another, have had an impact on this manuscript and are its rightful co-authors. All of the errors or omissions, however, are the complete and utter responsibility of the author.

Personally, the author thanks Michael Efroimsky, without whom this book would hardly have been published and Alex Zaslavsky, Oleg Orlyansky for numerous discussions. Special thanks to the translator Vladimir Slepkov.

Notation

a, **b**, **c**, ... — 3-dimensional vectors (bold font);
$\mathbf{a} \times \mathbf{b}$ — the vector product;
r — radius-vector with Cartesian coordinates x, y, z;
n — the unit vector ($\mathbf{n}^2 = 1$);
c — the speed of light;

$\gamma = 1/\sqrt{1-\mathbf{v}^2}$ — Lorentz-factor for the velocity **v**;
$\Gamma = (\gamma - 1)/v^2 = \gamma^2/(\gamma + 1)$ — the modified Lorentz-factor;

E, **p**, m — particle energy, momentum and mass;
L, **S** — angular momentum and spin;

s_α, c_α — abbreviations for $\sin\alpha$ and $\cos\alpha$;
$\operatorname{sh} x$, $\operatorname{ch} x$, $\operatorname{th} x$ — hyperbolic sine, cosine and tangent functions;
$\operatorname{ash} x$, $\operatorname{ach} x$, $\operatorname{ath} x$ — inverse hyperbolic functions;

$i, j, k, \ldots$ — the Latin indices run from 1 to 3;
$\alpha, \beta, \gamma, \ldots$ — the Greek indices run from 0 to 3;
$ds^2 = dt^2 - d\mathbf{r}^2$ — 4-dimensional interval;
A^α, $F^{\alpha\beta}$ — 4-vectors and tensors;
A, F — 4-vectors and tensors in index-free form;
δ_{ij}, δ^α_β — 3- and 4-dimensional Kronecker symbols;
$g_{\alpha\beta}$ — the metric tensor, $g_{\alpha\beta} = \operatorname{diag}\{1, -1, -1, -1\}$;
ε_{ijk}, $\varepsilon_{\alpha\beta\gamma\delta}$ — 3- and 4-dimensional Levi-Civita symbols, $\varepsilon_{0123} = 1$;

Restoring the constant *c*

We use units with $c = 1$. If it is necessary to restore the constant "c", then quantities containing the time taken to some power are multiplied by "c" to the same power. For the time (t), velocity (u) and acceleration (a) the following substitutions are performed:

$$t \mapsto c\,t, \qquad u \mapsto \frac{u}{c}, \qquad a \mapsto \frac{a}{c^2}.$$

For momentum (**p**), energy (E) and force (**F**):

$$\mathbf{p} \mapsto \frac{\mathbf{p}}{c}, \quad E \mapsto \frac{E}{c^2}, \quad \mathbf{F} \mapsto \frac{\mathbf{F}}{c^2},$$

and so on.

1 The Lorentz Transformations

This chapter considers the initial statements of the theory of relativity. We start from a discussion of the measurement procedures which must be performed in order for observers to agree upon units of length and time. Observers fixed relative to each other are considered, as well as observers located in different inertial frames of reference.

The key section of the chapter is "The Lorentz transformations". The derivation of these transformations used here is not well known, though it has already been known for over 100 years. The velocity-addition law will be derived with the help of the Lorentz transformations and the meaning of the fundamental constant c will be established. To make the picture complete, the traditional derivation is also considered; it is based on the postulate (in fact, a theorem) that the velocity of light is constant.

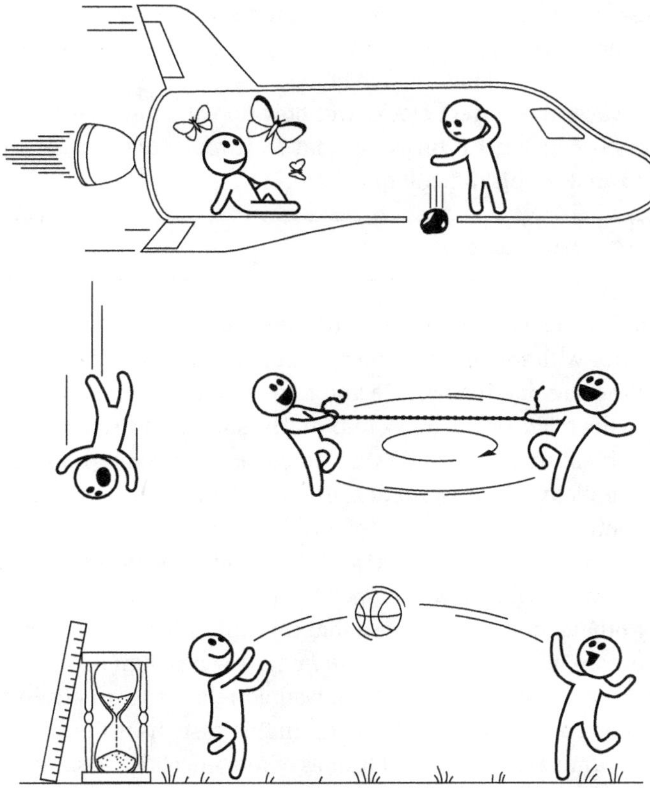

https://doi.org/10.1515/9783110515886-001

1.1 Stationary observers

Physical processes take place in space and time. Defining these terms necessitates determining measurement procedures that allow one to move from feelings to numbers. This is not easy and requires a number of explicit or implicit assumptions.

We will have to speak constantly of observers, active participants of the physical theory. Despite the surrounding world being objective, physics is ultimately being created to explain human perception, and to decrease its "subjectivity". So, the appearance of such projections of our "self" is more than likely unavoidable. Certainly, they may also be represented by some physical instruments or artificially constructed intelligent creatures.

▷ If a person did not have a memory, time would not exist for them. The perception of time is formed by our memory registering surrounding changes. These same changes are used to measure time changing. Periodic processes, the time durations of which are considered the same, are convenient for this purpose. People have been searching for the "correct" clock for all of human history.

What is however the correct clock, and how do we know that the duration of a periodic process is constant? It turns out that *the simplicity principle* [2] is the most practical way to answer this difficult question:

! Time is defined so that motion looks simple.

For example, it is not usually common to consider models explaining the complicated behavior of bodies with reference to the non-uniform rotation of the Earth. It is simpler to declare the sun dial inadequate and to search for another, more suitable uniform process (in the microworld, for example). The simplicity principle is a very strong statement, which assumes implicitly that time itself has some properties of uniformity, and the simplicity of our world description is "adjusted" to these properties for all kinds of motion.

The motion of a particle (an object) with constant velocity is the simplest kind of motion. Due to Newton's first law, it happens that when an object is moved far enough away from any others, one may suppose that this object feels no external influences. This *uniform* motion must agree with the *uniform* passage of time.

Besides this, the object's velocity is measured relative to an observer who must himself be unaffected by external influences. In this case, he is the *inertial* one. Newton's first law is formulated for inertial frames of reference. At the same time, it is also the *definition* of inertial frames of reference.

Non-uniformity of motion may occur because of an "inadequate" clock, non-inertiality of the observer, or some force affecting the object. Hence, time, motion, and the frame of reference are three sides of the same coin.

▷ We will not specify the clock mechanism used for time measurement. It would be convenient to declare it to be an atomic one; then we would remove the question about the agreement of the time units between the observers. Nevertheless, in this case we also have to use ***the identity principle*** for micro-objects, which is a postulate from another theory. A "mixture" of theories like this is not helpful from the axiomatic point of view. So, we will assume that the mechanisms of the different observers' clocks may differ. However, they are such that free bodies move with constant velocity relative to free (inertial) observers.

In what follows, we will consider events taking place at a given *moment* of time and at some *point* in space. Thus, clocks are supposed to be point-like and to be able to "tick" infinitely quickly. Of course, we preserve our previous assumption about having a uniform clock rate. This is an idealization, and we must accept that the concept of a "moment in time" will fail at the microscopic level. Nevertheless, this problem is ignored from here on. We also do not take into account any influence that making measurements might have on the objects (as usually happens in the quantum world). Finally, we ignore any converse influences on the observer during his interaction with the objects. For example, the observers may change the velocities of the objects and stay at "the same" point in space.

▷ By definition, spatial relations may be measured if there exists an etalon with constant length. As we will see, hard rulers will only give us an approximation. This creates a certain difficulty but we are not going to go into that now. Let us assume for now that rulers are always fixed, constant, and placed in close proximity to an observer.

With the ruler and clock, it is possible to measure the velocity and acceleration of a moving object. In order to measure the velocity it is necessary to perform two time measurements at two spatial points,

$$u = \frac{x_2 - x_1}{t_2 - t_1} = \frac{\Delta x}{\Delta t}.$$

Acceleration requires three such measurements, and so on. These measurements may also be performed with the help of one clock. For example, one can place a clock between two detectors that send a signal to it with *equal* velocities when an object flies past them. However, the establishment of the velocities' equality requires additional measuring procedures. From the mathematical point of view, velocity is a quantity that is defined at a given spatial point, and which appears as a result of a limit operation. It is one more idealization without which we cannot say anything about the uniformity of motion. In any case, we will assume that observers can measure the velocity of each object in their vicinity.

Considering several observers within one frame of reference is well established in cosmology. Without appealing to theological arguments, one can hardly speak about a single observer "spread over" the whole Universe. In the world of high velocities we will also consider observers in remote parts of space.

▷ Let us imagine that space is "filled" with plenty of observers. Some of them are fixed relative to each other and create *the frame of reference.* If the velocities of objects that are not affected by any external influences are constant relative to the observers, such a frame of reference is called an *inertial* frame.

! Due to the existence of the observers, a ruler and clock are placed at each point of the frame of reference.

In order to obtain a unified picture of the events occurring in space, the observers have to agree on their units of measurement. Let us start with the units of velocity. Let us suppose that in sequence, each of the observers measures the velocity of an object moving uniformly past them. The observers may agree that this velocity is equal, for example, to 1 m/sec. As a result, a common velocity unit will be obtained. Hence, an object that is not moving relative to the observer has zero velocity, by definition. It is assumed that in similar experiments where the objects have different velocities, the results of all observers will coincide. If this were not the case, it would be necessary either to reconsider the methods of length and time measurement or to seek the reason for the difference in the velocities. Now, imagine that the observers have agreed on a unit of velocity. How can they fix length and time etalons separately? It is possible to exchange copies of the etalons and to move them in space. But we would like to avoid such procedures, because the devices may potentially be deformed when their velocities change during this movement.

The simplest way to fix time units within mechanics is to send objects with the same velocities from one observer to another with some periodicity. The period of such a transfer may be set as the time unit (the left figure below):

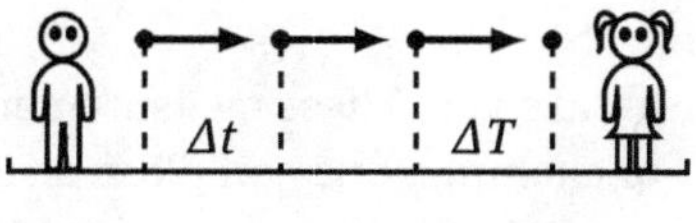

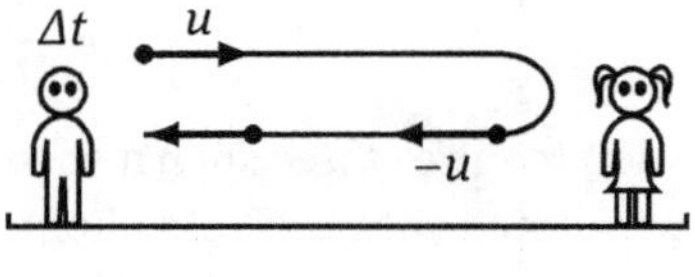

Once velocity and time units have been agreed upond, the observers obtain the same length units, too. Of course, they must ensure that they are in one frame of reference, so that the distance between them is constant. For this purpose, the "*radio-locating method*" may be used. The first observer sends an object towards the second one with constant velocity u (the right-hand figure above). The second observer, having received it, sends the object back with the *same* velocity (with the opposite sign). The first observer measures a time interval Δt between dispatching and receiving the object. If Δt does not change as this experiment is repeated, the distance between the observers is considered constant.

It is worth noting that the *isotropy* of space (that is, the equivalence of all directions) is not used in these experiments, though inertiality of the system is implied. The velocity of the object being sent back is controlled by both the first and the second observers. Thus, an additional hypothesis about the equality of the velocities going "forward" and "backward" is not necessary. Only their constancy is assumed.

It is possible to perform the same experiment "in reverse", having the second observer start, and the first one bounce the object back. The second observer must, (*by agreement*), obtain the same time duration

$$\Delta T = \Delta t$$

when sending the object as receiving it back. This is another way to agree upon time units that must yield the same results as the periodic sending of objects.

▷ Along with the length and time units, the observers must synchronize the time origin. For this purpose, it is possible to perform a radio-locating experiment which measures not only the time interval but also the absolute time values of events. Imagine that the object is sent at time t_1 by the first observer's clock. Having speed u, it will reach the second observer at time T (by his local clock). The object is returned back with the same velocity and arrives at the first observer at t_2. By the definition of velocity, and assuming that the observers are stationary relative to each other, the duration of the object's motion in each direction must be the same. So the event of reflection must be between the events of sending and receiving the object. The relation

$$T = \frac{t_1 + t_2}{2}$$

allows the observers to choose a time origin in the same way:

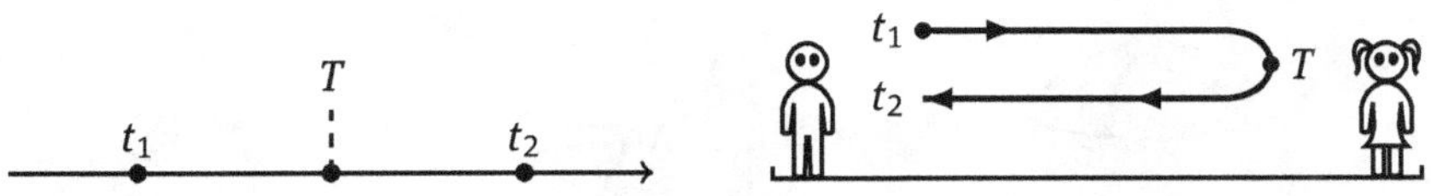

We should note that the *value* of the velocity u of the object used for clock synchronization ***does not matter***. According to Einstein, light signals are used for such a procedure. But if the observers can measure the velocities, this is not obligatory. Moreover, all of the described procedures only make sense if they yield the same result for an *arbitrary* object velocity.

It is assumed that the procedures for length and time unit agreement and the time origin synchronization procedure have a ***transitivity property***. This means the following. If observer A has calibrated their instruments to agree with those of B and B has reached the same agreement with C, then A and C are also in agreement. As a result, in a given frame of reference, all three observers have common units of time, length and velocity.

1.2 The properties of space

The observers each need a ruler to measure objects' velocities. We assume that they are not large (an observer is thought to be point-like in a given region of space). Thus, these rulers are not intended to measure distances to remote events and observers. It is possible to use the "radio-location method" to measure large distances (p. 4), when some object is being sent and returned back at a constant velocity u. Its magnitude does not matter, but must be known. If the total forward and backward journey takes time Δt, then the distance that passes as one direction is traversed equals, by definition, $L = u\,\Delta t/2$. Its constancy indicates that the observers are fixed relative to each other.

Different pairs of observers may measure the distances between each other in a similar way. A set of these distances is not arbitrary and is defined by the geometrical properties of space.

! The *dimensionality* of space is its key property.

Let us consider the set of points that are equidistant from a given one. It is a *surface* (a sphere) and, by definition, it has one dimension fewer than the whole space. It is possible to choose a fixed point on this surface and to consider the set of points *on the surface* that are equidistant from this central one. This yields a circle, which in its turn has one dimension fewer than the sphere. Usually, we assume that such a procedure eventually comes to an end even on a microscopic level, so that the final set of points is finite. The number of iterations performed is the number of dimensions of space (the first figure below):

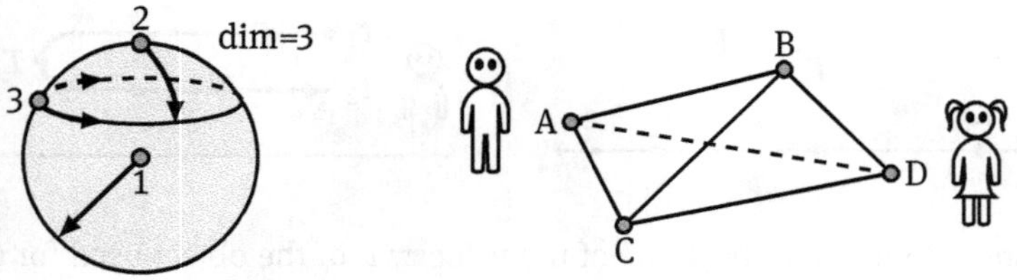

It is postulated that such a measurement of the dimensionality of space will yield the same results at different points in space. In other words, its dimensionality is a universal property of space. As we know, our space is three-dimensional.

There are other methods, apart from the one just described, by which it is possible to perform a simultaneous study of the geometry and dimensionality of space. For example, if there are four points on a Euclidean plane (2D space) that do not lie along a straight line, then the distances between them cannot be arbitrary. Above, the distance AD depends on the other five distances. However, if these points do not lie in a plane, there is no such connection, and all of the distances turn out to be independent.

The second important property of empty space is its *uniformity*. !

Any experiments held at different points in space (by different observers) must yield the same results.

The third property of space is *isotropy*. !

Measuring an object's velocity, observers obtain not only its magnitude, but also its direction (it is a vector). We will consider the case that there are no "preferred directions" in space, but rather that all directions are equivalent.

Because of the isotropy of space, a circle may be divided into an equal number of sectors and the concept of an angle may be introduced. The complete circle is taken to be 2π radians. Angles and distances allow for the study of the geometric properties of space. For example, if the sum of the angles in a triangle formed by three observers equals 2π, then the space they inhabit is most likely ***Euclidean***.

In fact, a uniform and isotropic space may only be of one of three geometry types, which differ in the sign of their ***space curvature***. If the sum of the angles in a triangle is larger than 2π, then the space has positive curvature; if it is smaller than 2π, then the space has negative curvature. Due to the uniformity and isotropy of space, triangles with equal sides will exhibit the same deviation from 2π, so that the curvature is constant in different directions and at different points in space. A concrete example of a uniform isotropic space with constant positive curvature is the 2D space on the surface of a sphere. In the final chapter we will consider spaces with constant curvature in more detail. We will see that the velocity space of special relativity is a space of constant negative curvature.

In the first book we will consider our space to be three-dimensional and Euclidean. In the second, the more general case will be discussed. In Euclidean space, a global coordinate system may be introduced. To do this, one of the observers is declared to be situated at the "origin" of the frame of reference. A triplet of numbers $\{x, y, z\}$ is associated with each observer. These numbers are called ***the coordinates***, and define the observer's position in space. They also "enumerate" the observers (or the "space points" where they are situated). In Euclidean space, this enumeration may be chosen in such a way that the squared distance between any two points with the coordinates $\{x_1, y_1, z_1\}$ and $\{x_2, y_2, z_2\}$ will be equal to $L^2 = (x_2 - x_1)^2 + (y_2 - y_1)^2 + (z_2 - z_1)^2$.

The values of the coordinates at the origin of the frame of reference are taken to be zero. Adding the moment in time to the coordinates, $\{t, x, y, z\}$, allows for the complete identification of any ***event***, taking place at a given instant t and at a given point in space $\{x, y, z\}$. When we study relativistic effects, what we usually consider are events. Actually, any event is always a process. However its duration for any observer is negligibly small.

1.3 Inertial frames of reference

Let us now turn to observers in different inertial frames of reference. For simplicity we will consider two of these systems, S and S', the latter moving with velocity v relative to the former. We are interested in the connection between the time and coordinates of some event which is observed from both frames of reference. Let us discuss the 1-dimensional case. In what follows, $\{t, x\}$ stands for the time and the coordinate of the event for an observer in S, and $\{t', x'\}$ represents the same for S'. If they are connected, then there must exist a functional dependence:

$$t' = f(t, x, v), \qquad x' = g(t, x, v). \tag{1.1}$$

The existence of this dependence is an axiom of the theory. This may be considered as an "Axiom of World Knowability".

! The results of observations performed in different systems must be connected to each other.

It is assumed that the functions $f(t, x, v)$ and $g(t, x, v)$ depend on the relative velocity v of the systems. This connection might also include, for example, the temperature of each of the observers or their heights. However, in kinematics, we believe that the time and the coordinate of an event are enough for its full description, and the relative velocity of two inertial frames of reference is the only parameter that makes them different.

In each system, all observers are fixed relative to each other. Characterizing the speed of S' relative to S with one number v, we suppose that all of the observers in S' are moving past S with the same velocity v. All observers in S move, in relation to the system S', at the same speed in the opposite direction. To make the results of the observations consistent, the observers in the different frames of reference (as well as in a single one) must agree upon their length and time units.

Let us start with the velocity units. The representatives of two systems of reference may agree to assume their ***relative velocity*** v to be the same. If the systems' axes are co-directional, then the velocity of S' will be equal to v for an observer in S (the left figure). Similarly, the velocity of S will be $-v$ for an observer in S' (the right figure).

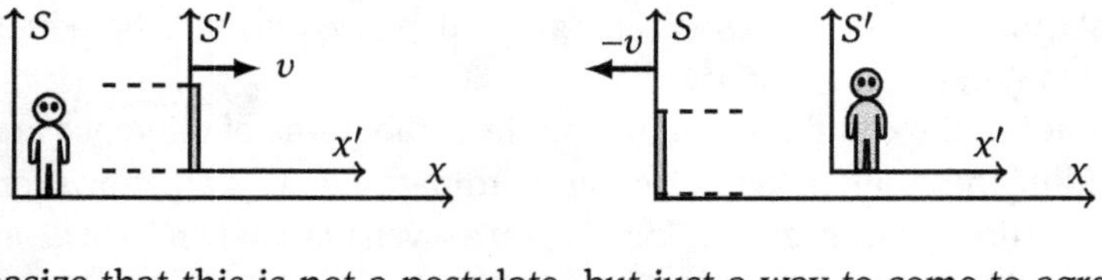

We emphasize that this is not a postulate, but just a way to come to agreement on the velocity units. At the same time it also contains an idea about the equivalence of inertial frames of reference.

▷ To agree on a unit of length, the observers may place their rulers perpendicularly to the direction of motion (along the y-axis) and ensure that they coincide. Figuratively speaking, this means that the observer in the system, say, S' is flying along a fence placed in the system S and draws two one-metre-high lines on it, which are parallel to the x-axis. In this way he gives information about his length unit.

Certainly, it is possible to use two particles flying along the x-axis instead of the fence. The shortest distance between their trajectories in both systems is considered unitary. It is worth noting that such a procedure is possible only in a space with more than one dimension. Hence the chosen orientation of the coordinate axes requires the extension of the transformations (1.1) by the additional relations

$$y' = y, \qquad z' = z.$$

If the length units perpendicular to the motion have been fixed, the observers may consider any other lengths to be fixed, too. This is a consequence of the *isotropy* of space in each frame of reference and the possibility of the "slow rotation" of the rulers without their deformation.

Agreeing on velocity and length units yields a mutual time unit. One may fix the time axis origin to be some event, for example, that at which the origins of the reference frames coincide, ($x = x' = 0$), assuming that at this moment $t = t' = 0$,

$$f(0, 0, v) = g(0, 0, v) = 0.$$

▷ Now, let us formulate an important property of the functions $f(t, x, v)$ and $g(t, x, v)$. If the equations (1.1) are solved with respect to t and x, an inverse relation (from the primed quantities to the non-primed ones) is obtained. Here, the functions turn out to be *the same*, while the velocity changes its sign,

$$\begin{cases} t = f(t', x', -v) \\ x = g(t', x', -v). \end{cases} \tag{1.2}$$

This is an important requirement which has a profound meaning. The initial transformations (1.1) may be interpreted from the point of view of an observer in the system S. They are measuring the coordinates and the time of some event $\{t, x\}$. With the help of the transformations they "figure out" the values of the same measurement in the system S' moving with the velocity v with respect to them.

The inverse transformations solve the same problem from the point of view of the observer in S'. However, the velocity of the system S equals "$-v$" for them, since the x- and x'-axes point in the same direction. So, the primed and unprimed quantities are swapped, and the replacement

$$v \longmapsto -v$$

is performed. As for the rest, the functional form of the transformations must be equivalent. This is a reflection of the principle of relativity .

▷ Galilean transformations are examples of transformations between two inertial frames of reference. In classical mechanics, it is assumed that time has the same tempo for all observers:

$$t' = t.$$

Let two frames of reference be situated so that their x-axes are parallel each to other and the origins of the systems coincide at time $t = t' = 0$.

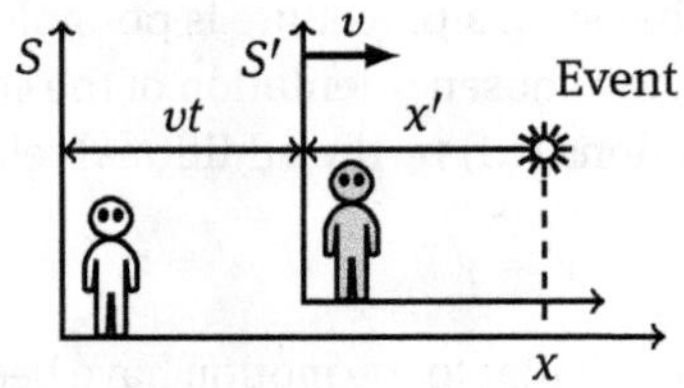

Then the coordinates and the time of some event observed from each system are related as follows,

$$\begin{cases} t' = t, \\ x' = x - vt. \end{cases} \tag{1.3}$$

Galileo did not introduce such transformations, they appeared much later. He is however an author of ***the principle of relativity***, formulated in the book "Dialogue Concerning the Two Chief World Systems: Ptolemaic and Copernican" (1632). We reproduce his formulation here, with some reduction:

> Shut yourself up with a friend in the main cabin below the decks of a large ship, and have with you there some flies, butterflies, and other small flying animals. Have a large bowl of water with some fish in it; hang up a bottle that empties drop by drop into a wide vessel beneath it. With the ship standing still, observe carefully how the little animals fly with equal speed to all sides of the cabin. The fish swim indifferently in all directions; the drops fall into the vessel beneath; and, in throwing something to your friend, you need throw it no more strongly in one direction than another, the distances being equal; jumping with your feet together, you pass equal spaces in every direction.
> When you have observed all these things carefully (though there is no doubt that when the ship is standing still everything must happen in this way), have the ship proceed with any speed you like, so long as the motion is uniform and not fluctuating this way and that. You will discover not the least change in all the effects named, nor could you tell from any of them whether the ship was moving or standing still.

It is usually said that Galileo's relativity principle is formulated for mechanical systems only, while Einstein extended it to cover all physical phenomena. This is not completely true. As we see, Galileo stated that in inertial frames of reference, all phenomena (even biological ones) pass in the same way. He did not divide physics into mechanics and other phenomena. Moreover, it seems he had no concept of physics. He was just thinking about the nature of the world.

▷ Transformations between frames of reference must be unambiguous, so that given values t, x give unique values t', x' always. Besides this, they should fulfil *the group properties*. This means that unity and inverse transformations exist, and the composition (sequence) of any two transformations is also a transformation. Let us illustrate this with the help of Galilean transformations.

The unity transformation corresponds to $v = 0$. In this case $t' = t$ and $x' = x$, so that the two frames of reference coincide. It is easy to write *the inverse transformation* if one inverts the equations (1.3):

$$\begin{cases} t = t', \\ x = x' + v t'. \end{cases}$$

As a result new Galilean transformations are obtained, as should be the case, but with the replacement

$$v \mapsto -v.$$

In order to define *the composition of transformations* let us consider three inertial frames of reference S_1, S_2, and S_3. Let S_2 move with velocity v_1 relative to S_1 and let S_3 move with velocity v_2 relative to S_2. Finally, the third system moves with some velocity v_3 relative to S_1:

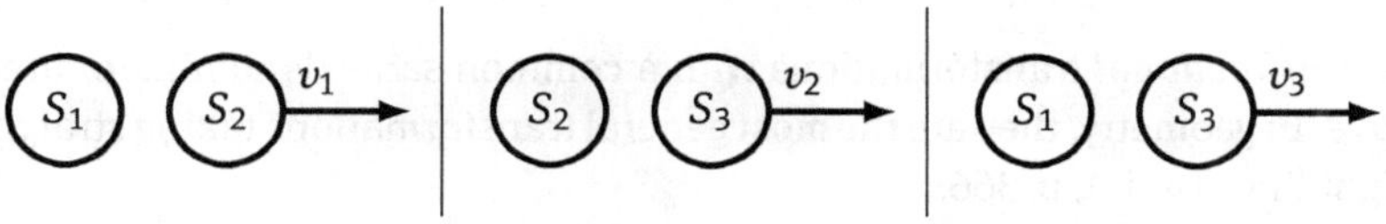

The Galilean transformations between the pairs of systems take the form:

$$\begin{cases} t_2 = t_1, \\ x_2 = x_1 - v_1 t_1, \end{cases} \qquad \begin{cases} t_3 = t_2, \\ x_3 = x_2 - v_2 t_2, \end{cases} \qquad \begin{cases} t_3 = t_1, \\ x_3 = x_1 - v_3 t_1. \end{cases}$$

All of these relations are the same due to the equivalence of all of the frames of reference. The relative velocity is the only value distinguishing them.

Substituting the first system into the second one and comparing it to the third one, it is easy to see that the velocity v_3 is not arbitrary. It is connected with v_1 and v_2 as follows:

$$v_3 = v_1 + v_2. \tag{1.4}$$

This simple velocity-addition rule is a result of the Galilean transformations. It may also be obtained by considering two systems and some object moving relative to them. In the above calculations, the third frame of reference acted as such an object.

The group axioms are very strong restrictions. With the help of some additional assumptions, they allow us to find an explicit form for the functions $f(t, x, v)$ and $g(t, x, v)$ in some generality.

1.4 The Lorentz transformations

Our current aim is to establish an explicit form for the functions f and g connecting the observation results for some event in two inertial frames of reference S and S',

$$t' = f(t, x, v), \qquad x' = g(t, x, v). \tag{1.5}$$

In order to do this, let us define a set of axioms that should hold for these functions. We assume that the transformations (1.5) are continuous, differentiable, and unambiguous. This mathematical requirement is very natural. Although it *narrows* the class of possible functions, it is still left more than wide enough.

Let us choose a definition of inertial frames of reference as the first physical axiom:

Axiom I.
If a particle moves steadily in a system S, then its motion will be steady any other system S', too, and vice versa.

In spite of the rather general character of this statement, it fixes the functional dependence of the transformations on the coordinates and time completely,

$$t' = \frac{A(v)\,t + B(v)\,x}{1 + a(v)\,t + b(v)\,x}, \qquad x' = \frac{D(v)\,x + E(v)\,t}{1 + a(v)\,t + b(v)\,x}. \tag{1.6}$$

Such *linear-fractional* transformations with a common same denominator are called *projective*. In geometry, they are the most general transformations taking straight lines to straight lines (< H_1), p. 366:

$$x' = x'_0 + u'\,t'.$$

The next axiom is:

Axiom II.
If the velocities of two free particles are equal in a system S, then they will also be equal in any other system S'.

This leads to the requirement that the coordinates and time transformations must be given by the linear functions (< H_2):

$$t' = A(v)\,t + B(v)\,x, \qquad x' = D(v)\,x + E(v)\,t, \tag{1.7}$$

where the coefficients A, B, D, E depend on the relative velocity v of the frames of reference, but do not depend on t and x. The derivation of the projective transformations (< H_3), p.363 may be omitted, assuming that linearity (1.7), which is a result of the first two axioms, is itself "natural".

Let us note that in (1.6), (1.7), the time origin is chosen so that the origins of the systems coincide at $t = t' = 0$, $x = x' = 0$.

▷ Appealing to the procedure for agreeing upon velocity units, we suppose that the point $x' = 0$ of the S' system moves along the trajectory $x = vt$ relative to S. Substituting $x' = 0$, $x = vt$ into the second equation (1.7), we obtain $E = -vD$. Similarly, $x = 0$ and $x' = -vt'$ yield $E = -vA$, or $A = D$. Let us introduce two functions of the velocity, $A = D = \gamma(v)$ and $\sigma(v) = -B/A$. In terms of these, the transformations may be rewritten as follows:

$$\begin{cases} t' = \gamma(v)\,[\,t - \sigma(v)\,x\,], \\ x' = \gamma(v)\,[\,x - vt\,]. \end{cases} \tag{1.8}$$

The third axiom expresses *the relativity principle*:

Axiom III.
All inertial frames of reference are equivalent.

Consider three frames of reference, S_1, S_2, and S_3. Let S_2 move with velocity v_1 relative to S_1 and S_3 move with velocity v_2 relative to S_2:

Let t_1 and x_1 stand for the time and coordinate of an event observed in S_1, and similarly for S_2 and S_3. The transformations are:

$$\begin{cases} t_2 = \gamma_1\,[t_1 - \sigma_1\,x_1], \\ x_2 = \gamma_1\,[x_1 - v_1\,t_1\,], \end{cases} \qquad \begin{cases} t_3 = \gamma_2\,[t_2 - \sigma_2\,x_2], \\ x_3 = \gamma_2\,[x_2 - v_2\,t_2\,], \end{cases}$$

where $\gamma_1 = \gamma(v_1)$, $\sigma_1 = \sigma(v_1)$, etc. Now we can substitute $\{t_2, x_2\}$ from the first system into the second one:

$$\begin{cases} t_3 = \gamma_1\gamma_2\,[(1 + v_1\sigma_2)\,t_1 - (\sigma_1 + \sigma_2)\,x_1] = \gamma_3\,[t_1 - \sigma_3\,x_1], \\ x_3 = \gamma_1\gamma_2\,[(1 + v_2\sigma_1)\,x_1 - (v_1 + v_2)\,t_1\,] = \gamma_3\,[x_1 - v_3\,t_1]. \end{cases}$$

The second equalities in the equations give the transformation between the systems S_1 and S_3 for a velocity v_3. These equations must hold at arbitrary t_1 and x_1. Let us equate the coefficients of t_1 in the first equation of the system and those of x_1 in the second equation. Then

$$\begin{cases} \gamma_3 = (1 + v_1\,\sigma_2)\,\gamma_1\gamma_2, \\ \gamma_3 = (1 + v_2\,\sigma_1)\,\gamma_1\gamma_2, \end{cases} \tag{1.9}$$

which yields $1 + v_1\,\sigma_2 = 1 + v_2\,\sigma_1$, or

$$\frac{\sigma(v_1)}{v_1} = \frac{\sigma(v_2)}{v_2} = \alpha = const. \tag{1.10}$$

Since the velocities v_1 and v_2 are arbitrary *independent* quantities, α is some constant that is the same for all inertial frames of reference.

▷ The equivalence of different systems means also that the transformation from S to S' (1.8) is the same as that from S' to S. In other words, the inverse transformation must coincide with the direct one, up to the replacement $v \mapsto -v$. For example, for the coordinate [see the 2nd equation (1.8)]

$$x = \gamma(-v)\,[\,x' + v\,t'\,] = \gamma(-v)\,\gamma(v)\,[\,1 - \alpha\,v^2\,]\,x,$$

where in the second equality, t', x' are substituted from the direct transformation (1.8) and we have taken into account that $\sigma(v) = \alpha\,v$. As a result,

$$\gamma(-v)\,\gamma(v) = \frac{1}{1 - \alpha\,v^2}. \tag{1.11}$$

Finally, in order to define $\gamma(v)$, we need one more axiom.

Axiom IV.
In inertial frames of reference, space is isotropic.

This means that the transformation (1.8) should not change under inversion of the system axes, i.e. $x \mapsto -x$ and $x' \mapsto -x'$. Under such an inversion, the velocity changes its sign, $v \mapsto -v$, hence for (1.8),

$$-x' = \gamma(-v)\,[-x + v\,t].$$

This equality turns into (1.8) again only if $\gamma(v)$ is an even function of the velocity,

$$\gamma(-v) = \gamma(v).$$

This allows us to to find that

$$\gamma(v) = \frac{1}{\sqrt{1 - \alpha\,v^2}}.$$

The positive sign of the square in (1.11) is chosen so that one obtains the identity transformations at zero velocity, i.e. so that $\gamma(0) = 1$.

In Chapter 3, the effect of a rod contraction is considered, in the case that its beginning and end coordinates are measured simultaneously ($\Delta t = 0$). The result of this measurement $\Delta x' = \gamma(v)\,\Delta x$ must not depend on the velocity direction. This observation also yields the evenness of the $\gamma(v)$ function.

The numerical value of α is constant and its sign cannot be fixed without additional axioms or experiments. In fact, three theories are possible, where $\alpha > 0$, $\alpha = 0$, and $\alpha < 0$. All these theories may exist and do not contain any inconsistencies, despite the fact that the case $\alpha < 0$ has quite unusual physical consequences. The case $\alpha \neq 0$ is more general than $\alpha = 0$ since it contains the latter as a limit of small α values.

Hence, the *functional form* of the transformations between the observers of two inertial frames of reference is completely determined up to the constant α. stablishment of its *magnitude* and *sign* is an experimental issue. The fundamental constant α could be zero, but in our world, it is positive.

▷ It is possible to express "α" through the constant "c":

$$\alpha = \frac{1}{c^2}.$$

As a result,

$$t' = \frac{t - vx/c^2}{\sqrt{1 - v^2/c^2}}, \qquad x' = \frac{x - vt}{\sqrt{1 - v^2/c^2}}. \tag{1.12}$$

These transformations fulfil all of the axioms formulated above and the choice we made,

$$\alpha > 0.$$

▷ For an axiomatic definition of the *fundamental velocity* constant "c" value additional axioms are necessary. Classical mechanics is based on the axioms (**I**)–(**IV**) but *adds* the following statement to them.

Axiom V.
If two events are simultaneous in one frame of reference, they will be simultaneous in any other frame as well.

This simultaneity of events ($\Delta t' = 0$, hence $\Delta t = 0$) yields $c = \infty$ (for more details, see p. 52) and the Galilean transformations, $t' = t$, $x' = x - vt$.

The axioms (**I**)–(**V**) define the functions $f(t, x, v)$ and $g(t, x, v)$ completely. If the 5th axiom is omitted, the quantity of information decreases, and our theory becomes incomplete. However, this incompleteness manifests itself only in the presence of an undefined constant "c", so it leads to a *parametrically incomplete theory* (see p. 31 in the next Chapter for details). In this sense, the 5th axiom contains a minimum of information.

▷ Let us note that we have not just derived the Lorentz transformations. We have also demonstrated that

---

The theory of relativity is self-consistent, if classical mechanics is. **!**

This is a result of the fact that the Lorentz transformations are derived from a subset of the axioms used to derive the Galilean transformations. Each theorem in the theory is derived from some group of axioms (p. 30). If none of the conclusions in classical mechanics lead to contradictions, then they will not lead to them in the theory of relativity, which uses fewer axioms. When a theory (classical mechanics) is augmented with a new independent axiom, *in a non-contradictory way*, there arises the possibility of obtaining new theorems, decreasing the arbitrariness in the initial constrained system of axioms (for example, the possibility of showing that $c = \infty$). Thus, relativity theory cannot contain any logical inconsistencies, since its initial postulates are non-contradictory.

1.5 Units

The fundamental velocity coincides numerically with the velocity of light,

$$c = 299\,792\,458 \; m/sec.$$

It is convenient to define time units so that $c = 1$. For example, it is possible to choose $1/299\,792\,458$ of the “usual second” in SI as a “new second”. All the formulas in the theory of relativity look simpler in this system of units. For example, if the Lorentz transformations are extended, assuming constancy of the coordinate perpendicular to the motion (“lines on a fence”, p. 9), they appear as follows,

i

$$t' = \frac{t - vx}{\sqrt{1 - v^2}}, \qquad x' = \frac{x - vt}{\sqrt{1 - v^2}}, \qquad y' = y, \qquad z' = z. \tag{1.13}$$

If it is necessary to restore the constant “c”, then the quantities containing time raised to some power are multiplied by “c” to the same power. For example, for the time t, velocity $u = dx/dt$ and acceleration $a = d^2x/dt^2$, the following substitutions are performed:

$$t \mapsto ct, \qquad u \mapsto \frac{u}{c}, \qquad a \mapsto \frac{a}{c^2}. \tag{1.14}$$

In what follows, we will use a system of units where $c = 1$. For all the physical quantities appearing in the theory there hold simple rules similar to (1.14). This convenient system of units will allows us to simplify the mathematics significantly without “losing” the fundamental constant c, since it may easily be restored.

One more simplification is achieved by introducing abbreviations for the relativistic factors,

$$\gamma = \frac{1}{\sqrt{1 - v^2}} \tag{1.15}$$

and

$$\Gamma = \frac{\gamma - 1}{v^2}. \tag{1.16}$$

At $v = 0$, the factor γ equals unity, while as $v \to 1$, it tends to infinity. For small velocities, γ and Γ may be expanded in Taylor series, ($\lessdot C_2$):

$$\gamma \approx 1 + \frac{v^2}{2} + \frac{3\,v^4}{8} + \ldots, \qquad \Gamma \approx \frac{1}{2} + \frac{3\,v^2}{8} + \frac{15\,v^4}{16} + \ldots$$

We will often need to perform different algebraic manipulations with the gamma factors, and so state some identities,

$$v^2 = 1 - \frac{1}{\gamma^2}, \qquad \Gamma = \frac{\gamma^2}{1 + \gamma}, \qquad \gamma - \Gamma = \frac{\gamma}{\gamma + 1}. \tag{1.17}$$

These identities can be proved as a simple exercise.

▷* Let us generalize the Lorentz transformations to the case of a relative velocity of arbitrary direction. Let the origin of the system S' move with velocity $\mathbf{v}$ relative to the inertial system S (the first figure),

The figure depicts a plane, but we generally assume that the motion is 3-dimensional. The observers obtain common time units by assuming the absolute value of the relative velocity to be constant; length units are then agreed upon by "comparing rulers" in a direction perpendicular to the motion. Fixing the components

$$\mathbf{v} = \{v_x, v_y, v_z\}$$

(the projections onto the axes) yields a choice of orientation for the coordinates axes (which is unique up to a rotation around $\mathbf{v} \lessdot C_3$). For an observer in S', the velocity components of the origin in S take the opposite sign.

In the third figure, the radius-vector $\mathbf{r}$ is expanded into a sum of two vectors, $\mathbf{r}_\parallel$ and $\mathbf{r}_\perp$. The first of them is directed along the vector $\mathbf{v}$ and the second one is perpendicular to it ($v = \sqrt{\mathbf{v}^2} = |\mathbf{v}|$ is the absolute value of the velocity vector),

$$\mathbf{r} = \mathbf{r}_\parallel + \mathbf{r}_\perp, \qquad \mathbf{r}_\parallel = \frac{(\mathbf{rv})}{v^2}\,\mathbf{v}.$$

The length of the vector $\mathbf{r}_\parallel$ is equal to the projection of $\mathbf{r}$ onto the unit vector $\mathbf{v}/v$ pointing in the velocity direction, which also defines the direction of $\mathbf{r}_\parallel$. Now the Lorentz transformations may be written for each component,

$$t' = \gamma\,(t - \mathbf{v}\mathbf{r}_\parallel), \qquad \mathbf{r}'_\parallel = \gamma\,(\mathbf{r}_\parallel - \mathbf{v}\,t), \qquad \mathbf{r}'_\perp = \mathbf{r}_\perp.$$

Really, $\mathbf{r}_\parallel$ is directed along $\mathbf{v}$ and is like x in the usual Lorentz transformations. Similarly $\mathbf{r}_\perp$ is perpendicular to the velocity and is like y. Taking into account that $\mathbf{r}' = \mathbf{r}'_\parallel + \mathbf{r}'_\perp$ and replacing $\mathbf{r}_\perp$ with $\mathbf{r} - \mathbf{r}_\parallel$, one may write the Lorentz transformations as follows:

$$t' = \gamma\,(t - \mathbf{vr}), \qquad \mathbf{r}' = \mathbf{r} - \gamma\,\mathbf{v}\,t + \Gamma\,\mathbf{v}\,(\mathbf{vr}). \tag{1.18}$$

The inverse transformation may be obtained by swapping the primed and unprimed quantities and substituting $\mathbf{v} \mapsto -\mathbf{v}$. If $\mathbf{v} = \{v, 0, 0\}$ and $\mathbf{r} = \{x, y, z\}$, then (1.18) yields (1.13). As we will see later, generally speaking, equality of the relative velocities does not mean that the coordinates axes of the two systems are parallel. That the Lorentz transformations take the form (1.18) only implies that the observers have agreed upon a system of units, as described above.

1.6 Velocity addition

Let us consider the inertial frames of reference S and S'. Let their x- and x'-axes be directed parallel to the relative velocity. The velocity of the S' system relative to S equals v, while the velocity of S relative to S' equals "$-v$".

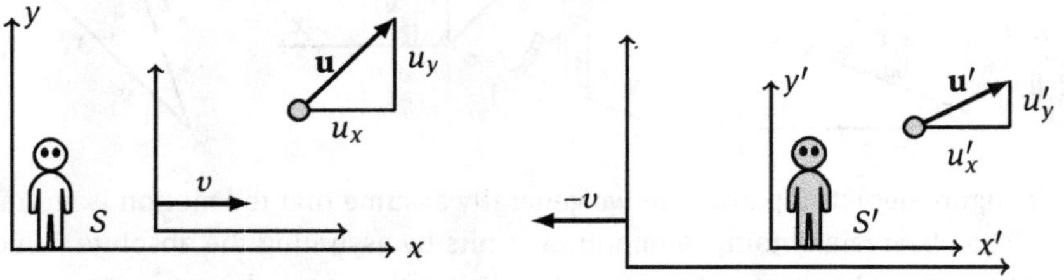

An event is defined by its time t and position $\{x, y\}$. In each frame of reference, the observers detect this event with the help of their instruments and obtain the values $\{t, x, y\}$ for S and $\{t', x', y'\}$ for S'.

We recall that observers can perform measurements only in their local vicinity, so we imagine each frame of reference to be "filled" with observers. An event is detected by two observers in S and S', respectively, located at the point where it has occurred. Due to the time synchronization procedure, the measurement they obtain will be accepted unambiguously by their "brothers" from *other* frames of reference. The effects of the theory of relativity manifest themselves at large velocities, so it will be necessary to study large distances in order to obtain noticeable differences from classical kinematics. Hence, introducing a set of observers is useful.

In a similar way, the results of the observations of some process can be compared. Let this process consist of two consecutive events, namely a beginning at the time moment t_1 (in the system S) and an end at the moment t_2. Its spatial location (in the plane) is also defined by two points $\{x_1, y_1\}$ and $\{x_2, y_2\}$. The time interval between the events and the difference in their coordinates equal

$$\Delta t = t_2 - t_1, \qquad \Delta x = x_2 - x_1, \qquad \Delta y = y_2 - y_1.$$

We take the Lorentz transformations (1.13) for each of the events and subtract them. Due to the linearity of the transformations and constancy of the velocity v, the following Lorentz-like transformations hold for the differences:

$$\Delta t' = \frac{\Delta t - v\Delta x}{\sqrt{1-v^2}}, \qquad \Delta x' = \frac{\Delta x - v\Delta t}{\sqrt{1-v^2}}, \qquad \Delta y' = \Delta y \tag{1.19}$$

(we will often consider the 2D space $\{x, y\}$, because the z-axis is similar to y due to symmetry).

▷ Let us consider a moving object. One may measure its location (i.e., its coordinates $\{x_1, y_1\}$) at the time moment t_1 and then at the time moment t_2. By definition, in the system S, the projections of its velocity $\mathbf{u} = \{u_x, u_y\}$ equal

$$u_x = \frac{\Delta x}{\Delta t}, \qquad u_y = \frac{\Delta y}{\Delta t},$$

and the same formulas hold for the primed quantities in S'. If the velocity of the object is constant, then the value Δt does not matter. In order to describe motion with varying velocity we assume that Δt is arbitrarily small, and the velocity is the derivative of the coordinates with respect to time.

From the Lorentz transformations for the differences (1.19) it is not difficult to find out the relation between the velocity of the object for observers in the systems S and S',

$$u'_x = \frac{u_x - v}{1 - u_x v}, \qquad u'_y = \frac{u_y \sqrt{1 - v^2}}{1 - u_x v}. \tag{1.20}$$

The inverse velocity transformations are obtained by direct calculation. However, due to the equivalence of the inertial frames of reference it suffices to simply change the sign of the velocity v and to swap the primed and non-primed quantities,

$$u_x = \frac{u'_x + v}{1 + u'_x v}, \qquad u_y = \frac{u'_y \sqrt{1 - v^2}}{1 + u'_x v}. \tag{1.21}$$

For example, if we stand on a railway platform and $\mathbf{u}' = \{u'_x, u'_y\}$ is a fly's velocity relative to a train moving with velocity v, then the fly's velocity relative to us is the result of the combined motion of the train and the fly. In classical mechanics, this composition takes the form

$$u_x \approx u'_x + v, \qquad u_y \approx u'_y.$$

In the theory of relativity, such relations are just an approximation which only holds if the velocities of the train and the fly are much less than the fundamental velocity $c = 1$. The faster the train or fly moves, the more the "addition" of their velocities (1.21) differs from the classical one.

Let us practice restoring the fundamental constant c. We have to make the replacement

$$v \mapsto \frac{v}{c}$$

for all the velocities, so the velocity transformation (along the x-axis, for example) may be written as

$$u'_x = \frac{u_x - v}{1 - u_x v/c^2}. \tag{1.22}$$

The classical velocity-addition law is obtained for $u_x v \ll c^2$.

▷ Let an object move along the x-axis with the fundamental velocity $u_x = c$. Then, in another system S', its velocity (1.22) equals

$$c' = \frac{c - v}{1 - c\,v/c^2} = c.$$

Hence, an object having the velocity "c" in one frame of reference has the same velocity in any other system. So, "c" is also called *the invariant velocity.*

▷ With the help of the transformations (1.20) it is possible to verify (< H_6) that the squared velocity $\mathbf{u}^2 = u_x^2 + u_y^2$ transforms as follows,

$$1 - \mathbf{u}'^2 = \frac{(1 - \mathbf{u}^2)\,(1 - \mathbf{v}^2)}{(1 - \mathbf{uv})^2}, \tag{1.23}$$

where $\mathbf{u}\,\mathbf{v} = u_x\,v$ is the projection of the object velocity $\mathbf{u}$ onto the system velocity $\mathbf{v}$. If the object moves in some arbitrary direction with the fundamental velocity $\mathbf{u}^2 = c^2 = 1$ in one frame of reference, then in any other inertial system $\mathbf{u}'^2 = 1$, so "c" is an invariant velocity regardless of its direction.

Usually this invariance is used in order to derive the Lorentz transformations (see. Sect. 1.8). However, it is in fact a *consequence* of the theory of relativity. Moreover, for our everyday practice, it is one of the most unusual and strange consequences of the theory. In the course of their daily lives, people get used to the Galilean velocity-addition rule. The impossibility of "catching" light looks almost like a contradiction of the theory. Certainly, there is no contradiction here; large velocities must simply be added in a completely different way than is done in classical mechanics.

▷* Let us write the velocity transformation in vector form with the help of the vector Lorentz transformations (1.18). Dividing $\Delta\mathbf{r}'$ by $\Delta t'$, we obtain

$$\mathbf{u}' = \frac{\mathbf{u} - \gamma\,\mathbf{v} + \Gamma\,\mathbf{v}\,(\mathbf{vu})}{\gamma\,(1 - \mathbf{uv})}. \tag{1.24}$$

With the help of the triple vector product (p. 334)

$$\mathbf{a} \times [\mathbf{b} \times \mathbf{c}] = \mathbf{b}(\mathbf{ac}) - \mathbf{c}(\mathbf{ab}),$$

this transformation may be re-written as follows (see the definition (1.16)),

$$\mathbf{u}' = \frac{\mathbf{u} - \mathbf{v} + [\mathbf{v} \times [\mathbf{v} \times \mathbf{u}]]\,\Gamma/\gamma}{1 - \mathbf{uv}}. \tag{1.25}$$

If the velocity of the frame of reference S' is parallel to the object velocity, then the product $\mathbf{v} \times \mathbf{u} = 0$, and (1.25) coincides with one-dimensional transformation of the velocity along the x-axis (1.20).

▷ The fundamental invariant velocity "c" is the maximum possible velocity of a "material" object in motion. Imagine that in the system S, an observer creates his clone and sends him into flight with velocity v (the system S_1). The first clone creates then creates a second and sends him with the same velocity ***relative to himself*** (the system S_2) and so on, up to infinity. In classical physics, the n-th clone would have the velocity $u_n = n\,v$ relative to the system S; this velocity would tend to infinity as $n \to \infty$. In the relativistic world, the velocities of the n-th and the $(n-1)$-th clones relative to the frame of reference S are connected as follows,

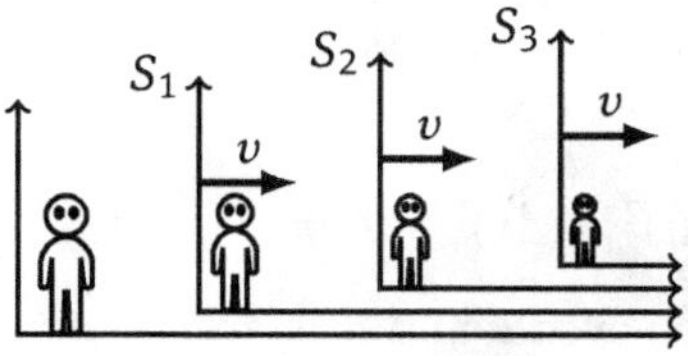

$$u_n = \frac{u_{n-1} + v}{1 + u_{n-1}\,v}$$

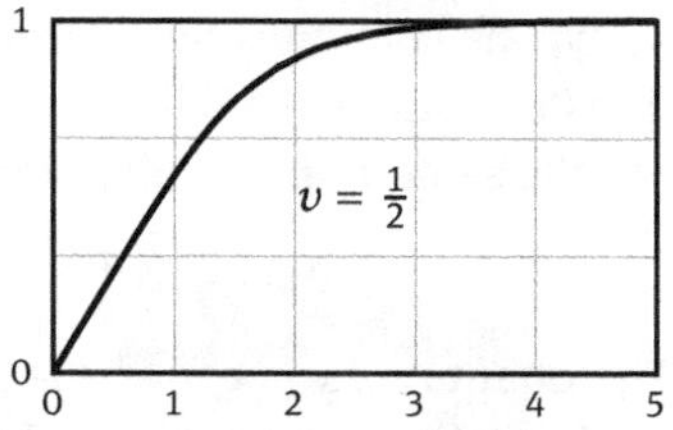

If one builds a graphical representation of this relation starting from $u_0 = 0$, $v = 1/2$, one obtains the plot shown in the right-hand figure. The velocity u_n tends to $c = 1$ as $n \to \infty$. Although u_n is permanently increasing, the differences between the consecutive velocities relative to the observer in S become smaller and smaller. As $n \to \infty$ it is possible to put $u_n = u_{n-1} = u_\infty$ and to obtain an asymptotic value

$$u_\infty = \frac{u_\infty + v}{1 + u_\infty\,v}$$

independent of v, which yields $u_\infty = 1$.

$\triangleright^*$ Let us find the explicit dependence of u_n on n. The velocity-addition law (1.21) yields the relation ($\lessdot H_7$)

$$\frac{1+u_x}{1-u_x} = \frac{1+u'_x}{1-u'_x}\,\frac{1+v}{1-v}. \tag{1.26}$$

Introducing the ***hyperbolic arctangent*** (p. 355), one obtains,

$$\operatorname{ath}(u_x) = \operatorname{ath}(u'_x) + \operatorname{ath}(v), \qquad \textit{where} \qquad \operatorname{ath}(v) = \frac{1}{2}\ln\frac{1+v}{1-v}.$$

Hence, $\operatorname{ath}(u_n) = n\,\operatorname{ath}(v)$, or

$$u_n = \frac{1-w^n}{1+w^n}, \qquad w = \frac{1-v}{1+v} < 1. \tag{1.27}$$

Of course, $u_n \to 1$ as $n \to \infty$. The velocity-addition formula written in terms of the hyperbolic arctangent has an important geometric meaning, which we will discuss when considering the Lobachevsky space in the last chapter of the book.

Apart from this imaginary experiment involving clones, there are other compelling reasons to believe that "c" must be the maximum possible velocity. These will be considered a bit later.

1.7 The interval

Let us consider two events $\{t_1, x_1, y_1\}$ and $\{t_2, x_2, y_2\}$ taking place at different moments in time and at different points in space. These events are detected by observers in different inertial reference frames S and S'. One can calculate the time difference between these events $\Delta t = t_2 - t_1$ in the system S and $\Delta t' = t'_2 - t'_1$ in the system S'. Similarly, one computes the coordinate differences Δx, $\Delta x'$, and so on. From the Lorentz transformations (1.19) it follows (⪡ H_8) that the following combination of differences has the same value for all observers,

$$(\Delta s)^2 = (\Delta t')^2 - (\Delta x')^2 - (\Delta y')^2 = (\Delta t)^2 - (\Delta x)^2 - (\Delta y)^2 .$$

The quantity Δs is called *the interval* between the events and it is an *invariant* of the Lorentz transformations. Interval invariance has a deep geometric meaning. The quantity Δs is a distance in the pseudo-Euclidean geometry of space and time. We will discuss it in more detail in Chapter 7.

If a light flash occurs at the point $\{x_1, y_1\}$ and propagates as a spherical wave (a circle in a plane) with velocity $c = 1$, then at time Δt its radius R will be

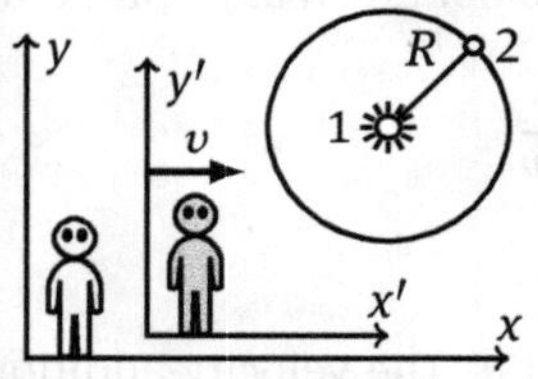

$$R^2 = (\Delta x)^2 + (\Delta y)^2 = (\Delta t)^2 .$$

Thus, $(\Delta s)^2 = 0$; such intervals are called *lightlike*. Lightlike intervals relate events that may be connected by a signal propagating with the fundamental velocity. A lightlike interval equals zero for all observers. So, a spherical light wave looks spherical from any inertial frame of reference.

If $(\Delta s)^2 > 0$, then the interval is called *timelike*. In particular, at

$$\Delta x = \Delta y = 0$$

the interval Δs equals the time interval Δt which has elapsed according to the clock fixed in the given system. Events related by a timelike interval may be connected by a signal propagating with velocity $\mathbf{u}$ less than $c = 1$,

$$(\Delta s)^2 = (\Delta t)^2 - (\Delta x)^2 - (\Delta y)^2 = (\Delta t)^2 (1 - \mathbf{u}^2) > 0,$$

where $\mathbf{u}^2$ is the squared velocity of the movement by Δx, Δy in the time Δt. The "timelikeness" of the interval is an invariant property of all the observers.

Finally, if $(\Delta s)^2 < 0$, then the interval is called *spacelike*. Two events for which $(\Delta s)^2 < 0$ cannot be connected by light signals or by "ordinary" particles having velocity less than the fundamental one.

▷ An object moving with velocity arbitrarily close to the fundamental velocity "c" differs significantly from objects having exactly the velocity "c". For example, the Lorentz transformations (1.12) turn yield infinity at $v = c$. As a result, it is impossible to relate an inertial frame of reference "filled" with observers, clocks, and rulers to objects having the velocity "c".

Our world might well have been constructed so that there were no objects moving with the velocity "c". In fact, the situation looks quite natural. In such a world, "c" would be a limit for the velocity, unreachable by any real object. But our world seems to be organized in a different way, and there are objects that are significantly different from other ones and which do move with the fundamental velocity. Light is the most important representative of them. It makes possible to us the study of remote bodies; moreover, life on our planet exists thanks to the light emitted by the Sun. At high energies, we perceive light as particles (photons), while at low energies it looks like a wave (electromagnetic radiation) to us.

We will often speak about light quanta or photons. In the classical (non-quantum) theory of relativity both of these entities are just another name for a light impulse that is negligibly short compared to the time of its propagation between two points.

Other than light, there may exist other entities which move with the fundamental velocity. Most likely, gravitational interaction propagates with the velocity of light. It was believed for a long time that neutrinos (weakly-interacting particles) are lightlike. However, not ver long ago, it was established that they are of small, but nonzero mass.

The significant difference between lightlike objects and "normal" ones is that they cannot change this property at any point during their "life". They cannot be slowed down or stopped without being destroyed. They do not change their velocity. Of course, here we are talking about motion in a vacuum. In a medium, the velocity of light becomes smaller. But, in fact, this velocity is the averaged velocity of *different* photons being re-emitted ("with a delay") by the atoms of the medium. Between the atoms, photons travel with velocity c. The "non-quantum" picture of the same process is built by summing up the set of secondary waves that appear when the charged particles of the medium are caused to oscillate by the electromagnetic wave.

Since it is possible to have lightlike objects that differ significantly from normal ones, one might suppose that ***tachyons***, which are able to move faster than the fundamental velocity "c", also exist. As with light, a tachyon remains a tachyon for its entire life. Its velocity may approach the velocity "c" from above but it will never reach this limit. Assuming the existence of tachyons yields quite unusual consequences. Besides that, there are no experimental signs of the tachyon's existence. Special difficulties (causality violations, and the like) appear when one tries to describe their interaction with "normal" matter.

1.8 Einstein's Axiomatics

A discussion of the logical principles of the theory relativity would be incomplete without considering the traditional approach provided by Einstein. In 1905, Einstein formulated his postulates in the paper "On the Electrodynamics of Moving Bodies" as follows [5].

---

1. The laws of physics are the same for all observers in uniform motion relative to one another (principle of relativity).

---

2. The speed of light in a vacuum is the same for all observers, regardless of their relative motion or of the motion of the light source.

The second postulate is usually somewhat misunderstood. Thus, we refer to the explanation proposed by Einstein himself in 1921 in the lectures "The meaning of relativity" [6].

> The consequence of the Maxwell-Lorentz equations that in a vacuum light is propagated with the velocity c, at least with respect to a definite inertial system K, must therefore be regarded as proved. According to the principle of special relativity, we must also assume the truth of this principle for every other inertial system.

The second postulate itself is quite "harmless". The theories of ether just assumed the existence of such a stationary frame of reference (ruhenden Koordinatensystem, by Einstein). An analogy with the propagation of sound in air was drawn. If an observer is fixed relative to the air, then the sound velocity is constant for him, and does not depend on the velocity of the sound source. Here, it is essential that preference is given to the frame of reference related to the medium. In any other system, the sound velocity will be different, and will depend on the direction of its propagation.

The existence of some medium (ether) was assumed wherever light (electromagnetic waves) was to propagate. It was believed that the Maxwell equations were related to the frame of reference fixed relative to this medium. Hence, the value of the velocity of light obtained from the Maxwell equations was also related to the stationary frame of reference.

It is the combination of the second postulate and the relativity principle which is the key point in Einstein's axiomatics. This yields completely new and very strong statement,

---

1 + 2. The velocity of light is the same *in all* frames of reference.

Thus, measurements of light impulse velocity made by different observers (moving relatively to each other) will always give the same value, independently of how this impulse has been created. This postulate contradicts "classical intuition" based on the Galilean addition law for velocities. Nevertheless, it is correct, and, as was shown above, it results from much more natural initial statements if we remove the axiom that time is absolute.

In fact, merging of the two postulates meant a rejection of the ether as the medium where the light is propagating. Really, the relativity principle has meaning only for "empty space" where it is impossible to mark some inertial frame of reference as preferred.

It is worth noting that with Einstein's axiomatics, it is not possible to speak about the velocity of light, but only about the existence of a maximum possible velocity for all objects in nature. If we denote this maximum by "c" and assume that this fundamental constant is the same for all ***equivalent*** inertial frames of reference, then we obtain Einstein's "constancy postulate" in its powerful statement 1 + 2.

Before Einstein, both Lorentz and Poincare obtained transformation laws between frames of reference based on the invariance of the Maxwell equations. The historical importance of Einstein's 1905 paper lies in the fact that he confined himself to the use of just one statement following from the theory of electromagnetism: that electromagnetic waves exist and propagate at the speed of light. In this sense, his arguments are more general. When we speak about the fundamental maximum velocity instead of the velocity of light, the arguments become even more general. We do not bind ourselves to the theory of a specific interaction or to a specific agent moving *exactly* with the fundamental velocity. Of course, it is assumed that such an agent must exist.

Let us note that the derivation of the Lorentz transformations described above did not require this agent. It might not even exist if there were no massless particles in the world. Actually, to derive the Lorentz transformations, only the first of Einstein's postulates is required. The second statement (in its strong form 1+2) is a corollary of the theory.

Let us now consider the derivation of the Lorentz transformations based on the strong version of the postulate about the contancy of the velocity of light 1 + 2. Following Einstein, we will assume that the transformations are linear, based on "*the properties of homogeneity which we attribute to space and time*" [5].

In order to establish consistent length units between two frames of reference, Einstein supposes that a copy of a stationary inelastic ruler in the S system accelerates to the velocity v and becomes a length etalon in the S' system. Certainly, perfectly inelastic rulers do not exist, and they may deform while accelerating. The problem is simplified a bit if the ruler is placed perpendicularly to the acceleration vector and all of its points gain the same acceleration *simultaneously*. After the necessary velocity is reached, the acceleration becomes zero. Due to the isotropy of space in an inertial frame of reference, the ruler may be rotated along any direction.

Nevertheless, Einstein supposes that the coordinates y' and y of the axes perpendicular to the velocity may differ,

$$y' = \eta(v)\, y,$$

where $\eta(v)$ is some even (due to isotropy) function of the velocity. Since the observers are equivalent, the same relation must also hold for the inverse transformation, so

$$y = \eta(-v)\, y' = \eta(-v)\, \eta(v)\, y$$

or

$$\eta(-v)\, \eta(v) = \eta^2(v) = 1.$$

Since we must have $\eta(0) = 1$ at $v = 0$, we obtain $\eta = 1$, and therefore that the coordinates perpendicular to the relative velocity of the frames of reference do not change.

Let us write the linear transformations as follows,

$$t' = \gamma\,(t - \sigma\, x), \qquad x' = \gamma\,(x - v\, t), \qquad y' = y,$$

where $\gamma = \gamma(v)$ and $\sigma = \sigma(v)$ are unknown functions of the velocity. As earlier, we assume that the origin of the S' system (the point $x' = 0$) is moving along the trajectory

$$x = v\, t,$$

and, correspondingly, that the origin of the S system is moving along the trajectory

$$x' = -v\, t'.$$

At the initial time moment $t' = t = 0$ the origins of the frames of reference coincide.

Let us suppose that at the moment $t' = t = 0$ a light flash occurs, for example, at the point $x = x' = 0$, $y = y' = 0$. In each the frame of reference, the spherical light wave is described by the equations

$$x^2 + y^2 = (ct)^2, \qquad x'^2 + y'^2 = (ct')^2.$$

Due to the postulate about the constancy of the velocity of light, the radii of both the spheres (in the plane, circles) increase with the the same velocity c.

Let us substitute the linear transformations of the coordinates and time into the equation of the light sphere in the system S'. Squaring them, we obtain,

$$\gamma^2 x^2 - 2\,\gamma^2\, v\, x\, t + \gamma^2\, v^2\, t^2 + y^2 = \gamma^2\,(ct)^2 - 2\,\gamma^2\, c^2\, \sigma\, x\, t + \gamma^2\, \sigma^2\, c^2\, x^2.$$

In order to obtain an equation describing a sphere in the system S, the cross terms xt must vanish.

Hence,

$$\sigma = \frac{v}{c^2}.$$

Collecting like terms, we have

$$\gamma^2 \cdot (1 - v^2/c^2)\, x^2 + y^2 = \gamma^2 \cdot (1 - v^2/c^2)\, (ct)^2 .$$

Taking the light sphere equation in S, we obtain the following relation for the function γ,

$$\gamma^2 \cdot (1 - v^2/c^2) = 1.$$

Since $\gamma(0) = 1$ at zero relative velocity, we must select the positive square root. This yields the desired Lorentz transformations,

$$t' = \frac{t - v\,x/c^2}{\sqrt{1 - v^2/c^2}}, \qquad x' = \frac{x - v\,t}{\sqrt{1 - v^2/c^2}}, \qquad y' = y, \qquad z' = z,$$

where we have added the transformation for the second direction perpendicular to the motion, z. Its derivation is completely similar to the one for y.

In such a derivation of the Lorentz transformations, accounting for at least two spatial dimensions is crucial. If only the motion of the light impulse along the x-axis were considered, instead of the light sphere (or the circle, as above), it would be possible to find the function $\sigma(v) = v/c^2$, but not $\gamma(v)$. Certainly, relativity theory may also be developed for 1-dimensional space, but in this case, using only the postulate of light velocity constancy does not suffice. According to the relativity principle, the group properties must be used, as was done in Sect. "*The Lorentz transformations*" (p. 12). In particular, one writes down an inverse transformation for the coordinates x and x' and substitutes in the direct transformation to obtain γ. However, instead of using the composition of transformations for three frames of reference that gave us the function σ in Sect. 1.3, it is possible to use the light velocity constancy postulate in all of the inertial frames of reference.

2 A bit of philosophy and history

In this chapter, the general issues of the axiomatic foundation of physical theories are considered, and the origin of the fundamental constants is discussed. The principle of *parametric incompleteness* will be formulated. This enables new physical theories to be obtained using existing ones.

To get a feel for the possible limitations of the theory of relativity, it is worth reading the section "*Out of the known*", although this is not necessary for our present purpose. The last three sections of the chapter are historical in nature.

https://doi.org/10.1515/9783110515886-002

2.1 The axiomatic development of theories

To prove some statement, it is necessary to argue using other statements (assumptions). Let us imagine each statement as a circle and let us mark with arrows those statements which have been used for its proof. Since there are a lot of possible statements (generally speaking, an infinite number), such a “network” of interactions may turn out to be very cumbersome.

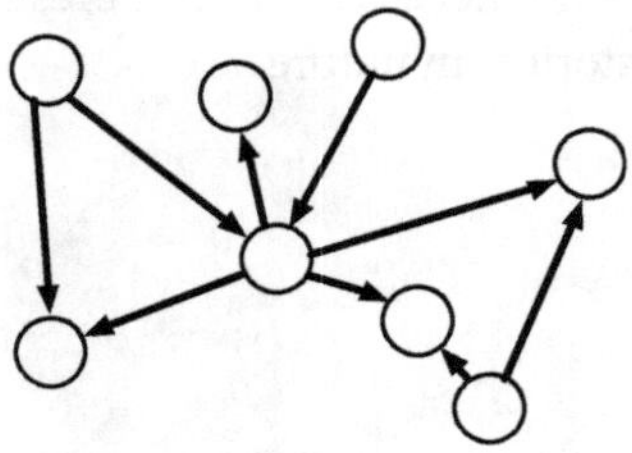

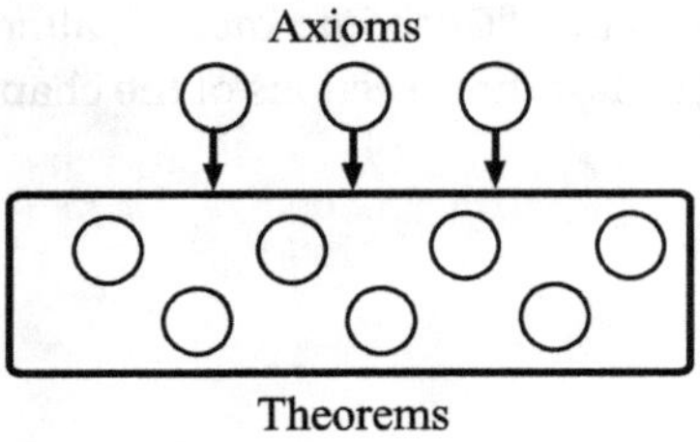

However, it is usually possible to put this chaos into some kind of order. It turns out that

---

! there exist several initial statements (axioms) from which all of the rest are derived.

In other words, one may succeed in obtaining an *infinite* number of statements from a *finite* set of initial statements. This wonderful approach to mathematical and physical theories is called the *axiomatic method.*

Euclidean geometry appears to be have been the first theory to be formulated axiomatically. At the beginning, certain “self-evident” statements, namely, the axioms, were stated. These were accepted without proof. Then, the other statements, the theorems, were obtained with the help of logical reasoning. It must be said that these logical derivations were quite informal, and were often based on intuitive evidence. In the figurative phrase of Vladimir Uspensky, the proofs were *psychological*, i.e., so convincing that you would be eager to convince other people with them.

After a while, there arose the necessity of formalizing a theory and proving it, so that no hidden “evidence” would be able to break the logical order. In particular, it was necessary to reject natural language, with its ambiguity, wherever possible. David Hilbert’s “Foundations of Geometry” and his subsequent works are a brilliant example of this kind of approach. All of the objects in a theory are denoted by formal symbols, and logical derivations reduce to the “permutation” of these symbols according to previously defined rules. Proofs that are built in this way can be tested, for example, using a computer. This makes formal mathematical theory practically flawless.

▷ The axioms of a "good" theory have the properties of independence, self-consistency, and completeness.

Independence means that no axiom can be derived from other ones. **!**

If one succeeds in deriving one axiom from some others, the system of axioms is reduced. This procedure is continued until only independent statements remain. Self-contradictory theories are of little use and should be avoided. Finally,

---

completeness of a system of axioms means that each statement within the theory may be proved using said axioms. **!**

If a statement (and its negation) cannot be derived from the axioms, it may be added to them. To have a "good" theory, the set of axioms must be finite.

A complete and self-consistent theory cannot be changed by adding a new axiom to the existing ones. If this occurs, the system will either not be independent anymore, or some self-contradictions will appear in the theory (if the negation of some statement that can be derived from the initial axioms is added).

▷ The theorems of a theory may be split into dependence classes by assigning all the theorems that may be derived from a given group of axioms to one class. For example, let us consider three axioms, A_1, A_2 and A_3. If the theorem T_1 may be derived from A_1, A_2:

$$A_1,\ A_2 \quad \Rightarrow \quad T_1,$$

and does not require the axiom A_3, then it is put into the class $\{A_1, A_2\}$. In some sense, the axioms themselves contain some "***axiomatic information***", which is needed for deriving a given theorem. The more "informative" a theorem is, the more axioms may be needed for its derivation. It may occur that a theorem belongs to two classes simultaneously, for example, $\{A_1, A_2\}$ and $\{A_1, A_3\}$. This means that in the axioms A_2 and A_3, the information needed for the theorem overlaps partially.

Basically, it is possible to enumerate a subset of the theorems that do not need any axiom for their proof. Generally speaking, such theorems lead to an incomplete theory. However, this theory may be quite meaningful. Moreover, as we have seen from the example of the Lorentz transformations, it may turn out to be more general than that which follows from the full set of axioms. The addition of unused axioms yields new theorems which are more restrictive. For example, the statement that time is absolute (i.e., that it elapses at the same rate in all frames of reference) is more restrictive than the assumption of different time passage for different observers. The generality of a theory just means that it allows for some phenomena, which are formulated in the form of statements that would be rejected in a specialised theory, as a result of a "restricting" addition of axioms .

2.2 The parametric incompleteness principle

The significance of the fundamental constants c and $\hbar$ in modern physics can hardly be overstated. They define the structure of the basic formulae in relativistic and quantum theories. Their numerical values

$$c = 299\,792\,458\ m/s, \qquad \hbar = 1.054\,571\,800(13) \times 10^{-34}\ J \cdot s,$$

specify the scales of the phenomena for which appropriate corrections to classical mechanics are essential.

There are a plenty of questions concerning the fundamental constants, the full answers for which are unknown. Here are some of them.

?

- ▷ Why is it impossible to calculate the values of the fundamental constants without turning to experiment?
- ▷ Why do the fundamental constants appear in more general physical theories but are absent in classical physics?
- ▷ Are there fundamental constants different to c and $\hbar$, and are generalizations of classical mechanics corresponding to them possible?
- ▷ Is the set of all possible fundamental constants finite?
- ▷ Do the "constants" change over time, or not?

Let us clarify what we mean by fundamental physical constants. Physics consists of three closely related phenomena.

1) *Structure*: electrons, quarks, atoms, ...;
2) *Interactions*: electromagnetic, strong, ...;
3) *Mechanics*: relativistic, quantum.

By "mechanics", we do not mean the "mechanistic" phenomena of classical physics, but the general properties of space and time, the influence of measurement on objects, etc., that is, everything that forms the basis of all of our physical theories. Mechanics defines the laws which all of the structural units and interactions must obey. It is the base on which the other two aspects of the building that is physics rest. For example, the principles of the theory of relativity restrict the class of possible interactions. The same interactions occur between different structural units, many of which form the basis of our world.

Let us agree that the fundamental constants are the physical parameters that define the structure of the theories' formulas, applicable to all forms of matter and types of interaction. These constants define the properties of mechanics.

In modern physics, three such constants are known, namely, the fundamental velocity c, Planck's constant $\hbar$ and, probably, the gravitation constant G_{grav}. The electron charge, and the masses and other important parameters of elementary particles, are not fundamental *in the sense described above*.

So, for example, when speaking about the constant "c", we usually use the term "speed of light". By doing so, we mix two essentially different constants, namely the velocity "c_{em}" of the propagation of electromagnetic waves in a vacuum, and the fundamental velocity "c" defining the structure of the theory of relativity.

The fact that $c_{em} = c$ is a property of one of the existing interactions. **!**

The constant "c" defines a relativistic theory that holds for all forms of matter. In particular, in order to measure the value of the fundamental velocity, there is no need to perform any electrodynamic experiments. It is enough to compare observation results in two frames of reference and to extract the value of "c" from the Lorentz transformations. Even if a photon had non-zero mass and there were no other massless particles, this would not alter the theory of relativity with the constant "c".

This is why the parameter "c" is considered a fundamental physical constant. At the same time, the parameter "c_{em}" is only a parameter of one of the interactions associated with the zero photon mass, despite the fact that $c_{em} = c$. In this sense, "c_{em}" is not something fundamental. Certainly, the definition of "c_{em}" as "non-fundamental" does not downgrade it.

In classical mechanics, the fundamental constants are absent. More precisely, their value is fixed in a trivial way (0 or ∞). The gravitational constant G_{grav} is present in classical mechanics, as is the velocity of light, but they acquire their fundamental meaning only in modern theories of space and time.

Mechanics that generalize the classical theory do not refute it, but rather restrict its scope of applicability.

If the fundamental constants tend to their limit values, then more general theories turn into classical physics. This fact is known as the *correspondence principle*. **!**

Of course, this does not mean that every phenomenon has a classical analogue. But the "backward compatibility" of new theories is quite high and it is always possible to find the conditions of a "smooth transition" from new concepts to old ones. The formal change of the values of the fundamental constants (decreasing or increasing of them) is the main instrument of such a transitions. This property of physical theories is also evidence of their consistency. The "transformation" of new theories into the old ones is smooth, and so the new theories are free from logical contradictions.

▷ So, why do the fundamental constants and more general mechanics corresponding to them occur at all? An answer to this question is related to the axiomatic analysis of the basis of physical theories. In mathematics, the axiomatic method has been used since the time of Euclid, but in fact axiomatic systems have only really garnered attention since the appearance of non-Euclidean geometry.

For more than two thousand years, people tried to prove "the fifth" axiom about parallel lines in Euclidean geometry. In the attempt to do this, many theorems were derived *independently* of this axiom. In Bolyai's terminology, having the fifth axiom resulted in an "ideal geometry". Consequently, a new theory appeared, namely non-Euclidean geometry. In contrast to Euclidean geometry, a constant λ appears in the formulas of non-Euclidean geometry. This is the curvature of space. Its value cannot be found from the initial axioms. As $\lambda \to 0$, the formulas of non-Euclidean geometry turn into the corresponding theorems of the Euclidean theory. Moreover, the addition of the fifth axiom automatically fixes the value $\lambda = 0$. Non-Euclidean geometry seems to be the first theory where the fundamental constant λ defining its structure appeared as a result of a reduction of the initial axiomatic information. The same situation also occurs in physics.

▷ The axiomatic system of any theory must fulfil three properties. It must be *independent*, *self-consistent*, and *complete* (p. 30). Completeness means that any statement may be proved or disproved with the help of the axioms. In these terms, classical physics is complete relative to the fundamental constants, while the theory of relativity is not. Really, such statements as c=299792458 m/sec or $c = \infty$ can be neither proved nor disproved by use of the deductive method (certainly, if an appropriate experiment is not performed). At the same time, in classical mechanics, the "theorem" $c = \infty$ follows from the axiom about the absoluteness of time (the fifth axiom from Sect. "*The Lorentz transformations*", p. 12).

! When a theory contains a constant, the value of which cannot be derived from the initial axioms, we call the theory *parametrically incomplete*.

Incompleteness appears because the reduced system of axioms contains less information than the initial one. Decreasing the amount of information leads to some unavoidable incompleteness in the theory. This incompleteness may be minimal, in the sense that all of the functional relations of the theory can be derived from the axioms until finally, only a set of constants is left undefined. Certainly, the term "axiomatic information" requires a more precise definition.

For example, rejecting the axiom of the absoluteness of time of classical mechanics, we arrive at a more general theory. The theory of relativity is historically the first manifestation of the principle of parametric incompleteness in physics. Quantum mechanics is another example of the same principle. We should expect that these two examples will not be the last.

▷ After an axiomatic description of basic concepts (space, time, mass, state etc.) classical mechanics becomes a rather formal mathematical theory, whose system of axioms must fulfil the completeness requirement.

On the basis of classical mechanics it is possible to develop a set of parametrically incomplete theories by use of the deductive method. In doing so, it is necessary to remove some subset of the classical mechanics axioms. In these theories (mechanics), the role of the fundamental physical constants will be played by the parameters, whose origin lies in the incompleteness that appears as a result of the reduction of the initial information. The limiting values of these constants again yield a classical theory. What we obtain is something like the correspondence principle in reverse,

---

it is possible to derive new, more general theories from classical physics by removing some axioms. **!**

This deductive way of deriving "new physics" is extremely attractive. Of course, not all of the possible theories have to be realised in our world. But, with the help of the parametric incompleteness principle, it is possible to derive those theories that *are already contained* in an axiomatic basis of classical physics (they arise from some subset of the classical mechanics axioms). It is possible that such "deformations" of initial logical structures must inevitably appear in reality, and that they will be discovered sooner or later if only precise enough measurements can be made. A theory that fulfils the correspondence principle cannot be disproved. Only experiment can confirm it. By changing the basic constant of such a theory, it is always possible to move its effects beyond the scope of current experimental precision.

In recent times, the slang term "the Theory of Everything" has become widespread. Usually this term refers to the creation of a unified theory that covers all of the known interactions in physics. Certainly, concurrently with this, it is necessary to develop "the Theory of Everything" relative to the basis of physical construction, i.e., the mechanics restricting the properties of these interactions. From the point of view of the parametric incompleteness principle, such

a Theory of Everything will be created from Nothing.

What is this "Nothing"? It is what is left in the axiomatics of classical physics when all of the axioms containing minimal axiomatic information are thrown away and all of the fundamental constants (due to parametric incompleteness) have appeared in this kind of generalized mechanics. We believe that such a Theory of Everything will necessarily be constructed, sooner or later.

2.3 Outside of the known*

A picture without its frame is just a rolled piece of linen. A more complete understanding of a physical theory emerges when its application and scope are clarified. We take a non-formal axiomatic approach to physical theories. Based on this, we can fit different frames on the wonderful picture called "the theory of relativity".

The observers described in the first section of the previous chapter are "bodyless". They perform their measurements without changing the properties of particles. In fact, this is impossible in practice. If the objects have very small mass, then their properties obey quantum (wave) laws defined by the Planck constant $\hbar$. Unification of these ideas with the world of large velocities yields *quantum field theory*, which forms a basis for modern methods in elementary particle physics analysis. Two fundamental constants are simultaneously at the heart of this theory, namely, $\hbar$ and c.

However, even within quantum field theory, the differentiability of quantities depending on x and t is taken to be a central principle. We assume that space and time are continuous and "smooth". Nevertheless, we have to be ready to challenge this basic principle at any moment. At the microscopic level, the properties of space and time may differ essentially from those we usually consider.

▷ The second axiom which we have used to derive the Lorentz transformations looks a bit artificial, despite the evidence for it. It was necessary in order to substantiate the linearity of the transformations. It is possible to neglect this axiom by placing linear-fractional (projective) transformations into the base of the theory,

$$t' = \frac{A(v)\,t + B(v)\,x}{1 + a(v)\,t + b(v)\,x}, \qquad x' = \frac{D(v)\,x + E(v)\,t}{1 + a(v)\,t + b(v)\,x}.$$

Sophus Lie noted the group character of the transformations, while Philipp Frank and Hermann Rothe applied them to a problem of inertial observers in 1911. The singularity which appears when the denominators vanishe, however, made them abandon further analysis of this theory. Now, the evolution of the universe and the existence of a particular moment in time at which "everything appeared", are the basis of cosmological ideas. This is why, in fact, the singularity of such transformations is an advantage.

▷ Comparatively recently [14],[15], explicit expressions have been obtained for the velocity functions entering the linear-fractional transformations, with the help of axioms similar to those used in Sect. 1.4. Their physical meaning has been analysed. As a result, in addition to the fundamental velocity c, one more constant λ has appeared,

$$t' = \frac{\gamma\,(t - vx/c^2)}{1 + \gamma\lambda\,xv/c - (\gamma - 1)\lambda\,ct}, \qquad x' = \frac{\gamma\,(x - vt)}{1 + \gamma\lambda\,xv/c - (\gamma - 1)\lambda\,ct},$$

where $\gamma = 1/\sqrt{1 - v^2/c^2}$. The Lorentz transformations are contained in the numerators of these expressions, while the denominator tends to unity as λ tends to zero.

The same occurs at comparatively small distances and times that are far enough from the point of initial synchronization

$$x = x' = 0, \qquad t = t' = 0.$$

However, when we turn to cosmological scales, the denominator becomes much more essential.

The physical meaning of such a generalization of the standard theory of relativity is quite simple. Up to now we have assumed that our space is flat (Euclidean). But the basic geometric principles of uniformity and isotropy also hold for spaces of constant curvature. The transformations between inertial observers in such a space turn out to be linear-fractional, and the constant λ is the curvature of the space [16]. The transformations between two observers in *one* frame of reference take the form

$$X = \frac{x - R}{1 - \lambda^2 R x}, \qquad Y = \frac{y\sqrt{1-(\lambda R)^2}}{1 - \lambda^2 R x}, \qquad \frac{1 + \lambda c T}{1 + \lambda c t} = \frac{\sqrt{1-(\lambda R)^2}}{1 - \lambda^2 R x}.$$

where (x, y, t) are the coordinates and time measure of some event, as measured by one observer, and (X, Y, T) are those measured by another; R is the distance between the observers along the x-axis. These transformations generalize the usual translational transformations of Euclidean space.

In a space of a constant curvature, the flow of time is different even for observers *fixed* relative to each other. As a result, the further a light source is located from an observer, the lower the frequency of the light it emits. Considering the well-known empirical Hubble's law of a red shift and some of the difficulties of standard cosmological theory, this simple property of projective relativity theory requires precise study. We will return to these question when discussing cosmology in the second part of the book.

▷ Even restricting to linear transformations, there are different ways to generalize the Lorentz transformations. For example, ignoring the isotropy axiom, one obtains (⋖ H_9), p. 366 the following relations,

$$x' = \left(\frac{c+v}{c-v}\right)^{\mu} \frac{x - vt}{\sqrt{1 - v^2/c^2}}, \qquad t' = \left(\frac{c+v}{c-v}\right)^{\mu} \frac{t - vx/c^2}{\sqrt{1 - v^2/c^2}},$$

where μ is a new fundamental dimensionless constant characterizing the non-isotropy of space. It arises again out of the reduction of the axiomatic information in the theory. Although the majority of observations have confirmed the isotropy of space, occasionally news about the possible weak non-isotropy of our universe appears. If this is confirmed, it will be necessary to analyse more precisely the parametrically incomplete generalizations of the Lorentz transformations appearing when the assumption of isotropy is removed. In particular, these transformations may be re-written in a covariant three-dimensional form with a preferred direction vector. Generally, considering 3D space extends the possibilities of deriving more general Lorentz transformations significantly.

▷ However, we have not yet exhausted all of the possibilities even in the one-dimensional case. For example, our procedure for fixing velocity units warrants attention. The requirement $v' = -v$ is natural, but is not the only possibility. Removing this assumption and using the isotropy of space yields the following transformations:

$$x' = \gamma\,(x - vt), \qquad t' = \gamma\,([1 + 2\varepsilon\, v/c]\, t - vx/c^2),$$

where

$$\gamma = \frac{1}{\sqrt{1 + 2\varepsilon\, v/c - v^2/c^2}} \left[\frac{c + (\varepsilon + \sqrt{1+\varepsilon^2})\, v}{c + (\varepsilon - \sqrt{1+\varepsilon^2})\, v} \right]^{\mu},$$

and ε is a third fundamental constant. If ε is non-zero, then the velocity of the system S relative to the system S' equals $v' = -v/(1 + 2\varepsilon v/c)$. This result was obtained by Frank and Rothe, as mentioned above. In such generalized transformaions, two preferred invariant velocities appear, c_1 and c_2. These equal $c_{1,2} = c\,(\varepsilon \pm \sqrt{1+\varepsilon^2})$ and coincide with "c" at $\varepsilon = 0$. Removing the condition $v' = v$ makes the observers "asymmetric", and so a new procedure for agreeing upon velocity units is required.

▷ Our world is relativistic, so the Lorentz transformations are definitively true in some approximation. Due to the correspondence principle, they turn into the Galilean transformations at small velocities or as $c \to \infty$. But, because of the same correspondence principle, more general transformations with corresponding fundamental constants are possible. The Lorentz transformations *must* appear when these constants tend to zero. Violation of the correspondence principle is a strong heuristic reason to reject certain theories. For example, the existence of a preferred frame of reference that is absolute in space is not impossible, but this "absoluteness" must be "small", and should manifest itself only in limiting cases. This point justifies analyses of the consequences that arise when the relativity principle (the third axiom) is thrown away or weakened. The main point is that the correspondence principle must hold, and that the Lorentz transformations (and then the Galilean transformations) must be obtained at a limit value of the corresponding fundamental constant.

▷ In its basic formulation, the theory of relativity theory is a theory of flat three-dimensional Euclidean space and one-dimensional linear time extended by the relativity principle. Curvature in a space (more precisely, curvature in a unified space and time) may be caused by the action of massive objects in it. In such a space, free particles follow the shortest path, as usual, but due to the presence of curvature these paths turn out to be "curved", and are called ***geodesic lines***. So, in a neighbourhood of a massive body, "probe" particles move along trajectories, the non-relativistic limits of which are described by Newton's theory of gravity. This beautiful theory, developed by Albert Einstein and, independently, David Hilbert, is known as "general relativity theory" (GRT), and will be discussed in more detail in the second part of the book. Before its emergence, our space had been considered to be flat Euclidean space, and time assumed to be one-dimensional, uniform, and isotropic.

2.4 A bit of history

Fundamental historical facts regarding the story of relativity may be found in many books, such as [8]. There are also collections of the basic works on relativity theory [9],[5]. So, our historical overview will be very short. Apart from the commonly known facts, we will touch upon those questions that are generally poorly described in the literature .

The appearance and development of the theory of relativity grew out of questions about the nature of light. Clearly, this is because light is the object that propagates the most quickly out of all those which can be directly perceived. For a long time, it was believed that its velocity was infinite, and there were reasonable arguments to support this. The trajectory of an arrow flying rapidly is more “straight” than the “curved” trajectory of a thrown stone. Light rays look like ideal straight lines, so their velocity was thought to be very large and, very possibly, infinite.

Nevertheless, in 1638, Galileo Galilei attempted to measure (without success, however) the velocity of light. Only in 1676 di Ole Römer succeed, by observing an apparent change in the orbital periods of the moons of Jupiter during the Earth’s motion around the Sun. In fact, this was the first demonstration of a Doppler effect, where the change of the satellites’ periods acted like a signal frequency. In 1727, James Bradley obtained a similar value for the velocity of light by analysing the aberration (displacement) of a star’s location as a result of the same motion of the Earth. The aberration effect and the Doppler effect will be considered in more detail in Chapter 3. In 1849, Armand Hippolyte Louis Fizeau was the first to measure the velocity of light. His experiment will be discussed below.

There was also the problem of understanding the “internal structure” of a light signal. The corpuscular and wave hypotheses were concurrent for a long time. The emission of “light corpuscles” at the velocity of light by luminous bodies looked very pretty within the framework of geometric optics. A light ray was thought of as the “trajectory” of a large number of such corpuscles. In 1818, however, Augustin Fresnel explained the phenomenon of diffraction using the wave hypothesis and the earlier results of Christiaan Huygens (1969) and Thomas Yung (1800). This phenomenon could hardly be explained by the corpuscular theory. It was shown that the geometric laws of optics could also be obtained using the concept of light waves.

Fifty years later, in 1861, James Maxwell expressed the experimental results of Michael Faraday (1831) in mathematical form. It followed from his equations that different phenomena (such as the electrostatic charging of a hairbrush or the attraction of metals by a magnetic ore) are just different manifestations of one electromagnetic interaction. Maxwell’s equations admit solutions in the form of electromagnetic waves propagating with the velocity of light. This unification of such different theories and the explanation of the nature of light in terms of electromagnetic field oscillations was the most beautiful theoretical acheivement in physics since Isaac Newton’s works on the theory of gravity. The wave nature of light became generally accepted.

The propagation of waves in water and the form of sound oscillations in air were well-known and studied. The velocity of a wave “signal” relative to the medium (water, air) is constant and is defined by the physical properties of the medium only. This velocity does not depend on the signal source velocity, but does depend on the motion of an observer (receiver of the signal) *relative to the medium*. This fact distinguishes the propagation of waves in a medium from corpuscular signals, the velocity of which would depend on the motion of both source and receiver.

These analogies resulted in the appearance of the concept of a ***world aether***, that is, a medium through which light propagates. In analogy with the apparent wind felt during motion through air, an aether wind was expected to be detected as a result of the motion of the Earth through this substance. Numerous experiments conducted by Albert Michaelson and Edward Morley failed to detect this aether wind. This was thought to be strange, since the Earth moves quite quickly in space as it rotates around the Sun.

Many attempts were made to explain these results. For example, it was suggested that the aether could be trapped by the Earth during its motion through space, like the air inside an aircraft, and so appear fixed relative to the experimentalists on its surface. In order to check this idea, experiments were conducted on mountains, but not even a light aether wind was detected. The concept of world aether is closely related to the existence of a preferred frame of reference. On Earth, we have a criterion of "absolute rest" due to the air. A preferred fixed frame of reference may also be bound to the world aether. However, giving preference to the air is possible only because of the existence of the Earth itself, relative to which it is fixed. It is not clear what the preference of the aether should be, i.e. *relative to what*, and why the aether is fixed ($\lessdot C_1$).

▷ Besides the experiments and philosophical thinking regarding the nature of the aether, work started on the mathematical analysis of Maxwell's equations. Any physics equations (for example, the equations of electromagnetic interaction) depend on the properties of space and time. Of course, this fact was realised only very gradually. Joseph Larmor in 1900 and Hendrik Lorentz in 1904 derived the transformations of space, time, and the electromagnetic field that leave Maxwell's equations unchanged (covariant). At the suggestion of Henri Poincare, these transformations were called ***Lorentz transformations***. Lorentz himself was an adherent of the aether theory. For example, he interpreted the shortening of length that followed from his theory as a real ruler contraction caused by the electromagnetic interaction of its atoms with the aether. Poincare seems to have been the first to realise the physical meaning of these transformations and stated the principle of relativity and invariance of the velocity of light relative to different inertial frames of reference. He also interpreted the transformations as rotations in four-dimensional spacetime before Hermann Minkowsky introduced this concept in 1908. Nevertheless, Poincare's works were just improvements of Lorentz's electromagnetic theory, and they were rather difficult from a mathematical point of view.

In 1905, Einstein's famous work "On the Electrodynamics of Moving Bodies" appeared. Although "electrodynamics" appeared in the title, the Lorentz transformations were derived in the first section "The Kinematic Part" without explicit reference to Maxwell's equations. The derivation was very simple and took an axiomatic approach: "The following reflections are based on the principle of relativity and on the principle of the constancy of the velocity of light. [9]."

The simplicity and generality of Einstein's arguments meant that they were readily accepted by many of his contemporaries. In fact, his work gave birth to a new theory of space and time. Relativistic physics was the first example of a generalization of Newton's mechanics. After 1905, the theory of relativity developed rapidly within the paradigm proposed by Einstein. In time, the fundamental character of the velocity of light $c = 299792458$ m/sec became so evident that since 1983 its value, has been set as exact. Nowadays, our unit of length is defined as the distance over which light propagates in a single unit of time, which is in turn related to the radiation frequency in a caesium atom.

▷ Now let us tell a story that is not so widely known. On September 21 1910 at the meeting of German naturalists and doctors, Vladimir Ignatovskiy spoke about the derivation of the Lorentz transformations using their group properties [10]. The next year, a similar work by Philipp Frank and Hermann Rothe [12] generalizing Ignatovskiy's results was published in Annalen der Physik. They noticed that the most general transformations between two inertial observers have a linear-fractional form, and obtained an explicit form of the generalized transformations lying in the class of linear functions. By 1910, Einstein's authority had grown significantly and works in which, in fact, correct axiomatics of the theory of relativity were established went almost unnoticed. The only mention of them in popular manuals may be found in "Theory of Relativity" by Wolfgang Pauli [17]. But he plays down the significance of Ignatovskiy's, Frank's, and Rothe's works and writes,

> Nothing can, naturally, be said about the sign, magnitude and physical meaning of α. From group-theoretical assumptions it is only possible to derive the general form of the transformation formulas, but not their physical content.

Of course, this is not true, and, starting from the velocity-addition law obtained from the transformations, it is not difficult to establish the meaning of the constant $c = 1/\sqrt{\alpha}$ as an invariant maximum possible velocity. The fact that this velocity is also the velocity of light does not follow from the group assumption, and actually, *it need not be the velocity of light.* The sign of α and its magnitude are the subject of experimental investigations. In fact, two possibilities exist, $\alpha > 0$ and $\alpha < 0$. In our world, the first possibility has been realised, but the second one does not contain any self-contradictions, and might also occur.

2.5 Theory and experiment

Physical theories are created in order to describe the real world. No matter how beautiful a theory may be, without experimental confirmation, it remains only a theory. In its early days, the theory of relativity was confirmed by a number of experiments designed to measurement of the velocity of light under different conditions. These experiments had to involve electromagnetic waves, since the velocities achievable with other objects were too small to confirm the effects of the new theory.

Today, the situation is different.

! In fact, the theory of relativity has become an engineering science.

Due to the development of accelerator apparatus, it is now possible to accelerate very light particles, such as electrons, protons, atomic nucleii, etc. to ultrarelativistic velocities. These experiments are routine and are performed using the accelerators in many scientific centres every day. None of them has detected any deviations from the main consequences of the theory.

Firstly, these experiments pertain to relativistic mechanics. The equations for energy and momentum hold with huge accuracy. With their help, the tracks (trajectories) of the microparticles in electromagnetic fields are calculated. Based on these calculations, the technical parameters of accelerators and the masses, and other characteristics, of particles are established. Relativistic conservation laws explain many of the reactions that take place between elementary particles.

Such accelerators are also a platform for the testing and practical application of the direct consequences of the Lorentz transformations. At velocities as large as those reached in the accelerators, the unusual conclusions of relativity theory such as time dilation in a moving reference frame can be confirmed with high accuracy (p. 53). Many particles have a very short life time, but this increases precisely as predicted by the theory if the particles are moving fast [13]. Time dilation (relativistic and gravitational) must be taken into account in satellite navigation. So, in fact, further verification of the theory of relativity takes place each time we turn on the GPS positioning systems in our phones.

We should also not forget the scientific symbol of the 20th century that is the formula $E = mc^2$. All of energetics is ultimately based on this relation. The enormous energy "stored" in the mass of particles admits both peaceful and military implementation in the form of nuclear and thermonuclear weapons.

It is important to understand that there is no "main", crucial experiment that confirms the theory of relativity. The historically important experiments of the end of the 19th and beginning of the 20th centuries are not such experiments. Our belief in the truth of the theory of relativity is a consequence of the self-consistency of a huge number of experiments and theories that have arisen based on relativistic physics.

For example, the combining of relativity theory, quantum mechanics, and Maxwell's electrodynamics yielded *quantum electrodynamics*. Its properties are defined by three physical constants, which are the velocity of light "c", Planck's constant "$\hbar$" and the electron charge "e". Their dimensionless relation is called the *fine structure constant*,

$$\alpha = \frac{e^2}{\hbar c} = 1/137.035\,999\,679(94).$$

The predictions of these theories have been confirmed experimentally with incredible accuracy. Let us give only one example. The majority of elementary particles have spin, that is, their own rotational moment. If a particle is charged, then it acquires a magnetic moment, so that it becomes a small magnet. This moment can be measured by observing its interaction with an external magnetic field. Without taking quantum electrodynamic effects into account, the value of the magnetic moment may be calculated quite easily. But the result of this calculation differs from the measured value a little, so the term *anomalous magnetic moment* appeared. Quantum electrodynamics allows for more precise theoretical calculations with the help of the perturbation theory containing the constant α. Then, the relative error between the theoretical and experimental results is on the order of 10^{-12}. This is actually a success of the theory of relativity, which lies at the base of quantum electrodynamics! Similarly, astrophysics and cosmology are being developed due to the theory of relativity.

Certainly, all this does not mean that relativity theory is some ultimate truth. On the contrary, it would be very interesting to determine the scope of its applicability, beyond which, probably, new and even more unusual physics is hidden. However, the theory of relativity works wonderfully in the range of situations currently accessible by experiment, and no other theory exists that is able to explain *the whole variety* of experimental data accumulated until now. Let us consider some experiments that have played an essential historical role in the creation and development of the theory of relativity.

▷ One of the most famous experiments performed in the early days of the theory of relativity was that performed by Albert Abraham Michelson (1881). At the time, the aether theory dominated, and a major task was to establish the velocity of the Earth relative to the rest frame of the aether. In Michelson's experiment, a light ray is split into two rays using a semi-transparent mirror M.

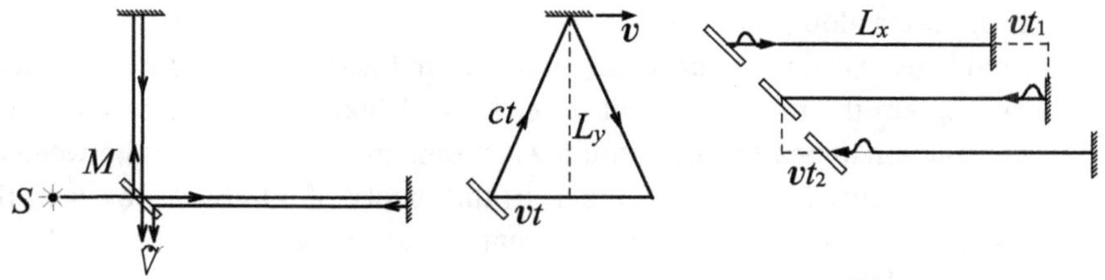

The rays then cover approximately equal distances in two perpendicular directions and turn back after being reflected by mirrors. After that, they are combined again.

Let us carry out the calculations of this experiment from the point of view of the system bound to the "aether". Let c be the velocity of light and let the experimental set-up (*Michelson interferometer*) move through the aether with the velocity v in a horizontal direction. In accordance with Pythagoras' theorem, during the time interval t_y, in a horizontal direction ("forward-backward"), the distance covered by the light ray is given by (see the second figure above),

$$(ct)^2 = L_y^2 + (vt)^2 \qquad \Rightarrow \qquad t_y = 2t = \frac{2L_y/c}{\sqrt{1 - v^2/c^2}}.$$

While moving horizontally, the light ray travels a longer path parallel to the Earth's motion (the mirror "recedes" from it), and travels a shorter path in the opposite direction,

$$\begin{aligned} ct_1 &= L_x + vt_1, \\ ct_2 &= L_x - vt_2 \end{aligned} \qquad \Rightarrow \qquad t_x = t_1 + t_2 = \frac{2L_x/c}{1 - v^2/c^2}.$$

The Earth moves around the Sun with velocity $v = 30\ km/sec$. Thus, the quantity $v/c \sim 10^{-4}$ is small, and the denominators in the expressions for t_x and t_y may be expanded in series. Since $\sqrt{1-x} \approx 1-x/2$ and $1/(1-x) \approx 1+x$, for the time differences $\Delta t = t_x - t_y$ we obtain

$$\Delta t \approx \frac{2L_x}{c}\left(1 + \frac{v^2}{c^2}\right) - \frac{2L_y}{c}\left(1 + \frac{v^2}{2c^2}\right) = \frac{2(L_x - L_y)}{c} + \frac{2L_x - L_y}{c}\frac{v^2}{c^2}.$$

The "horizontal" part of the light wave reaches the observer Δt seconds later than the "vertical" part. Their addition yields an interference picture.

If the apparatus is rotated slowly so that the vertical and horizontal arms are becoming swapped, the interference picture must change. This change allows us to calculate the velocity v. The experiment, however, yielded negative results - the velocity v turned out to be zero within the experimental uncertainties.

From the point of view of the theory of relativity, the results are explained as follows. As we will see in the next chapter, the horizontal (parallel to the motion) arm is shortened by a factor of $\sqrt{1 - v^2/c^2}$ relative to the "aether" (more precisely, to a stationary observer), while the length of the perpendicular arm does not change. After the replacement $L_x \mapsto L_x\sqrt{1 - v^2/c^2}$ has been made, the arm's passing time will pick up a similar dependence on the velocity and a rotation will not change the interference picture. But, in the reference frame associated with the interferometer, the light velocity is the same along each path.

Fitzgerald suggested the contraction of bodies in the direction of motion as a hypothesis to explain the negative result of Michelson's experiment. In Lorentz's theory, this contraction appeared as a result of electromagnetic interactions between the charged medium particles and the aether. In the theory of relativity, however, this property is just a kinematic effect which appears as a result of the very general properties of space and time.

The experiment described above was repeated further by Michelson and Morley (1887), and then Morley and Miller (1902-1904), and gave a negative result each time. In 1928, however, Miller declared the existence of "aether wind", but its velocity $v = 10\ m/sec$ *did not depend* on the velocity of the Earth's motion and was orientated perpendicularly to the orbital plane. Later, Kennedy, Joos, and others repeated his experiment and did not discover any effect.

The negative results of the Michelson-Morley experiment might be explained with the help of a corpuscular (ballistic) light model, where the velocity of the "light corpuscles" is added to the velocity of the set-up and has a constant value relative to it. But this model contradicts observations of binary stars (this is an argument of Willem de Sitter, 1913). If one adds velocities classically, then stars rotating around a common centre emit light "corpuscles" with the velocity $c + v$ if they are coming closer and with the velocity $c - v$ if they are moving away. At large distances, a later, but fast "picture" may outpace an earlier, slower one. If this were the case, then the visible behaviour of binary stars would be very strange, but this is not what is observed.

▷ The experiment by Fizeau (1851) was the first observation of the relativistic velocity-addition law. The scheme of this experiment is presented in the figure below. The light ray from the source S is split into two rays with the help of the semi-transparent mirror M. One of them travels the path AB with the water current and the second of them travels along CD against the current. The water runs through a glass tube with a velocity v of about 7 m/sec.

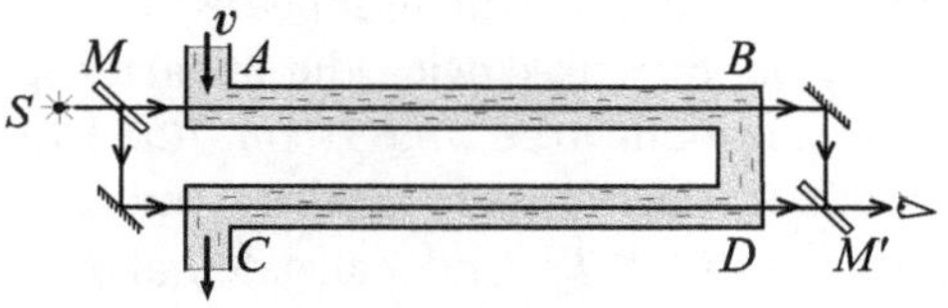

In stationary water, the velocity of light is lower than in a vacuum, $u' = c/n$, where $n > 1$ is a refraction index. This is due to the constant absorption and re-emittance of the light by the atoms of the medium. The velocity of the light is defined by the relativistic velocity-addition law relative to the laboratory reference frame; it increases with the current and decreases against it. Up to first order in v, this gives

$$u = \frac{u' \pm v}{1 \pm u'v/c^2} \approx (u' \pm v)(1 \mp \frac{u'v}{c^2}) \approx u' \pm (1 - \frac{u'^2}{c^2})v = \frac{c}{n} \pm (1 - \frac{1}{n^2})\, v,$$

where the approximate equality is obtained by expanding the denominator ($1/(1+x) \approx 1 - x$), since the water velocity $v/c \approx 2 \cdot 10^{-8}$ is very small. Exactly this dependence of the velocity of light on n and v was observed in the experiment. Certainly, it was not the velocity itself that was measured. Any peak of the light wave is split into two in M. Furthermore, these peaks travel with different velocities. After crossing the same path, one of the peaks recedes a little from the other. This effect is observed in the interference picture after the rays are re-combined by the semi-transparent mirror M'.

Fizeau himself, along with other scientists, believed for a long time that this experiment was evidence that the water ***partially*** captured the light aether, and that if the classical velocity-addition rule held, then the light's velocity c/n increased not by v but by $(1 - 1/n^2)v$. This property of the aether certainly would have explained Fizeau's experiment, but raised another question immediately. Namely, that of explaining the entrapment of the aether by the water. We should note that Fresnel's theory, and further Lorentz's electron theory, went a long way towards answering this question. In these theories, the situation is actually standard and the experiments can be "explained" *without* the theory of relativity. In any case, it is possible to invent some explanatory theory or another, but, sooner or later, this is contradicted by other experiments and requires more and more "patches".

▷ Fizeau's name is also associated with the first laboratory measurement of the velocity of light (1849). He used a rapidly rotating toothed wheel. A light ray passed through a hole between the teeth, was reflected off a mirror, and was returned back.

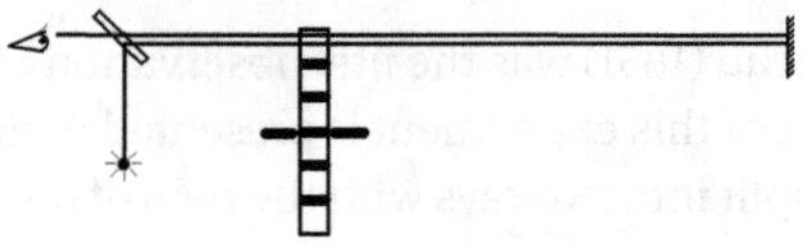

If the wheel did not have enough time to rotate by half a tooth during this motion, the light did not reach the receiver. In Fizeau's experiment, the distance to the mirror was 8.633 km. The wheel had 720 teeth. The light disappeared at the wheel velocity 12.6 rps. When the rotation velocity was increased twice, the signal reappeared, and so on. As a result, the velocity of light was found to be $2{\cdot}8.633\ km\ {\cdot}720{\cdot}2{\cdot}12.6/s \approx 313000\ km/s$.

In similar modern set-ups, electric light "breakers" are used rather than mechanical ones. These are known as *Kerr cell's*. Under an external electric field, some liquids transform a plane-polarized wave travelling through them into an elliptically polarized one. If no field is present, the polarization does not change. Imagine there are two polarizers with perpendicular axes in the path of the light, before its entrance into the liquid, and after its escape. In the absence of an external field, such a system is opaque, because the first polarizer cuts off all of the waves, excluding only those which are vertically polarized, for example. The second polarizer transmits only horizontally polarized waves. When the external field is turned on, however, the plane-polarized wave becomes elliptically polarized, part of it passes through the second polarizer, and the Kerr cell becomes transparent. Unlike a mechanical wheel, a light breaker like this may have very high frequency, up to $10^9 \div 10^{12}$ breaks per second.

If in Fizeau's experiment the relative error was 5%, 100 years later this experimental uncertainty was decreased by five orders of magnitude with the help of the Kerr cell. In absolute magnitude this uncertainty is not more than 1 km/sec. Using modern frequency standards, the uncertainty in the velocity of light has been decreased by a further three orders of magnitude, down to 1 m/sec.

2.6 Newtonian space and time*

Modern physics rejects the concepts of absolute space and time of classical Newtonian physics. Relativistic theory demonstrated that space and time are relative. Perhaps no other phrase appears more often in works on the history and philosophy of physics. But the situation is not all that simple, and such statements require additional clarifications (to be honest, quite linguistic ones). Nevertheless, in order to understand the state of modern science, it is often necessary to appeal to its foundations.

We know that time can be measured with the help of some uniform periodic process. But how do we know whether a process is ***uniform*** without a measure of time? The logical shortcomings in the definitions of such primary concepts are evident. The uniformity of the clock rate must be postulated and called the uniform time passage. For example, by defining time via steady motion, we in fact turn Newton's first law into a definition of uniform time flow. The clock rate is uniform if a body on which no external force is acting moves uniformly (by this clock) in a straight line. In this case, motion is considered relative to an inertial frame of reference which also requires Newton's first law and a uniform clock for its definition.

Another shortcoming originates from the fact that two processes which are uniform to a given precision may be revealed to be non-uniform when a more precise measurement is made. We continue to come up with increasingly more reliable etalons of time uniformity.

As mentioned above, a process is considered uniform and an acceptable measure of time only if it allows that all other phenomena be described as simply as possible. Obviously, some degree of abstraction is necessary in such a definition of time. The permanent search for the "right" clock is a result of our belief that time should have some sort of objective uniformity.

Newton understood the existence of these difficulties excellently. Moreover, in his "Principles", he introduced the concepts of absolute and relative time in order to emphasize the necessity for abstraction, and for the definition of some mathematical model of absolute time based on a concept of relative time as an "ordinary", measurable quantity. ***Here***, his understanding of time does not differ from the modern one, though some confusion has appeared because of ***differences in terminology***.

Let us turn to "The Mathematical Principles of Natural Philosophy" (1687) [4]. In simplified form, the Newtonian definitions of absolute and relative time are as follows.

> Absolute time, true and mathematical, flows equably in itself and by its nature without a relation to anything external, and by another name is called duration. Relative, apparent, and common time is any sensible external measure of duration you please (whether with accurate or with unequal intervals) which is commonly used in place of true time.

The relation between these two concepts, as well as their necessity, is clear from the following explanation.

> Absolute time is distinguished from relative time in astronomy by the common equation of time. For the natural days are unequal, which commonly may be taken as equal for the measure of time. Astronomers correct this inequality, so that they measure the motion of the heavens from the truer time. It is possible that there is no uniform motion, by which time may be measured accurately. All motions can be accelerated and decelerated, but the flow of absolute time is unable to change.

Relative Newtonian time is measurable time, while absolute time is its mathematical model, with properties that are obtained from relative time via abstraction. In general, speaking of time, space, and motion, Newton emphasizes repeatedly that they are perceived by our feelings, and thus are ordinary (relative).

> Relative quantities are not therefore these quantities themselves, the names of which they bear, but those perceptible measures (true or mistaken) of them which are used by ordinary people in place of the measured quantities.

Modelling these concepts requires the introduction of mathematical (absolute) objects that are ideal and independent of hardware inaccuracies. Newton's statement that "absolute time, true and mathematical, flows equably in itself and by its nature without a relation to anything external" is usually understood as a statement about the independence of time and motion. But it can be seen from the quotes above that Newton speaks of the necessity of abstracting away from the possible inaccuracies in the uniform rate of any clock. For him, absolute and mathematical time are synonymous!

Newton does not discuss anywhere the question that the rate of time flow may differ in different relative spaces (reference frames). Without any doubt, classical mechanics means the same uniformity of time in all frames of reference. But this property of time seems so evident that Newton, who was very precise in his statements, does not discuss it, and does not formulate it as one of the definitions or laws of his mechanics. Precisely this property of time is thrown away in the theory of relativity. Absolute time, however (*as Newton meant it*), still exists in the paradigm of modern physics.

Now, let us turn to Newtonian physical space. If we understand the existence of some preferred, privileged frame of reference to mean that space is absolute, there is no need for a reminder that such a frame is absent in classical mechanics. Galileo's brilliant description of the impossibility of establishing the absolute motion of a ship is a spectacular example of this. The theory of relativity did not refute classical mechanics on the non-existence of an absolute reference frame, which is rather absent from both theories. Nevertheless, the question of the relationship between absolute and relative space is not quite clear in Newton's approach. On the one hand, for both space and time, the term "relative" is used to mean "measurable quantity" (that is, perceived by our feelings) and the term "absolute" refers to "its mathematical model".

> Absolute space, by its own nature without relation to anything external, always remains similar and immovable: relative [space] is some mobile measure or dimension of this [absolute] space, which is defined by its position to bodies according to our senses, and by ordinary people is taken for an unmoving space

On the other hand, the discussion of the sailor on the ship is placed in the text. It may be read as a description of a preferred reference frame.

> If the earth also is moving; there is the true and absolute motion of the body, partially from the true motion of the ship in an unmoving space, partially from the motion of the ship relative to the earth: and if the body is moving relatively in the ship, the true motion of this arises partially from the true motion of the earth in motionless space,and partially from the relative motion both of the ship to the Earth as well as of the body to the ship; and from these relative motions the motion of the body relative to the Earth arises.

Hence, a concept of absolute motion is introduced that contradicts the Galilean relativity principle.

Nevertheless, absolute space and motion are introduced only in order to immediately place a question mark over their existence.

> In truth since these parts of space cannot be seen and distinguished from each other by our senses; we use in turn perceptible measures of these. For we define all places from the positions and distances of things from some body, which we regard as fixed... moreover it is impossible to know in turn from the situation of bodies in our regions, whether or not any of these at a given remote position may serve [to determine true rest in the absolute space for local bodies]; true rest cannot be defined from the situation of these bodies between themselves.

It is possible that the necessity of considering absolute space and absolute motion in it is related to the analysis of the relation between inertial and non-inertial frames of reference. Discussing the experiment with a rotating bucket, Newton shows that the motion is absolute in the sense that it can be identified within the bucket-water and water systems by the concave shape of the water's surface. Here, his point of view coincides with the modern one, too.

In the phrases quoted at the beginning of this section, there arose confusion as a result of the noticeable difference in usage of the terms "absolute" and "relative" by Newton on the one hand, and modern physicists on the other. Nowadays, by an absolute quantity, we mean one that is the same for different observers. As for relative quantities, they may be different for different observers. Instead of the *Newtonian* "absolute space and time", now we refer to a "mathematical model of space and time".

> Therefore, those who interpret these words truly violate the meaning of the Holy Scripture.

The mathematical structure of classical mechanics and relativity theory is well-known. The properties which these theories confer on space and time follow unambiguously from this structure. But the misty (philosophical) reasoning of deprecated "absoluteness" and revolutionary "relativity" hardly bring us nearer to a solution of the Main Mystery. The theory of relativity has its name arguably because it really has demonstrated that many things which seem absolute at small velocities are, at large velocities, not.

3 Kinematics

This chapter deals with the main kinematic effects of the theory of relativity. We will begin with the relativity of simultaneity and time dilation. Following convention, we will discuss a thought experiment involving two twins, which is sometimes called the twin paradox.

Observers can make measurements only within their immediate neighbourhood. This means that they can only receive information about remote events using various signals, most often with the help of light. Information received in this way must be corrected, in view of the finite value of the speed of light and some other effects of the theory of relativity. The surrounding universe does not "really" appear exactly as we observe it. First, we will consider the Doppler effect, distortions of moving objects, and the aberration effect. Then we will obtain relationships for motion under uniform acceleration.

https://doi.org/10.1515/9783110515886-003

3.1 Time

Consider two simultaneous events (e.g. flashes of light) in the reference frame S'. Suppose they happen at different points in the frame. Then the Lorentz transformations for the time and space increments are (see p. 19):

$$\Delta t' = \frac{\Delta t - v\Delta x}{\sqrt{1-v^2}}, \qquad \Delta x' = \frac{\Delta x - v\Delta t}{\sqrt{1-v^2}}. \tag{3.1}$$

Simultaneity means that $\Delta t' = 0$ ($t_1' = t_2'$), so from the first equation we get: $\Delta t = v\Delta x$. If $\Delta x = x_2 - x_1 > 0$, then $\Delta t = t_2 - t_1 > 0$. Therefore, from the perspective of an observer "at rest" in S, the *left* event happens *before* the right one ($t_2 > t_1$):

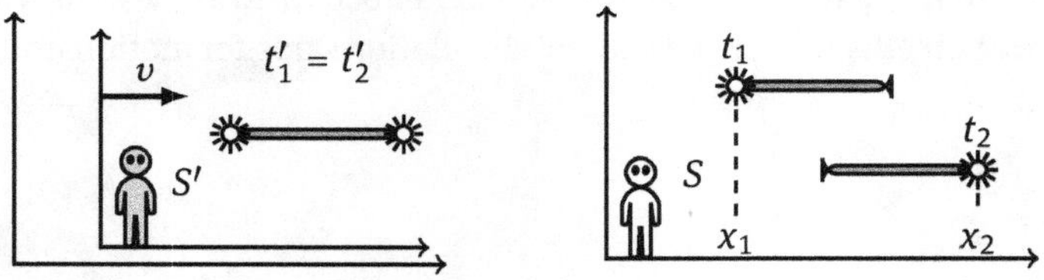

The concept of simultaneity is relative; events which are simultaneous for one observer will not appear simultaneous to another.

In particular, a spherical wave, as considered on p. 22, will look different to each observer, since a sphere is a set of *simultaneously* observed points equidistant from its centre.

According to the principle of relativity, all inertial frames are equivalent. Therefore, simultaneous events in the frame S will not appear to be simultaneous to observers in S'. It follows from the Lorentz transformations that $\Delta t' = -v\Delta x'$ for $\Delta t = 0$.

Because of the relativity of simultaneity, clocks cannot be synchronized between different inertial frames over the whole of space. Time synchronization within one inertial frame implies that all of its clocks show the same time *simultaneously*. However, from the perspective of another inertial frame, the same clocks show different times (compared to those synchronized within this frame).

This difference increases with the velocity of frame the S' and the distance between the events. To recover the constant "c" in the equation $\Delta t = v\Delta x$, we use the substitution given above (p. 16):

$$c\Delta t = \frac{v}{c}\Delta x, \qquad or \qquad \Delta t = \frac{v}{c^2}\Delta x.$$

In classical mechanics we have $\Delta t = \Delta t' = 0$, i.e. that simultaneity is an absolute concept, which immediately gives $c = \infty$.

▷ Another interesting effect is that of time dilation in moving frames. Let $\Delta x' = 0$, i.e., suppose the clock is at rest in S' and is moving relative to S: $\Delta x = v\Delta t$. In this case, we obtain from (3.1) that:

$$\Delta t' = \frac{\Delta t - v(v\Delta t)}{\sqrt{1-v^2}} = \Delta t\sqrt{1-v^2}$$

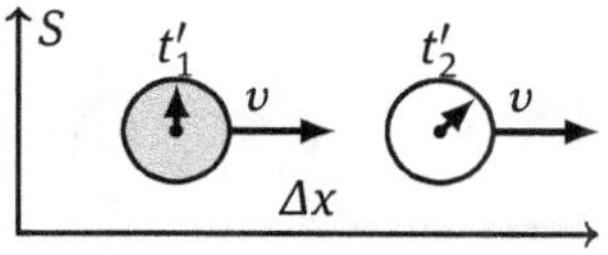

The time *interval* $\tau_0 = \Delta t'$ in S' is commonly denoted by the index zero, while the same interval from the perspective of the observer "at rest" has no index ($\tau = \Delta t$):

$$\tau = \frac{\tau_0}{\sqrt{1-v^2}}. \tag{3.2}$$

The time interval τ_0 measured by the observer at rest relative to the clock is called the clock's *proper time*. Since $v < 1$, we have $\tau > \tau_0$, and the travelling clock appears to run slower than the clock at rest.

▷ As an example, imagine a "light clock" which periodically reflects a flash of light between two mirrors. According to the proper time of the clock, its height and the time of one "tick" are related by $L = c\tau_0$.

For the observer at rest, this "tick" goes along the hypotenuse *at the same speed*, the speed of light. By the Pythagorean theorem:

$$L^2 = (c\tau_0)^2 = (c\tau)^2 - (v\tau)^2.$$

For $c = 1$ we obtain (3.2). For the observer at rest, each tick τ_0 of the moving clock lasts longer, since the light has to pass a longer distance $c\tau$ at the same speed c. All processes in S' appear to occur at this rate, so all of them will appear slower from the perspective of the observer in S.

▷ Is this time dilation real? Yes, just as real as our methods of measuring times and our procedures for coordinating units of measurement. A travelling object will "live longer" from the perspective of resting observers. This effect is observed for short-lived elementary particles. Of course, time dilation is relative. All clocks in S will run slower for the observers in S'. Inertial reference frames are equivalent, and all processes inside them look symmetric. Indeed, this variety of opinions is not surprising. No more so than opinions about the taste of Roquefort cheese.

▷ Let us consider in more detail what the observer sees in the frame of the travelling clock. For clarity, imagine that there are spaceports in S at *great* distances from each other along the x-axis. Each of them is equipped with a clock synchronised with all of the others in S. The figure below shows the observers in S:

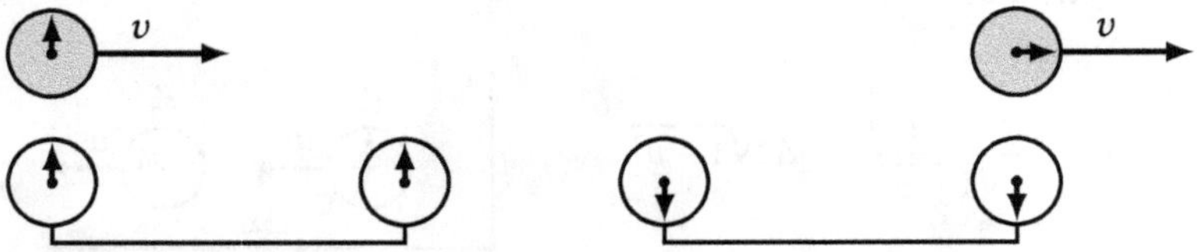

When passing the first clock at the point $x = 0$, a spaceship synchronizes its proper time with this clock. As a result, the clocks in the ship and in the spaceport show the same time, which we assume to be midnight (the first figure). When travelling past each other, the observers in the ship and in the spaceports will each see that "the other" clock is ticking more slowly than their own clock. Generally speaking, the *finite* "tick" of a moving clock can be measured only with a pair of fixed synchronized clocks. We will assume that there are such clocks on the ship.

? What happens when the ship reaches the next spaceport?

The Earth clock still runs *slower* for the astronaut. However, it will show a later time than the clock on the ship. Since the comparison of clocks at the same point in space is absolute (in contrast to their rate, i.e. the interval), the staff of the spaceport will also observe that the ship's clock is slow.

Let's consider the mathematics of this effect. Suppose that the ship's coordinate is $x' = 0$ in the Lorentz transformations (1.13), p. 16. Then the equation of its motion in S is

$$x = vt.$$

The time elapsed since the origins coincided at $x = x' = 0$ is shorter in S' than in S:

$$t' = \frac{t - v\,(vt)}{\sqrt{1 - v^2}} = t\,\sqrt{1 - v^2} < t. \tag{3.3}$$

On the other hand, the astronaut will see that the clock at rest in S ($\Delta x = 0$) is slow (see (3.1)):

$$\Delta t' = \frac{\Delta t}{\sqrt{1 - v^2}} > \Delta t.$$

Thus, although every specific clock in S is slow for the observer in S', different clocks *along his trajectory* show later times. If the astronaut suddenly stops and gets off at some station, he will see that the clock in the spaceport is ticking synchronously with his own clock, but still shows a later time for him. In other words, the astronaut will land in the "future" of the rest frame S.

▷ What is the physical reason behind the strange (from the astronaut's perspective) behaviour of this clock in the "resting" reference frame? To answer this question, consider a squadron of spaceships travelling one after another along the x-axis. Let the central ship $x' = 0$ synchronize its clock with the spaceport clock $x = 0$ (the figure is drawn from the S' perspective):

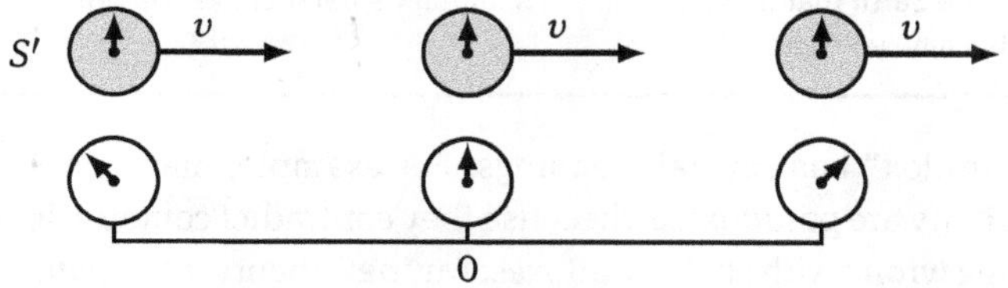

Simultaneously with this event (in S'), the astronauts on the other ships of the squadron see different pictures depending on whether they are travelling ahead of or behind the central ship. Since time is synchronized for the squadron, the origins of the two frames will coincide at the same instant $t' = 0$ for all ships. From the Lorentz transformations we obtain

$$t = vx.$$

Therefore, the spaceports behind ($x < 0$) will appear in the past ($t < 0$), and the spaceports ahead will appear in the future for the observers in S'.

When the central ship reaches a new spaceport, it discovers a future time there, even though its clocks are ticking more slowly. Their slow rate is not enough to compensate for the initial "jump to the future" (3.3). Time dilation always "goes together" with the relativity of simultaneity. Even though all of the observers in some frame appear to be in their present, they simultaneously exist in the past and in the future for observers in other frames.

▷ An interesting question is what happens when the squadron stops *very quickly*, at $t' = 0$, for example. Obviously, this "simultaneous stop" will not look as such in the rest reference frame S: $t = vx$. In fact, the last ship will be the first to slow down, then the central ship stops, and the flagship will be the last one to run its brake engines.

Let the distance between the ships of the squadron remain constant. If the speed of the central ship is zero relative to S, it should also be zero for all of the other ships, which are *at rest relative to it*. When the ships stop, they find themselves in the frame S, and should see reality as it appears to the observers in S, for whom some of the ships have already stopped, while others are still moving.

To explain this "*stopping paradox*", one has to consider accelerated frames, which are covered in Chapter 6. For now, we will discuss the simpler and better known "twin paradox".

3.2 The twin paradox

The twin paradox is wrapped in the romance of interstellar flight and the mist of misinterpretation. It became widely known due to Paul Langevin's formulation (1911), which in the popular phrasing reads as follows:

? One twin brother stays on Earth, while the other travels in space at near the speed of light. For the stay-at-home observer, the traveller's flow of time is slow. Therefore, he will be younger *after returning*. However, it was the Earth that moved, from the astronaut's perspective. Therefore, his stay-at-home brother has to become younger.

The word "paradox" has several meanings. For example, many implications of the theory of relativity are paradoxical, because they contradict common ideas. Of course, there is nothing wrong with such paradoxes. Any new theory is "*unusual*" and requires that old views be changed. However, the "paradox" in this story of twins is a "*logical contradiction*". Two different ways of thinking about the same event (the meeting of the brothers) give different results. This should not happen in a consistent theory.

Opponents of the theory of relativity often consider the twin paradox as proof of its logical inconsistency. Of course, this is not true. As mentioned in the first chapter, there are fewer axioms in the background of the theory of relativity than in the background of classical mechanics. Therefore, if classical mechanics is not contradictory, then the theory of relativity cannot be either. The paradox of the twins' story is due to incorrect reasoning rather than logical contradictions within the theory of relativity. We need to uncover the logical error.

There is a vast literature devoted to the twin paradox. The usual explanation is as follows. The brothers can only *directly* compare their ages if one of them (the traveller) comes back and, therefore, experiences accelerated motion associated with a non-inertial reference frame. Hence, the brothers are not completely symmetric. However, this solution does not explain why it is the astronaut who becomes younger. Moreover, we have the following immediate objection:

? "If it is all about acceleration, then the periods of acceleration and deceleration can be made arbitrarily short (for each observer!) compared to *arbitrarily* long and *symmetric* periods of a uniform motion".

A common answer is that calculations within the general theory of relativity give the same result. In fact, gravity has nothing to do with these calculations, and differential geometry is just a mathematical tool used to describe non-inertial frames. Even though such calculations are absolutely correct (see chapter 6), the physical reasons for what happens to the brothers often remain unclear.

To begin with, there is no need for the traveller to come back. He can just stop, and thus enter the Earth's frame. Although they are at a great distance, the brothers are not moving relative to each other and can synchronize their clocks to discover how much they have diverged (physically and biologically).

There is also the possibility that the spaceship starts moving backwards and returns to Earth. However, this will have no other effect than that of multiplying all times by a factor of two. Generally speaking, there is not even any need to assume an accelerated start from Earth. We can instead assume that the brothers were simultaneously born in two different inertial frames at the moment they were passing each other. Leaving aside the physiological details of such a birth, note that the brothers can easily synchronize the initial time (the event of their birth) at the same point of space, even though they appear in different frames at this moment of time.

Due to the relativity of simultaneity, the observer at rest will see those parts of the moving frame which have already passed him in the direction of travel as being "in the past", while those parts which have not yet reached him will appear to be "in the future". The size of this effect increases with the distance separating these regions from the point of the brothers' birth $x = x' = 0$ (p. 54):

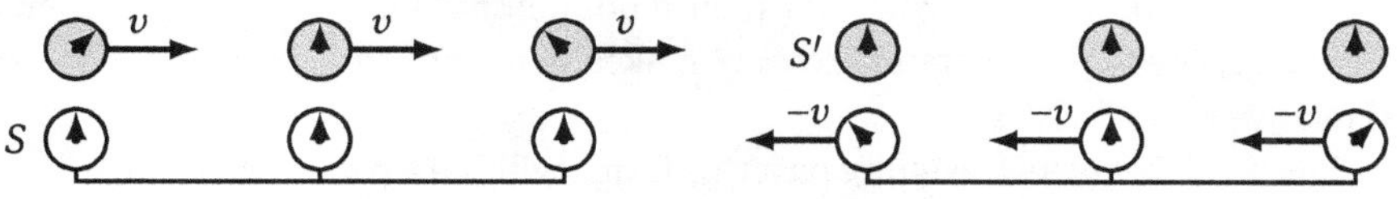

$t = 0, \quad t' = -vx'$ $\qquad\qquad$ $t' = 0, \quad t = vx$

An astronaut travelling past any "stationary" clock $\Delta x = 0$ sees that it runs slower than his own clock,

$$\Delta t' = \frac{\Delta t}{\sqrt{1 - v^2}}.$$

However, all clocks which he ***meets on his way*** will show a future time for him:

$$t' = t\sqrt{1 - v^2}$$

at $x' = 0$. Similarly, the staff of the spaceports see a younger him as he passes them. And if "the same-age nephews" (on the last ships of the squadron) are passing the stay-at-home brother at the same instant of time, they look older than the earthman. These effects are absolute for observers from different frames if they are at the same point in space, so they ***will not change after the stop.*** In fact, there is no need to consider non-inertial reference frames to understand the twin paradox! If the astronaut stops, he will "fall into the future" of the Earth's frame and will be younger there. Similarly, if the earthman accelerates, he will fall into the future of the astronaut's frame and will be younger there.

▷ The most paradoxical thing about the twin paradox is that sometimes it can be more easily explained than formulated. This paradox is often understood superficially, so let us consider the following reasoning:

OK, suppose the twins are not equivalent, and that the astronaut changes his frame of reference. This still does not resolve the paradox if it is formulated as follows. When travelling past *any clock* which is at rest relative to Earth, the astronaut sees that it goes slower than his clock. He is a "former earthman" and knows that all of these clocks are identical. Therefore, he must conclude that the time of his stay-at-home brother goes more slowly. **?**

In contrast to the lengths of rulers, time intervals accumulate. Therefore, the clocks cannot show equal times after the stop. Moreover, if the stop is very fast compared with the time of the uniform motion, it cannot make the slow clock of the Earth brother jump beyond the clock of the spaceship.

Therefore, the time on Earth should (from the astronaut's perspective) go slower, and the Earth brother should become younger. This, however, contradicts the similar reasoning of the earthman, who sees that all processes go slower for the moving objects. So, we cannot say what will happen when the traveller returns (to compare the clocks directly) ...

This *incorrect* reasoning misses something: along with time dilation, there is one more effect to consider, the relativity of simultaneity. In classical mechanics, all observers have one common present, whatever their motion might be. But this is not so in the theory of relativity. A "common present" makes sense only for observers who are at rest relative to each other.

However, observers travelling past this frame will see a continuous range of past, present, and future. Observers far ahead in the direction of the motion will see the distant future of the rest frame, while those moving at the rear will see its past.

! All clocks that the astronauts pass by are slower than their own clocks. However, this does not mean that they show a smaller "accumulated" time! Though slower, the clocks are in the future of the Earth's frame, and when the astronaut reaches them, they do not "have time" to compensate for this difference. As a result, when passing the next clock, the astronaut sees it in the future. Still, the observers of the "rest" frame register a slower time flow for the astronaut's clock.

We will conclude the story about the twin paradox with a fairy tale.

A tale of stem cells

Once upon a time there lived stem cells in some galaxy. One day, these cells were separated and taken to different star systems. The stars in the galaxy moved very slowly, so all of the planets inhabited by sentient beings had a common galactic time and obeyed the cult of political correctness. It had been 13 700 002 010 since the creation of the universe. The stem cells were prompted to divide in this year, and nine months later a girl was born next to each star. These "multiplet" sisters never saw each other and were not particularly worried about this. Except for one of them. Her name was Unity.

The time came for a quest, and Unity decided to take a journey to see her sisters. At that time anyone could take a spaceship across the galaxy, even a half-educated sophomore. So Unity took her dresses, a purse, her notes on physics, boarded a big red ship, and accelerated it to near the speed of light.

Travelling past the first star, Unity joyfully greeted her sister, who was conscientiously studying for her exams. The sisters were as much alike as two peas in a pod. However, Unity's sister's greeting was very sluggish. She slowly waved her hand and dropped her physics textbook in surprise. It started to fall as slowly as if it imagined itself a snowflake.

The same story repeated with the other sisters by which Unity passed. All of them lived at a slow pace. The second hands of their graceful clocks barely crawled over the dial, and the sisters' movements were slow and graceful.

However, while moving away from her home planet, Unity noticed that although her sisters were still slow, they grew older, had children, a dog, and a husband. While she herself was full of strength and youthful temper, she met more and more mature sisters. At last, passing one of the stars, she was heartbroken to see her sister pass, honourably and very slowly, into the other world. After that, Unity never met another sister.

Believing that her mission was over, Unity stopped near one cute star system with a pink spectrum and entered the Department of Physics at the local University. And there, from a textbook written by her sister, a famous scientist in the remote past of this planet, she learned the theory of relativity, and finally realized that the twin paradox appears only if the relativity of simultaneity is forgotten.

* * *

3.3 The Doppler effect

In our consideration of time dilation, we used a clock travelling at a velocity v past an observer in S. Similarly, the rate of *remote* clocks can be measured by receiving periodic signals from them. First, let's consider the one-dimensional situation where the clock approaches a "stationary" observer **A** in the frame S. Assume that the clock is moving with velocity v and periodically emits a signal with velocity $u > v$ (relative to S), with a period of one "second" according to its *proper time*:

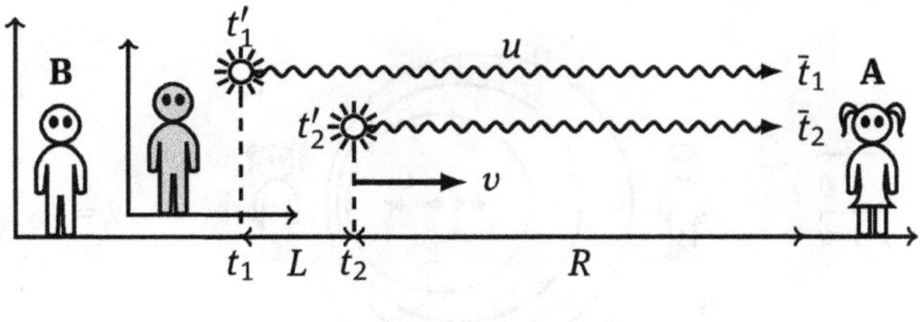

As is customary, the figure above depicts the signals as light waves, though they can be any objects, e.g. pellets periodically shot from a paintball gun with velocity $u < 1$. Suppose the observer **B** is next to the clock in S, and that for him each shot happens at time t_i. However, a remote observer **A** gets the information about the shots (in the form of paint pellets) at later times $\bar{t}_i$.

Consider two shots, the first made at a distance R from the stationary observer **A**, and the second made after the "gun" has traveled a distance L:

$$\bar{t}_1 = t_1 + \frac{R}{u}, \qquad \bar{t}_2 = t_2 + \frac{R-L}{u}.$$

Given that the distance $L = v\Delta t$ in S, we have:

$$\Delta\bar{t} = \bar{t}_2 - \bar{t}_1 = \Delta t - \frac{L}{u} = \Delta t\left(1 - \frac{v}{u}\right).$$

The clock in S' goes slower for the observer **B** in S who is measuring Δt, and

$$\Delta t' = \Delta t\sqrt{1 - v^2}.$$

Therefore, the times of the shots measured by the clock in S' and the times at which they are *detected* by the remote observer in S are related as follows:

$$\Delta\bar{t} = \Delta t'\,\frac{1 - v/u}{\sqrt{1 - v^2}}. \tag{3.4}$$

The numerator in this equation also appears in classical mechanics, whereas the denominator has its origins in relativity and is due to time dilation in the moving frame.

▷ The relationship just obtained plays an important role in determining the velocity v of the remote object. Let's imagine that a source is moving at velocity v and is emitting a light wave with frequency $\nu_0 = 1/\Delta t'$ (by the clock of the source). In this case, the two "gun shots" are two successive maxima of the amplitude of the electromagnetic field strength. The remote observer in S will receive this signal with frequency $\nu = 1/\Delta\bar{t}$. The light wave propagates with fundamental velocity $u = c = 1$.

Consider the two situations where the source is moving away from and towards the observer. The velocity of the signal and the velocity of the source have opposite signs $u = -1$ in the first case, and the same sign $u = 1$ in the second case. Therefore, we obtain from (3.4) the relation:

$$\nu = \nu_0\sqrt{\frac{1 - v}{1 + v}}$$

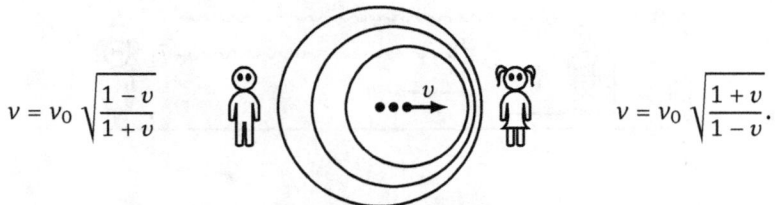

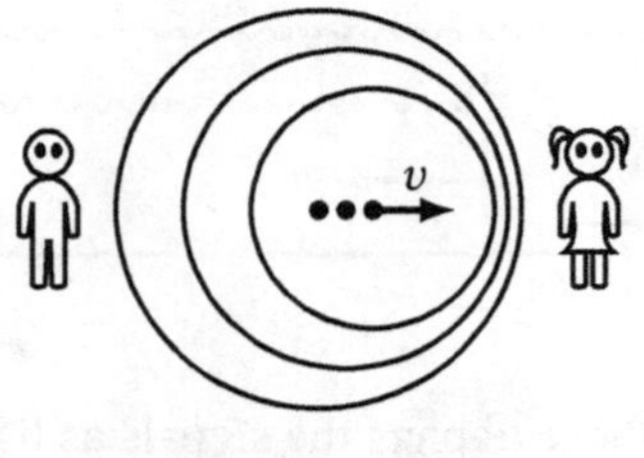

$$\nu = \nu_0\sqrt{\frac{1 + v}{1 - v}}.$$

This frequency change in the light emitted by a moving source is called the *longitudinal Doppler effect*. For the observer, the approaching source has a higher wave frequency than the proper frequency in the frame of the source. On the other hand, a receding source will have a lower frequency for the observer. Red light waves have lower frequencies relative to blue waves. Therefore, the emission spectrum of the receding source shifts to the red area (*redshift*), while the spectrum of the approaching source shifts to the blue area (*blueshift*).

The figure above gives a visual idea of the Doppler effect. A light source emits a spherical wave from each new position. New waves are "squeezed up" against the old ones in the direction of the motion, so that their length λ decreases and their frequency increases. For the receding source, the opposite is true.

Suppose now that the source is moving right "above the observer " so that for some short time it neither approaches nor recedes. Then, its frequency changes only due to the effect of time dilation $\Delta t' = \Delta\bar{t}\,\sqrt{1-v^2}$, and therefore:

$$\nu = \nu_0\,\sqrt{1-v^2}.$$

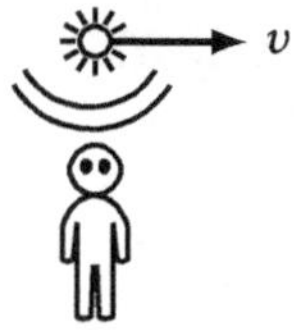

This is a case of the so-called ***transverse Doppler effect.*** In contrast to the longitudinal effect, it is purely relativistic in nature. As an exercise (< H_{10}), recover the constant "c" in the equations for the longitudinal and the transverse Doppler effects.

▷ Now let us combine the equations for the longitudinal and transverse Doppler effects. Assuming that the period between the emission of the pulses is much shorter than the time of their travel to the observer, we have:

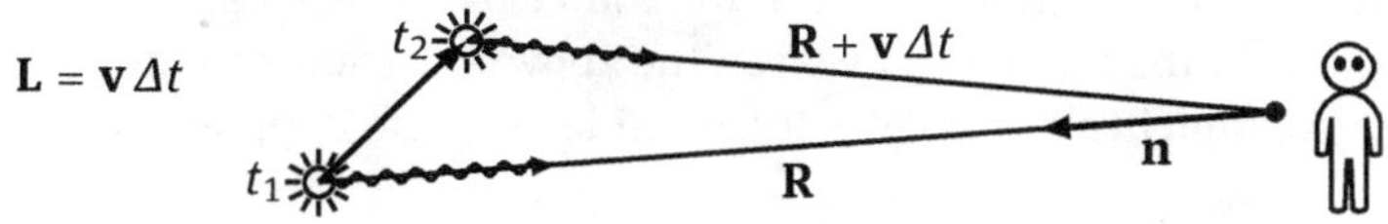

Let the radius vector $\mathbf{R}$ be directed from the observer to the source at the time the first signal is emitted. Since $\sqrt{1+x} \approx 1 + x/2$ (< C_2), we can expand the distance between the observer and the point of the emission of the second signal into a series with respect to small values of Δt:

$$\sqrt{(\mathbf{R}+\mathbf{v}\Delta t)^2} \;\approx\; \sqrt{\mathbf{R}^2 + 2\,\mathbf{R}\mathbf{v}\Delta t} \;\approx\; R\left(1 + \frac{\mathbf{R}\mathbf{v}}{R^2}\Delta t\right) \;=\; R + \mathbf{n}\mathbf{v}\Delta t,$$

where $R = |\mathbf{R}|$ is the distance to the source and $\mathbf{n} = \mathbf{R}/R$ is a unit vector in its direction. Let the velocity of the signals be equal to the fundamental velocity $u = c = 1$. Then the remote observer will measure the following time between their arrival:

$$\Delta\bar{t} = \bar{t}_2 - \bar{t}_1 = (t_2 + R + \mathbf{n}\mathbf{v}\Delta t) - (t_1 + R) = \Delta t\,(1 + \mathbf{n}\mathbf{v}). \tag{3.5}$$

Given that time dilates according to $\Delta t = \Delta t'/\sqrt{1-v^2}$, we obtain the following equations relating the time interval $\Delta\bar{t}$ between signals received in the rest frame to the interval $\Delta t'$ between the signals emitted from the moving frame (and similar relations for the frequencies):

$$\Delta\bar{t} = \Delta t'\,\frac{1+\mathbf{n}\mathbf{v}}{\sqrt{1-\mathbf{v}^2}}, \qquad \nu = \nu_0\,\frac{\sqrt{1-\mathbf{v}^2}}{1+\mathbf{n}\mathbf{v}}. \tag{3.6}$$

Clearly we have $\mathbf{nv} = -v$ if the source is approaching the observer, $\mathbf{nv} = v$ if it is receding, and $\mathbf{nv} = 0$ in case of transverse motion.

▷ There is one interesting manifestation of the Doppler effect. Suppose that there are two stationary markers, that an observer knows that the distance between them is L, and that the distance between the observer and the markers is R. A moving object emits light signals at t_1 and t_2 (by its local clock) when passing the first and the second marker, respectively. The observer will receive the signals at $\bar{t}_1$ and $\bar{t}_2$. Using these values to find the velocity of the object, we obtain from (3.5):

$$\bar{\mathbf{v}} = \frac{\mathbf{L}}{\Delta\bar{t}} = \frac{\mathbf{L}}{\Delta t\,(1+\mathbf{n}\mathbf{v})} = \frac{\mathbf{v}}{1+\mathbf{n}\mathbf{v}}.$$

Thus, the "apparent" velocity $\bar{\mathbf{v}}$ differs from the "real" velocity $\mathbf{v} = \mathbf{L}/\Delta t$. The word "real" means that the velocity is registered by observers near the markers. If the object is approaching the observer, the magnitude of its apparent velocity $\bar{v} = v/(1-v)$ can turn out to be arbitrarily larger than unity (the speed of light).

▷* The observer (receiver) was at rest in the above example. According to the principle of relativity, only the *relative* velocity between the source and the receiver matters. In the equations for the light signals ($u = 1$), the velocity $\mathbf{v}$ is this relative velocity, no matter who "believes" that he is at rest, the source or the receiver.

To emphasize the difference between the above situation and the propagation of signals in a medium in classical physics, consider how the Doppler effect is derived in acoustics.

Let the speed of *sound* relative to the air be c. Let w and u denote the velocities of the receiver and the source relative to the air, respectively (the triangle and the square below). Suppose both of them move along the x-axis (in the figure, from left to right):

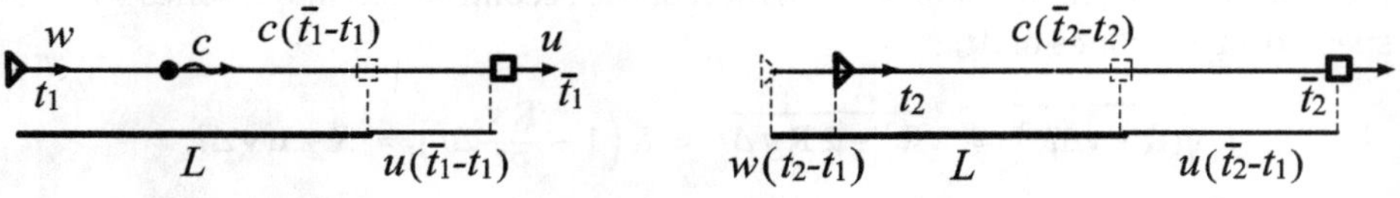

Consider two successive "claps" produced by the source at time instants t_1 and t_2. The signals reach the receiver at $\bar{t}_1$ and $\bar{t}_2$. After the first clap, the source has time to shift $w\,(t_2-t_1)$ to the right, and by the time of the second clap, the has shifted $u\,(\bar{t}_2-t_1)$ from the initial position. As a result, we obtain (right figure above):

$$L - w\,(t_2-t_1) + u\,(\bar{t}_2-t_1) = c\,(\bar{t}_2-t_2).$$

Subtracting the similar relation

$$L + u\,(\bar{t}_1-t_1) = c\,(\bar{t}_1-t_1)$$

for the first signal (the left figure above), we have:

$$\frac{t_2-t_1}{\bar{t}_2-\bar{t}_1} = \frac{\Delta t}{\Delta\bar{t}} = \frac{\nu}{\nu_0} = \frac{1-u/c}{1-w/c} \approx 1 - \frac{u-w}{c} + \ldots,$$

where the last approximate equality is obtained by expanding the denominator in a series ($1/(1+x) \approx 1-x$), and ν_0 is the proper frequency of the radiation (time dilation is not taken into account).

Therefore, if the velocities of the source and the receiver are small relative to the air, then only their relative velocity

$$v = u - w$$

is important for the consideration of effects up to first order. However, this is not true for velocities close to the speed of sound. Here, it is important to know who (source or receiver) is moving relative to the environment, and how. This difference is essential when considering both the relativistic Doppler effect and the classical effects of signal propagation in a medium.

3.4 The Doppler effect and the twin "paradox"

There are various manifestations of the Doppler effect. Besides the changes to the radiation spectra of moving objects, it also distorts our perception of the duration of remote processes. For example, when Ryomer first measured the speed of light in 1676 by observing the orbital periods of Jupiter's moons, he exploited a manifestation of the Doppler effect. The periods decreased when Earth was approaching Jupiter and increased when it was receding. The relativistic component is hard to measure at such velocities. However, the classical component is a first-order effect.

Consider again the twin "paradox" from the perspective of the Doppler effect. Suppose the brothers started transmitting their images to one another immediately after they parted. The traveller sees his brother sitting in an armchair next to the chimney, under the clock. In turn, the earth-dweller sees on his monitor the cabin of the spaceship equipped with an electronic clock over the steering wheel held by his brave travelling brother. The spaceship has to get to the nearest star and come back. Here are some extracts from the ship's flight logbook.

Travel diary.
After rapidly accelerating I have reached a speed approaching the speed of light. The g-forces were enormous, but thanks to the latest achievements in biocybernetics, I withstood them easily enough. The time at which I started to travel is the same by my clock and that of my stay-at-home brother. However, the frequency of the signal I'm receiving has markedly decreased as the Earth rapidly "recedes". The situation is not changing as the distance from Earth grows. The second hand of the clock above my brother's chimney is barely creeping and shows a time considerably behind my own. This is a result of the Doppler effect, combined with the delay in the video transmission due to the finiteness of the speed of light. The very bright stars up ahead look so densely crowded, while the stars behind seem notably scarce and more red in colour. The distances between the automatic beacons along my way have decreased ⟨...⟩.

After reaching my destination, I slowed down and took some memorable photos with the star in the background. When I stopped, the hand of my brother's clock above the fireplace immediately resumed its natural run, though the total time since the beginning of the flight has not changed and is far behind my time. I had nothing to do near that lonely star, and so quickly gathered speed in the opposite direction. Now, having recovered after the acceleration, I can see that my brother's clock has accelerated noticeably, and its second hand is running like crazy.

Earth is not far away. During the return journey, my brother's clock has managed to close the gap and, moreover, has outrun my chronometer. Tomorrow is my landing and the day of our long-awaited reunion. However, I have no doubt that it's me who is now the younger brother.

▷ Consider the effects observed by the traveller and his stay-at-home brother. Let the brothers send precisely timed signals to each other with a period of one second $\nu_0 = 1$ (by their respective clocks). Assume that the periods when the ship is accelerating are much shorter (from the perspective of both brothers) than the time of the whole journey. While the spaceship is moving away from Earth, each brother, due to the Doppler effect, observes a decrease in the frequency (increase in the period) of the signals being received. After the traveller stops near the star, he also stops "running away" from the terrestrial signals, and their period *immediately* becomes equal to his second. However, due to the signal delay, he sees his brother "in the past". After the traveller turns around and accelerates, he starts "running into" the approaching signals, and their frequency increases (their period decreases). The travel time by his clock is t'_1 in both directions. The *number* of "terrestrial second" signals received over the whole time of travel $t' = 2t'_1$ is equal to the frequency ν multiplied by the time:

$$t = t'_1\sqrt{\frac{1-v}{1+v}} + t'_1\sqrt{\frac{1+v}{1-v}} = \frac{2t'_1}{\sqrt{1-v^2}}.$$

The astronaut counted fewer seconds (first term) when moving away from Earth, and more seconds on his way back (second term). According to the time dilation equation, the total number of seconds sent from Earth is greater than the number of seconds sent to it.

The calculations of the earthman are somewhat different. While his brother is receding, the earthman also sees that the time signals from the spaceship come with longer periods. However, unlike his brother, the earthman will observe this effect for a *longer* time. The flight time to the star is t_1 by the Earth clock. The earthman will see the traveller stopping next to the star after some extra time $t_2 = L/c = L$ $(c = 1)$, once some light has covered the distance from the star. This is why a time $t_1 + t_2$ has to pass from the beginning of the journey before his monitor shows that his returning brother's clock has accelerated:

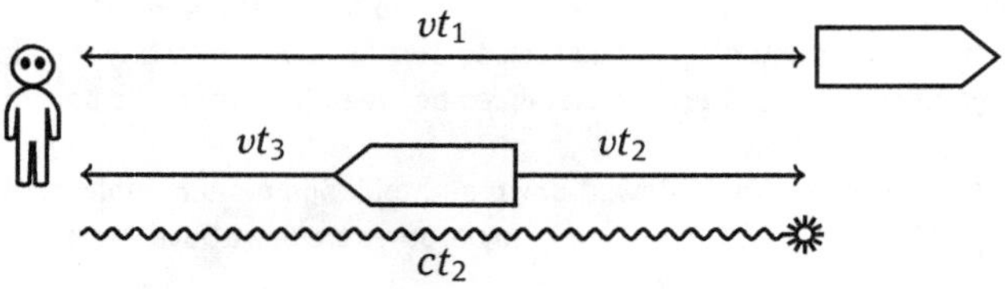

Given that the times are $t_2 = L = vt_1$ and $t_3 = t_1 - t_2$, we obtain:

$$t' = (t_1 + t_2)\sqrt{\frac{1-v}{1+v}} + t_3\sqrt{\frac{1+v}{1-v}} = 2t_1\sqrt{1-v^2}.$$

Therefore, the effect of time dilation for the brother who has changed his reference frame *is absolute* (equal for both brothers).

3.5 Lorentz contraction

In the first chapter we discussed how observers compare their length units in different frames. Two rulers in S and S' are considered identical if oriented perpendicular to the direction of their relative motion. Let's see how these rulers "appear" when oriented along the motion.

Consider a resting ruler in S' which passes an observer in S with velocity v. To measure the length of the ruler, the observer can *simultaneously* ($\Delta t = 0$) mark the coordinates of its beginning and the end. Setting $L_0 = \Delta x'$ and $L = \Delta x$, we obtain from the Lorentz transformations for increments (3.1), p. 52:

$$\Delta x' = \frac{\Delta x - v\Delta t}{\sqrt{1-v^2}} \quad \Rightarrow \quad L = L_0\sqrt{1-v^2}$$

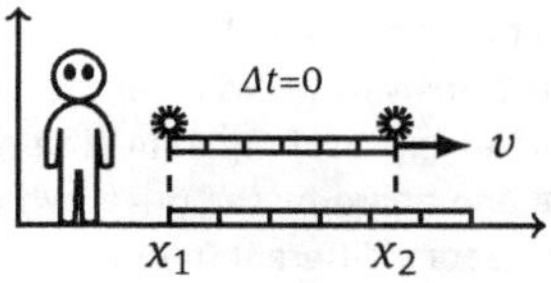

The passing ruler looks shorter than its doppelganger oriented perpendicular to the motion.

However, the observer in S' will be "dissatisfied" with such a measurement. Indeed, the difference between the coordinates of its ends is the ruler's length

$$L_0 = x'_2 - x'_1$$

according to the observer in S' (*proper length of the ruler*). Similarly, due to the simultaneity of the measurements, the observer in S can see that its length is

$$L = x_2 - x_1.$$

However, these measurements will not be simultaneous for the moving observer! Since

$$\Delta t' = t'_2 - t'_1 = -v\Delta x' < 0,$$

the moving observer will see that the left end of the ruler x'_1 matched the ruler of the "stationary" observer later than its right end x'_2.

▷ There is also another way to measure lengths. Suppose the observer in S marks the time t_1 when the right end of the ruler passes him, and then the time t_2 for the left end. Knowing the velocity of the ruler, he can find its length: $L = v\Delta t$. All of this occurs at one point of his reference frame $\Delta x = 0$, and is again associated with a length contraction:

$$L_0 = -\Delta x' = \frac{v\Delta t}{\sqrt{1-v^2}} = \frac{L}{\sqrt{1-v^2}}.$$

However, the observer in the frame of the moving ruler will again be "dissatisfied", because the clock in S goes slower for him, and the length $L = v\Delta t$ is incorrect from the perspective of S'.

Nevertheless, Lorentz contraction in moving frames as just as real as time dilation. Of course, the word "real" can be understood in a variety of ways. Lorentz believed that this contraction was a real consequence of the charged particles of the ruler interacting with the aether. The theory of relativity does not require an aether or any "force" to affect the rulers in different inertial frames. However, the contraction is objective and is registered by the observer. Unlike clocks, a ruler can recover its initial state (certain conditions being met) after being smoothly accelerated and then slowed down. A clock does not behave like this and will always remain "behind".

! **Does the ruler "really" contract?**

To give the answer, we will use the following analogy. Due to the Doppler effect, the frequency the whistle of an approaching train differs from that of a receding train. The "real" frequency of the whistle is the one heard by the driver who is at rest relative to the cab of the train. The observer on the platform hears a different frequency. And though his perception differs from that of the driver, it is no less objective and is not "false". These are not "mind games", these effects are actually registered by appropriate measuring instruments. Such relativity often occurs in classical physics. The Doppler effect and the relativity of velocity are just two examples. The situation is the same for relativistic effects such as length contraction, time dilation, simultaneity of events, etc.

Of course, the contraction which follows from the Lorentz transformations is relative. All inertial frames are equivalent. Just the same, the observer in the "moving" frame S' will see that all rulers at rest in S contract if they are oriented along his motion.

Length contraction, like any other effect of the theory of relativity, requires a clear "accompanying description" of the experimental conditions under which the effect is observed. Many "paradoxes" arise due to discrepancies between the results of observation and these descriptions. As a simple exercise, we invite the reader to consider the following "pencil and pencil case paradox"

? **Pencil and pencil case paradox.**

A quickly moving pencil flies into a stationary pencil case. In the frame of the pencil case, the length of the pencil is contracted, and the pencil case has enough room for it. In the frame of the pencil, the pencil case is contracted, and there is no room for the pencil. The question is: is there "really" enough room for the pencil in the pencil case?

Both of the above measurement methods will register that the ruler's length has contracted. Also, the object can be photographed with a very short exposure time. In this case, the results will depend on the model of the camera and the shape of the object (p. 71).

3.6 Contraction and rotation *

Like a ruler, any object is contracted in the direction of its motion if the coordinates of its surface points are registered ***simultaneously*** at a some instant in time. For example, a square (in its proper frame of reference) will "look like" a rectangle if it is moving along one of its sides. If the velocity vector is directed along the diagonal of the square, the square will be flattened along this diagonal to "become" a rhombus, but the diagonal oriented across the motion will not change its length.

Assume that the sides of the square are the coordinate axes (x', y') of a moving frame. Then, these axes will only be normal to each other for certain velocity directions. Moreover, they will rotate relative to the rest frame. The dashed and solid lines below show the coordinate grids of the "rest" and the moving frame (for observers in S) respectively. The magnitude of the velocity is $v = 0.8$:

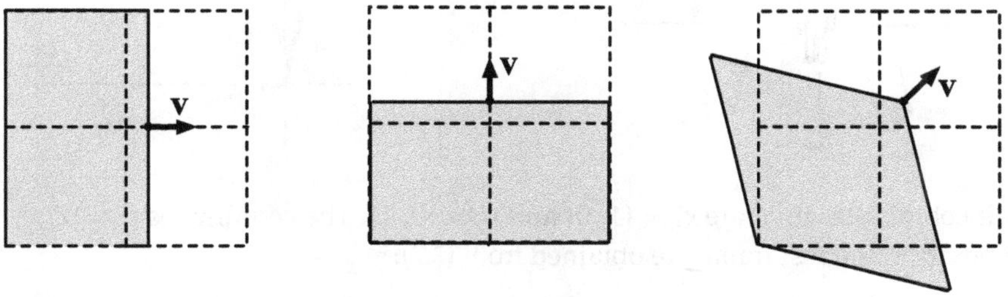

The coordinates of the points $\mathbf{r}$ in the rest frame S can be found from the coordinates $\mathbf{r}'$ of these points in S' (associated with the body) using the Lorentz transformations. Let us find the ***instantaneous*** shape of the moving body at time t in the frame S. To do this, we rearrange the Lorentz transformations so that the radius vector $\mathbf{r}$ depends on t and $\mathbf{r}'$. The inverse Lorentz transformations ($\mathbf{v} \mapsto -\mathbf{v}$), p. 17, are

$$\begin{cases} t &= \gamma\,(t' + \mathbf{v}\mathbf{r}'), \\ \mathbf{r} &= \mathbf{r}' + \gamma\mathbf{v}t' + \Gamma\,\mathbf{v}\,(\mathbf{v}\mathbf{r}'), \end{cases}$$

where

$$\Gamma = \frac{\gamma - 1}{v^2} = \frac{\gamma^2}{\gamma + 1},$$

so rewriting t', we have:

$$\mathbf{r} = \mathbf{v}t + \mathbf{r}' - \frac{\gamma}{\gamma + 1}\,\mathbf{v}\,(\mathbf{v}\mathbf{r}'). \tag{3.7}$$

This equation gives the positions of the points of a moving body in S at a given time t. The first term $\mathbf{v}t$ in (3.7) indicates that all points move parallel to each other with constant velocity $\mathbf{v}$. At $t = 0$, the origins of both frames coincide, and the shape of the moving body is determined by the second and the third terms of (3.7).

▷ At $t = 0$ we obtain from (3.7):

$$\mathbf{vr} = \mathbf{vr}'/\gamma, \qquad \mathbf{r}^2 = \mathbf{r}'^2 - (\mathbf{vr}')^2. \tag{3.8}$$

That is, longitudinal dimensions are subject to Lorentz contraction, and transverse dimensions ($\mathbf{vr}' = 0$) do not change. Consider two fixed points of the body. Using (3.7), we obtain expressions for each of their position vectors, whose scalar product is

$$\mathbf{r}_1\mathbf{r}_2 = \mathbf{r}'_1\mathbf{r}'_2 - (\mathbf{vr}'_1)(\mathbf{vr}'_2). \tag{3.9}$$

Let the body move within the (x, y)-plane. Let one point lie on the x'-axis and the other on the y'-axis:

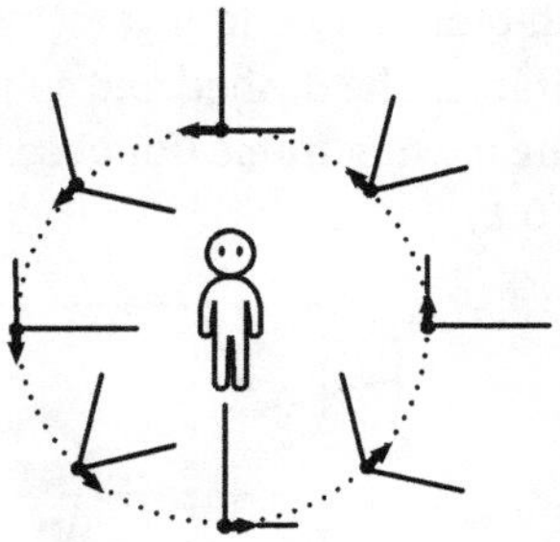

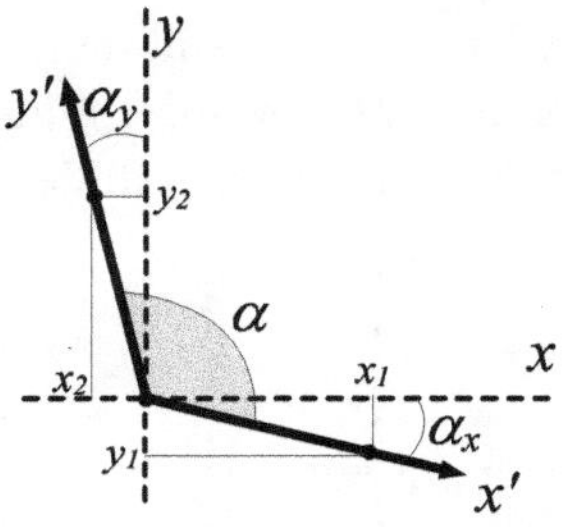

Their coordinates in S' are $\mathbf{r}'_1 = \{1, 0\}$ and $\mathbf{r}'_2 = \{0, 1\}$. The coordinates $\mathbf{r}_i = \{x_i, y_i\}$ of the points in the rest frame are obtained from (3.7):

$$x_1 = 1 - \frac{\gamma v_x^2}{\gamma + 1}, \quad y_1 = -\frac{\gamma v_x v_y}{\gamma + 1}; \qquad x_2 = -\frac{\gamma v_x v_y}{\gamma + 1}, \quad y_2 = 1 - \frac{\gamma v_y^2}{\gamma + 1}, \tag{3.10}$$

where

$$\gamma = \frac{1}{\sqrt{1 - \mathbf{v}^2}} = \frac{1}{\sqrt{1 - v_x^2 - v_y^2}}.$$

As a result, the sine of the angle α_x between the x'- and x-axes and, similarly, the sine of the angle α_y between the y'- and y-axes are given by

$$\sin\alpha_x = \frac{\gamma}{\gamma + 1}\frac{v_x v_y}{\sqrt{1 - v_x^2}}, \qquad \sin\alpha_y = \frac{\gamma}{\gamma + 1}\frac{v_x v_y}{\sqrt{1 - v_y^2}}, \tag{3.11}$$

where the magnitudes of $\mathbf{r}_1$ and $\mathbf{r}_2$ are obtained from the second equation of (3.8). The cosine of the angle between the axes of the moving frame is found from (3.9),

$$\cos\alpha = -\frac{v_x v_y}{\sqrt{(1 - v_x^2)(1 - v_y^2)}}. \tag{3.12}$$

Therefore, the coordinate axes of S' will be normal to the observers in S only if one of the velocity components is zero. This happens if the object is moving along a coordinate axis.

▷ Let us consider another interesting effect. Suppose the frame S' moves relative to the "rest" frame S along the x-axis with velocity v. If an observer in S' studies the motion of a square *along the* y'*-axis*, they will see it flattened in the direction of its motion (the second figure below). Let a third frame S'' be associated with the square.

What does the square look like to observers in the rest frame S?

The square is moving at an angle to the x-axis and, naturally, will turn into a rhombus. However, only the "horizontal" sides of the rhombus will be inclined, while the vertical sides will remain parallel to the y- and y'-axes:

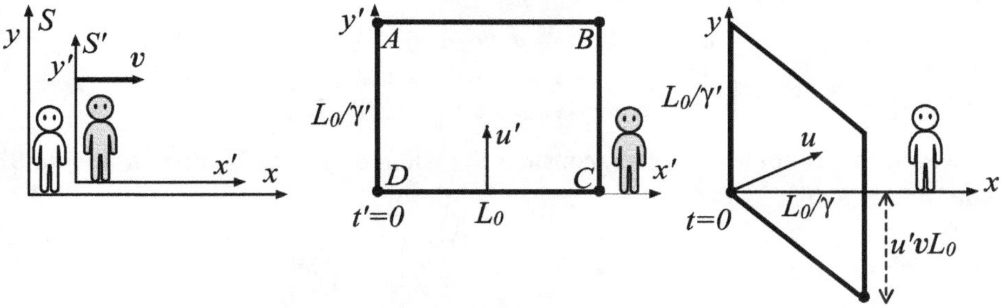

When moving parallel to the y'-axis in S', all points of the square move upwards *simultaneously*, in particular its lower vertices (points D, C) (the second figure). If point D was at the origin ($x' = y' = 0$) at time $t' = 0$, then point C will also lie on the x'-axis ($y' = 0$). However, this simultaneity will be true only in S'.

Simultaneous events in S' ($\Delta t' = 0$) are separated in time by

$$\Delta t = v\Delta x$$

in S (p. 52). Therefore, although both vertices in S' intersect the x'-axis simultaneously, the right vertex C will do it *later* than the left vertex D in the frame S. Since the square moves together with S' with velocity v along the x-axis, it is flattened by a factor of γ along x. It is as if the sides of the square slide along two vertical guides in S'. The distance between the guides in S is L_0/γ. Vertical rulers have the same length in S and S'. Therefore, since the side of the square AD (for example) gets shortened by a factor of

$$\gamma' = \frac{1}{\sqrt{1 - u'^2}}$$

when moving with velocity u' in S', it will have the same length in S. However, the side BC will fall behind AD. As a result, we will have a rhombus which is skewed vertically rather than squeezed along the diagonal (< H_{11}). In particular, if a horizontal rod is lifted up by an observer in S', it will appear *rotated* to an observer in S.

▷* The projections of the velocity of the frame S'' onto S' are $u'_x = 0$ and $u'_y = u'$. Adding these velocities (1.21), p. 19, we obtain the velocity of S'' relative to S:

$$\mathbf{u} = \{u_x, u_y\} = \{v,\ u'\ \sqrt{1-v^2}\}$$

and

$$\gamma_u = \frac{1}{\sqrt{1-\mathbf{u}^2}} = \gamma\gamma',$$

where $\gamma = 1/\sqrt{1-v^2}$, $\gamma' = 1/\sqrt{1-u'^2}$. It might seem that the ordinary Lorentz transformations should be true for the frames S'' and S:

$$\begin{cases} t'' &= \gamma_u\,(t - \mathbf{ur}), \\ \mathbf{r}'' &= \mathbf{r} - \gamma_u \mathbf{u}\, t + \Gamma_u\, \mathbf{u}\,(\mathbf{ur}), \end{cases}$$

or, in component form:

$$\begin{cases} t'' &= \gamma_u\,(t - u_x x - u_y y), \\ x'' &= x - \gamma_u u_x\, t + \Gamma_u u_x\,(u_x x + u_y y), \\ y'' &= y - \gamma_u u_y\, t + \Gamma_u u_y\,(u_x x + u_y y). \end{cases} \tag{3.13}$$

However, this is not true! In fact, consider the chain of transformations from S to S' and from S' to S'':

$$\begin{array}{lllllllll} t' &=& \gamma\,(t - vx), & x' &=& \gamma\,(x - vt), & y' &=& y, \\ t'' &=& \gamma'\,(t' - u'y'), & x'' &=& x', & y'' &=& \gamma'\,(y' - u't'). \end{array}$$

The square is moving along the y'-axis rather than the x'-axis, therefore, x' and y' are interchanged in the transformation from S' to S'' (second line). If t', x', y' are substituted from the first line into the second line ($u' = u_y\gamma$), we obtain the same transformation for t'' as in (3.13). By contrast, the relationship between x'', y'' and t, x, y is different:

$$\begin{cases} x'' &= \gamma\,(x - u_y\, t) \\ y'' &= \gamma' y - \gamma\gamma_u\, u_y\,(t - u_x\, x). \end{cases}$$

This may seem strange, especially in view of the fact that the Lorentz transformations in the one-dimensional case were obtained (p. 12) using a composition of transformations. So why a does a chain of Lorentz transformations give a Lorentz transformation only in the one-dimensional case, and not in the case of a plane?

The answer hints at quite sophisticated issues concerning the mathematical nature of the theory of relativity. As we will see later, space and time make up a single four-dimensional pseudo-Euclidean space. This space allows for various transformations of space and time, as long as they preserve the interval Δs^2. In particular, in addition to the pure Lorentz transformations (so-called ***boosts***) there are also the usual three-dimensional rotations of the Cartesian axes. It turns out that in the general case, a sequence of two boosts is equivalent to a boost followed by a three-dimensional rotation of the x''- and y''-axes through some angle ϕ.

3.7 Photographing objects *

What does a relativistic object look like when photographed or observed "visually"? If its different points are at different distances from us, we will see them at *different* times in the past due to the finiteness of the speed of light. When photographed, a point in *three-dimensional* space (x, y, z) is projected onto the *two-dimensional* surface of the photograph (X, Y) in the (x, y)-plane. We first consider the *orthogonal projection*, when only rays normal to the photograph are registered. Such a "camera" can be designed, for example, by placing a thick plate with cylindrical holes in front of the film. Their walls absorb skew rays and let vertical rays go. The orthogonal transform neglects the information about the coordinate z and has the form: $X = x$ and $Y = y$.

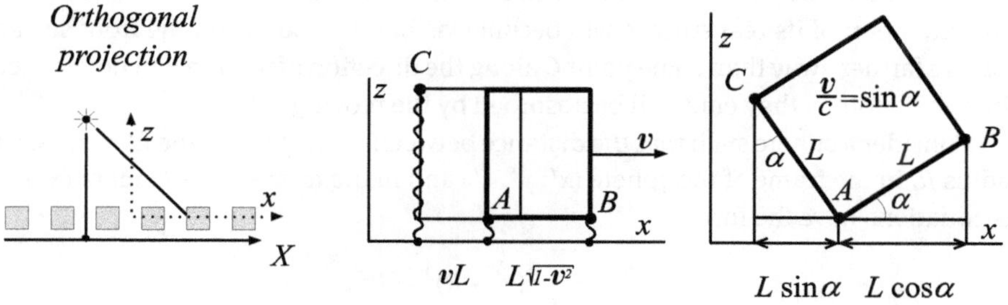

▷ Consider a moving cube with edges of length L (in its proper reference frame), and suppose its picture is taken with the "orthogonal" camera. The camera will capture light quanta that came to the film *at a given point* of time (the exposure time is very short). However, these quanta will have been emitted at different times in the past. The signals from points A and B will pass the same distance (the central figure above). The side of the cube facing the film has length $L\sqrt{1 - v^2}$ and looks contracted. However, from the point C the photon passes an additional distance along the cube's edge (normal to the film), so it is emitted

$$t = \frac{L}{c}$$

time units earlier (further, as usual, $c = 1$). At this instant of time the cube was

$$vt = vL$$

to the left. In the nonrelativistic case ($c = \infty$) the orthogonal projection of the cube onto the photograph will form a square, the image of one its sides. In the relativistic case this side is contracted, but we will also see the left side, which will be v times smaller. The projection will look like a picture of a resting cube rotated through the angle

$$\alpha = \arcsin v$$

(the last figure above).

▷ The situation is similar for a picture of a moving ball:

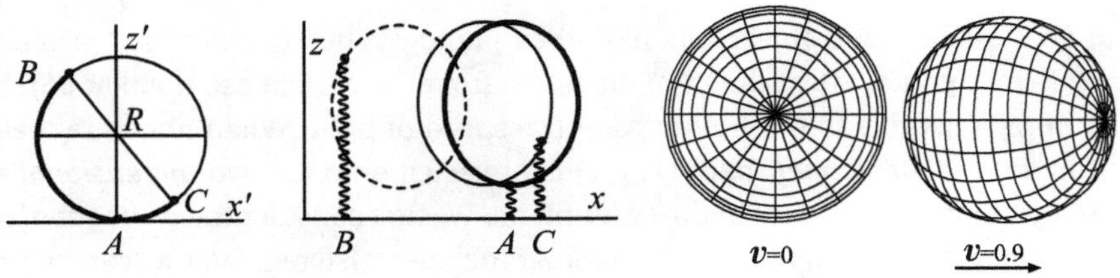

The picture of a stationary sphere represents only those points of the hemisphere which face the camera. A moving sphere can slip to the right "from under the photon" emitted by the point B, so that it will not stand the way of it reaching the film. As a result, a part of its rear surface will become visible. The points on the front surface that are farther away than some point C along the direction of motion will not be seen, since the photons they emit will be absorbed by the moving sphere.

Consider a sphere such that the distance between its center and the film equals its radius R. In the frame of the sphere (x', y', z') and in the frame of the camera (x, y, z) its equations have the form:

$$x'^2 + y'^2 + (z' - R)^2 = R^2,$$

or

$$\gamma^2 (x - vt)^2 + y^2 + (z - R)^2 = R^2, \tag{3.14}$$

where in the second equation we have performed a Lorentz transformation. Consider a beam of light falling on the film ($z = 0$) at the point $X = x$, $Y = 0$ at $t = 0$. Its trajectory along the z-axis is $z = -t$. Substituting this into the equation of the moving sphere (3.14), we find the time of light emission:

$$t = v(X + vR) - R \pm \frac{1}{\gamma} \sqrt{R^2 - (X + vR)^2}. \tag{3.15}$$

A solution exists if the determinant (the expression under the root) is greater than zero. Therefore, we obtain the limiting values of the coordinates X on the film:

$$X_{min} = -R(1 + v), \qquad X_{max} = R(1 - v). \tag{3.16}$$

The difference between these two points is $2R$, or the diameter of the resting sphere. These points emit beams of light at $t = -R(1 + v)$ and $t = -R(1 - v)$. Since $z = -t$, we obtain the coordinates of the extreme visible points on the surface of the sphere:

$$z = R(1 + v) \quad and \quad z = R(1 - v).$$

Under the orthogonal projection, both the cube and the sphere appear rotated without a Lorentz contraction. This effect is called the *Terrell-Penrose rotation* and is typical of "orthogonal" pictures of *three-dimensional* objects. By contrast, a flat ruler moving parallel to the film still shrinks due to Lorentz contraction.

▷ *Projective transformation* is a more typical way to take photos. Suppose there is a single small hole (a diaphragm) in a thin screen. Light beams fall through the hole onto the photograph to create an image of the object (z is measured from the screen):

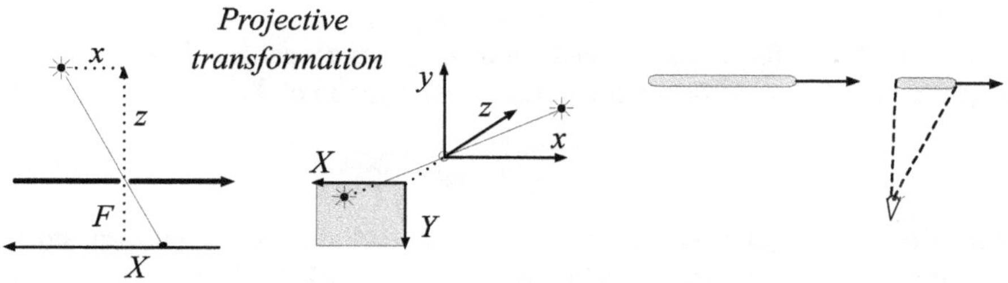

Using the similarity of the triangles we obtain the relationship between the two- and three-dimensional coordinates:

$$X = F\frac{x}{z}, \qquad Y = F\frac{y}{z}, \tag{3.17}$$

where F is the *focal length* between the screen and the film. The image in such "*camera obscura*" is inverted, so the figure represents a rotated image of the photograph's plane (coordinates X, Y).

▷ Let the light source move along the path

$$x = x_0 + vt, \quad y = 0, \quad z = z_0.$$

At some point in time $\bar{t}$ the light emitted in the past enters the diaphragm:

$$t = \bar{t} - \sqrt{x^2 + z^2}.$$

Substituting t into the trajectory to find x, and then using (3.17), we obtain the law for the motion of the source on the film:

$$X(\bar{t}) = \frac{F}{z_0}\gamma^2\left(x_0 + v\bar{t} - v\sqrt{(x_0 + v\bar{t})^2 + z_0^2/\gamma^2}\right). \tag{3.18}$$

Consider a rod of proper length L_0 moving parallel to the film. Relative to the camera, it is subject to Lorentz contraction and has length $L = L_0/\gamma$. Assume $x_{01} = -L_0/2\gamma$ and $x_{02} = L_0/2\gamma$ for two light sources (the beginning and the end of the rod). Then the length of the rod $\Delta X = X_2(\bar{t}) - X_1(\bar{t})$ *on the film* changes with time as:

$$\Delta X = \frac{F}{z_0}\gamma\left[L_0 - v\sqrt{\left(\frac{L_0}{2} + v\gamma\,\bar{t}\right)^2 + z_0^2} + v\sqrt{\left(\frac{L_0}{2} - v\gamma\,\bar{t}\right)^2 + z_0^2}\,\right].$$

For $\bar{t} \to -\infty$ the length of the rod looks larger at $(F/z_0)\, L_0 \gamma\, (1+v)$. Gradually, it contracts to the Lorentz value $(F/z_0)\, L_0/\gamma$, and then gets even shorter; as $\bar{t} \to \infty$ its length approaches $(F/z_0)\, L_0\, \gamma(1-v)$.

▷ Consider also how photos represent three-dimensional objects. Let

$$f(x', y', z') = 0$$

be the equation of the surface of such an object in its proper frame. Relative to the current time $\bar{t}$, the time of emission of the photons from point (x, y, z) is somewhere in the past

$$t = \bar{t} - \sqrt{x^2 + y^2 + z^2},$$

depending on the distance from the diaphragm. Substituting the Lorentz transformations and then projective transformations $x = zX/F$, $y = zY/F$ into the equation of the surface, we obtain:

$$f\Big(z\gamma\Big[\frac{X}{F} + v\sqrt{1 + (X^2 + Y^2)/F^2}\Big] - v\gamma\bar{t},\ z\frac{Y}{F},\ z\Big) = 0. \tag{3.19}$$

The photograph is taken by a rapid opening and closing of the diaphragm. The equation should be solved with respect to z for each point (X, Y) of the photo. If there are several solutions, we take the one corresponding to the minimum distance from the diaphragm (other points of the surface will be shielded by the body). Knowing the time

$$t = \bar{t} - z\sqrt{1 + (X^2 + Y^2)/F^2},$$

we can use the Lorentz transformations to obtain the image of the point (x', y', z') on the surface under projection onto the photograph (X, Y).

The figures below show such a "photo" in the cases of a stationary ($v = 0$, the first figure) and a moving cube ($v = 0.9$, the second figure):

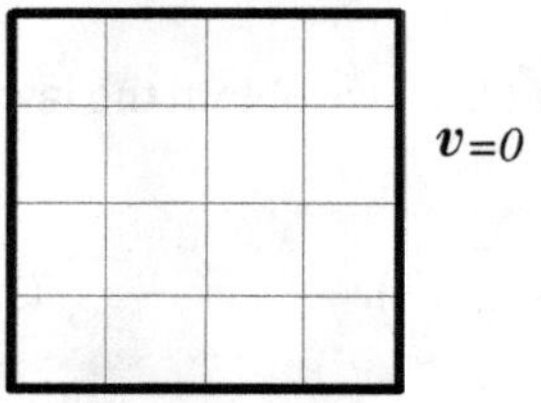

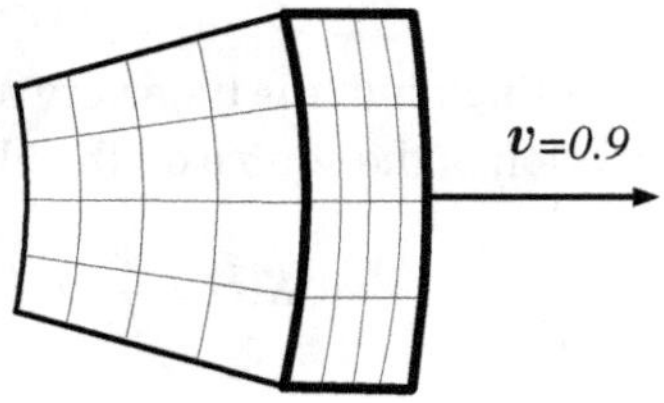

Note that any vertical rod moving in the horizontal direction appears curved, since the images of its center and of its edges are emitted at different points in time in the past, so the ends of the rod will bend "backwards". A moving sphere looks rotated in the photograph, just the same as in the case of the orthogonal projection.

Thus, the appearance of the relativistic world depends essentially on how we observe it. To get a realistic model of a photograph, we have to take into account the lighting properties and correct it according to the Doppler effect (p. 59) and the aberration effect, which also changes the luminosity of objects (p. 82). Note that the above signal delay can also be interpreted as an aberration which "distorts" the position of the object. Let us now consider this effect in more detail.

3.8 Aberration

Aberration is similar to the Doppler effect, but the "distortion" here affects the apparent position of the object, rather than the radiation frequency. Like the Doppler effect, aberration has a classical component and some corrections due to relativistic effects. Aberration was first observed as a change in the position of the stars during Earth's motion along its orbit around the Sun. So, we will start with this example.

Let an observer S be at rest relative to the Sun. Suppose he sees a stationary star in the direction θ from the plane of Earth's orbit (this angle is called its *declination*). Another observer S' travels together with Earth relative to the first observer with velocity v (the left figure). For now, we will assume that the motion is straightforward:

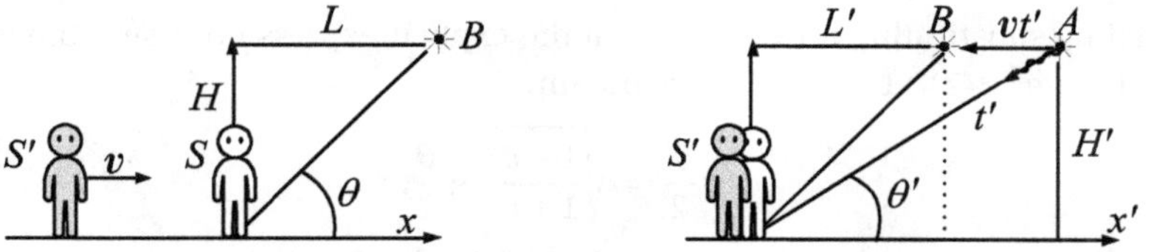

When the observers find themselves at one point, the earthman S' will see the star at an angle θ' (the second figure). The star is travelling towards him with velocity v, so it is seen at point A, which was passed some time t' ago. The time is needed for the light to travel the length of the hypotenuse ($c = 1$). The "true" position of the star corresponds to the point B. The observer S resting relative to the star also sees it in the past, but always in the same direction (at an angle θ). Decomposing the hypotenuse t' into the catheti, we obtain the relation between the angles:

$$\begin{cases} t' \sin\theta' = H', \\ t' \cos\theta' = vt' + L'. \end{cases}$$

Combining these equations, we obtain

$$\frac{\sin\theta'}{\cos\theta' - v} = \frac{H'}{L'} = \frac{H}{L\sqrt{1-v^2}} = \frac{\operatorname{tg}\theta}{\sqrt{1-v^2}},$$

where we have used the fact that $\operatorname{tg}\theta = H/L$ for the observer at rest relative to the star, that the horizontal distance to the star is reduced $L' = L\sqrt{1-v^2}$ for the earthman (Lorentz contraction), and and $H' = H$. Since $\cos\theta = 1/\sqrt{1+\operatorname{tg}^2\theta}$, we have:

$$\cos\theta = \frac{\cos\theta' - v}{1 - v\cos\theta'}, \qquad \sin\theta = \frac{\sqrt{1-v^2}\,\sin\theta'}{1 - v\cos\theta'}. \tag{3.20}$$

The distortions in the shape of the photos of moving objects considered in the previous section were also an example of aberration.

▷ The equations for aberration can also be obtained from the velocity-addition law (p. 18). In this case, the object moving with velocity $\mathbf{u}$ relative to the frame S and velocity $\mathbf{u}'$ relative to S' is a light signal that propagates from the source to the observers:

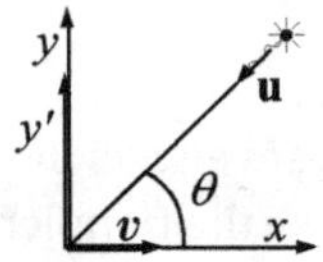

$$u'_x = \frac{u_x - v}{1 - u_x v}, \qquad u'_y = \frac{u_y\sqrt{1-v^2}}{1 - u_x v}.$$

As is seen from the figure, the projections of the speed of light are $u_x = -\cos\theta$ and $u_y = -\sin\theta$. This is also true for the moving observer S', since the magnitude of the speed of light $c = 1$ is the same in both frames. Substituting u_x, u_y into the velocity-addition law, we obtain:

$$\cos\theta' = \frac{\cos\theta + v}{1 + v\cos\theta}, \qquad \sin\theta' = \frac{\sqrt{1-v^2}\,\sin\theta}{1 + v\cos\theta}. \tag{3.21}$$

These equations are inverse relative to those discussed above. As usual, they can be obtained by substituting $v \mapsto -v$ or from the explicit expression. Using the identity $\cos\theta = (1 - \mathrm{tg}^2\,\theta/2)/(1 + \mathrm{tg}^2\,\theta/2)$, we obtain:

$$\mathrm{tg}\,\frac{\theta'}{2} = \sqrt{\frac{1-v}{1+v}}\;\mathrm{tg}\,\frac{\theta}{2}. \tag{3.22}$$

The difference between the observation angles for the stationary and the moving observer is the sine of the difference between the angles $\alpha = \theta - \theta'$:

$$\sin\alpha = \sin\theta\cos\theta' - \cos\theta\sin\theta' = \frac{v + \left(1 - \sqrt{1-v^2}\right)\cos\theta}{1 + v\cos\theta}\,\sin\theta.$$

At low velocities, the following approximation is true: $\sin\alpha \approx v\sin\theta$. When $v \ll 1$, the angle is also small and $\sin\alpha \approx \alpha \approx v\sin\theta$. The difference between the observations is maximal for $\theta = \pi/2$ (when the source is directly above the stationary observer).

In the theory of relativity, it is not important which is moving, the receiver or the source. All motion is relative. However, the situation is different when, for example, the propagation of sound is considered. In this case, there is a special reference frame associated with the air. The sound always propagates with constant speed relative to this frame, independently of the speed of the source. However, the speed of sound depends on the speed of the receiver relative to the air. This gives rise to various effects, depending on whether the receiver or the source is moving relative to the surroundings.

Naturally, moving in one direction, it is impossible to detect the aberration of a star. However, when revolving around the Sun, the velocity of the Earth changes, and the positions of the stars therefore change as well. Let's consider this effect in more detail.

3.9 Parallax *

We now express the effect of aberration in vector form. To do this, we begin with the relationship between the velocities of an object as measured by observers in S and S', see (1.24), p. 20:

$$\mathbf{u}' = \frac{\mathbf{u} - \gamma\mathbf{v} + \Gamma\,\mathbf{v}\,(\mathbf{vu})}{\gamma(1 - \mathbf{uv})}, \qquad \Gamma = \frac{\gamma^2}{\gamma + 1} = \frac{\gamma - 1}{v^2}.$$

Denote by $\mathbf{n}$ and $\mathbf{n}'$ the *unit* vectors directed from the two obervers toward the light source. Naturally, the light signal propagates towards the observers, i.e., in the direction opposite to these vectors: $\mathbf{u} = -\mathbf{n}$ and $\mathbf{u}' = -\mathbf{n}'$. Therefore,

$$\mathbf{n}' = \frac{\mathbf{n} + \gamma\mathbf{v} + \Gamma\,\mathbf{v}\,(\mathbf{vn})}{\gamma(1 + \mathbf{vn})}. \tag{3.23}$$

For the angle θ between the direction to the object and the velocity vector, we have

$$\mathbf{nv} = v\cos\theta,$$

and the same relation holds for the primed values. Multiplying the left and right sides of (3.23) by $\mathbf{v}$, we easily obtain the relation between the cosines of θ and θ' (3.21). The relation for the sines gives the vector product $\mathbf{n}' \times \mathbf{v}$, which is of magnitude $v\sin\theta'$.

At low velocities we can disregard the term $\mathbf{v}\,(\mathbf{vn})$ in the equation for aberration (3.23) (order v^2, since $\Gamma \approx 1/2$, $\gamma \approx 1$, see p. 16). Expanding the denominator as a series ($1/(1 + x) \approx 1 - x$), we obtain a linear approximation of $\mathbf{n}'$ in terms of v:

$$\mathbf{n}' \approx \frac{\mathbf{n} + \mathbf{v}}{1 + \mathbf{vn}} \approx \mathbf{n} + \mathbf{v} - \mathbf{n}\,(\mathbf{nv}) = \mathbf{n} - \mathbf{n} \times [\mathbf{n} \times \mathbf{v}], \tag{3.24}$$

where in the last equation we have used the formula for the vector triple product (p. 334). As an exercise, we recommend verifying that this ***approximate formula*** implies, up to first order with respect to v, that the primed vector has unit length,

$$\mathbf{n}'^2 \approx 1.$$

Since the vector product between the directions is

$$\mathbf{n}' \times \mathbf{n} = [\mathbf{v} \times \mathbf{n}]\,\frac{1 + (\mathbf{vn})\,\Gamma/\gamma}{1 + \mathbf{vn}}, \tag{3.25}$$

we can find the sine of the angle between them, which for unit vectors is just the magnitude $|\mathbf{n}' \times \mathbf{n}|$. If the velocity is small, then the aberration angle is also small:

$$\alpha \approx |\mathbf{v} \times \mathbf{n}| = v\sin\theta.$$

No aberration occurs if the position of the source is in line with the motion of the reference frame, and it is maximal if $\mathbf{n}$ and $\mathbf{v}$ are normal to each other. In this case, the angle is $\alpha = v$, and is directed along the motion.

▷ In fact, the Earth does not travel along a straight line. Rather, it revolves around the Sun on an elliptic orbit. The eccentricity (oblateness) of this orbit is not great (there is only a 3% difference between the minimum and the maximum distances from the Sun). Therefore, to simplify the coming calculations, we assume the Earth's orbit to be circular with radius

$$R = 149597870700\ m \approx 1.496 \cdot 10^8\ km = 1\ AU.$$

The distance R is called the *astronomical unit* (AU).

Let's put aberration aside for a moment. Even without this effect, the stars nearest to Earth visibly move against the celestial sphere, due to the revolution of Earth around the Sun. This is called the *annual parallax.* To an observer on Earth, a star "above the Sun" will move in a circle:

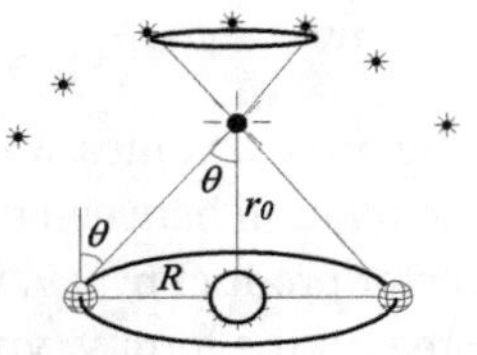

This circle is seen against a background of "static" stars, which are much further from the Sun than the star for which the parallax effect is observed. In fact, all stars traverse a circle (or ellipse, in the general case) on the celestial sphere, if they are above the Sun. The farther the star is from the Sun, the smaller the ellipse. The dimensions of the ellipses can be used to find the distances to the nearest stars using geometrical methods.

A *parsec* is the distance from which the average radius of the Earth's orbit is seen at an angle of one second.

In the figure above, the proportions are strongly distorted. Actually, $r_0 \gg R$, and therefore, $R/r_0 = \operatorname{tg}\theta \approx \theta$. One angular second is 1/3600 of a degree, so:

$$1\ pc = \frac{1\ AU}{1''} = \frac{1\ AU}{2\pi/(360 \cdot 3600)} = 206\,265\ AU = 3.0857 \cdot 10^{13}\ km.$$

The word "parsec" is formed from two words: "parallax" and "second". Light covers one parsec in 3.26 years. For comparison, 1 AU corresponds to 500 light seconds (8 min. 20 sec.). The distance to the nearest star system Alpha Centauri is 1.3 pc. Parallaxes measured by orbital telescopes make it possible to calculate distances up to 500 pc. This is the most direct way of measuring distances to relatively close stars.

▷ To describe the parallax effect, consider a unit vector $\mathbf{n}_0 = \mathbf{r}_0/r_0$ directed to the star in question, where $\mathbf{r}_0$ is the radius vector from the Sun to the star. For an observer on Earth the direction to the star is $\mathbf{n} = \mathbf{r}/r$. The radius vectors of the observers are related by $\mathbf{r} = \mathbf{r}_0 - \mathbf{R}$, where $\mathbf{R}$ is the radius vector directed from the Sun to the Earth:

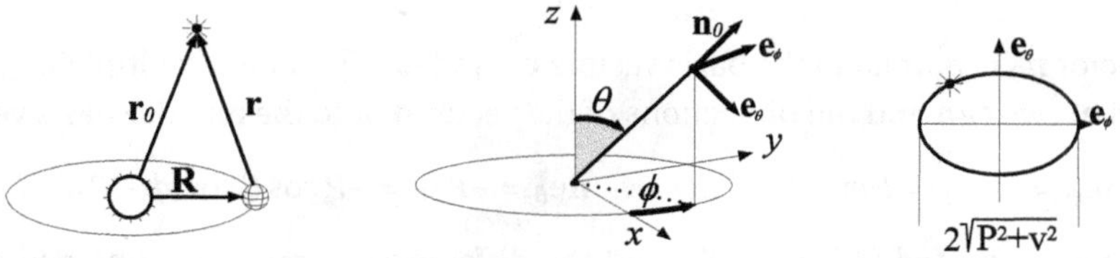

Assuming $r_0 \gg R$, we can express the distance from the Earth to the star in the same way that we did for the Doppler effect (p. 61):

$$r = \sqrt{(\mathbf{r}_0 - \mathbf{R})^2} \approx \sqrt{\mathbf{r}_0^2 - 2\mathbf{r}_0\mathbf{R}} \approx r_0 - \mathbf{n}_0\mathbf{R} = r_0\,(1 - \mathbf{n}_0\mathbf{P}),$$

where $\mathbf{P} = \mathbf{R}/r_0$ is the *parallax vector*. Therefore:

$$\mathbf{n} = \frac{\mathbf{r}}{r} \approx \frac{\mathbf{r}_0 - \mathbf{R}}{r_0(1 - \mathbf{n}_0\mathbf{P})} = \frac{\mathbf{n}_0 - \mathbf{P}}{1 - \mathbf{n}_0\mathbf{P}}.$$

Expanding the denominator in terms of small P, we have:

$$\mathbf{n} \approx (\mathbf{n}_0 - \mathbf{P})(1 + \mathbf{n}_0\mathbf{P}) \approx \mathbf{n}_0 + \mathbf{n}_0(\mathbf{n}_0\mathbf{P}) - \mathbf{P}.$$

Using the mnemonic "BAC-CAB" (p. 334), the relationship between the unit vectors directed toward the star from the Sun ($\mathbf{n}_0$) and from the Earth ($\mathbf{n}$) can be written using the vector triple product:

$$\mathbf{n} \approx \mathbf{n}_0 + \mathbf{n}_0 \times [\mathbf{n}_0 \times \mathbf{P}]. \tag{3.26}$$

We now express the components of the unit vector $\mathbf{n}_0$ in terms of the spherical angles (θ, ϕ) (see the second figure above):

$$\mathbf{n}_0 = (s_\theta c_\phi,\ s_\theta s_\phi,\ c_\theta),$$

where $s_\theta = \sin\theta$, $c_\theta = \cos\theta$. On the surface of the sphere we can define two orthogonal unit vectors $\mathbf{e}_\phi \sim \partial\mathbf{n}_0/\partial\phi$, $\mathbf{e}_\theta = \partial\mathbf{n}_0/\partial\theta$ normal to $\mathbf{n}_0$ (< C_5):

$$\mathbf{e}_\phi = (-s_\phi,\ c_\phi,\ 0), \qquad \mathbf{e}_\theta = (c_\theta c_\phi,\ c_\theta s_\phi,\ -s_\theta). \tag{3.27}$$

As is easily verified, $\mathbf{e}_\phi\mathbf{e}_\theta = 0$ and $\mathbf{e}_\theta^2 = \mathbf{e}_\phi^2 = 1$. These vectors indicate the direction of a small change in the angular coordinates. It would also be useful to verify that the vectors are normal to $\mathbf{n}_0$: $\mathbf{n}_0\mathbf{e}_\phi = \mathbf{n}_0\mathbf{e}_\theta = 0$.

The Earth revolves around the Sun in a circle, and the components of the radius vector $\mathbf{R}$ are:

$$\mathbf{R} = \{R\cos(2\pi t/T),\ R\sin(2\pi t/T),\ 0\},$$

where the time T is the orbital period. When t changes from 0 to T, the radius vector returns to its original position. Since **R** rotates, the components of the parallax vector also change periodically:

$$\mathbf{P} = \frac{\mathbf{R}}{r_0} = \{P\cos(2\pi t/T),\ P\sin(2\pi t/T),\ 0\}. \tag{3.28}$$

The vector $\mathbf{n}_0$ is normal to the basis vectors $\mathbf{e}_\phi$ and $\mathbf{e}_\theta$. Therefore, multiplying (3.9) by $\mathbf{e}_\phi$ and $\mathbf{e}_\theta$, we can find the projections of the vector **n** onto the angular basis vectors:

$$\mathbf{n}\mathbf{e}_\phi = -\mathbf{P}\mathbf{e}_\phi = P\sin(\phi - 2\pi t/T),\quad \mathbf{n}\mathbf{e}_\theta = -\mathbf{P}\mathbf{e}_\theta = -P\cos\theta\cos(\phi - 2\pi t/T),$$

where we have used (3.27), (3.28), and the difference formulas for sine and cosine. Since the ellipse of revolution is small, in a neighbourhood of it, the vectors $\mathbf{e}_\phi$, $\mathbf{e}_\theta$ on the sphere can be used as perpendicular Cartesian axes (because a sphere is a plane in small scales). The values $\mathbf{n}\mathbf{e}_\phi$ and $\mathbf{n}\mathbf{e}_\theta$ are projections onto these axes ($\lessdot$ C$_6$). As a result, the star traverses an ellipse with semiaxes P and $P\cos\theta$ on the surface of the celestial sphere:

$$\frac{(\mathbf{n}\mathbf{e}_\phi)^2}{P^2} + \frac{(\mathbf{n}\mathbf{e}_\theta)^2}{P^2\cos^2\theta} = 1.$$

For $\theta = 0$ (the star is above the Sun), we obtain a circle with radius equal to the parallax P.

▷ Consider now the aberration due to the motion of Earth. Substituting the relation (3.9) into (3.24) and neglecting terms of order $P\,v$, we can write:

$$\mathbf{n}' \approx \mathbf{n}_0 + \mathbf{n}_0 \times [\mathbf{n}_0 \times (\mathbf{P} - \mathbf{v})]. \tag{3.29}$$

The velocity of the Earth **v** along its circular orbit is normal to **R**, and its magnitude is 30 km/s, or $v \sim 10^{-4}$ in the units of the speed of light. Differentiating the radius vector **R** with respect to time, we obtain the velocity components

$$\mathbf{v} = \{-v\sin(2\pi t/T),\ v\cos(2\pi t/T),\ 0\}.$$

It is a good exercise to verify that the semiaxis of the ellipse is $\sqrt{P^2 + v^2}$.

Even for the nearest star, the parallax is as small as $P = R/r_0 \sim 4\cdot 10^{-6}$, i.e., it is 27 times smaller than the dimensionless velocity v. This is why the effect of aberration on the elliptic motion of stars during the year is much more significant than that due to their parallax. Note that equation (3.29) does not take into account relativistic effects of order v^2.

3.10 Starry sky

Aberration, like the Doppler effect, has various manifestations. Let us assume that some stars are uniformly distributed in space in some reference frame and then consider how these stars appear from a fast-moving spaceship.

When observed from a stationary point, a starry sky looks like stars projected onto the "celestial sphere" (< C$_7$). If the stars are uniformly distributed, their density is, on average, constant as a function of direction. Define spherical angles (θ, ϕ) and consider a small area on the sphere bounded by small increments $(d\theta, d\phi)$:

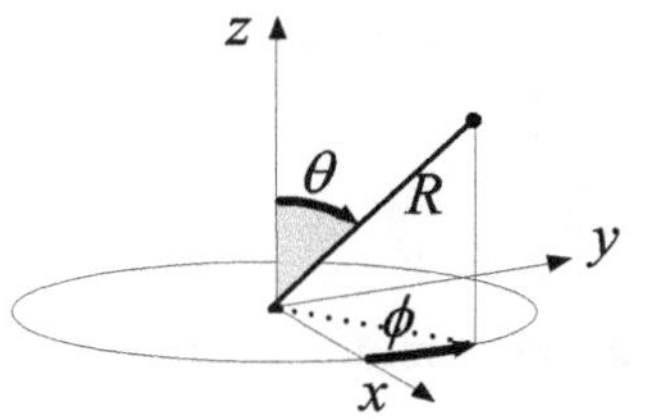

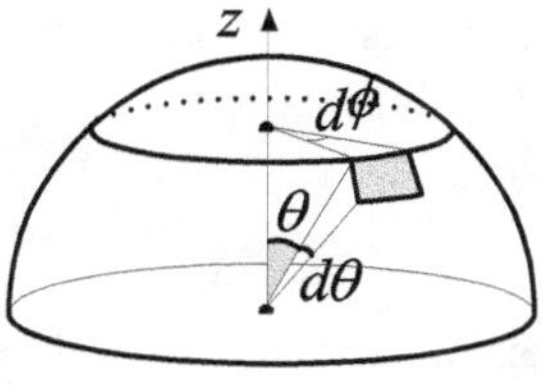

If $\phi = const$, the angle increment $d\theta$ describes an arc of length $Rd\theta$ on the sphere of radius R. If θ is constant, the angle increment $d\phi$ describes an arc $R\sin\theta\, d\phi$ where $R\sin\theta$ is the distance from the arc to the z-axis. As a result, the area bounded by the increments $(d\theta, d\phi)$ is equal to $R^2\sin\theta\, d\theta\, d\phi$. Its ratio to the squared radius of the sphere is called a ***solid angle***

$$d\Omega = \sin\theta\, d\theta\, d\phi.$$

Since the area of the sphere is $4\pi R^2$, integration of the solid angle over (θ, ϕ) gives 4π:

$$\int d\Omega = \int_0^{2\pi} d\phi \int_0^{\pi} \sin\theta\, d\theta = 2\pi \int_{-1}^{1} d(\cos\theta) = 4\pi.$$

A solid angle $d\Omega$ is a small area on the surface of the unit sphere. However, for fixed $d\theta$ and $d\phi$, the area is different for different θ. For this reason, it is more convenient to work with relative values. The ratio of the number of stars dN seen in the solid angle $d\Omega$ to the magnitude of this angle is called the density of ***angular distribution*** $n(\theta, \phi)$. In the general case, this density is a function of both angles:

$$dN = n(\theta, \phi)\, d\Omega = n(\theta, \phi)\, \sin\theta\, d\theta\, d\phi = -n(\theta, \phi)\, d(\cos\theta)\, d\phi.$$

For a uniform distribution the density is constant:

$$n(\theta, \phi) = \frac{N}{4\pi},$$

where N is the total number of visible stars.

▷ Now consider a (primed) reference frame attached to a spaceship. Let its velocity v be directed along the z-axis. The aberration of the angle θ' between the velocity and the direction to the star (3.20) is:

$$\cos\theta = \frac{\cos\theta' - v}{1 - v\cos\theta'}. \tag{3.30}$$

We can write $s_\theta\, d\theta = -dc_\theta = -(\partial c_\theta/\partial c_{\theta'})\, dc_{\theta'}$ and take the derivative (3.30) with respect to $\cos\theta'$. Since $\phi' = \phi$, for $n(\theta,\phi) = N/4\pi$ we obtain

$$dN = -\frac{N}{4\pi}\frac{\partial(\cos\theta)}{\partial(\cos\theta')}\,d(\cos\theta')\,d\phi = -\frac{N}{4\pi}\frac{1-v^2}{(1-v\cos\theta')^2}\,d(\cos\theta')\,d\phi'.$$

As a result, the distribution density of the stars on the "celestial sphere", as seen from the spaceship, has the form

$$n'(\theta',\phi') = \frac{N}{4\pi}\frac{1-v^2}{(1-v\cos\theta')^2}. \tag{3.31}$$

This expression reaches its maximum at $\theta' = 0$ and its minimum at $\theta' = \pi$:

$$n'(0,\phi') = \frac{N}{4\pi}\frac{1+v}{1-v}, \qquad n'(\pi,\phi') = \frac{N}{4\pi}\frac{1-v}{1+v}.$$

It is a useful exercise to verify (< H_{12}) that integration of $n(\theta',\phi')$ over the entire surface of the sphere $0 \leqslant \theta' \leqslant \pi$ and $0 \leqslant \phi' < 2\pi$ gives, as expected, exactly N visible stars.

The "photos" below show a starry sky. The first picture corresponds to the stationary reference frame. The second and the third pictures were taken from a spaceship moving with velocity $v = 0.9$ (from the front and back of the ship, respectively):

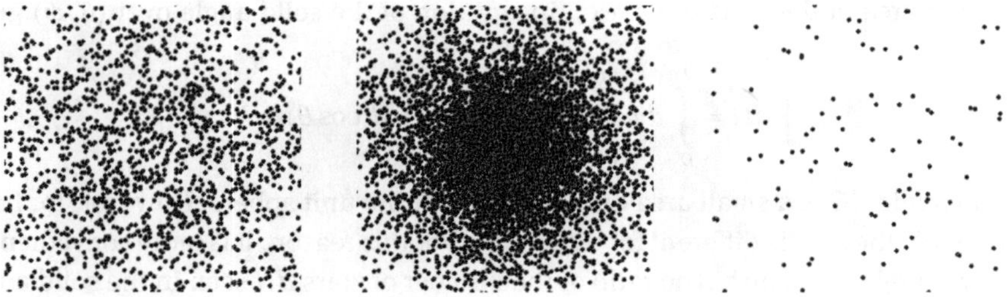

Remember that the Doppler effect makes the spectra of the stars in front of and behind the ship shift toward the blue (frequencies increase) and toward the red (frequencies decrease) regions, respectively. However, these are not the only distortions of the starry sky, there are also aberrative effects to consider.

⊳ Consider a light source (a star) that emits light energy uniformly in all directions (in its proper frame). *The photon flux* J in a given direction is the *number* of photons dN emitted per unit time dt into the solid angle $d\Omega$. Similarly, define the *stellar luminosity* I using the energy flux E brought by these photons:

$$J = \frac{dN}{d\Omega\,dt}, \qquad I = \frac{dE}{d\Omega\,dt} = h\nu J, \tag{3.32}$$

where, for simplicity, we have assumed that the spectrum of the star is concentrated in the vicinity of one frequency ν, so that the energy is $dE = h\nu\,dN$, where $h = 2\pi\hbar$ is the *Planck constant*.

Let the star move with velocity v along the x-axis relative to an observer. In the chosen coordinates (the figure below), the velocity component of the photons is $u_x = \cos\theta$, and the sum of velocities $u'_x = (u_x - v)/(1 - u_x v)$ gives the aberration formula:

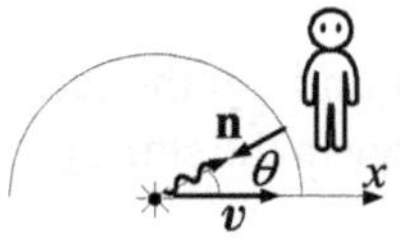

$$\cos\theta' = \frac{\cos\theta - v}{1 - v\cos\theta}$$

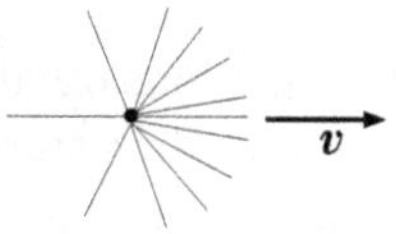

The flux in the frame attached to the star is $J_0 = const$, and it does not depend on θ', since the radiation was assumed to be isotropic. Using the formula for aberration, we can express the solid angle $d\Omega' = -d(\cos\theta')\,d\phi'$ in terms of $d\Omega$, which is attached to the stationary observer. We will also use equation (3.6), p. 61, which describes the relationship between the periods over which the photons are emitted dt' and received dt. After carrying out calculations similar to those for the distribution of stars in the sky above, we obtain for the photon flux:

$$dN = J_0\,d\Omega'\,dt' = J_0\,\frac{1 - v^2}{(1 - v\cos\theta)^2}\,d\Omega\,\frac{\sqrt{1 - v^2}}{1 - v\cos\theta}\,d\bar{t}.$$

Note that sometimes the equation $dt' = \sqrt{1 - v^2}\,dt$ is used. This is not correct, since the number of photons is measured by a single observer, who meets the photons, rather than by numerous observers "distributed along the trajectory of the star". Therefore, it is necessary to use the Doppler formula.

As a result, the photon flux for a stationary observer depends on the observation angle relative to the velocity of the star,

$$J = J_0\,\frac{(1 - v^2)^{3/2}}{(1 - v\cos\theta)^3}. \tag{3.33}$$

This expression is maximal at $\theta = 0$ and minimal at $\theta = \pi$. As a result, the "rays" of the moving star "bunch up" going forward, and the largest number of photons is emitted in the direction of motion (the second figure above).

▷ Calculating the luminous flux requires another application of the Doppler effect (3.6), p. 61. The projection of the unit vector **n** onto the velocity **v** is determined by the angle θ, $\mathbf{nv} = -v\cos\theta$, and

$$\nu = \nu_0\,\frac{\sqrt{1 - v^2}}{1 - v\cos\theta},$$

where ν_0 is the natural emission frequency of the photons in the frame of the star. Given that $I = h\nu J$, we find that

$$I = I_0\,\frac{(1 - v^2)^2}{(1 - v\cos\theta)^4}. \tag{3.34}$$

In addition to aberration, the Doppler effect also distorts the energy distribution of the radiation, due to the higher frequency (and therefore energy) of the photons travelling towards the observer.

Returning to the starry sky observed from the spaceship, we conclude that not only the number of stars, but also their brightness, will significantly increase up ahead on the ship's course. This is due to the Doppler effect and, ***doubly***, to aberration (the displaced position of the stars and the redistributed intensity of their radiation).

⊳ The equations obtained demonstrate one more effect. We can find out the ***total energy*** radiated by a star before it "burns out". If the total light energy in the star's proper frame is E_0, then the total number of radiated photons is

$$N = \frac{E_0}{h\nu_0}, \qquad \text{and} \qquad dN = \frac{N}{4\pi}\, d\Omega'$$

of them will be emitted into a unit solid angle. For the observer at rest,

$$dE = h\nu\, dN = \frac{h\nu N}{4\pi}\, d\Omega' = \frac{E_0}{4\pi}\frac{\nu}{\nu_0}\, d\Omega' = \frac{E_0}{4\pi}\frac{\sqrt{1-v^2}}{1-v\cos\theta}\frac{1-v^2}{(1-v\cos\theta)^2}\, d\Omega.$$

Integrating over the whole sphere, we obtain:

$$E = \int dE = \frac{E_0}{4\pi}\int \frac{(1-v^2)^{3/2}}{(1-v\cos\theta)^3}\, d\Omega = \frac{E_0}{2}\int_{-1}^{1} \frac{(1-v^2)^{3/2}}{(1-vz)^3}\, dz = \frac{E_0}{\sqrt{1-v^2}}.$$

A moving star has higher energy E in a stationary frame than its own energy E_0. As we will see in the next chapter, this relationship between the energy and the velocity

$$E = \frac{E_0}{\sqrt{1-v^2}}, \tag{3.35}$$

is of a quite general nature.

3.11 Accelerated motion

Uniformly accelerated motion, in the sense of classical mechanics, is impossible in the theory of relativity. If the velocity of an object increases at a constant rate $u(t) = a\,t$, sooner or later it will exceed the fundamental speed c. This is impossible due to energy limitations, as discussed in Chapter 4. Let us consider one type of accelerated motion, where the velocity grows, but never exceeds unity ($c = 1$).

For the sake of clarity, let us assume that a spaceship moves with a constantly increasing velocity $u = u(t)$ past a stationary observer. Consider the reference frame associated with the ship (the right figure):

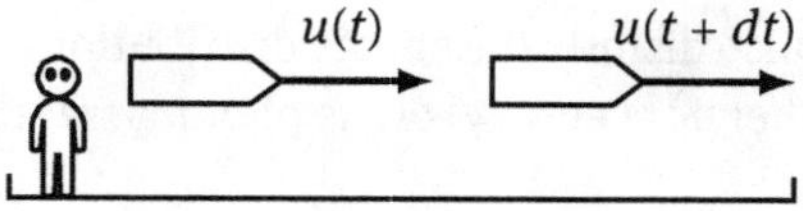

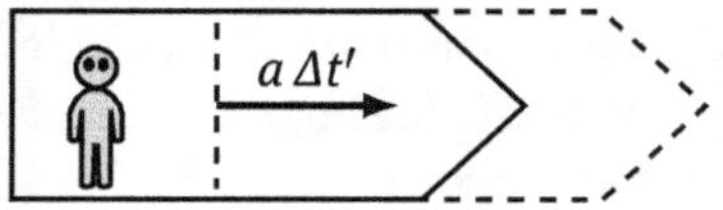

Since over the time increment dt' the velocity of the ship changes only slightly, its increment *relative to the previous state* can be calculated as $a\,dt'$, where a is some constant. From the velocity-addition formula (1.21), p. 19, we obtain that the new velocity relative to the stationary observer is ($v = u(t)$, $u'_x = a\,dt'$)

$$u(t+dt) = \frac{a\,dt' + u(t)}{1 + u(t)\,a\,dt'}.$$

Since dt' is small, we can expand the denominator in a series $[1/(1+x) \approx 1 - x]$ and multiply it by the numerator, preserving the order of smallness with respect to $a\,dt'$:

$$u(t+dt) \approx (u + a\,dt')(1 - u\,a\,dt') \approx u + (1-u^2)\,a\,dt',$$

where $u = u(t)$. Given that by the ship's clock time satisfies

$$dt' = dt\sqrt{1-u^2},$$

we can write the derivative of velocity $du/dt = [u(t+dt) - u(t)]/dt$ as

$$\frac{du}{dt} = a\,(1-u^2)^{3/2},$$

or

$$\frac{d}{dt}\left(\frac{u}{\sqrt{1-u^2}}\right) = a. \qquad (3.36)$$

This differential equation describes the velocity of an object for a stationary observer if the object, from its own "perspective", is "trying" to move with uniform acceleration.

Taking as initial condition $u_0 = u(0)$ and integrating equation (3.36), we obtain

$$\frac{u(t)}{\sqrt{1-u^2(t)}} = \pi(t) = \pi_0 + a\,t, \qquad \text{where } \pi_0 = \frac{u_0}{\sqrt{1-u_0^2}}. \qquad (3.37)$$

The function $\pi(t) = \pi_0 + at$ is linear with respect to time and can be arbitrarily large. However, the relativistic velocity $u(t)$ never exceeds unity:

$$u(t) = \frac{\pi_0 + at}{\sqrt{1 + (\pi_0 + at)^2}}. \qquad (3.38)$$

Given that $u(t) = dx/dt$, integrating ($\lessdot H_{13}$) once more with the initial condition $x(0) = x_0$, we obtain the law of motion

$$x(t) = x_0 + \frac{1}{a}\left(\sqrt{1 + (\pi_0 + a\,t)^2} - \sqrt{1+\pi_0^2}\right). \qquad (3.39)$$

This can be expressed in terms of the velocity:

$$x(t) = x_0 + \frac{1}{a}\left(\frac{1}{\sqrt{1-u^2(t)}} - \frac{1}{\sqrt{1-u_0^2}}\right) \approx x_0 + \frac{u^2(t) - u_0^2}{2a},$$

where the approximate equality holds for small velocities. To reintroduce the fundamental velocity c in the above equations, one makes the substitutions $t \mapsto ct$, $u \mapsto u/c$, and $a \mapsto a/c^2$ ($\lessdot$ H$_{14}$) (see p. 16). This results in both $a\,t$ and π_0 being divided by c. Passing to the limit $c \to \infty$ and expanding (3.39) into a series with respect to small π_0 and $a\,t$, we obtain the classical equations for motion under uniform acceleration:

$$u(t) \approx u_0 + at + \dots, \qquad x(t) \approx x_0 + u_0 t + \frac{at^2}{2} + \dots.$$

The graphs below demonstrate the time dependence of the speed and coordinates of an object which moves with unit acceleration for a unit of time and then starts slowing down.

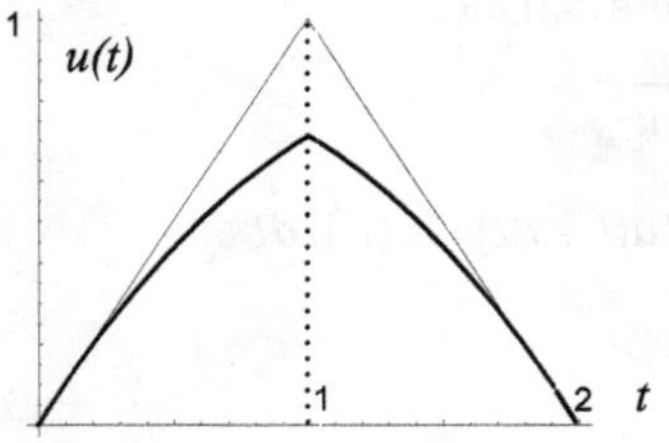

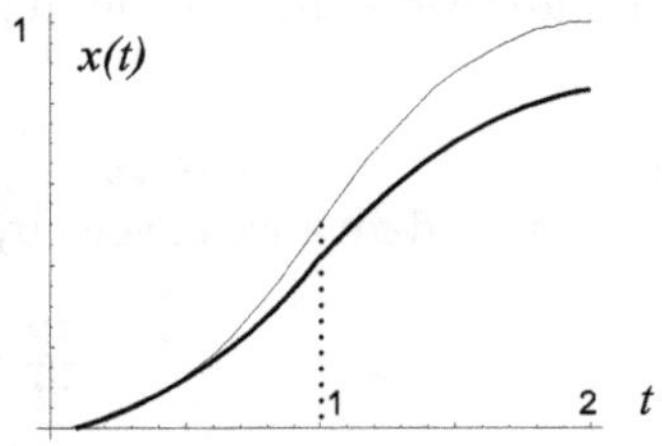

The thin upper lines on the graphs correspond to the classical dynamics of uniformly accelerated motion.

▷ Time dilation takes place for both a uniformly moving and an accelerating ship. Consider these effect from the perspective of an observer on Earth. Suppose the ship accelerates for some time τ_1, then moves uniformly for a time τ_2, and then spends τ_1 time units slowing down:

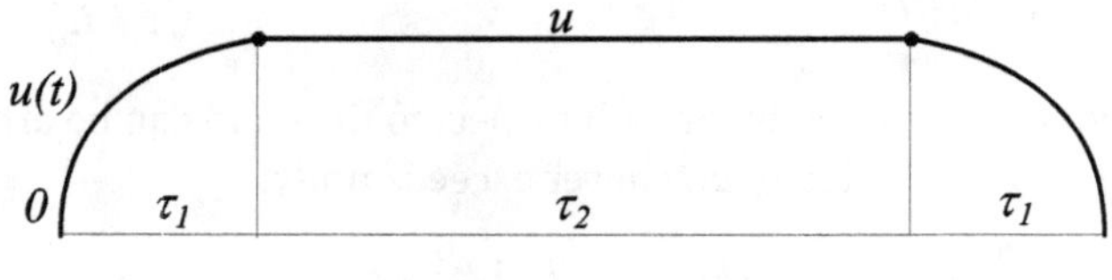

Let us calculate the proper time of the travel as perceived by the ship. Since the velocity of the ship changes insignificantly for a short time interval dt, it can be attached to a locally inertial frame. The time counted by a moving clock is related to the time of the stationary observer by [see (3.2), p. 53]:

$$dt' = \sqrt{1 - u^2(t)}\, dt \qquad \Rightarrow \qquad t'_2 - t'_1 = \int_{t_1}^{t_2} \sqrt{1 - u^2(t)}\, dt.$$

As a result of the integration, small time intervals are summed up on Earth and on the ship. We also assume that acceleration does not affect the course of time. In Chapter 6, we will examine this assumption in more detail.

As before, time intervals for the astronaut will be marked with the index zero. For the first stage of acceleration we have

$$\tau_{01} = \int_0^{\tau_1} \sqrt{1-u^2(t)}\, dt = \int_0^{\tau_1} \frac{dt}{\sqrt{1+(at)^2}} = \frac{1}{a}\, \mathrm{ash}(a\,\tau_1), \tag{3.40}$$

where ash(x) is hyperbolic arcsine, which is the inverse function of the hyperbolic sine (see p. 355):

$$\mathrm{sh}(x) = \frac{e^x - e^{-x}}{2}, \qquad \mathrm{ash}(x) = \ln(x + \sqrt{1+x^2}) \approx x - \frac{x^3}{6} + \ldots.$$

Over the acceleration period τ_1, the velocity of the ship increases to (3.38)

$$u = \frac{a\tau_1}{\sqrt{1+(a\tau_1)^2}},$$

and the ship then moves uniformly. Therefore, in the second stage:

$$\tau_{02} = \tau_2 \sqrt{1-u^2} = \frac{\tau_2}{\sqrt{1+(a\tau_1)^2}},$$

and it is during this period that the dilation of time is most pronounced, since the velocity is at its largest.

For the final period of slowing down,

$$\tau_{03} = \int_{\tau_1+\tau_2}^{2\tau_1+\tau_2} \sqrt{1-u^2(t)}\, dt = \int_{\tau_1+\tau_2}^{2\tau_1+\tau_2} \frac{dt}{\sqrt{1+(\pi_0+at)^2}} = \frac{1}{a}\, \mathrm{ash}(a\,\tau_1).$$

From symmetry considerations, we could have just used the result obtained for the acceleration stage (3.40). Summing up the time intervals of each stage, we finally obtain:

$$\tau_0 = \frac{2}{a}\, \mathrm{ash}(a\tau_1) + \frac{\tau_2}{\sqrt{1+(a\tau_1)^2}}.$$

The first term is smaller than τ_1, and the second term is smaller than τ_2. They can ($\lessdot$ H_{15}) be expanded in the Taylor series

$$\tau_0 \approx 2\tau_1 + \tau_2 - \left(\frac{\tau_1}{3} + \frac{\tau_2}{2}\right)(a\tau_1)^2.$$

The time that passed in the stationary frame is $2\tau_1 + \tau_2$. The travel time by the ship's clock τ_0 is smaller. In the acceleration stages, time slows down more slowly than during the stage of uniform motion.

Let us estimate how long it would take to fly to the Alpha Centauri star system, which is 4.3 light-years away from Earth. A *light-year* is the distance that light travels in one year:

$$1\, ly = (299792458\, m/s) \cdot (365.25 \cdot 24 \cdot 3600\, s) \approx 0.9461 \cdot 10^{16}\, m = 0.307\, pc.$$

When measuring distances in light years and time in usual years, we are still in the system $c = 1$. In this system, unit acceleration $1\ ly/yr^2 = 9.5\ m/s^2$ is close to the rate of free fall acceleration on the surface of the Earth. Assume for the sake of comfort (artificial gravity) that the spaceship is moving with acceleration $a = 1$ for half of the journey, and then immediately starts slowing down ($\tau_2 = 0$). From the perspective of the Earth [see(3.39)], at the halfway point $x = 4.3/2$ ly there holds

$$x = \frac{1}{a}\left[\sqrt{1 + (a\tau_1)^2} - 1\right],$$

so the first half of the journey takes

$$\tau_1 = \frac{1}{a}\sqrt{(1 + ax)^2 - 1} \approx 3\ yr.$$

Accordingly, the total back and forth flight time will take 12 years. The astronaut's proper time at the moment of return will be

$$4\,\mathrm{ash}(a\tau_1)/a = 7.3\ years,$$

which is 40 % smaller. In 64 years of their proper time the astronaut can travel to and from Andromeda galaxy, which is 2.5 million light years away from Earth. On Earth the journey will take 5 million years. Unfortunately, things are not so simple, and in the next chapter we will consider the technical limitations of such flights.

4 Dynamics

This chapter is concerned with questions relating to the dynamics of particles. The laws of conservation of energy and momentum are universal tools for obtaining information on the motion of particles after their collision, even if the details of their interaction are unknown.

Expressions for relativistic energy and momentum will be obtained from quite general considerations using an "informal" axiomatic method. We will then calculate the outcomes of various particle collisions and decays.

Conservation laws can also be used to analyse the details of interstellar flights and their possible limitations which, unfortunately, show them to be unrealistic.

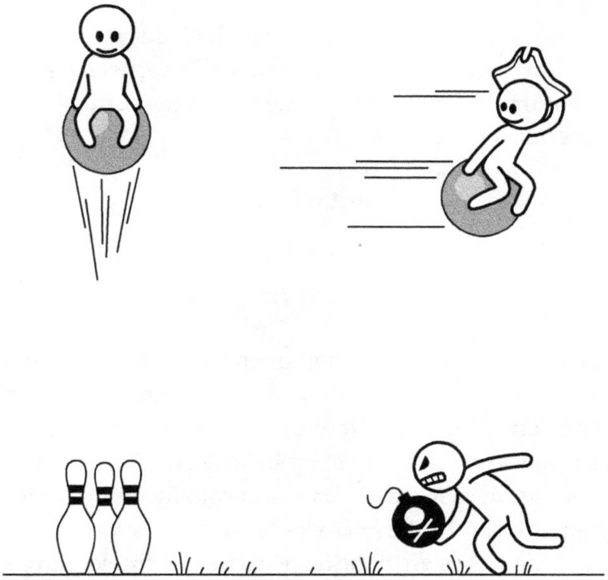

https://doi.org/10.1515/9783110515886-004

4.1 Inertial mass

After space and time, mass is the most mysterious property used to describe objects in the world around us. This is how it was defined by Isaac Newton:

> The quantity of matter is the measure of the same, arising from its density and bulk conjunctly. [4]

This definition has been criticized repeatedly, because it seems more natural to use mass and volume to define density: $\rho = m/V$. However, the calculation of mass from *geometric* parameters seems quite acceptable for bodies which are of equal density. If there were fundamental particles of equal mass, a definition similar to that of Newton would be quite appropriate for classical mechanics.

For example, the situation is exactly the same as that in the case of charge. However, there are still no consistent theories of mass quantization for particles similar to those available for the charge of particles (though things are not that straightforward with the latter, either).

Many of the equations of classical mechanics include mass as a factor. For example, if $\mathbf{u}$ and $\mathbf{a}$ are the velocity and the acceleration of a particle, E and $\mathbf{p}$ are its kinetic energy and momentum, and $\mathbf{F}$ is the force exerted on the particle, then the following "definitions" of *inertial mass* are possible:

$$m\,\mathbf{a} = \mathbf{F},$$

$$m\,\mathbf{u} = \mathbf{p},$$

$$m = 2E/\mathbf{u}^2.$$

Velocity and acceleration come from kinematics and thus reduce to measurements of length and time. Therefore, they are well-defined quantities in dynamics. However, the same cannot be said of force, momentum, and energy. Rather, *they* are defined in terms of mass. We could try to exclude dynamic quantities from the definition of mass. Suppose some standard force acts equally on two different masses m_1 and m_2. For instance, it may be the force exerted by the first particle upon the second, which is equal to the inverse sign of the force exerted by the second particle upon the first (Newton's third law). As a result of Newton's laws, we have that

$$m_1\mathbf{a}_1 = -m_2\mathbf{a}_2,$$

which is both a law of dynamics and a definition of a particle mass.

We will define mass using another method via the concept of elastic collision and some symmetry considerations. The inertial properties of mass are revealed when something changes the velocity of an object. This requires some impact from another object such as a collision.

Suppose two different particles collide and then move apart along the same straight line. If the particles stay "the same" after the collision, the collision is called *elastic*.

Axiom I.
If the collision is elastic, there is a reference frame such that the particles' velocities change in sign but not in magnitude as a result of it.

Indeed, suppose that a particle A has velocity $\bar{u}_1$ in some frame before the collision, and velocity $-\bar{u}_2$ after the collision. Moving with a suitable velocity v with respect to this frame, we can make the velocities equal in magnitude:

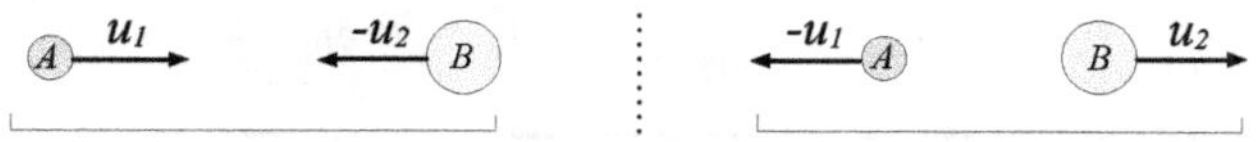

Axiom I claims that the velocity of neither particle A nor B will change its magnitude in such a frame. This symmetry of elastic collisions can be related to *time reversibility* (if seen in the opposite direction, the film of such a collision will be the same, as far as the velocity or the "appearance" of the particles are concerned).

Axiom II.
The i-th particle is characterized by a positive scalar parameter (its mass) m_i. These parameters determine the velocities u_1, u_2 after the symmetric elastic collision:

$$\begin{array}{lll} \text{If } m_1 < m_2, & \text{then} & u_1 > u_2; \\ \text{If } m_1 = m_2, & \text{then} & u_1 = u_2; \\ \text{If } m_1 > m_2, & \text{then} & u_1 < u_2. \end{array}$$

This *definition* does not give instructions on how to calculate the absolute mass. To do this, some function

$$\frac{m_2}{m_1} = F(u_1, u_2)$$

must be specified to obtain the ratio of the masses. The fact that the masses are given in the form of a ratio m_2/m_1 can be substantiated (though not proven) by the following thought experiment:

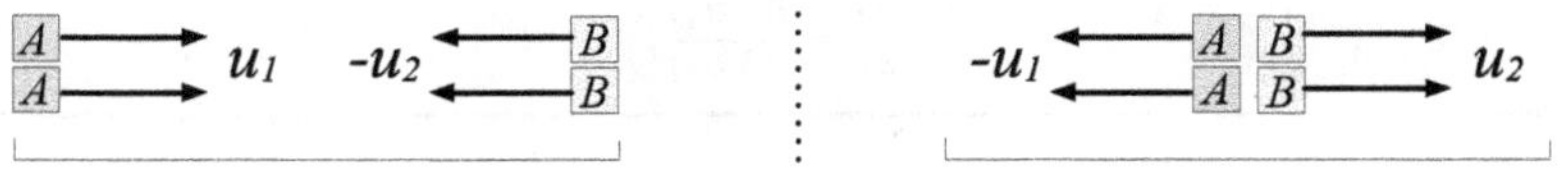

Let the particles A be equivalent to each other and different to the particles B (these properties are given by the ordering rules of Axiom II). Upper particles will collide the same way as lower particles. On the other hand, the pairs of particles AA and BB can be considered as a single object containing twice the amount of "matter".

Therefore, we assume that the following axiom is true:

Axiom III.
A proportional increase in the particle masses does not change the velocities of the particles after the collision.

The ratio between two particle masses should comply with that obtained from collisions with other particles. Therefore, we require the *transitivity axiom* to be true:

Axiom IV.
The operation of measuring the mass of a particle is *transitive*:

$$\textit{If}\ \frac{m_2}{m_1} = F(u_1, u_2), \quad \textit{and}\ \frac{m_3}{m_2} = F(u_2, u_3), \quad \textit{then}\ \frac{m_3}{m_1} = F(u_1, u_3).$$

The first two relations define the function F. The elocity u_3 in the third relation is *the same* as that in the second ratio m_3/m_2. This is a very strong requirement (< C_4) which immediately gives

$$F(u_1, u_2)\, F(u_2, u_3) = F(u_1, u_3).$$

This equation does not include masses, and the velocity u_3 is arbitrary (it corresponds to an arbitrary mass m_3). Since it does not depend on u_1 or u_2, it can take any fixed value, e.g. $u_3 = 0$. Therefore, function $F(u_1, u_2)$ is a ratio of two identical functions $F(u_1, 0)/F(u_2, 0)$ of the variables u_1 and u_2:

$$\frac{m_2}{m_1} = F(u_1, u_2) = \frac{f(u_1)}{f(u_2)}. \tag{4.1}$$

In classical mechanics, the function $f(u)$ can be found from the requirement that mass should not depend on the units used to measure velocity. For example, equation (4.1) should give the same ratio m_2/m_1 for any playback speed of the film of the particle collision. Therefore, if the function $f(u)$ does not include any other parameters including the dimension of time (for example, fundamental velocity c!), then the following axiom is true:

Axiom V.

$$\frac{m_2}{m_1} = \frac{f(\lambda\, u_1)}{f(\lambda\, u_2)},$$

---

where λ is an arbitrary parameter independent of the velocities and the masses. This condition completely determines the form of the function $f(u)$.

▷ To find the function $f(u)$, let us differentiate the equation

$$m_1\, f(\lambda\, u_1) = m_2\, f(\lambda\, u_2)$$

with respect to λ and divide it by the initial equation:

$$\frac{f'(\lambda u_1)\, u_1}{f(\lambda u_1)} = \frac{f'(\lambda u_2)\, u_2}{f(\lambda u_2)}.$$

Let us set $\lambda = 1/u_2$ and set

$$x = \frac{u_1}{u_2}, \qquad a = \frac{f'(1)}{f(1)},$$

to obtain

$$\frac{f'(x)}{f(x)} = \frac{a}{x} \qquad \Rightarrow \qquad f(x) = f_0\, x^a,$$

where f_0 is the integration constant. Consequently,

$$\frac{m_2}{m_1} = \frac{u_1^a}{u_2^a}, \qquad or \qquad \frac{m_2^{1/a}}{m_1^{1/a}} = \frac{u_1}{u_2}.$$

The parameter a is arbitrary and basically determines the *deformation of the mass scale*:

$$m \mapsto m^a.$$

In the simplest case $a = 1$, we obtain the following definition:

$$\frac{m_2}{m_1} = \frac{u_1}{u_2}. \tag{4.2}$$

Therefore, if the operation of mass measurement is transitive and independent of the units of velocity, then in classical physics it determines (up to the deformation power) the relation between the particles' velocities and the ratio of their masses. The mass of some particle can be taken as a standard. The masses of other particles can then be expressed in terms of the standard mass using symmetric collisions and the relation (4.2).

The fact that any particle "resists" attempts to change its velocity is a fundamental property of our world. As far as we currently understand its laws, we take this property as an experimental fact. It is just as fundamental as the existence of space and time. Without mass, mechanics would be quite a poor mathematical theory of non-interacting material points.

Definition (4.2) is a special case of some more general conservation laws which are of great importance in both classical and relativistic physics. Before considering relativistic dynamics, let us recall some of the main facts concerning conservation laws for classical energy and momentum.

4.2 Classical energy and momentum

Consider a symmetric collision of two *identical* particles:

Generally speaking, symmetry does not prohibit the simultaneous increase of the particles' velocities after the collision. Besides time reversibility (Axiom I), there is another way to "explain" why this does not happen. Suppose there is some scalar velocity function which does not change after an elastic collision. When it was first introduced, this function was poetically referred to as a "living force" (vis viva), but later received the prosaic name *energy*. Due to the isotropy of space, energy must depend on the squared velocity vector $E = E(\mathbf{u}^2)$. *If* the total "living force" of the bodies remains constant during a symmetric elastic collision, the magnitudes of the final particle velocities will not change.

Taking the definition

$$m_1 u_1 = m_2 u_2$$

and *assuming* that the energy of a particle is proportional to its mass $E = m\, g(\mathbf{u}^2)$, we can obtain the explicit form of the function g using Galilean transformations and the following axiom:

Axiom VI.
Mass is a proper characteristic of a particle. It is identical for all inertial observers.

Consider the law of conservation of energy for a symmetric elastic collision of particles with masses m_1 and m_2:

$$m_1\, g(u_1^2) + m_2\, g((-u_2)^2) = m_1\, g((-u_1)^2) + m_2\, g(u_2^2).$$

In a frame travelling with an *arbitrary* velocity $\mathbf{v}$, the velocities of the particles change as $\mathbf{u} \mapsto \mathbf{u} - \mathbf{v}$, but their masses do not change. Therefore, the law of conservation of energy has the following form:

$$m_1\, g((u_1 - v)^2) + m_2\, g((-u_2 - v)^2) = m_1\, g((-u_1 - v)^2) + m_2\, g((u_2 - v)^2).$$

We differentiate with respect to v and set $v = 0$. Assuming $m_1 u_1 = m_2 u_2$, we obtain:

$$m_1\, g'(u_1^2)\, u_1 = m_2\, g'(u_2^2)\, u_2 \quad => \quad g'(u_1^2) = g'(u_2^2) = const.$$

Since the masses were reduced, we conclude from their arbitrary nature that u_1 and u_2 are also arbitrary and independent quantities. Therefore, the latter relation can be true only if it is constant. Therefore, the function $g(\mathbf{u}^2)$ is linear with respect to the squared velocity.

▷ The symmetry of collisions of identical bodies cannot be fully "explained" by conservation of energy alone. Since E depends on the square of the velocity, both particles could potentially travel in the same direction after the collision. This is why we need another vector quantity, the ***momentum***, to be conserved. Its form can be obtained from the law of conservation of energy using Galilean transformations and the invariance of mass. Consider the total energy of several particles moving with velocities $\mathbf{u}_i$:

$$\sum_i m_i \frac{\mathbf{u}_i^2}{2} = const. \tag{4.3}$$

If the collision is elastic, the factor in front of $m\mathbf{u}^2$ is arbitrary, so we take it to be $1/2$. For an observer travelling with velocity $\mathbf{v}$ relative to the "stationary" frame where (4.3) was derived, all particles change their velocities by $\mathbf{v}$. Since all inertial observers are equivalent, energy should be conserved in all frames, and therefore:

$$\sum_i m_i \frac{(\mathbf{u}_i - \mathbf{v})^2}{2} = \sum_i m_i \frac{\mathbf{u}_i^2}{2} - \mathbf{v}\sum_i m_i \mathbf{u}_i + \frac{\mathbf{v}^2}{2}\sum_i m_i = const. \tag{4.4}$$

The first term after the equality sign must be constant if energy is to be conserved. Since the velocity $\mathbf{v}$ is arbitrary, the total momentum and the total mass of the system should also be conserved:

$$\sum_i m_i \mathbf{u}_i = const, \qquad \sum_i m_i = const.$$

To find them, differentiate the conservation law (4.4) with respect to $\mathbf{v}$. The condition $\mathbf{v} = 0$ expresses the conservation of total momentum. Another differentiation with respect to $\mathbf{v}$ gives conservation of mass. If the energy were non-linear with respect to the square of the velocity, ***given*** that the linear Galilean transformations are true, there would be more than three conservation laws.

All conservation laws include the same parameter, the mass m, to characterize different particles. In particular, conservation of momentum for a symmetric elastic collision gives the above definition of classical mass:

$$m_1 u_1 - m_2 u_2 = -m_1 u_1 + m_2 u_2,$$

or

$$\frac{m_2}{m_1} = \frac{u_1}{u_2}.$$

The law of conservation of energy is fulfilled automatically and ***independently*** in the explicit form of function $E(\mathbf{u}^2)$.

4.3 Relativistic energy

Now consider the case of relativistic dynamics. We will first answer the question:

? "What happens if the collision is *inelastic*?"

Consider as an example two identical particles with masses m moving with velocities u and $-u$ which collide to form a particle with mass M:

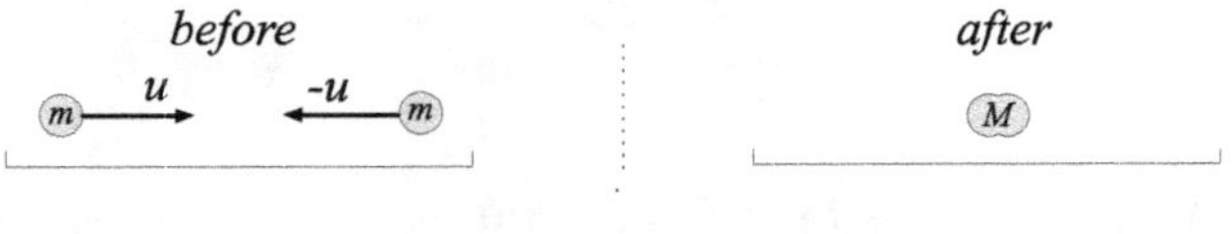

Symmetry considerations suggest that the velocity of the resulting particle M is zero. There is no reason to believe that it should move left or right. The same follows from the law of conservation of momentum in the form

$$\mathbf{p} = m\,\mathbf{u}\,g(\mathbf{u}^2),$$

where g is an arbitrary function.

According to classical mechanics, the mass $M = 2m$ should be also conserved. However, since the resulting particle is stationary, the classical kinetic energy is not conserved. The initial particles are moving and have "living force" in them, in contrast to the resulting stationary particle. In such a case, the energy of motion of the initial particles is said to be converted into energy "internal" to the body M. This internal energy can manifest itself in the degree of heating or deformation of the resulting object. For a more precise description we need to build some ***structure model*** for physical objects. For example, we may assume that they are composed of numerous constantly moving small particles (atoms). As a result of collision, the velocity of this motion increases and the initial "living force" is conserved due to the increased "living forces" of the atoms.

Even though this picture is absolutely correct for most objects in our world, it is desirable that our reasoning be independent of such details concerning particle structure. Mechanics and conservation laws are universal, and should be applicable also to unstructured objects such as ***point particles***. In this case, the original point objects are assumed to disappear as a result of the interaction, and a new point particle M is produced instead. But where does the living force go in this case?

The resulting particle with mass M can then be involved in further collisions, wher the energy accumulated from the original components should somehow manifest itself. Until now we have assumed that a particle's energy grows together with its velocity and mass. Mathematically, this statement is expressed by the equation $E = m\,f(\mathbf{u}^2)$, where f is a monotonically increasing function.

▷ A massive particle has greater energy. So why should life force "lost" not transform into the mass of the body? We now have to take a decisive step and ask:

"Whence does it follow that masses are conserved after inelastic collisions?" ?

No general considerations suggest this. Indeed, the resulting particle cannot be considered a simple sum of the two initial particles. Rather, a new object is created, and its mass *may* differ from the total mass of the original particles.

If so, the law of conservation of energy for inelastic collisions can be "saved" by stating that kinetic energy does not disappear; instead, it increases the mass, which is manifested in further experiments with the new particle. Note that we have made no "a priori" assumptions about the energy function. It was introduced to explain the invariance of the velocity magnitudes after symmetric elastic collisions of particles. This is quite an arbitrary function of the square of the velocity:

$$E = m\,f(\mathbf{u}^2).$$

In particular, nothing prohibits it from having a *nonzero value* when $\mathbf{u} = 0$. In this case, since the units of measurement of energy and mass are arbitrary, we can take $f(0) = c^2$. In our system of units we have $c = 1$, and the energy of a resting particle is equal to its mass $E = m$. This ***rest energy*** is the energy created from the "living forces" of original particles. The equation

$$E = mc^2$$

became the icon of science of the twentieth century. Its consequences are immense. All power engineering (not just atomic) ultimately follows from this equation.

We will further assume that our world is built in such a way that particles increase in mass when produced after inelastic collisions. In other words, the energy of motion does not disappear. Rather, it reinforces the inertial properties of a new object. Currently, this looks only natural. Of course, it is easy to be smart when the right answer is already known. It is more difficult if nobody knows it. But the real power of thought is manifested when the answer is known, generally accepted, absolutely habitual, and yet incorrect. However, the equation $E = m\,c^2$ has been proved by numerous experiments and is undoubtedly true.

Let us find the dependence $E = mf(\mathbf{u}^2)$ of a particle's energy on its velocity from the viewpoint of the theory of relativity. The only requirement for the function f is that it is not zero at zero velocity:

$$f(0) = 1.$$

Also, the relativistic velocity-addition formula and the axiom of mass *invariance* will be assumed true. Remember that the parameter m is a characteristic of a particle and has the same value in all inertial frames. Let us use several thought experiments to find the form of the function f.

▷ Consider an inelastic collision of identical particles with masses m moving towards each other with velocity u. Let this motion take place along the the y-axis (the first figure below). Consider the same collision in the frame S' moving *to the left* with velocity "$-v$" (the second figure):

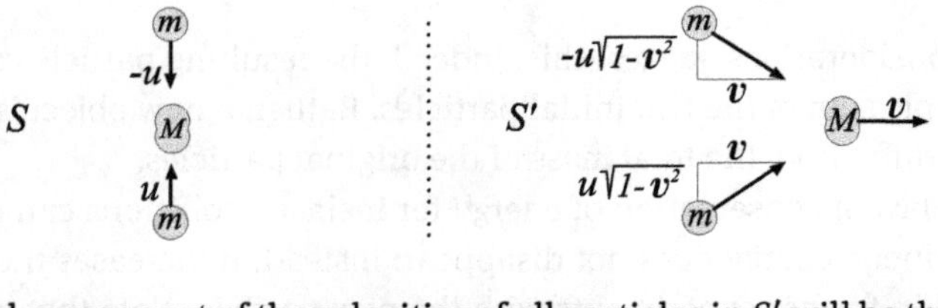

The horizontal component of the velocities of all particles in S' will be the same as the velocity v, while their vertical components change according to the velocity-addition formula (1.20), p. 19. Since $u_x = 0$, $u_y = \pm u$ in the stationary frame, we have

$$u'_y = \pm u\,\sqrt{1 - v^2} \quad \text{and} \quad u'_x = v.$$

Suppose there is a conserved quantity proportional to the mass of the body multiplied by some function of the squared velocity: $E = m\,f(\mathbf{u}^2)$. Since $\mathbf{u}^2 = u_x^2 + u_y^2$, the law of conservation of energy in the moving frame S' is as follows:

$$2m\,f(v^2 + u^2(1 - v^2)) = M\,f(v^2).$$

In the rest frame ($v = 0$), we have $M = 2mf(u^2)$. Substituting in this expression for the mass M and writing $x = u^2$, $y = v^2$, we obtain the following functional equation:

$$f(x + y - xy) = f(x)f(y). \tag{4.5}$$

To solve it, we can derive with respect to y and set $y = 0$. Introducing the constant $a = f'(0)$, we obtain a differential equation:

$$f'(x)\,(1 - x) = a\,f(x),$$

or

$$f(x) = \frac{1}{(1 - x)^a},$$

where the integration constant was found from the condition $f(0) = 1$. Therefore, as a function of velocity, the energy has the form:

$$E = \frac{m}{(1 - \mathbf{u}^2)^a}. \tag{4.6}$$

Let us again consider colliding particles to find the value of a.

▷ Suppose the particle velocities are directed along the x-axis, and that the frame S' moves to the right with velocity $v = -u$:

The velocity of the particle on the right is zero in S'. Setting $v = -u$ in the equation

$$u'_x = \frac{u - v}{1 - uv},$$

we obtain $2u/(1 + u^2)$ for the velocity of the particle on the left. Using (4.6), we obtain the law of conservation of energy in S':

$$\frac{m}{[1 - 4u^2/(1 + u^2)^2]^a} + m = \frac{M}{(1 - u^2)^a} = \frac{2m}{(1 - u^2)^{2a}},$$

where the conservation law

$$M = 2mf(u^2)$$

for the rest frame was used in the second equation. After some elementary algebraic manipulation, we obtain:

$$(1 + u^2)^{2a} + (1 - u^2)^{2a} = 2.$$

This expression can only hold for an arbitrary velocity u if $a = 1/2$ (it is enough to set $u = 1$).

Thus, the final expression for energy is:

$$E = \frac{m}{\sqrt{1 - \mathbf{u}^2}}. \tag{4.7}$$

Note that the energy of the body becomes equal to its mass for $\mathbf{u} = 0$. If, however, the velocity tends to unity (the fundamental velocity), the energy tends to infinity. This is the way in which energy considerations explain why an "ordinary" particle can never reach the speed of light. In classical mechanics, the energy of motion

$$E = m\,\frac{\mathbf{u}^2}{2}$$

grows together with the velocity without limit. However, in reality, it always remains finite. The relativistic expression for the energy, like the Lorentz transformations, is singular (infinite) for $|\mathbf{u}| = 1$. This means that infinite energy is required to accelerate a massive particle to the fundamental velocity.

To obtain (4.7), we considered an inelastic collision in two frames moving along the x- and x-axes. As an exercise (< H_{16}) , we suggest considering one-dimensional motion along the x-axis and solving the functional equation in this case.

4.4 Relativistic momentum

Using equation (1.23), p. 20, write the energy in the frame S', which moves with velocity $\mathbf{v}$:

$$E' = \frac{m}{\sqrt{1-\mathbf{u}'^2}} = \frac{m\,(1-\mathbf{u}\,\mathbf{v})}{\sqrt{1-\mathbf{u}^2}\,\sqrt{1-\mathbf{v}^2}}.$$

Using the notation

$$\mathbf{p} = \frac{m\,\mathbf{u}}{\sqrt{1-\mathbf{u}^2}}, \tag{4.8}$$

we rewrite the energy in a more compact form:

$$E' = \frac{E-\mathbf{v}\mathbf{p}}{\sqrt{1-\mathbf{v}^2}}. \tag{4.9}$$

Consider a set of colliding particles. According to the principle of relativity, the energy is conserved in any inertial frame:

$$\sum_i E'_i = \frac{1}{\sqrt{1-\mathbf{v}^2}}\sum_i E_i - \frac{\mathbf{v}}{\sqrt{1-\mathbf{v}^2}}\sum_i \mathbf{p}_i = const.$$

The value of this expression does not change before or after the collision. The sum of the energies E_i in the frame S is also conserved. Since the velocity $\mathbf{v}$ is arbitrary, these two laws can be be fulfilled simultaneously only if the following law of conservation of *momentum* is true:

$$\sum_i \mathbf{p}_i = const. \tag{4.10}$$

Therefore, conservation of momentum (4.8) follows from the law of conservation of energy and from the Lorentz transformations.

For the symmetric elastic collisions considered above, conservation of momentum leads to the following relativistic generalization of the definition of mass (4.2):

$$\frac{m_2}{m_1} = \frac{u_1/\sqrt{1-u_1^2}}{u_2/\sqrt{1-u_2^2}}. \tag{4.11}$$

Note that once again, the mass of a particle is its "personal" parameter, identical for all observers regardless of their relative velocity.

There is no law of mass conservation during collisions in the relativistic world. It is "contained" in the law of conservation of energy and in the concept of mass as the energy the body has in its proper frame. Naturally, there can still be elastic collisions which do not change particle masses. However, it is only due to inelastic collisions that the energy of colliding particles can transform into the masses of resulting particles.

▷ A transformation similar to (4.9) can also be written for the momentum. Consider a frame S' moving along the x-axis so that $\mathbf{v} = (v, 0, 0)$. Energy (4.7) and momentum

(4.8) are related by a simple equation $\mathbf{p} = E\,\mathbf{u}$. Therefore, we can use (4.9) to find the x-components of the momentum:

$$p'_x = E'\,u'_x = \frac{E - vp_x}{\sqrt{1-v^2}}\,\frac{u_x - v}{1 - u_x v} = E\,\frac{1 - vu_x}{\sqrt{1-v^2}}\,\frac{u_x - v}{1 - u_x v} = \frac{p_x - vE}{\sqrt{1-v^2}},$$

where we have also used the transformation of the x-component of the velocity and the equation $p_x = E\,u_x$, which we applied twice. The the transformation of the momentum y-components is found similarly:

$$p'_y = E'\,u'_y = \frac{E - vp_x}{\sqrt{1-v^2}}\,\frac{u_y\sqrt{1-v^2}}{1 - u_x v} = E\,\frac{1 - vu_x}{1 - vu_x}\,u_y = p_y.$$

Therefore, we have the following transformations of energy and momentum for the relative motion of the frames S and S' along parallel x- and x'-axes:

$$E' = \frac{E - vp_x}{\sqrt{1-v^2}}, \qquad p'_x = \frac{p_x - vE}{\sqrt{1-v^2}}, \qquad p'_y = p_y, \qquad p'_z = p_z. \tag{4.12}$$

Compare these with the Lorentz transformations:

$$t' = \frac{t - vx}{\sqrt{1-v^2}}, \qquad x' = \frac{x - vt}{\sqrt{1-v^2}}, \qquad y' = y, \qquad z' = z.$$

As can be seen, there is a remarkable symmetry in the fact that the quadruple $\{E, p_x, p_y, p_z\}$ is transformed in the same way as $\{t, x, y, z\}$. This is not an accidental coincidence. Energy and momentum are components of the four-vector

$$p^\nu = \{E,\ \mathbf{p}\},$$

and the event is described by the four-vector

$$x^\nu = \{t,\ \mathbf{x}\}.$$

We consider these questions in more detail in chapters 7 and 8.

Similarly to the vector Lorentz transformations (1.18), p. 17, a more general transformation can be written for energy and momentum:

$$E' = \gamma\,(E - \mathbf{v}\mathbf{p}), \qquad \mathbf{p}' = \mathbf{p} - \gamma\mathbf{v}E + \Gamma\,\mathbf{v}\,(\mathbf{v}\mathbf{p}). \tag{4.13}$$

Assuming that $\mathbf{p} = m\mathbf{u}\,f(\mathbf{u}^2)$ and that momentum is conserved, the function $f(\mathbf{u}^2)$ can also be obtained from the relativistic velocity-addition formula. To see this, consider two identical particles stuck together after a collision and write down the law of conservation of momentum in a frame moving perpendicular to the velocities of the initial particles with arbitrary velocity v (< H_{17}).

4.5 Some examples

Consider again the energy and the momentum of a particle with mass m moving with velocity $\mathbf{u}$,

$$E = \frac{m}{\sqrt{1-\mathbf{u}^2}}, \qquad \mathbf{p} = \frac{m\mathbf{u}}{\sqrt{1-\mathbf{u}^2}}. \tag{4.14}$$

Squaring and subtracting them from each other, we can remove the velocity:

$$E^2 - \mathbf{p}^2 = m^2. \tag{4.15}$$

Dividing the momentum by the energy, the mass can also be removed:

$$\mathbf{p} = E\mathbf{u}. \tag{4.16}$$

In this form, the relationship between energy and momentum is also true for massless particles such as photons. In this case $c = 1$, so the velocity can be written as a unit vector $\mathbf{u} = \mathbf{n}$ where $\mathbf{n}^2 = 1$. Using Planck's formula $E = h\nu$, we have:

$$\mathbf{p} = E\mathbf{n} = h\nu\,\mathbf{n} = \frac{h}{\lambda}\mathbf{n}, \tag{4.17}$$

where $\lambda = 1/\nu$ is a wavelength associated with the quantum properties of the photon (more precisely, with an ensemble of photons).

Remember that we have to multiply all quantities with dimensions equal to some power of time by "c" to the same power to recover the fundamental velocity "c" in the above equations. Therefore, for the velocity ($\mathbf{u}$), momentum ($\mathbf{p}$), energy (E), force ($\mathbf{F}$), and acceleration ($\mathbf{a}$) the following substitutions are necessary:

$$t \mapsto tc, \quad \mathbf{u} \mapsto \frac{\mathbf{u}}{c}, \quad \mathbf{p} \mapsto \frac{\mathbf{p}}{c}, \quad E \mapsto \frac{E}{c^2}, \quad \mathbf{F} \mapsto \frac{\mathbf{F}}{c^2}, \quad \mathbf{a} \mapsto \frac{\mathbf{a}}{c^2}.$$

With the constant "c" present, equations (4.14) have the form:

$$E = \frac{mc^2}{\sqrt{1-\mathbf{u}^2/c^2}}, \qquad \mathbf{p} = \frac{m\mathbf{u}}{\sqrt{1-\mathbf{u}^2/c^2}}, \tag{4.18}$$

while the relations (4.15), (4.16) are written:

$$E^2 - \mathbf{p}^2c^2 = m^2c^4, \qquad \mathbf{p} = \frac{E}{c^2}\mathbf{u}.$$

We can decompose the energy and momentum (4.18) into series with respect to $1/c$:

$$E = mc^2 + \frac{m\mathbf{u}^2}{2} + \frac{3}{8}\frac{m\mathbf{u}^4}{c^2} + \dots, \qquad \mathbf{p} = m\mathbf{u} + m\mathbf{u}\frac{\mathbf{u}^2}{2c^2} + \dots$$

Except for the rest energy mc^2, the leading terms correspond to the classical expressions $E = m\mathbf{u}^2/2$ and $\mathbf{p} = m\mathbf{u}$.

▷ The energy in atomic physics and in the physics of elementary particles is measured in *electron-volts* rather than in SI units. This is the energy that an electron with charge "e" acquires when passing a potential difference of one volt. An electron-volt is a very small amount of energy as compared to the energies we usually meet:

$$1\, eV = 1.602\,176\,53(14) \cdot 10^{-19}\, J.$$

In fact, it is small even for typical tasks on the microscopic scale. That is why powers of it, with prefixes

$$\text{kilo } (10^3), \quad \text{mega } (10^6), \quad \text{giga } (10^9), \quad \text{tera } (10^{12}),$$

etc., are usually used. Note the number in parentheses in the definition $1\, eV$. This gives a typical range for the last two significant digits, which are subject to experimental error.

In this book we use the system of units $c = 1$, so mass is equal to the rest energy of the particle. Therefore, the masses of the electron and the proton are expressed in electron-volts as:

$$m_e \approx 0.511\, MeV, \qquad m_p \approx 938\, MeV.$$

Quantum laws determine certain characteristic dimensions and energies of atoms, such as the Bohr radius r_B and the Rydberg constant E_R:

$$r_B = \frac{\hbar^2}{m_e e^2} \approx 0.529 \cdot 10^{-10}\, m, \qquad E_R = \frac{e^2}{2r_B} \approx 13.6\, eV.$$

The energy E_R measures the mass defect of the hydrogen atom, i.e. the difference between the total "electron + proton" mass and the mass of the hydrogen atom. The mass of the proton is much larger than this energy; the relative change in the mass of the atom due to the binding energy is of order 10^{-8}.

However, the whole mass of a particle can also transform into energy via radiation. This happens, for example, when an electron and an antielectron (positron) annihilate

$$e^+ + e^- \mapsto 2\gamma,$$

releasing $2\, m_e \approx 1 MeV$ of radiation energy.

Changes in the energy of normal objects around us produce very small changes in their masses. If a clothes iron with mass m is dropped from a height H of one meter and all of its potential energy is converted into internal energy, then the relative change in its mass will be:

$$\frac{\Delta m}{m} = \frac{mgH/c^2}{m} = \frac{gH}{c^2} \sim 10^{-16}.$$

If the same iron is used for its intended purpose and is heated up to 200 degrees, then it will receive an additional 10^5 J/kg of energy per unit mass, given that its specific heat (Fe) is about 500 J/(kg K). The relative change in its mass is of order 10^{-12} in this case.

4.6 Kinetic energy

The energy required to accelerate a particle to velocity $\mathbf{u}$,

$$T = E - m = \frac{m}{\sqrt{1-\mathbf{u}^2}} - m,$$

is called the *kinetic* energy. This energy is quoted among the other parameters of particle accelerators. For small velocities it tends to the classical expression $T \approx m\mathbf{u}^2/2$. If $T \gg m$, then the kinetic energy approximately equals the total energy ($T \approx E$). For example, the Large Hadron Collider (LHC) at CERN accelerates protons $m_p \approx 0.9$ GeV up to kinetic energies $T \sim 7$ TeV = 7000 GeV (2016).

Fast moving particles can collide either with a fixed *target* or with a crossed beam of accelerated particles. In the latter case, there is a noticeable energy gain. Assume that the particles are identical and that the total energy of each beam is $E = m/\sqrt{1-u^2}$. This means that the particles' velocities are $u^2 = 1 - m^2/E^2$:

crossed beam
u
-u
m
m
E
E
fixed target
u'
m
E'
m

Suppose there is a reference frame that moves to the left (the second figure) with velocity $v = -u$ in which the particles of the right beam are at rest. To calculate the energy of the left beam, we use the transformation law for the energy (4.9), p. 100, with $\mathbf{v} = -\mathbf{u}$ and the equation $\mathbf{p} = E\mathbf{u}$:

$$E' = \frac{E + \mathbf{u}\mathbf{p}}{\sqrt{1-\mathbf{u}^2}} = \frac{1+u^2}{\sqrt{1-u^2}}\,E.$$

Substitution of $u^2 = 1 - m^2/E^2$ gives the relationships between the total and the kinetic energies in the two frames:

$$E' = \frac{2E^2}{m} - m, \qquad \frac{T'}{m} = 2\left(1 + \frac{T}{m}\right)^2 - 2. \tag{4.19}$$

These relationships can be used to compare the collision of two beams with energies E, T with the scattering of a single beam with energy E', T' by a stationary target. For high energies $T \gg m$, we have

$$T' \approx \frac{2T^2}{m}.$$

Due to this quadratic dependence, energies equivalent to those for a collision with a stationary target are reached very rapidly. For example, the energy 7 TeV of each proton beam in the Large Hadron Collider (LHC) will produce the same result as the energy 133 TeV of a single beam in a fixed-target accelerator. Therefore, a 19-fold gain is obtained. At low velocities ($T \ll m$), the kinetic energy of scattering by a target is $T' \approx 4T$, i.e. there is only a fourfold gain of using two beams instead of one beam (< C_8).

▷ Almost all power engineering draws upon solar energy, either currently available or accumulated in the past. This, in turn,is produced by gravitational forces which squeeze and heat up a huge plasma ball until it reaches the pressures and temperatures required for thermonuclear reactions. ***Fusion reactions*** produce deuterium ^{2}D from two hydrogen nuclei (protons):

$$p + p \mapsto {}^2\mathrm{D} + e^+ + \nu_e. \tag{4.20}$$

This reaction is accompanied by an "energy release". This is not a very good term, since the total energy of any reaction is constant. What is meant is the difference between the ***kinetic energies*** of the particles before and after the collision:

$$\sum_i (T_i + m_i) = \sum_j (T_i' + m_j') \qquad => \qquad \Delta T = (\sum_i m_i) - (\sum_i m_i').$$

To calculate the released energy of the ***motion***, the difference between the particle masses before and after the interaction must be found. Thus, for the reaction (4.20) underlying the thermonuclear energy of the Sun, we have ($m_\nu \approx 0$):

$$\Delta T = 2m_p - (m_D + m_e) = 2 \cdot 938.272 - (1875.613 + 0.511) = 0.42\ MeV$$

or

$$\Delta T / 2m_p = 2 \cdot 10^{-4}.$$

This balance of kinetic energy is important; it means that the energy can be transformed further, into other "useful" forms of energy (e.g., electrical).

Proton-proton synthesis is slow (because of its small probability), and almost half of the reaction energy is carried away by neutrinos ν_e, which interact weakly with matter. A positron e^+ annihilates with an electron e^-, and deuterium ^2D fuses with a proton p to produce helium-3, with a higher gain of energy. Still another fusion reaction produces helium-4. As a result, four protons and two electrons transform into the nucleus of a helium atom, two neutrinos, and six photons:

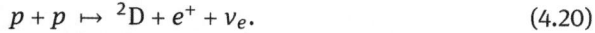

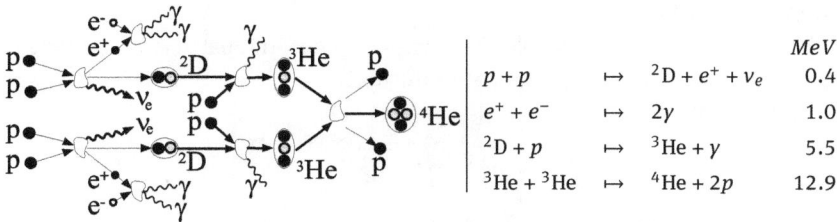

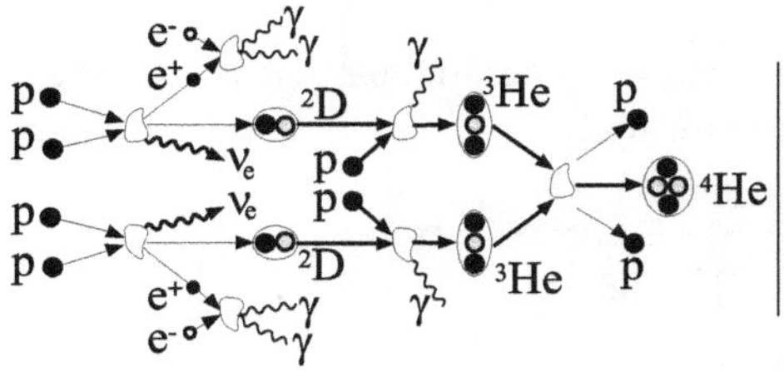

			MeV
$p + p$	$\mapsto$	$^2\mathrm{D} + e^+ + \nu_e$	0.4
$e^+ + e^-$	$\mapsto$	2γ	1.0
$^2\mathrm{D} + p$	$\mapsto$	$^3\mathrm{He} + \gamma$	5.5
$^3\mathrm{He} + {}^3\mathrm{He}$	$\mapsto$	$^4\mathrm{He} + 2p$	12.9

The total energy 26.7 MeV is equivalent to just 0.7% of the mass of four protons. Because of radiation, the Sun loses about 10^{17} kg a year, or $5 \cdot 10^{-14}$ % of its mass.

4.7 Decays and inelastic collisions

Suppose that, due to some internal reason, a resting particle of mass M decays into two particles with masses m_1, m_2 and velocities u_1, u_2, respectively:

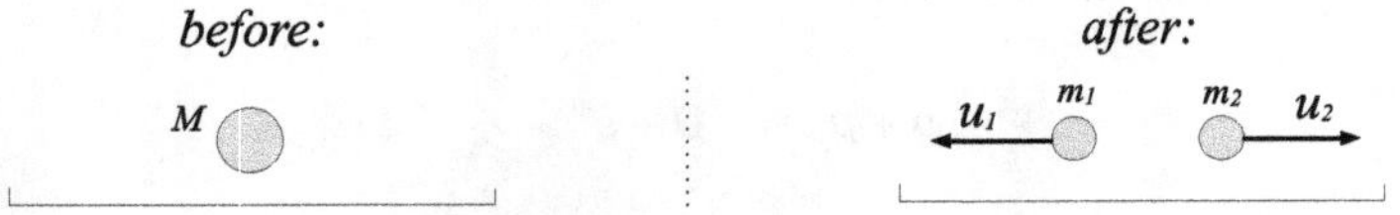

We can use the conservation laws of energy and momentum to find the energies of the decay products:

$$E = E_1 + E_2, \qquad p = p_1 + p_2. \tag{4.21}$$

A standard approach to such problems involves substituting for one of the unknowns (energy or momentum) using the equation

$$E^2 - p^2 = m^2.$$

Taking E_1 and p_1 over to the left of the conservation laws, squaring them, and subtracting them from one from another, we have:

$$(E - E_1)^2 - (p - p_1)^2 = E_2^2 - p_2^2 = m_2^2.$$

By expanding the brackets and again using the expression for mass squared, we obtain

$$M^2 + m_1^2 - 2EE_1 + 2pp_1 = m_2^2.$$

Since the initial particle is at rest, for it we have $E = M$ and $p = 0$. Therefore, the energy E_1 of the first decay product is easily obtained. To find E_2, the indices must be interchanged:

$$E_1 = \frac{M^2 + m_1^2 - m_2^2}{2M}, \qquad E_2 = \frac{M^2 + m_2^2 - m_1^2}{2M}. \tag{4.22}$$

Thus, the energies are fully determined by the mass of the initial particle and the masses of its decay products. If desired, also their velocities can be found from the equation $E = m/\sqrt{1 - u^2}$.

When is such a decay possible? The mass of the initial particle is converted into the energy of the particles produced after the decay:

$$M = \frac{m_1}{\sqrt{1 - u_1^2}} + \frac{m_2}{\sqrt{1 - u_2^2}} > m_1 + m_2.$$

If $M < m_1 + m_2$, such ***spontaneous*** decay is, in light of energy considerations, impossible, and some external impact is required to break the ***binding energy*** equal to $m_1 + m_2 - M$.

▷ The requirement $M > m_1 + m_2$ prohibits a particle from emitting another particle while maintaining its original mass. For example, a uniformly moving electron cannot emit a photon and still remain an electron. In fact, nor can it turn into any other particle in such a process of decay, but that is another story.

Why do particles decay? This happens as a result of processes taking place "inside" the initial particle. An explosion as a result of chemical reactions is the simplest example. In the quantum world, decay is usually associated with the *tunnelling* of a quantum particle through a potential barrier that it cannot surmount in classical physics. In the relativistic quantum world of elementary particles the situation is even less intuitive. For example, a free neutron decays over an average of 15 minutes into a proton, an electron, and an antineutrino, though they were not "its *constituents*" before that:

$$n \mapsto p + e^- + \bar{\nu}_e.$$

The products of this decay are "produced" at exactly the moment it takes place. To describe such physics, a quantum field theory is needed.

▷ Decay is the opposite of "fusion", when a third particle is produced by two colliding particles. Consider a *laboratory reference frame* where the first particle has mass m_1, velocity v, and energy

$$E_1 = \frac{m_1}{\sqrt{1 - v^2}},$$

and the second particle with mass m_2 is at rest. The particle resulting from their collision has mass M and velocity u:

Writing the conservation laws of energy and momentum,

$$E_1 + m_2 = E, \qquad p_1 = p,$$

squaring these, and subtracting them from each other, we can find the squared mass of the resulting particle, which, naturally, is greater than the sum of the masses of the initial particles:

$$M^2 = m_1^2 + m_2^2 + \frac{2m_1 m_2}{\sqrt{1 - v^2}} = (m_1 + m_2)^2 + 2m_1 m_2 \left(\frac{1}{\sqrt{1 - v^2}} - 1 \right).$$

Its velocity can be found from the law of conservation of momentum $p_1 = p$. However, this can be achieved more quickly using the relation between momentum, energy, and velocity:

$$u = \frac{p}{E} = \frac{p_1}{E_1 + m_2} = \frac{v}{1 + (m_2/m_1)\sqrt{1 - v^2}}.$$

If $v \to 1$, the velocity of the produced particle also tends to the speed of light.

4.8 Elastic collisions

Consider a particle with energy E_1 and mass m_1 which collides into a resting particle of mass m_2. Suppose that the particle masses do not change after the interaction (the collision is *elastic*):

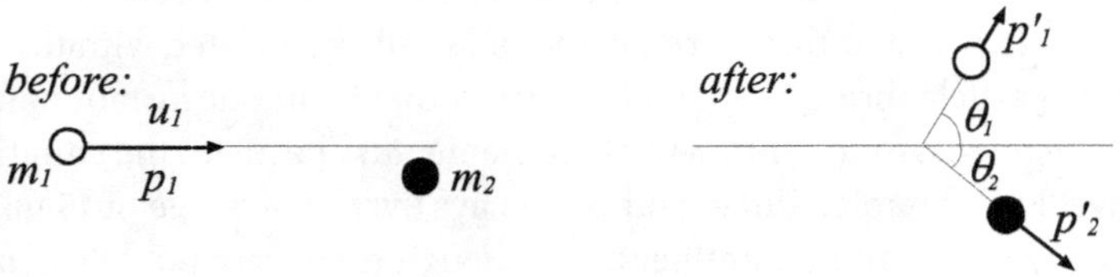

If the collision is not "head-on", then the particles scatter at some angles to the velocity $\mathbf{u}_1$ of the colliding particle. Use primes to denote all quantities after the collision and consider the conservation laws

$$E_1 + E_2 = E_1' + E_2', \qquad \mathbf{p}_1 + \mathbf{p}_2 = \mathbf{p}_1' + \mathbf{p}_2', \tag{4.23}$$

where $E_2 = m_2$ and $\mathbf{p}_2 = 0$. Isolating the energy and the momentum of one of the particles

$$(E_1 + m_2 - E_1')^2 - (\mathbf{p}_1 - \mathbf{p}_1')^2 = m_2^2,$$

we have

$$\mathbf{p}_1\mathbf{p}_1' = (E_1 + m_2)\,E_1' - m_1^2 - E_1 m_2.$$

The scalar product of the momenta is expressed in terms of their magnitudes and the angle between them: $\mathbf{p}_1\mathbf{p}_1' = p_1 p_1' \cos\theta_1$. Therefore, the emission angle of the particles is:

$$\cos\theta_1 = \frac{(E_1 + m_2)\,E_1' - m_1^2 - E_1 m_2}{p_1 p_1'}, \qquad \cos\theta_2 = \frac{(E_1 + m_2)(E_2' - m_2)}{p_1 p_2'}$$

(The second angle is obtained similarly, by isolating the energy and the momentum of the second particle after the collision).

The dispersion angle depends on the impact parameter and the nature of the interaction between the particles. However, regardless of the latter, the energy of the scattered particle is related to this angle via the conservation laws. The greater the energy loss of the first particle, the greater the scattering angle. Since $E_2' > m_2$ and $E_2 = m_2$, it follows from the law of conservation of energy (4.23) that $E_1' < E_1$, i.e., the energy decreases when passing to the resting particle.

Substituting $p' = \sqrt{E_1'^2 - m_1^2}$ into the expression for $\cos\theta_1$ and differentiating it with respect to E_1', we ($\lessdot$ H$_{18}$) find the maximum value of the angle:

$$(\sin\theta_1)_{max} = \frac{m_2}{m_1}. \tag{4.24}$$

A maximum other than $\theta = \pi/2$ exists only if the mass of the incident particle m_1 is greater than the mass of the target m_2.

▷ An elastic collision looks simpler if the incident particle has no mass, $m_1 = 0$. For example, in the *Compton Effect* (1923) a photon is scattered by an electron. In this case $p_1 = E_1$, $p'_1 = E'_1$, and the energy of the photon after the collision is:

$$\frac{m_2}{E'_1} = \frac{m_2}{E_1} + 1 - \cos\theta_1.$$

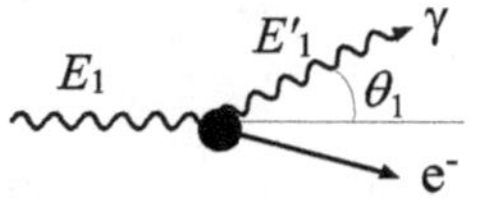

A photon (according to Planck's formula) is characterised by its frequency $E = h\nu$ and wavelength $\lambda = 1/\nu$. The relative change $\Delta\lambda = \lambda' - \lambda$ of the latter is:

$$\frac{\Delta\lambda}{\lambda} = \frac{E_1}{m_2}(1 - \cos\theta_1).$$

Thus, the larger the scattering angle, the longer the wavelength (and the smaller the energy). Therefore, the relative change of the wavelength becomes noticeable when the energy E_1 of the photon is comparable to the mass of the electron $E_1 \sim m_2$. The Planck constant in electron-volts is:

$$\hbar = \frac{h}{2\pi} = 6.582\,118\,99(16) \cdot 10^{-22}\ MeV \cdot s.$$

Therefore, the electron mass $m_e = 0.511$ MeV corresponds to the photon frequency 10^{20} Hertz. The wavelength is $\sim 10^{-12}$ m in this case. These photons are of hard *X-radiation*, which can also be obtained in an X-ray tube, where electrons are accelerated by an electric field to collide into the anode and emit X-rays (bremsstrahlung). X-rays also knock electrons out of inner electron shells of the atoms. Electrons from upper energy levels replace them, also emitting photons in the X-ray spectrum. The value

$$\frac{h}{mc} = 2.4263 \cdot 10^{-12}\ m$$

is called the *Compton wavelength of the electron*. The change of a photons's wavelength (or frequency) due to scattering by an electron is a quantum-relativistic effect. In classical electrodynamics, electromagnetic waves do not change their frequency when scattered by a charged particle.

We considered above a collision in the laboratory frame, one particle being at rest and the other colliding it with the impulse p_1. We have used the term "*laboratory frame*" several times by now. It comes from the experiments where one type of particle is propelled by the accelerator and directed to the stationary target built of particles of another type. Indeed, this is quite an arbitrary term, since the same laboratories carry out experiments with multiple colliding beams propelled in accelerators.

▷ Consider now an elastic collision of two particles in the frame where their total momentum is zero.

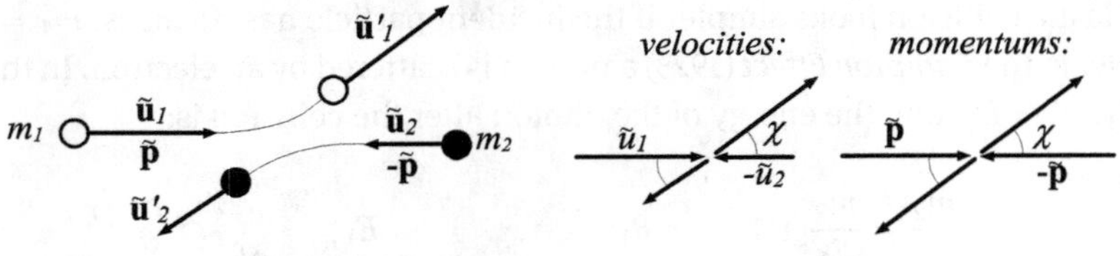

The first figure shows the dynamics of such a collision and the trajectories of the particles. As they approach each other, the interaction between them increases, and their trajectories get curved. When the particles are sufficiently far from each other, they again become free and move at some constant velocity.

The frame where the momenta of the colliding particles have equal magnitudes is called the *center-of-mass* frame or the *center of inertia* frame. By the law of conservation of momentum, both particles have the same scattering angle. Let us denote it by χ. The energies of the particles will not change after the collision. Indeed, it follows from the conservation law for equal initial momenta that:

$$\sqrt{\tilde{\mathbf{p}}^2 + m_1^2} + \sqrt{\tilde{\mathbf{p}}^2 + m_2^2} = \sqrt{\tilde{\mathbf{p}}'^2 + m_1^2} + \sqrt{\tilde{\mathbf{p}}'^2 + m_2^2},$$

whence $\tilde{\mathbf{p}}^2 = \tilde{\mathbf{p}}'^2$ and, consequently, $\tilde{E}_i = \tilde{E}'_i$. The energy and the momentum in the center-of-mass frame are designated with a tilde sign to distinguish them from the laboratory frame. The velocities for the same momenta obey the relation:

$$|\tilde{\mathbf{p}}| = \tilde{E}_1\tilde{u}_1 = \tilde{E}_2\tilde{u}_2 \qquad => \qquad \frac{m_2}{m_1} = \frac{\tilde{u}_1/\sqrt{1-\tilde{u}_1^2}}{\tilde{u}_2/\sqrt{1-\tilde{u}_1^2}},$$

and are also different for different masses.

▷* Now let us express the energy in the *laboratory* frame in terms of the angle χ defined in the *center-of-mass* frame. To do this, consider an inertial reference frame moving *to the left* with velocity $\mathbf{v} = \tilde{\mathbf{u}}_2$. The second particle is at rest in this frame before the collision. Using the energy conversion formula (4.9), we see that the magnitudes of the energy and the momentum of the first particle do not change in the center-of-mass frame before and after the collision:

$$\tilde{E}_1 = \tilde{E}'_1 = \tilde{E}, \qquad |\tilde{\mathbf{p}}_1| = |\tilde{\mathbf{p}}'_1| = \tilde{p}.$$

The energies in the laboratory frame are:

$$E_1 = \frac{\tilde{E}_1 + \tilde{u}_2\tilde{p}_1}{\sqrt{1-\tilde{u}_2^2}}, \qquad E'_1 = \frac{\tilde{E}_1 + \tilde{u}_2\tilde{p}_1\cos\chi}{\sqrt{1-\tilde{u}_2^2}}, \tag{4.25}$$

where we have applied the fact that the energy does not change after the collision and that the scalar product **vp** is equal to the product of the magnitudes and the cosine of the angle χ.

Therefore, the energy transferred in the laboratory frame is:

$$E_1' - E_1 = -\frac{\tilde{p}_1 \tilde{u}_2}{\sqrt{1-\tilde{u}_2^2}} (1-\cos\chi) = -\frac{\tilde{p}^2}{m_2} (1-\cos\chi),$$

where $\tilde{u}_2$ is expressed in terms of momentum $\tilde{p}_2$ equal to $\tilde{p}$ (in the center-of-mass frame). To find it, we substitute the relationships between the energy and the momentum $\tilde{E}_1 = \sqrt{\tilde{p}^2 + m_1^2}$ and between the momentum and the velocity

$$1 + \frac{\tilde{p}^2}{m_2^2} = \frac{1}{1-\tilde{u}_2^2}$$

into the first equation of (4.25):

$$E_1 = \frac{\tilde{E}_1}{\sqrt{1-\tilde{u}_2^2}} + \frac{\tilde{p}^2}{m_2} = \sqrt{\tilde{p}^2 + m_1^2}\,\sqrt{1+\tilde{p}^2/m_2^2} + \frac{\tilde{p}^2}{m_2}.$$

Solving for $\tilde{p}$ we obtain:

$$\frac{\tilde{p}^2}{m_2} = \varepsilon = \frac{(E_1^2 - m_1^2)\, m_2}{m_1^2 + m_2^2 + 2m_2\, E_1},$$

which gives E_1', and using the law of conservation and equation $E_2' = E_1 + m_2 - E_1'$, we have:

$$E_1' = E_1 - \varepsilon\,(1-\cos\chi), \qquad E_2' = m_2 + \varepsilon\,(1-\cos\chi).$$

Thus, the energies of the scattered particles in the laboratory frame are, in quite a simple manner, related to the rotation of the momenta in the center-of-mass frame given by the angle χ. The quantity $\varepsilon\,(1-\cos\chi)$ expresses the energy transferred by the first particle and grows together with χ. In the case of a "head-on" collision ($\chi = \pi$), this energy reaches it maximum value 2ε.

▷ To conclude, let us consider one important aspect related to conservation laws [20]. When interacting, particles change their velocity. Therefore, their energy and momentum must be calculated at a particular instant of time. When writing a conservation law, e.g. for the momentum, we sum up particle momenta calculated in a specific frame at a given point of time. However, since the particles occupy different points of space, such a sum will correspond to different times (relativity of simultaneity, p. 52) in another frame. Conservation laws can only be unambiguous either before or after the interaction. In this case the particles are moving with constant velocities (momenta), and even non-simultaneous values of their momenta will give the same sum. We will consider this question in more detail in Chapter 10.

4.9 Jet propulsion

Interstellar distances are enormous, and relativistic velocities are needed to cover them. In *empty space* a vehicle can accelerate only by losing a part of its mass. Let us consider some of the details of relativistic jet propulsion. Let a rocket of mass M travel with velocity v, and for a short time interval dt emit something with energy ε and velocity $-u$ in the direction opposite to its motion:

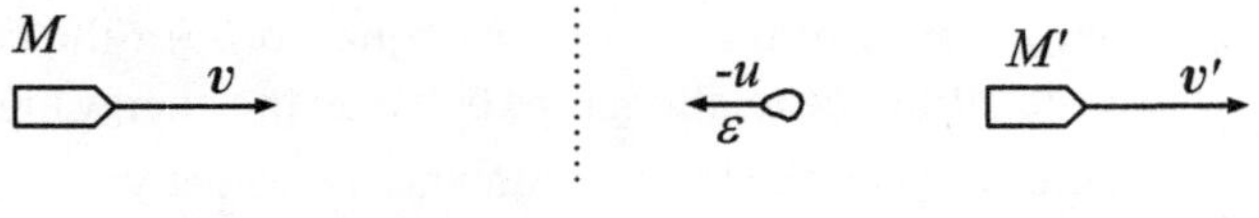

As a result, the velocity of the rocket increases and its mass decreases. The laws of conservation of energy and momentum have the form:

$$\frac{M}{\sqrt{1-v^2}} = \frac{M'}{\sqrt{1-v'^2}} + \varepsilon, \qquad \frac{M\,v}{\sqrt{1-v^2}} = \frac{M'\,v'}{\sqrt{1-v'^2}} - u\,\varepsilon,$$

where M' and v' are the mass and the velocity of the rocket after the emission, and the relation

$$p = -u\,\varepsilon$$

has been substituted for the momentum of the emitted object (the velocity u is directed opposite to v). Assuming that the changes in energy and momentum are small, we can write them as differentials ($df = f' - f$):

$$d\left(\frac{M}{\sqrt{1-v^2}}\right) = -\varepsilon, \qquad d\left(\frac{v\,M}{\sqrt{1-v^2}}\right) = u\,\varepsilon,$$

and therefore

$$d\left(\frac{v\,M}{\sqrt{1-v^2}}\right) = -u\,d\left(\frac{M}{\sqrt{1-v^2}}\right). \tag{4.26}$$

The exhaust velocity $u_x = -u$, as well as the velocity of the rocket, were above measured relative to the "rest" reference frame. Assume that the exhaust velocity is constant ***relative to the rocket***

$$u'_x = -u_0 = const.$$

Then, according to the relativistic velocity-addition formula, the velocity for the observer "at rest" is:

$$u_x = \frac{u'_x + v}{1 + u_x v} = \frac{-u_0 + v}{1 - u_0 v} = -u.$$

Obviously, the current velocity u in the resting frame depends on the velocity of the rocket v and decreases when the latter increases ($u_0 < 1$).

The mass of the rocket $M = M(t)$ and its velocity $v = v(t)$ are functions of time. By rearranging, we can assume that the remaining mass of the rocket depends on its current velocity $M = M(v)$. Dividing (4.26) by dv, we obtain the differential equation:

$$\frac{d(vf)}{dv} = \frac{v - u_0}{1 - u_0 v}\frac{df}{dv}, \qquad where \qquad f = f(v) = \frac{M(v)}{\sqrt{1 - v^2}}.$$

Expanding the derivative of the product and dividing the differentials leads to

$$\frac{df}{f} = \frac{v - 1/u_0}{1 - v^2}\,dv = \frac{vdv}{1 - v^2} - \frac{1}{2u_0}\left(\frac{dv}{1 - v} + \frac{dv}{1 + v}\right).$$

This equation is easily integrated (< H_{19}). Returning from thr function f to the mass $M = f\sqrt{1 - v^2}$, we finally obtain:

$$\frac{M}{M_0} = \left(\frac{1 - v}{1 + v}\right)^{1/(2u_0)}, \tag{4.27}$$

where $M_0 = M(0)$ is the mass of the rocket at the beginning of the acceleration ($v_0 = v(0) = 0$). Restoring the fundamental velocity $v \mapsto v/c$ (< H_{20}), we obtain the well-known ***Tsiolkovsky equation*** in the nonrelativistic limit:

$$M = M_0\, e^{-v/u_0}.$$

▷ When deriving the equation for jet propulsion, we assumed that the flow of matter ejected from the rocket was constant both in velocity and in intensity. In fact, accelerating in this way is quite inefficient, since both the useful cargo and the fuel need to be accelerated in the initial stages. It is more rational to get rid of as much matter as possible from the very beginning, although the resulting acceleration will be very large. So, if some matter is emitted “instantly” rather than gradually, e.g. as a result of an explosion, it will change the ratio M/M_0 between the useful and the initial mass of the rocket. Let us write the conservation laws for energy and momentum for a rocket that was initially at rest and had mass M_0. Suppose it emits some mass m with velocity u_0:

$$M_0 = \frac{M}{\sqrt{1 - v^2}} + \frac{m}{\sqrt{1 - u_0^2}}, \qquad \frac{Mv}{\sqrt{1 - v^2}} = \frac{mu_0}{\sqrt{1 - u_0^2}}.$$

Rearranging, we obtain

$$\frac{M}{M_0} = \frac{\sqrt{1 - v^2}}{1 + v/u_0}.$$

When $u_0 = 1$ (fundamental velocity), the result coincides with that of (4.27). If, however, $u_0 < 1$, the equations are different, and instead of exponential mass decrease we have a much softer dependence on u_0.

4.10 Space flights

As follows from (4.27), the higher the exhaust velocity u_0 the smaller the amount of initial mass lost by the rocket. In chemical rocket engines, the exhaust velocity is about

$u = 10^{-5}$ in the units of the fundamental velocity (or 3 km/s). Therefore, after being accelerated up to half the speed of light $v = 1/2$, the rocket will have only 10^{-23856} of its mass. Obviously, this kind of relativistic acceleration is absolutely unrealistic.

An engine with exhaust velocity close to unity $u_0 \sim 1$ is more economical. In this case, it will spend 40% of the initial mass to reach half the speed of light. Such engines are called *photonic*, but jets can also work with charged particles accelerated up to near-light velocities by electromagnetic fields.

Imagine that a ship has the form of a large cyclic accelerator which gradually propels charged particles with an electric field and keeps their orbit circular or spiral with a magnetic field. Having reached near-light speeds, they are ejected in the form of a jet stream. Another possibility is to have a long linear accelerator propelling quite heavy ionized atoms. Such engines are often called *ionic*. They have a very small reactive power, but at the stage of cruise acceleration (gaining speed at the start, e.g. from orbit around the Earth), can gradually and fairly effectively increase the speed of the rocket. Hybrid cases are also possible, where several pairs of linear accelerators are cycled by reversing rings with a magnetic field. In this case, the engine can work in both linear and cyclic modes.

Besides chemical fuels or nuclear reactions, rockets in the solar system can also use mirror sails to travel with the help of sunlight pressure. Also, a spaceship with no engines can travel due to acceleration by external sources, and so can move away from Earth by receiving pulses (for example, laser pulses) from a chain of accelerating installations rotating at different distances around the Earth. The recoil from such impulses will also gradually increase the speed of the installations. This untwisting and growing "sling" can receive the energy for its pulses from solar panels and constantly launch many small apparatus to study the solar system. Jules Verne's idea (electromagnetic guns, etc.) could also possibly be developed.

Travelling at near-light speeds carries a number of risks. Besides non-humanoid space pirates, hydrogen atoms are the main enemy of spaceships. According to the latest estimates, each cubic centimetre of interstellar space contains approximately one such atom. Even if they do not move quickly, these atoms will cause strong radioactive radiation and heating when colliding with the skin of a fast travelling ship.

One possible (though yet technically unclear) way to protect ships from these particles would be electromagnetic fields which could also capture them to replenish the substance exhausted during the jet propulsion. Dust and micrometeorites are also big problems which can be fatal even for automatic probes.

The second problem is related to the energy. Even if we neglect the mass lost by the rocket as given by the Tsiolkovsky equation, we would need $15 \cdot (3 \cdot 10^8)^2 \approx 10^{18}$ J to accelerate a 100 kg probe up to half the speed of light . By current standards, this is a huge amount of energy. However, the development of controlled thermonuclear synthesis will make such energies more realistic. The physics of elementary particles is also by no means standing still. The deeper we go down the structural stairs, the greater the bound-state energies that become available.

Besides the technological problems that arise in studies of deep space, there are also problems associated with the appropriateness of such expeditions. The farther an expedition goes into the space, the later mankind will learn about the results of its research. At some point, its value will be completely cancelled out by the delay in obtaining new information. The benefits of manned space exploration are also doubtful, even within the solar system. Computer technologies are rapidly developing and in the very near future they will be comparable to the human intellect, and may even exceed it. Therefore, weak biological organisms which need protection from radiation, air, and water make manned expeditions senseless. The only reason to equip spaceships with biological crews would be possible problems with our Sun. At the moment there are no reasons to worry that its current calm regime will suddenly come to an end and that it will suddenly become, for example, a supernova star. However, we know very little about its physics. In any case, we have to think about the possibility of "arks" that could be used to save biological creatures and human knowledge.

5 Force and equations of motion

Although much information about particle interactions can be derived from the conservation laws, equations of motion are required for a more detailed description.

We will study the concept of force and a generalization of Newton's second law to describe particles moving in an external force field. We will then find out which properties a force should have so that the total energy and the total momentum of a particle are conserved. In general, such a force can be a function of the particle's velocity.

We consider the motion of relativistic particles subject to a constant force, a central field, and a spherically symmetric field. If in the spherically symmetric case the force is a function of the test particle's velocity, the particle's orbit is subject to additional precession in accordance with Einstein's theory of gravity.

https://doi.org/10.1515/9783110515886-005

5.1 Definition of force

Predicting the future is the main task of physics. In classical mechanics, the dynamics of particles are described by second-order differential equations. Therefore, their coordinates can be found as functions of time from their initial positions and velocities. The situation is much more complex in relativistic mechanics, so a new dynamic quantity, the interaction field, must be defined in order to describe a system of interacting particles. Nevertheless, the classical approach involving second-order differential equations can be used for "small" particles moving in an external field. Particles which have no counter-impact on the sources of the force field are called *test* particles.

Newton's second law is in some sense a definition of force in classical mechanics:

$$\mathbf{F} = m\mathbf{a}$$

(i.e. force is a function of the particle's position and velocity and is proportional to its acceleration). The same definition can be written as $\mathbf{F} = d\mathbf{p}/dt$, where $\mathbf{p} = m\mathbf{u}$ is the non-relativistic momentum of the particle. These two relations are not equivalent in relativistic theory where the following *definition* is used for the force:

$$\mathbf{F} = \frac{d\mathbf{p}}{dt} = \frac{m\,\mathbf{a}}{\sqrt{1-\mathbf{u}^2}} + \frac{m\,\mathbf{u}\,(\mathbf{ua})}{(1-\mathbf{u}^2)^{3/2}}, \tag{5.1}$$

where $\mathbf{a} = d\mathbf{u}/dt$, and the second equation is obtained by direct differentiation of the relativistic momentum

$$\mathbf{p} = \frac{m\mathbf{u}}{\sqrt{1-\mathbf{u}^2}}.$$

For example, *Coulomb's law* is true for a small charge q moving near a stationary charge Q:

$$\frac{d\mathbf{p}}{dt} = qQ\,\frac{\mathbf{r}}{r^3}. \tag{5.2}$$

We could possibly define force as

$$m\mathbf{a} \quad \text{or} \quad m\mathbf{a}/\sqrt{1-\mathbf{u}^2}.$$

However, it would then follow from Coulomb's empirical equation of motion (5.2) that the force depends not only on the position of the test charge, but also on its velocity and, even worse, on its acceleration. Therefore, definition (5.1) was chosen to make the force vector (from which the empirical equations of motion are derived) as *simple* as possible. The question is: why does "simplicity" arise when the force is the derivative of the momentum? Leaving this aside, we will take (5.1) as a definition and consider general dynamical properties as independent of the specific form of interaction $\mathbf{F}$.

▷ The scalar product of the velocity with the force is the change of the particle's energy of motion $E = m/\sqrt{1-\mathbf{u}^2}$:

$$\frac{dE}{dt} = \mathbf{uF}. \tag{5.3}$$

Besides direct verification, this equation is easily obtained by differentiating the relationship between energy, momentum, and mass

$$E^2 = \mathbf{p}^2 + m^2$$

and substituting in

$$\mathbf{p} = \mathbf{u}E.$$

As follows from (5.3), if a particle is moving normal to the force vector, the energy of its motion and, consequently, the magnitude of its velocity do not change.

▷ The acceleration of a test body is thus

$$\mathbf{a} = \frac{d\mathbf{u}}{dt} = \frac{d}{dt}\left(\frac{\mathbf{p}}{E}\right) = \frac{\mathbf{F} - \mathbf{u}\,(\mathbf{uF})}{E}. \tag{5.4}$$

Since the mass of the particle is not explicitly present in the above equation, the equation can be also applied to massless objects such as photons. The trajectory of light deviates from a straight line in gravitational fields. As a result, the apparent position of light sources differs from their "real" positions. During a solar eclipse, the positions of stars near the edge of the Sun's disk deviate from their positions in the absence of the Sun:

Calculating this effect became the triumph of Einstein's theory of gravitation. If the force of gravitational attraction depends on the particle's energy E rather than its mass m, the curvature of its trajectory can be described by the special theory of relativity (p. 133).

▷ From (5.4) we can obtain the time dependence for the square of the velocity:

$$\frac{d\mathbf{u}^2}{dt} = \frac{2(\mathbf{uF})}{E}\,(1 - \mathbf{u}^2). \tag{5.5}$$

As follows from this equation, when the velocity $\mathbf{u}$ approaches the speed of light, its magnitude tends to a constant value, since the factor $(1 - \mathbf{u}^2)$ "freezes up" the dynamics. In particular, "light-like" objects whose velocity initially had unit *magnitude* $\mathbf{u}^2 = 1$, will never change this value. Of course, the direction of their velocity can change due to some force $\mathbf{F}$ (if it depends on energy E of the object).

5.2 Force in different reference frames

Let us find the relationship between force vectors for observers in two inertial frames. To do this, consider the Lorentz transformations (p. 17) for the time interval dt between two successive positions of an object in space:

$$dt' = \frac{dt - \mathbf{v}\,d\mathbf{r}}{\sqrt{1-\mathbf{v}^2}} = \frac{1 - \mathbf{v}\mathbf{u}}{\sqrt{1-\mathbf{v}^2}}\,dt,$$

where $\mathbf{v}$ is the relative velocity of the frames and $\mathbf{u} = d\mathbf{r}/dt$ is the velocity of the object. Using the following equation for the squared velocity (1.23), p. 20,

$$\frac{\sqrt{1-\mathbf{u}'^2}}{\sqrt{1-\mathbf{u}^2}} = \frac{\sqrt{1-\mathbf{v}^2}}{1-\mathbf{v}\mathbf{u}}, \tag{5.6}$$

we can rewrite this equation in a more symmetric form:

$$dt'\,\sqrt{1-\mathbf{u}'^2} = dt\,\sqrt{1-\mathbf{u}^2}. \tag{5.7}$$

The same expression appears on both the left- and the right-hand sides, written for each inertial frame. This is why it is called the *invariant* of the transformation. In this specific case, the infinitesimal proper time dt_0 is the same in both frames S and S'. Indeed, if the clock travels at velocity $\mathbf{u}$, it will show that the time

$$dt_0 = dt\sqrt{1-\mathbf{u}^2}$$

has passed relative to the frame S (p. 53). The same formula can also be written for the frame S', with the primed velocity and time appearing on the right-hand side. In both cases, the left-hand side contains the same quantity, the proper time of the clock.

Taking momentum and energy differentials (p. 101) and assuming that the relative velocity $\mathbf{v}$ is constant, we have:

$$dE' = \frac{dE - v\,dp_x}{\sqrt{1-v^2}}, \qquad dp'_x = \frac{dp_x - v\,dE}{\sqrt{1-v^2}}, \qquad dp'_y = dp_y.$$

Dividing these relationships by the time differentials (5.7) and using that $dE/dt = \mathbf{u}\mathbf{F}$, we obtain:

$$\frac{\mathbf{u}'\mathbf{F}'}{\sqrt{1-\mathbf{u}'^2}} = \frac{\mathbf{u}\mathbf{F} - v\,F_x}{\sqrt{1-\mathbf{u}^2}\,\sqrt{1-v^2}}, \qquad \frac{F'_x}{\sqrt{1-\mathbf{u}'^2}} = \frac{F_x - v\,(\mathbf{u}\mathbf{F})}{\sqrt{1-\mathbf{u}^2}\,\sqrt{1-v^2}}.$$

We thus have relationships between the force components for the observers in the two inertial frames S' and S. Since $dp'_y = dp_y$ and $dp'_z = dp_z$, the expressions $F_y/\sqrt{1-\mathbf{u}^2}$ and $F_z/\sqrt{1-\mathbf{u}^2}$ are invariant, if the force projections are normal to the direction of velocity v.

▷ As mentioned above (p. 101), the quadruple of energy and momentum components $\{E, \mathbf{p}\}$ is transformed the same way as the quadruple of time with radius vector $\{t, \mathbf{r}\}$. The situation is similar for the force. In this case, the corresponding four quantities of the (*four-force*) are:

$$f^\alpha = \left\{\frac{\mathbf{u}\mathbf{F}}{\sqrt{1-\mathbf{u}^2}}, \frac{\mathbf{F}}{\sqrt{1-\mathbf{u}^2}}\right\}. \tag{5.8}$$

In the first chapter we obtained transformations for the coordinates and time in vector form (p.17):

$$\mathbf{r}' = \mathbf{r} - \gamma \mathbf{v} t + \Gamma \mathbf{v}\,(\mathbf{v}\mathbf{r}), \tag{5.9}$$

where

$$\Gamma = \frac{\gamma - 1}{v^2}.$$

Similar vector transformations are also true for the force:

$$\frac{\mathbf{F}'}{\sqrt{1-\mathbf{u}'^2}} = \frac{\mathbf{F} - \gamma \mathbf{v}\,(\mathbf{u}\mathbf{F}) + \Gamma \mathbf{v}\,(\mathbf{v}\mathbf{F})}{\sqrt{1-\mathbf{u}^2}}. \tag{5.10}$$

Note also the relationship

$$\mathbf{u}'\mathbf{F}' = \frac{(\mathbf{u}-\mathbf{v})\,\mathbf{F}}{1-\mathbf{u}\mathbf{v}}, \tag{5.11}$$

which follows from the transformation of the four-force zero components and the relation (5.6).

Furthermore, we will need the inverse transformation for the force. As usual, this is obtained by exchanging the primed quantities quantities with the unprimed, and making the substitution $\mathbf{v} \mapsto -\mathbf{v}$:

$$\frac{\sqrt{1-\mathbf{v}^2}}{1-\mathbf{v}\mathbf{u}}\,\mathbf{F} = \mathbf{F}' + \gamma \mathbf{v}\,(\mathbf{u}'\mathbf{F}') + \Gamma \mathbf{v}\,(\mathbf{v}\,\mathbf{F}'). \tag{5.12}$$

Here we have also used equation (5.6).

The transformations of the force projections between inertial frames can be used to find the interaction between two moving objects, provided that the “static force” is known when one of the objects is at rest. We will return to this fact in the next volume when using Coulomb’s law (5.2) to find the force exerted by a charge moving with velocity $\mathbf{v}$. We will also discover a specific force component which can be identified as a magnetic field. Finally we will obtain Maxwell’s equations and deduce classical electrodynamics, which is really based “only” on Coulomb’s law and the Lorentz transformations.

5.3 Motion in electric fields

Consider the following dynamic equation describing a relativistic particle moving in a constant force field:

$$\frac{d}{dt}\frac{m\mathbf{u}}{\sqrt{1-\mathbf{u}^2}} = \mathbf{F} = const.$$

In particular, such a force acts on a charge q moving in a constant electric field (Volume 2). Despite its simplicity, this equation has rather cumbersome solutions in relativistic dynamics.

After integrating the equation, we obtain:

$$\frac{\mathbf{u}}{\sqrt{1-\mathbf{u}^2}} = \mathbf{w}_0 + \mathbf{a}\,t, \tag{5.13}$$

where $\mathbf{a} = \mathbf{F}/m$ is a constant vector (simply a constant, not a three-acceleration!) and $\mathbf{w}_0$ is an integration constant. The latter can be expressed in terms of the initial velocity $\mathbf{u}_0$ of the object at $t = 0$:

$$\mathbf{w}_0 = \frac{\mathbf{u}_0}{\sqrt{1-\mathbf{u}_0^2}}.$$

Squaring (5.13) and expressing $\mathbf{u}^2$ in terms of the time, we have

$$\mathbf{u}(t) = \frac{\mathbf{w}_0 + \mathbf{a}\,t}{\sqrt{1+(\mathbf{w}_0 + \mathbf{a}\,t)^2}}. \tag{5.14}$$

By definition, $\mathbf{u}(t) = d\mathbf{r}/dt$, so the law of motion of a particle is obtained by integrating the function $\mathbf{u}(t)$:

$$\mathbf{r}(t) = \int \mathbf{u}(t)\,dt = \frac{\sigma}{a}\int \frac{\mathbf{w}_0 + \mathbf{a}\,t}{\sqrt{1+\sigma^2\left(t+\frac{\mathbf{w}_0\mathbf{a}}{a^2}\right)^2}}\,dt,$$

where we have completed the square in $(\mathbf{w}_0 + \mathbf{a}\,t)^2 = \mathbf{w}_0^2 + 2\mathbf{w}_0\mathbf{a}\,t + \mathbf{a}^2\,t^2$ ($\lessdot$ H$_{21}$) with respect to t, $a = |\mathbf{a}|$, and the following constant was introduced

$$\sigma = \frac{a}{\sqrt{1+\frac{[\mathbf{w}_0 \times \mathbf{a}]^2}{a^2}}}.$$

Making the change of variables $\tau = t + (\mathbf{w}_0\mathbf{a})/a^2$, we obtain:

$$\mathbf{r}(t) = \frac{\sigma}{a}\int \frac{\mathbf{a}\,\tau + [\mathbf{a}\times[\mathbf{w}_0\times\mathbf{a}]]/a^2}{\sqrt{1+\sigma^2\tau^2}}\,d\tau,$$

where a vector triple product formula was used.

Using the tabulated integrals

$$\int \frac{\tau\,d\tau}{\sqrt{1+\sigma^2\tau^2}} = \frac{1}{\sigma^2}\sqrt{1+\sigma^2\tau^2}, \qquad \int \frac{d\tau}{\sqrt{1+\sigma^2\tau^2}} = \frac{1}{\sigma}\ln\left(\sigma\tau + \sqrt{1+\sigma^2\tau^2}\right),$$

which can be verified directly by differentiation, we have:

$$\mathbf{r}(t) = \mathbf{r}_0 + \frac{\mathbf{a}}{a^2}\left(\sqrt{1+(\mathbf{w}_0+\mathbf{a}\,t)^2} - \sqrt{1+\mathbf{w}_0^2}\right) + \frac{\mathbf{a}\times[\mathbf{w}_0\times\mathbf{a}]}{a^2}\,\tau_0(t),$$

where $\mathbf{r}_0$ is the position of the body at $t = 0$. Integration constants were chosen to make $\mathbf{r}(0) = \mathbf{r}_0$. The quantity

$$\tau_0(t) = \frac{1}{a}\ln\frac{\sqrt{1+(\mathbf{w}_0+\mathbf{a}t)^2} + at + (\mathbf{w}_0\mathbf{a})/a}{\sqrt{1+\mathbf{w}_0^2} + (\mathbf{w}_0\mathbf{a})/a} \tag{5.15}$$

can be interpreted as the object's proper time, which by definition is

$$\tau_0(t) = \int_0^t \sqrt{1 - \mathbf{u}^2(t)}\, dt = \int_0^t \frac{dt}{\sqrt{1 + (\mathbf{w}_0 + \mathbf{a}\, t)^2}}.$$

If a clock is moving together with the particle, it will show that a time interval $\tau_0(t)$ has passed at time t. Given that the initial velocity and the "acceleration" of the particle are known, we can use the expression for $\tau_0(t)$ to calculate the time of the moving clock. In the special case that $\mathbf{u}_0 = \mathbf{w}_0 = 0$, the proper time is (p. 87):

$$\tau_0(t) = \frac{1}{a} \ln\left(\sqrt{1 + (at)^2} + at\right).$$

If the ***proper acceleration*** $\mathbf{a}$ (p. 85) and the initial velocity $\mathbf{u}_0$ are parallel to each other, the vector product $\mathbf{a} \times [\mathbf{w}_0 \times \mathbf{a}]$ is zero and the expression for the trajectory is considerably simplified. The body is moving along a straight line. If the x-axis is directed along this line, we have the expression (3.39) mentioned above on p. 85:

$$x(t) = x_0 + \frac{1}{a}\left(\sqrt{1 + (w_0 + a\, t)^2} - \sqrt{1 + w_0^2}\right).$$

Since this trajectory is a hyperbola in the (x, t)-plane, such motion is often called ***hyperbolic***. The general solution has a more complex time dependence, so the term "hyperbolic motion" is used only for motion in a straight line. In general, it's better to use the term ***uniformly accelerated relativistic motion***.

5.4 Motion in magnetic fields

Consider another example of how relativistic dynamical equations are solved. Let the force depend on the velocity as follows:

$$\frac{d\mathbf{p}}{dt} = \mathbf{u} \times \mathbf{B}, \tag{5.16}$$

where $\mathbf{u}$ is the velocity of the particle, and $\mathbf{B}$ is some constant vector. As we will see in the next volume, this equation describes a unit charge moving in a constant homogeneous magnetic field.

The derivative of energy with respect to time is the product of the velocity with the force $dE/dt = \mathbf{uF}$, and the scalar product is $\mathbf{u}[\mathbf{u} \times \mathbf{B}] = [\mathbf{u} \times \mathbf{u}]\mathbf{B} = 0$. Therefore, the energy of the motion $E = m/\sqrt{1 - \mathbf{u}^2}$ and, consequently, the magnitude of the velocity do not change in such a force field. We can use this to write the relationship between the momentum, the energy, and the velocity: $\mathbf{p} = E\mathbf{u}$. Taking the constant energy out of the derivative with respect to time, we have the following equation of motion:

$$\frac{d\mathbf{u}}{dt} = \omega\, [\mathbf{u} \times \mathbf{b}], \tag{5.17}$$

where $\omega = |\mathbf{B}|/E$ is a constant unit vector and $\mathbf{b} = \mathbf{B}/|\mathbf{B}| = const$. Multiplying the equation by **b**, we conclude that the velocity component in the direction of **b** does not change:

$$\frac{d(\mathbf{bu})}{dt} = 0. \tag{5.18}$$

The particle is moving uniformly along the vector **b**. The direction of the velocity changes only in the plane normal to **b**.

Equation (5.17) is easily solved in a suitable coordinate system. To do this, we direct the z-axis along vector **b** to obtain a system of three differential equations (< H_{22}) . We will solve (5.17) in vector notation. We expand the velocity vector into a sum $\mathbf{u} = \mathbf{u}_\parallel + \mathbf{u}_\perp$, where

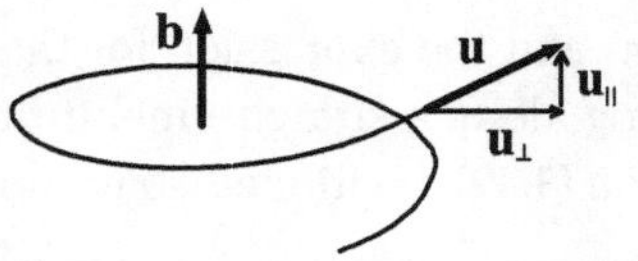

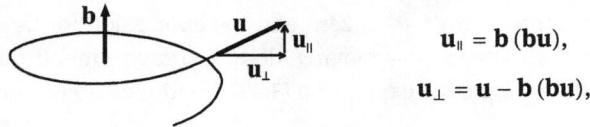

$$\mathbf{u}_\parallel = \mathbf{b}\,(\mathbf{bu}),$$
$$\mathbf{u}_\perp = \mathbf{u} - \mathbf{b}\,(\mathbf{bu}),$$

where $\mathbf{u}_\parallel$ and $\mathbf{u}_\perp$ are the *longitudinal* and the *transverse* velocity components with respect to **b**, respectively. In fact, **bu** is the projection of the velocity onto **b**. Multiplying it by the *unit* vector **b**, the longitudinal component is obtained. Subtracting it from the velocity vector gives us the transverse component.

Since $\mathbf{b} \times \mathbf{b} = 0$ and $\mathbf{bu} = const$, the transverse component satisfies equation (5.17):

$$\frac{d\mathbf{u}_\perp}{dt} = \omega\,[\mathbf{u}_\perp \times \mathbf{b}].$$

Using that $\mathbf{u}_\perp \mathbf{b} = 0$, we differentiate this equation with respect to time and expand the vector triple product to obtain

$$\frac{d^2\mathbf{u}_\perp}{dt^2} = \omega\,\frac{d\mathbf{u}_\perp}{dt} \times \mathbf{b} = \omega^2\,[\mathbf{u}_\perp \times \mathbf{b}] \times \mathbf{b} = -\omega^2\,\mathbf{u}_\perp.$$

As a result, we obtain the oscillator equation:

$$\frac{d^2\mathbf{u}_\perp}{dt^2} + \omega^2\mathbf{u}_\perp = 0. \tag{5.19}$$

Its solution has the form

$$\mathbf{u}_\perp(t) = \boldsymbol{\alpha}\,\cos(\omega t) + \boldsymbol{\beta}\,\sin(\omega t), \tag{5.20}$$

where $\boldsymbol{\alpha}$ and $\boldsymbol{\beta}$ are two constant vectors normal to **b**. They are found from the initial conditions:

$$\mathbf{u}_\perp(0) = \mathbf{u}_0 - \mathbf{b}\,(\mathbf{bu}_0) = \mathbf{b} \times [\mathbf{u}_0 \times \mathbf{b}] = \boldsymbol{\alpha}, \qquad \frac{d\mathbf{u}_\perp(0)}{dt} = \omega\,[\mathbf{u}_0 \times \mathbf{b}] = \omega\boldsymbol{\beta},$$

where $\mathbf{u}(0) = \mathbf{u}_0$ is the initial velocity value. As a result:

$$\mathbf{u}(t) = \mathbf{b}\,(\mathbf{b}\mathbf{u}_0) + \mathbf{b}\times[\mathbf{u}_0\times\mathbf{b}]\,\cos(\omega t) + [\mathbf{u}_0\times\mathbf{b}]\,\sin(\omega t). \quad (5.21)$$

Here the transverse velocity $\mathbf{u}_\perp$ is expressed in terms of the full velocity $\mathbf{u}$, and we have also used that $\mathbf{u}\mathbf{b} = \mathbf{u}_0\mathbf{b} = const.$

Integrating the velocity over time gives the trajectory of the particle:

$$\mathbf{r} = \mathbf{r}_0 + \mathbf{b}\,(\mathbf{b}\mathbf{u}_0)\,t + \frac{\mathbf{b}\times[\mathbf{u}_0\times\mathbf{b}]}{\omega}\,\sin(\omega t) - \frac{\mathbf{u}_0\times\mathbf{b}}{\omega}\,(\cos(\omega t) - 1),$$

where $\mathbf{r}_0 = \mathbf{r}(0)$ is the initial position of the particle at $t = 0$. This is the equation of a spiral. The particle revolves in the plane normal to $\mathbf{b}$ with angular velocity ω. The vectors weighted by the sine and the cosine lie within this plane. They are normal to each other and have the same magnitude, which determines the radius R of the spiral that "unwinds" along $\mathbf{b}$:

$$R = \frac{|\mathbf{u}_0\times\mathbf{b}|}{\omega} = \frac{|\mathbf{u}_0|}{\omega}\,\sin\phi, \quad (5.22)$$

where ϕ is the angle between the vectors $\mathbf{u}_0$ and $\mathbf{b}$. If $\phi = \pi/2$ (the initial velocity is normal to $\mathbf{b}$), the radius is at its maximum and is proportional to the initial momentum of the particle $R = |\mathbf{u}_0|E/|\mathbf{B}| = |\mathbf{p}_0|/|\mathbf{B}|$.

5.5 General expression for the force

▷ Let us find the expression of the force which would be consistent with the laws of conservation of energy and angular momentum. Suppose that *space is isotropic* around some fixed "point" source of force. Then the force acting on the test particle can depend only on its radius vector $\mathbf{r}$, which originates from the centre of the field, and on the velocity of the test particle $\mathbf{u}$:

$$\mathbf{F} = f_1\,\mathbf{r} + f_2\,\mathbf{u}\,(\mathbf{r}\mathbf{u}) + f_3\,[\mathbf{r}\times\mathbf{u}],$$

where f_i are scalar functions. They can depend on the distance $r = \sqrt{\mathbf{r}^2}$ between the point and the force source, on the magnitude of the velocity $u = \sqrt{\mathbf{u}^2}$, and on the scalar product $\mathbf{r}\mathbf{u}$. The factor $(\mathbf{r}\mathbf{u})$ with which f_2 is multiplied appears out of convenience.

For such a force, a scalar function of the coordinates and velocity called the *total energy* is conserved (does not change with time). In its simplest form, it is the sum of the energy of motion $E = m/\sqrt{1-\mathbf{u}^2}$ and the *potential* energy $V(r)$:

$$\mathcal{E} = E + V(r) = const. \quad (5.23)$$

Differentiating (< H_{23}) with respect to time using (5.3), we have

$$\frac{d\mathcal{E}}{dt} = \mathbf{u}\mathbf{F} + V'(r)\,\frac{\mathbf{r}\mathbf{u}}{r} = f_1\,(\mathbf{r}\mathbf{u}) + f_2\,(\mathbf{r}\mathbf{u})\,\mathbf{u}^2 + V'(r)\,\frac{(\mathbf{r}\mathbf{u})}{r} = 0.$$

From this equation it follows that

$$f_1 = -f_2\,\mathbf{u}^2 - V'(r)/r.$$

Therefore, the *additive energy* (5.23) is conserved *if* the force has the following form:

$$\mathbf{F} = -\frac{V'(r)}{r}\,\mathbf{r} - f_2(r, u, \mathbf{ru})\,[\mathbf{u} \times [\mathbf{r} \times \mathbf{u}]] + f_3(r, u, \mathbf{ru})\,[\mathbf{u} \times \mathbf{r}], \tag{5.24}$$

where the "BAC - CAB" rule was used:

$$\mathbf{u} \times [\mathbf{r} \times \mathbf{u}] = \mathbf{r}\mathbf{u}^2 - \mathbf{u}(\mathbf{ru}),$$

p.334. The second and the third components of the force are always normal to the velocity. Therefore, they change its direction, but not its magnitude (i.e., the energy of motion).

If the force depends on the velocity, the total energy does not have to be the *sum* of the functions $E(\mathbf{u})$ and $V(r)$. As an exercise (< H_{24}) we propose verifying that the *multiplicative energy*

$$\mathcal{E} = E\,e^{V(r)} = const \tag{5.25}$$

is conserved, *if* the force has the form

$$\mathbf{F} = -E\,\frac{V'(r)}{r}\,\mathbf{r} - f_2(r, u, \mathbf{ru})\,[\mathbf{u} \times [\mathbf{r} \times \mathbf{u}]] + f_3(r, u, \mathbf{ru})\,[\mathbf{u} \times \mathbf{r}]. \tag{5.26}$$

Note that the first term of the equation now contains the energy of motion.

▷ Besides the total energy, the angular momentum vector **L** normal to the velocity **u** and to the radius vector **r** can also be conserved. If this happens, the trajectory of the particle always remains in the same plane, which passes through the initial values of the vectors **r** and **u**. The classical formula

$$\mathbf{L} = m\,[\mathbf{r} \times \mathbf{u}]$$

can be generalized to the relativistic setting in various ways. The simplest expression for the momentum

$$\mathbf{L} = \mathbf{r} \times \mathbf{p} = E\,[\mathbf{r} \times \mathbf{u}] \tag{5.27}$$

is conserved

$$\frac{d\mathbf{L}}{dt} = \mathbf{u} \times \mathbf{p} + \mathbf{r} \times \frac{d\mathbf{p}}{dt} = \mathbf{r} \times \mathbf{F} = f_2\,(\mathbf{ru})\,[\mathbf{r} \times \mathbf{u}] + f_3\,(\mathbf{r}\,(\mathbf{ru}) - \mathbf{u}\,\mathbf{r}^2) = 0,$$

if $f_2 = f_3 = 0$ (in the general case, the vectors **u**, **r** point in arbitrary directions, and $\mathbf{r} \times \mathbf{u}$ is normal to them).

However, this does not mean that a vector integral of motion cannot be constructed for $f_2 \neq 0$. Note that the vector $\mathbf{u} \times [\mathbf{r} \times \mathbf{u}]$ in the expressions (5.24) and (5.26) always stays within the $(\mathbf{r}, \mathbf{u})$-plane and never takes the particle's trajectory out of it (unlike the third component of the force f_3). Therefore, if f_2 is non-zero, the following vector is conserved:

$$\mathbf{L} = g(E)\,[\mathbf{r} \times \mathbf{p}], \tag{5.28}$$

where $g(E)$ is some function of the energy of motion. For example, for (5.26) with $f_3 = 0$, the equality

$$\frac{d\mathbf{L}}{dt} = g'(E)\,(\mathbf{uF})\,[\mathbf{r} \times \mathbf{p}] + g(E)\,[\mathbf{r} \times \mathbf{F}] = 0$$

holds ($\lessdot\, H_{25}$) if

$$f_2 = E^2\,\frac{g'(E)}{g(E)}\,\frac{V'(r)}{r}. \tag{5.29}$$

In the simplest case of a linear dependence $g(E) = E$, the ***"modified" angular momentum*** $\mathbf{L} = E\,[\mathbf{r} \times \mathbf{p}]$ (motion within the plane) is conserved under the action of the force:

$$\mathbf{F} = -E\,\frac{V'(r)}{r}\,\{\mathbf{r} + \mathbf{u} \times [\mathbf{r} \times \mathbf{u}]\}. \tag{5.30}$$

There is a common factor E in the multiplicative total energy (5.25) and in the power law $g(E) = E^{\nu}$. For gravitational interaction at low velocities the force must transform into Newton's law of universal gravitation, therefore $V(r) = -GM/r$, $E \approx m$, where M is the mass of the field source, m is the mass of the particle, G is the gravitational constant. This force gives correct values for the advance of the perihelion precession of Mercury and for the deviation of light (p. 131).

5.6 Motion in a field $V(r) = -\alpha/r$

Consider a particle of mass m in an attractive field with potential $V(r) = -\alpha/r$ corresponding to the force

$$\mathbf{F} = -\alpha\,\frac{\mathbf{r}}{r^3}, \qquad \alpha > 0.$$

In the case of interacting charges, α is their product. For gravitational interaction, the constant is the product of the masses and the gravitational constant $\alpha = GmM$. Since α has the dimensions of energy squared times distance, the substitution $\alpha \mapsto \alpha/c^2$ is necessary to recover the fundamental speed.

The simplest case is circular motion, when the source of the force is at the center of the circle with radius r_0. The magnitude of the velocity of such motion is constant, and the change in its direction causes centripetal acceleration $a = u^2/r_0$ directed to the centre of the circle. The expressions for the force (5.1) and for the magnitude L of angular momentum $\mathbf{L} = \mathbf{r} \times \mathbf{p}$ are:

$$\frac{m\,u^2/r_0}{\sqrt{1-u^2}} = \frac{\alpha}{r_0^2}, \qquad L = \frac{mur_0}{\sqrt{1-u^2}}.$$

The first equation gives the relationship between the velocity of the particle and the radius of its orbit. The higher the speed, the smaller the radius, and for $u \to 1$, $r_0 \to$

0. Combining these relations, we have $Lu = \alpha$, and since $u < 1$, we conclude that $L > \alpha$. This inequality is a condition for the possibility of stationary circular motion. This is typical of relativistic dynamics, which, unlike classical mechanics, implies that velocity is limited. For $L < \alpha$, circular motion is impossible.

⊳ Now consider the case of non-circular motion. Designate again $c_\phi = \cos\phi$, etc. The radius vector $\mathbf{r}$ of the particle and its velocity $\mathbf{u} = d\mathbf{r}/dt$ are written in polar coordinates as:

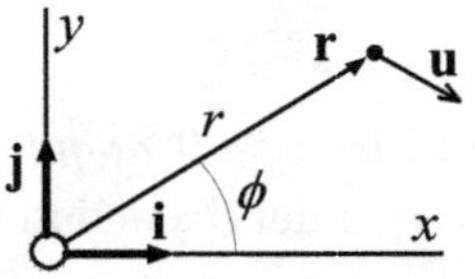

$$\mathbf{r} = r\,c_\phi\,\mathbf{i} + r\,s_\phi\,\mathbf{j},$$

$$\mathbf{u} = (\dot{r}\,c_\phi - r\,\dot{\phi}\,s_\phi)\,\mathbf{i} + (\dot{r}\,s_\phi + r\,\dot{\phi}\,c_\phi)\,\mathbf{j},$$

where $\mathbf{i}, \mathbf{j}$ are orthogonal unit basis vectors in the Cartesian space used to decompose the radius vector $\mathbf{r}$, and the point denotes differentiation in time. Multiplying the vectors $\mathbf{r}$ and $\mathbf{u}$, we obtain:

$$\mathbf{ur} = r\dot{r}, \qquad \mathbf{u}^2 = \dot{r}^2 + r^2\dot{\phi}^2, \qquad \mathbf{r}\times\mathbf{u} = r^2\dot{\phi}\,\mathbf{k}, \tag{5.31}$$

where $\mathbf{k} = \mathbf{i}\times\mathbf{j}$ is a unit vector directed along the z-axis normal to the particle's plane of motion.

These relations give the magnitude of the angular momentum vector:

$$L = \frac{m\,r^2\dot{\phi}}{\sqrt{1 - \dot{r}^2 - r^2\dot{\phi}^2}} = const. \tag{5.32}$$

Besides this, the additive total energy $\mathcal{E} = E + V(r)$ is also conserved:

$$\mathcal{E} = \frac{m}{\sqrt{1 - \dot{r}^2 - r^2\dot{\phi}^2}} - \frac{\alpha}{r} = const. \tag{5.33}$$

A particle's motion can be either finite (limited within some region of space) or infinite (starting at infinity and returning there). In the latter case $\alpha/r \to 0$ for $r \to \infty$ and, consequently, $\mathcal{E} > m$ (if the velocity is nonzero at infinity). If, however, $\mathcal{E} < m$, the particle's motion is finite.

Combining the energy and the angular momentum, we obtain the angular velocity:

$$\dot{\phi} = \frac{L/r^2}{\mathcal{E} + \alpha/r}. \tag{5.34}$$

Expressing the derivative of the distance using (5.33) and substituting in $\dot{\phi}$, we have:

$$1 - \dot{r}^2 = \frac{m^2 + L^2/r^2}{(\mathcal{E} + \alpha/r)^2}. \tag{5.35}$$

For $\dot{r} = 0$ we have an equation quadratic in r for the "turning" points, where the distance stops increasing and starts decreasing (or vice versa):

$$\frac{L^2 - \alpha^2}{r^2} - \frac{2\mathcal{E}\alpha}{r} + m^2 - \mathcal{E}^2 = 0.$$

Introducing three constants:

$$\beta = \frac{\alpha^2}{L^2}, \qquad r_0 = \frac{L^2}{\alpha\mathcal{E}}(1-\beta), \qquad 1 - \epsilon^2 = \frac{m^2 - \mathcal{E}^2}{\alpha\mathcal{E}}\, r_0, \tag{5.36}$$

let us write this equation for $\beta \neq 1$ in the form:

$$\frac{1}{r^2} - \frac{2}{r_0 r} + \frac{1-\epsilon^2}{r_0^2} = 0. \tag{5.37}$$

If $\beta < 1$ and $0 < \epsilon < 1$ (finite motion), we have the following two solutions:

$$r_{1,2} = \frac{r_0}{1 \pm \epsilon}.$$

The parameter ϵ is called the *eccentricity.* A circular orbit with radius $r_1 = r_2 = r_0$ corresponds to the value $\epsilon = 0$. For $\epsilon > 1$ there is only one positive solution $r_1 = r_0/(1+\epsilon)$, which is the distance at which the trajectory turns around during infinite motion.

▷ The particle's trajectory is described by the functions $r(t)$ and $\phi(t)$. Suppressing time, we obtain the equation of the trajectory $r = r(\phi)$. Since $d\phi/dr = (d\phi/dt)(dr/dt) = \dot{\phi}/\dot{r}$, from (5.34) and (5.35) we have

$$\frac{d\phi}{dr} = \pm\frac{L/r^2}{\sqrt{(\mathcal{E}+\alpha/r)^2 - m^2 - L^2/r^2}},$$

where the plus-minus sign is due to the square root taken in (5.35). Using the constants introduced above (5.36), this relation can be rewritten as follows:

$$\sqrt{1-\beta}\, d\phi = \pm\frac{(1/r^2)\, dr}{\sqrt{\dfrac{\epsilon^2-1}{r_0^2} + \dfrac{2}{r_0 r} - \dfrac{1}{r^2}}} = \pm\frac{-d\rho}{\sqrt{\dfrac{\epsilon^2}{r_0^2} - \left(\rho - \dfrac{1}{r_0}\right)^2}},$$

where we have substituted $r = 1/\rho$ and completed the square in the second equation. By a standard integral, we obtain:

$$\frac{r_0}{r} = 1 + \epsilon\cos(\sqrt{1-\beta}\,\phi). \tag{5.38}$$

If it were not for the multiplier $\sqrt{1-\beta}$ under the cosine, the equation would coincide with the trajectory of a non-relativistic particle in this field (verify that $\beta \mapsto \beta/c^2$). In Cartesian coordinates, $x = r\cos\phi$, the trajectory for $\beta = 0$ has the form $r_0 = r + \epsilon x$. Since $r = \sqrt{x^2+y^2}$, we easily obtain:

$$\left(\frac{x}{a} + \epsilon\right)^2 + \frac{y^2/a^2}{1-\epsilon^2} = 1, \qquad a = \frac{r_0}{1-\epsilon^2}.$$

For $0 < \epsilon < 1$, this is the equation of an ellipse with semi-major axis $a = (r_1 + r_2)/2$:

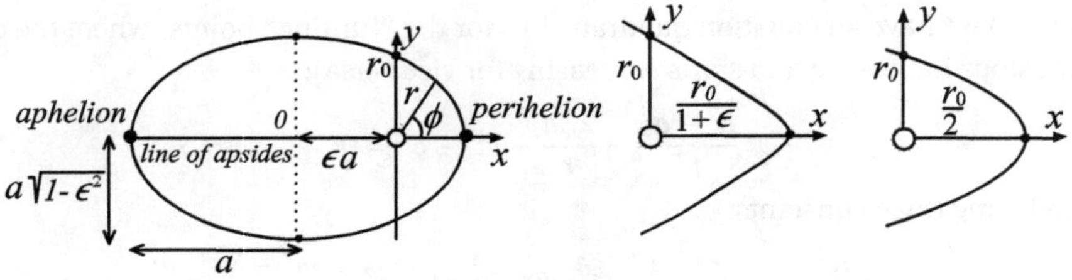

If $\varepsilon = 0$, the ellipse degenerates into a circle. For $\epsilon > 1$, $a < 0$, a hyperbola is obtained (the second figure). Finally, if $\epsilon = 1$, the term with x^2 in $r_0 = r + \epsilon x$ is reduced, and we obtain a parabolic trajectory: $y^2 = r_0^2 - 2r_0\,x$ (the third figure above). The parabola and the hyperbola look similar, though hyperbolae tend more quickly to a straight line with increasing distance from the centre. Both curves come from infinity and return there. Parabolas correspond to a stationary particle at infinity ($\mathcal{E} = m$).

▷ If $\beta \neq 0$, the trajectory changes fundamentally. Assume that a particle is at the point of perihelion at the minimum distance $r_1 = r_0/(1+\epsilon)$ from the centre (5.38) at $\phi = 0$. The next time the particle comes to the perihelion will be when the angle is $\phi = 2\pi + \delta\phi$. At this moment of time, the cosine is 1 and its argument is 2π, therefore:

$$\delta\phi = 2\pi\Big(\frac{1}{\sqrt{1-\beta}} - 1\Big) \approx \pi\beta, \tag{5.39}$$

where the approximate equality is written for small β. The first figure below shows the trajectory for ϕ ranging from 0 up to 18π:

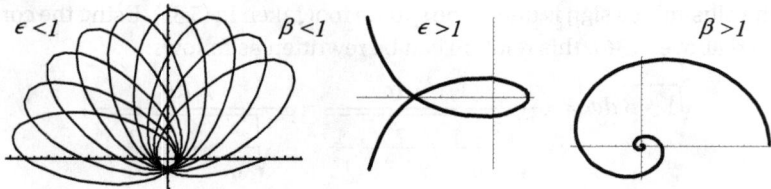

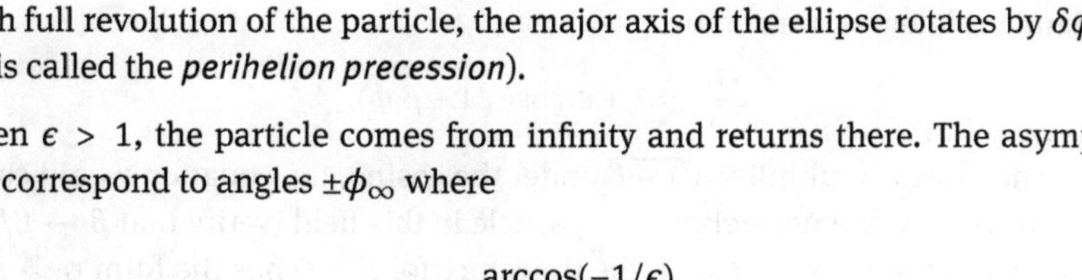

At each full revolution of the particle, the major axis of the ellipse rotates by $\delta\phi$ (this effect is called the *perihelion precession*).

▷ When $\epsilon > 1$, the particle comes from infinity and returns there. The asymptotes $r = \infty$ correspond to angles $\pm\phi_\infty$ where

$$\phi_\infty = \frac{\arccos(-1/\epsilon)}{\sqrt{1-\beta}}. \tag{5.40}$$

For large ϕ_∞ the particle can make several revolutions around the centre of the field before it goes to infinity (the second figure above).

▷ Finally, for the case $\beta > 1$, we have:

$$\sqrt{\beta - 1}\, d\phi = \pm \frac{(1/r^2)\, dr}{\sqrt{\frac{1-\epsilon^2}{r_0^2} + \frac{2}{r_0 r} + \frac{1}{r^2}}} = \pm \frac{-d\rho}{\sqrt{\left(\rho + \frac{1}{r_0}\right)^2 - \frac{\epsilon^2}{r_0^2}}},$$

where $r_0 = L^2(\beta - 1)/\alpha\mathcal{E}$ and $1 - \epsilon^2 = (\mathcal{E}^2 - m^2)\, r_0/\alpha\mathcal{E}$. Integrating this (assuming $r = r_0$ for $\phi = 0$), we obtain:

$$\frac{r_0}{r} = \epsilon\, \mathrm{ch}(\sqrt{\beta - 1}\,\phi) - 1. \tag{5.41}$$

The cases $\mathcal{E} > m$ ($\epsilon < 1$) and $\mathcal{E} < m$ ($\epsilon > 1$) correspond to infinite and finite motion, respectively. Finite motion has the form of a spiral approaching the centre of the field. A particle demonstrating such behaviour is said to be *falling to the centre* (the third figure above).

5.7 Velocity-dependent gravity *

As shown above (p. 127), if a force is proportional to the energy of motion E, if the multiplicative total energy together with the analogue of the angular momentum are conserved, the force should have the form:

$$\mathbf{F} = -E\,\frac{\alpha}{r^3}\,\{\mathbf{r} + \mathbf{u} \times [\mathbf{r} \times \mathbf{u}]\}. \tag{5.42}$$

We assume that for $\alpha = GM$ this force is the force of gravity created by a mass M and that for low velocities it gives Newton's law of gravitation. If the constant α is divided by the speed of light squared, we obtain a quantity with the dimensions of length. Twice its value is usually called the *gravitational radius* of the body with mass M. So, for the Sun the gravitational radius is:

$$\frac{2\alpha}{c^2} = \frac{2GM}{c^2} = 2.953\,250\,077\,0(2)\ km, \tag{5.43}$$

where $G = 6.6743(7)\cdot 10^{-11}\ \mathrm{m^3\,kg^{-1}\,s^{-2}}$ is the gravitational constant. For comparison, Earth's gravitational radius is 8.87 millimetres.

Taking into account the relation (5.4), the equation of motion of a particle under the influence of this force can be written in the following form:

$$\frac{d\mathbf{u}}{dt} = -\frac{\alpha}{r^3}\,\{(1 + \mathbf{u}^2)\,\mathbf{r} - 2\,(\mathbf{r}\mathbf{u})\,\mathbf{u}\}. \tag{5.44}$$

In this case, the total energy is conserved (remember that $dE/dt = \mathbf{u}\mathbf{F}$):

$$\tilde{\mathcal{E}} = E\, e^{-\alpha/r} = const, \tag{5.45}$$

and so is the modified angular momentum:

$$\tilde{\mathbf{L}} = E\,[\mathbf{r} \times \mathbf{p}] = E^2\,[\mathbf{r} \times \mathbf{u}]. \qquad (5.46)$$

For circular motion, the velocity and the radius vector are normal to each other. Substituting the centripetal acceleration $a = u^2/r$ into (5.44), we obtain the relationship between the velocity and the radius of the orbit:

$$r_0 = \alpha\,\frac{1+u^2}{u^2}.$$

If $u \to 1$, we have $r_0 \to 2\alpha$ (in contrast to zero, like in the model of the previous section). A particle moving along a circle can never be closer to the centre of the field than the gravitational radius 2α.

However, the escape velocity is always smaller than the speed of light. As is easily obtained from the law of conservation (5.45), if a particle starts from distance $R > 2\alpha$ with velocity $u = (1 - e^{-2\alpha/R})^{1/2}$, it will have zero velocity at infinity.

▷ Let us square the modified angular momentum (5.46), $\tilde{L}^2 = E^4\,\{\mathbf{r}^2\mathbf{u}^2 - (\mathbf{r}\mathbf{u})^2\}$, and substitute into it $\mathbf{u}^2 = 1 - m^2/E^2$ and $\mathbf{r}\mathbf{u} = r\dot{r}$:

$$\dot{r}^2 = 1 - \frac{m^2}{E^2} - \frac{\tilde{L}^2/r^2}{E^4}. \qquad (5.47)$$

Directing the z-axis of the coordinate system along the angular momentum, we have its magnitude $\tilde{L} = E^2\,r^2\dot{\phi}$ in polar coordinates. Repeating the reasoning of the previous section and replacing E by $\tilde{\mathcal{E}}\,e^{\alpha/r}$, we obtain:

$$d\phi = \frac{\pm(\tilde{L}/r^2)\,dr}{\sqrt{\tilde{\mathcal{E}}^4\,e^{4\alpha/r} - \tilde{\mathcal{E}}^2 m^2\,e^{2\alpha/r} - \dfrac{\tilde{L}^2}{r^2}}} \approx \frac{(1/r^2)\,dr \,/\, \sqrt{1-\tilde{\beta}}}{\sqrt{\dfrac{\tilde{\varepsilon}^2-1}{\tilde{r}_0^2} + \dfrac{2}{\tilde{r}_0 r} - \dfrac{1}{r^2}}}.$$

In the second approximate equality, the exponentials are expanded in series with respect to $\alpha \ll r$ up to second order, $e^x = 1 + x + x^2/2$, and the following parameters are introduced:

$$\tilde{\beta} = 2\tilde{\mathcal{E}}^2(4\tilde{\mathcal{E}}^2 - m^2)\,\frac{\alpha^2}{\tilde{L}^2}, \qquad \tilde{r}_0 = \frac{\tilde{L}^2\,(1-\tilde{\beta})}{\alpha\,\tilde{\mathcal{E}}^2\,(2\tilde{\mathcal{E}}^2 - m^2)}, \qquad 1 - \tilde{\varepsilon}^2 = \frac{\tilde{\mathcal{E}}^2\,(m^2 - \mathcal{E}^2)}{1-\tilde{\beta}}\,\frac{\tilde{r}_0^2}{\tilde{L}^2}.$$

Note that the dimensions of α and $\tilde{L}$ are different from those of the previous model. In a non-relativistic approximation, the particle still moves along an ellipse in the case of finite motion. Since the parameters $\tilde{\beta}$ and $\tilde{r}_0$ are multiplied by α, which is proportional to $1/c^2$, in the major approximation with respect to α we can take a non-relativistic limit for $\tilde{\mathcal{E}}$ equal to m. Therefore, $\tilde{\beta} \approx 6\,\alpha^2\,m^4/\tilde{L}^2$ and $\tilde{r}_0 \approx \tilde{L}^2/\alpha m^4$. Using (5.39), we obtain the following advance of the orbital perihelion:

$$\delta\phi = \frac{6\pi\alpha}{r_0}. \qquad (5.48)$$

For Mercury the parameter r_0 is $5.546 \cdot 10^7$ km, and its period of revolution around the Sun is 88 days. Therefore, the advance of the perihelion of Mercury is 43″ angular seconds per century:

$$\frac{3 \cdot 2.953}{5.546 \cdot 10^7} \cdot \frac{360 \cdot 60 \cdot 60}{2\pi} \cdot \frac{365.24 \cdot 100}{88} = 42.96''.$$

The first fraction is the relation (5.48), where (5.43) was used. It is equal to the change of the perihelion in radians. The second fraction converts radians into angular seconds, and the last fraction shows the time of the perihelion turn. This result is six times greater than that in the previously discussed model with force $\mathbf{F} = -m\,\alpha\,\mathbf{r}/r^3$. Einstein's theory of gravitation leads to the same expression. However, the physics of the effect is much more difficult in this case.

▷ Consider now how a light beam is deflected by the gravitational field of the Sun when moving in accordance with equation (5.44) with $\mathbf{u}^2 = 1$:

$$\frac{d\mathbf{u}}{dt} = -\frac{2\alpha}{r^3}\{\mathbf{r} - (\mathbf{r}\mathbf{u})\mathbf{u}\}. \tag{5.49}$$

Let the light propagate along the x-axis. In the system of units $c = 1$ the *small* angle θ of the deviation of the beam is equal to the y-component of the velocity after passing the Sun (the ratio of a cathetus to the unit hypotenuse):

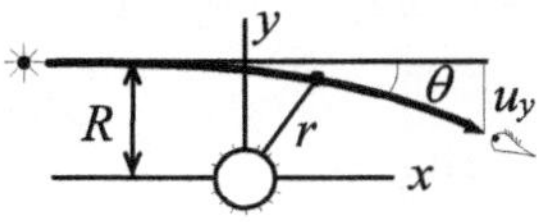

$$\theta \approx \operatorname{tg}\theta = |u_y|.$$

Before the beam passes the Sun, its y-component is $u_y(-\infty) = 0$. We integrate the y-projection of equation (5.49) over time:

$$u_y = u_y(\infty) - u_y(-\infty) = -2\alpha \int_{-\infty}^{\infty} \frac{y - (\mathbf{r}\mathbf{u})\,u_y}{r^3}\,\frac{dx}{u_x},$$

where we have made the substitution $dt = dx/u_x$. In the zeroth-order approximation with respect to α, the light propagates along a straight line. Let us expand the trajectory and the velocity in a series with respect to α. Since the integral is already multiplied by α, the values in the major approximation with respect to α can be replaced by "unperturbed" values:

$$u_y = 0, \qquad u_x = 1, \qquad y = R,$$

and the squared distance to the trajectory is

$$r^2 = R^2 + x^2.$$

Therefore:

$$u_y = -2\alpha \int_{-\infty}^{\infty} \frac{R\,dx}{(R^2 + x^2)^{3/2}} = -\frac{4\alpha}{R}. \tag{5.50}$$

Thus, the deviation angle of the light is $4\alpha/R$. If the visible position of the star is near the surface of the Sun, then R is approximately equal to its radius $6.955 \cdot 10^5$ km. Since the gravitational radius

$$2\,\alpha = 2.953\ km,$$

we obtain 1.75″ angular seconds for the deviation angle. This conclusion is confirmed by observations of the displacement of the positions of stars during a solar eclipse.

When the deviation of a light beam is calculated using Einstein's theory of gravity, the same result is obtained. However, in this case it is related to the curvature of space-time in a neighbourhood of a massive object, which curves the trajectory of any object.

▷ There is another interesting effect usually attached to the theory of gravity, which can be described in this simple model. Consider two points at distances r and r_0 from the centre of the force field. From the law of conservation of energy (5.45) it follows that

$$\exp\left(\frac{\alpha}{r} - \frac{\alpha}{r_0}\right) = \frac{E}{E_0} = \frac{\nu}{\nu_0}, \tag{5.51}$$

where E and E_0 are the energies of the particles' motion at these points. Let the particle be a quantum of light. According to quantum theory, its energy is proportional to its frequency

$$E = h\nu,$$

where h is the Planck constant. Therefore, the frequency ν_0 of light radiated from a distance r_0 will differ from the frequency ν of the same light received at a distance r, which can be seen from the second equality. If the light source is closer to the centre than the receiver ($r_0 < r$), the frequency decreases $\nu < \nu_0$ (the quantum of light loses energy) and the emission spectrum shifts to the red region. If the gravitational field is strong and the source is close to the centre, the frequency of light can decrease so much that the source will become invisible.

▷ The escape velocity

$$u = (1 - e^{-2\alpha/R})^{1/2}$$

exists only if $R > 2\alpha$. Neither particles nor light can escape to infinity from below the gravitational radius 2α. Therefore, an object with radius smaller than 2α is invisible. Light cannot escape the horizon. Objects whose radius is smaller than the gravitational radius are called ***black holes***. In practice, they are observed as invisible super-massive formations. One of these objects is quite likely in the centre of our galaxy and is called ***Sagittarius A****. The area around it radiates in the infrared and X-ray ranges. This radiation originates from ultra-relativistic particles trapped by the black hole at its low orbits (but not absorbed by it). From the parameters of nearby stars moving around the black hole in elliptical orbits it has been possible to determine accurately the mass of the black hole. It is about four million solar masses, and is concentrated in a space no more than six light hours across, which is comparable with its gravitational radius $2.953 \cdot M/M_{\odot}$, where $M_{\odot}$ is the mass of the Sun.

The density of supermassive black holes is relatively small. For the radius

$$r_g = \frac{2GM}{c^2},$$

the density is

$$\rho = \frac{M}{4\pi r_g^3/3} = \frac{3c^6}{32\pi M^2 G^3}$$

and it becomes smaller when the mass of the object grows. For Sagittarius A* it is about 10^6 kg/m^3, which is comparable with the density of matter in the centre of the Sun.

6 Non-inertial frames

This chapter concerns non-inertial reference frames. First, we will study rigid, uniformly accelerating frames and show that the rate of their time flow depends on the relative positions of observers. We will then consider the physics of a rotating disc. The most unusual property of such a frame is the impossibility of synchronizing clocks along certain closed curves, despite the fact that all of their points experience a common flow of time. The third example of a "non-rigid" frame will give the general definition of a reference frame. In conclusion, we will consider some "paradoxes" related to non-inertial frames.

The idea of non-inertial frames will be considered again in chapter 9, from a more general mathematical point of view.

https://doi.org/10.1515/9783110515886-006

6.1 Uniformly accelerating frames

Consider a frame $S\ :\ \{t,\ x\}$. Let the point with coordinate $x\ =\ 0$ in it move with a constant *proper acceleration*

$$a = const$$

relative to the inertial (laboratory) frame $S_0\ :\ \{T,\ X\}$. We assume that the axes labeled x and X are parallel to each other and have the same direction. The trajectory of a uniformly accelerated relativistic particle (p. 85) has the form:

$$X(T) = \frac{1}{a}\left[\sqrt{1+(aT)^2}-1\right]. \tag{6.1}$$

The particle is at rest at $T\ =\ 0$, and then starts moving with increasing velocity (but decreasing acceleration):

$$U(T) = \frac{dX}{dT} = \frac{aT}{\sqrt{1+(aT)^2}}. \tag{6.2}$$

Let the origin of the frame S (and the observer *at this* point) move relative to S_0 according to equation (6.1):

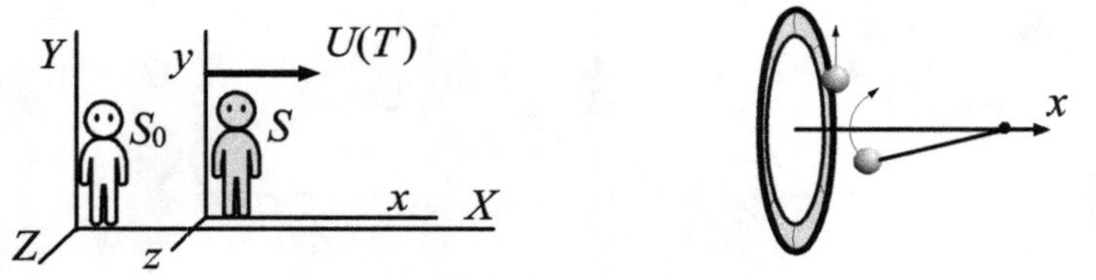

Space is not isotropic inside the non-inertial frame S. More precisely, it is isotropic inside the (y, z)-plane normal to the acceleration, but not isotropic along the x-axis. In particular, free particles will not undergo uniform linear motion.

! We will assume that the proper acceleration a is *small*, although the velocity U of the frame S relative to S_0 can be large.

This means that the observer at the origin of S can neglect, at least locally, the anisotropy of space with respect to their standards of length. This means that their deformation is ignored, or that some corrections are made due to the elasticity of the materials the rulers are made of. As a result, the rulers are assumed to maintain a constant length when reoriented from the isotropic (y, z)-plane into the direction of acceleration. The situation is the same when "rigid" rulers are used in our immediate neighbourhood on the Earth's surface under $g = 9.8\ m/s^2$ acceleration.

As well as the rulers, the observer needs a clock. To avoid discussions of how the clock is affected by inertia forces, we assume that only processes inside the isotropic

(y, z)-plane are used to measure the time. For example, one might use a friction-free ball rolling around a circular groove normal to the motion $\lessdot$ C_9. In addition, these processes are chosen so as to make all other motion appear as simple as possible relative to them (p. 2). The rate of such a clock is assumed to be the same as that in the inertial frame $S'_0 : \{T', X'\}$ ***comoving*** relative to the observer. Such a frame moves with respect to S_0 with the same velocity as the origin of S (at any given moment of time). For the time being, when talking about the frames S and S'_0 we mean only two observers (non-inertial and inertial) at rest relative to each other and having the same coordinate $x = 0$. They have common clocks and rulers. Hence,

---

we assume that the rate of time flow of the moving clock relative to the resting clock depends only on its velocity and *not on its acceleration*. **!**

Therefore, we can use the standard formula for time dilation

$$dt = dT\,\sqrt{1 - U^2},$$

p. 53, to calculate the time t shown by the clock moving with velocity $U(T)$ in the laboratory frame. If the velocity changes with time $U = U(T)$, integration is required:

$$t = \int_0^T \sqrt{1 - U^2(T)}\,dT. \tag{6.3}$$

The experimental evidence for equation (6.3) is profound; it has been observed that the lifetime of muons in a circular accelerator [13] increases according to (6.3) to within a relative error of 10^{-3}. For muons with speed $U = 0.9994$, time slows down by a factor of 29. A ring with radius $R = 7$ metres creates an extremely high acceleration ($U^2/R \sim 10^{15} \cdot g$, where $g = 9.8\ m/s^2$). However, the lifetimes of particles still depend only on their velocities.

Calculating the integral (6.3) of the velocity (6.2), we obtain the relationship between the times:

$$t = \frac{1}{a}\,\mathrm{ash}(aT),$$

or

$$aT = \mathrm{sh}(at) = \frac{e^{at} - e^{-at}}{2}, \tag{6.4}$$

where $\mathrm{ash}(aT)$ is the hyperbolic arcsine. Note that t is the time shown by a single moving clock (at the origin of S), while T is the time shown by a set of resting synchronized clocks placed *along the trajectory of the motion* in the laboratory frame S_0 (see also p. 87).

▷ Now consider two observers in a non-inertial frame S; assume, for clarity, that these observers are spaceships. Suppose that the distance between them before the acceleration was x_0. While the spaceships were in the spaceports, their crews synchronized their clocks with each other and with other observers in S_0. Denote by t and t' the times on the clocks of the first and the second ships, which started from

$$X = 0 \quad \text{and} \quad X = x_0,$$

respectively. All observers have a common time T in the inertial frame.

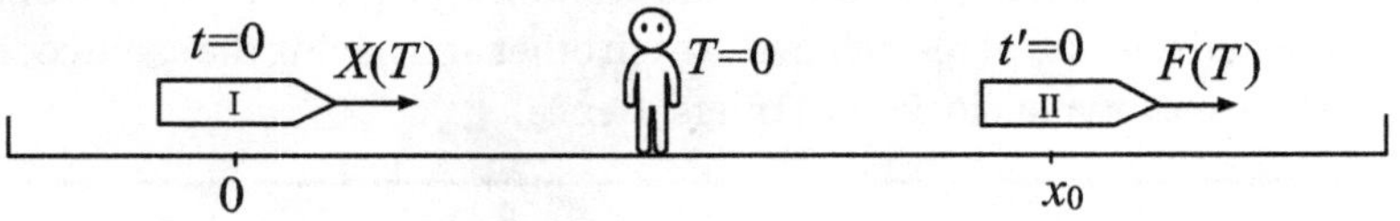

The coordinate of the first (left) ship at the origin of S changes with time as (6.1):

$$X(T) = \frac{1}{a}\left[\sqrt{1 + (aT)^2} - 1\right] = \frac{1}{a}\left[\mathrm{ch}(at) - 1\right], \tag{6.5}$$

where we have subsituted the ship's proper time (6.4) into the second equation.

! A *rigid reference frame* is a set of observers at rest relative to each other.

Above we defined the trajectory $X(T)$ of the origin of S. How must the second spaceship move in order to keep the distance between the ships constant from *their point of view*? The answer "in the same way" is not true.

Events which are simultaneous in one frame will not be in another frame. Ships which are accelerating in synchrony from the perspective of S_0 will not be such in S, and vice versa. Therefore, the constancy of distances between points in a reference frame is, generally speaking, a *relative concept*. As we will see, if the frame S is "kept" rigid by its observers, it will appear contracted in the direction of motion for observers in the inertial frame S_0.

Let us assume that the squadron uses *radiolocation* to keep the distance between its ships constant. One ship sends a light signal towards the other. When reflected, the signal comes back. The time needed for the light to travel in both directions should be the same by the *local clock* of the ship. Let us find out what trajectory $F(T)$ the second ship should have *relative to* S_0 so that frame S appears rigid *to the observers in* S_0. The calculations will be carried out in the rest frame S_0.

▷ Suppose the first ship sends forward a light signal at T_1 which is reflected from the second ship at T, and arrives back at the first ship at T_2. The distance travelled by the signal equals the duration of its motion ($c = 1$):

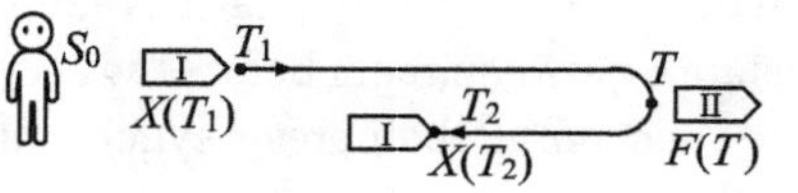

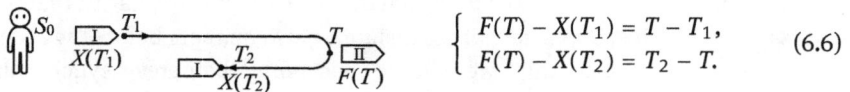

$$\begin{cases} F(T) - X(T_1) = T - T_1, \\ F(T) - X(T_2) = T_2 - T. \end{cases} \tag{6.6}$$

According to (6.5), the coordinate of the first ship is $X(T)$ and the coordinate of the second ship is $F(T)$. Expressing the time $T = \mathrm{sh}(at)/a$ of the laboratory frame S_0 in terms of the proper time of the first ship t, we easily obtain (< H_{26}) from (6.5):

$$1 + aX(T) \pm aT = \mathrm{ch}(at) \pm \mathrm{sh}(at) = e^{\pm at}.$$

Furthermore, from the system of equations (6.6) we have the relationships

$$1 + aF(T) - aT = e^{-at_1}, \qquad 1 + aF(T) + aT = e^{at_2}, \tag{6.7}$$

where t_1 and t_2 are the times when the signal was sent and received, respectively, by the clock of the ***first ship***. If the distance between the ships is constant, the time $\tau_0 = t_2 - t_1$ needed for the signal to go "there and back" is also constant and does not depend on the time t_2 it was received. Subtracting equations (6.7), we have

$$2aT = e^{at_2} - e^{-a(t_2-\tau_0)} \quad \Rightarrow \quad e^{at_2} = aT + \sqrt{e^{a\tau_0} + (aT)^2}. \tag{6.8}$$

Substituting e^{at_2} into the second equation (6.7), we obtain the trajectory $F(T)$:

$$1 + a\,F(T) = \sqrt{e^{a\tau_0} + (aT)^2} = \sqrt{(1 + ax_0)^2 + (aT)^2}, \tag{6.9}$$

where the initial condition $F(0) = x_0$ or $e^{a\tau_0/2} = 1 + ax_0$ was used in the second equation. Call the ***radar distance*** l_0 half of the time τ_0 taken by the signal to cover the path in both directions:

$$l_0 = \frac{t_2 - t_1}{2} = \frac{\tau_0}{2} = \frac{1}{a}\ln(1 + ax_0). \tag{6.10}$$

The velocity of the second ship $U_2(T) = dF(T)/dT$ relative to the resting frame is

$$U_2(T) = \frac{a_2 T}{\sqrt{1 + (a_2 T)^2}}, \qquad a_2 = \frac{a}{1 + ax_0} = a\,e^{-al_0}. \tag{6.11}$$

Therefore, to keep the frame S rigid, the second ship must also be uniformly accelerating, though with a smaller proper acceleration a_2.

6.2 Times and distances

Knowing the velocities of the ships, we can find the times shown by their clocks at the time T of frame S_0:

$$T = \frac{1}{a}\,\mathrm{sh}(at), \qquad T = \frac{1}{a\,e^{-al_0}}\,\mathrm{sh}\left(a\,e^{-al_0}\,t'\right). \tag{6.12}$$

We can express the coordinates of the second ship (6.9) in the frame S_0 in terms of its proper time:

$$F = \frac{1}{a\,e^{-al_0}}\,\mathrm{ch}\left(ae^{-al_0}\,t'\right) - \frac{1}{a}. \tag{6.13}$$

Using (6.8), (6.12), (6.13) and the distance l_0 (6.10), we can $(\lessdot H_{27})$ find the time t' when the signal reaches the second ship:

$$t' = (t_1 + l_0)\, e^{al_0} = (t_2 - l_0)\, e^{al_0} = \frac{t_1 + t_2}{2}\, e^{al_0}. \tag{6.14}$$

Note that

$$t' \neq \frac{t_1 + t_2}{2},$$

so the initial synchronization of the clocks (p. 4) has been lost. To understand why, let us find the relationship between the times on the ships.

▷ In the relations (6.12), the times t and t' which have elapsed on the first and the second ships since the moment of their start are compared with *different* clocks T synchronized in S_0. Therefore, we cannot equate via T in (6.12) to obtain the relationship between t and t', since simultaneous events in S_0 (the same T) will not be such in S.

Imagine that the second ship sends a "time signal" to the first ship. It starts at t' and arrives at the first ship at t_2. Using equation (6.14), we have

$$t_2 = t'\, e^{-al_0} + l_0. \tag{6.15}$$

If sent at regular intervals with frequency $\Delta\nu' = 1/\Delta t'$ by the clock of the second ship, the signals will also arrive at regular intervals, but with a different frequency $\nu = 1/\Delta t$:

$$\nu = \nu'\, e^{al_0} \approx \nu'\,(1 + al_0), \tag{6.16}$$

where the approximate equality is written for

$$al_0 \ll 1.$$

The farther away the source of the signal is in the direction of travel, the higher the frequency ν of the received signal. Therefore, if the distance between the observers in the non-inertial reference frame remains constant, the flow of time *is different* for them. As a result, initially synchronized clocks will desynchronize over time.

▷ Observers in the accelerating frame feel as they would in the Earth's homogeneous gravity field. For them, all "untouched" objects will move with acceleration a, *regardless of their mass.*

If, following Einstein, we assume that physics in a uniformly accelerating frame is the same as in a homogeneous gravitational field, the equation for the frequencies (6.16)

will be also true under Earth's conditions. In particular, a receiver placed below the source of some standard radiation ν' should get a signal of higher frequency

$$\nu \approx \nu' \,(1 + gl_0/c^2),$$

where $a = g = 9.8\ m/s^2$, and l_0 is the distance between the source and the receiver (we also restored the constant "c" here ($g \mapsto g/c^2$)). This ratio was confirmed to a high degree of accuracy in the Pound-Rebka experiment (1960). Their device was $l_0 = 22.5\ m$ high and gave a relative change of frequency

$$\frac{gl_0}{c^2} = 2.5 \cdot 10^{-15}.$$

Nowadays, this effect is taken into account in satellite GPS navigation systems, where l_0 is 10^7 times greater.

▷ Length and time standards for the non-inertial observer coincide with those in the co-moving inertial frame. In particular, the speed of light is isotropic for them, and has the same value whether moving in the direction of the x-axis or against it. Therefore, the observer can calculate the distance to the second ship

$$l_0 = \ln(1 + ax_0)/a$$

as half of the time $\tau_0/2$ of the radar experiment. Taking from (6.15) the time required to cover this distance, the observers can compare the times of their clocks:

$$t = t'\, e^{-al_0}. \tag{6.17}$$

If some event occurs on the second ship at time t', the observer on the first ship *can take it to be simultaneous* to the moment t by his clock, since information about this event will reach his ship after a time equal to the distance l_0.

Remember that the distance between the ships is x_0 before they start. However, space then becomes anisotropic due to acceleration, so the distance to the second ship *decreases* ($x_0 \mapsto l_0$) and then remains constant afterwards. In fact, the observers will not discover this jump of distance until enough time to carry out the radar measurement has passed. At the moment the engines are turned off, the non-inertial frame becomes inertial and space becomes isotropic again.

▷ The distance between the ships in the uniformly accelerating frame is kept constant. For observers at rest in S_0, the distance between the ships decreases. This is not surprising, if we think of Lorentz contraction, because the velocity of the ships increases. Imagine that two ships are connected with a "ruler". Its length in S_0 is the difference between the coordinates of the ships:

$$L = F(T) - X(T) = \frac{1}{a}\sqrt{e^{2al_0} + (aT)^2} - \frac{1}{a}\sqrt{(1 + (aT)^2}.$$

The observer in S on the first ship believes that the length of the ruler (the distance between the ships) is

$$l_0 = \ln(1 + ax_0)/a.$$

If we consider the velocity of the first ship (6.2) relative to the inertial frame S_0 and use the Lorentz factor corresponding to this velocity

$$\gamma = \frac{1}{\sqrt{1 - U^2(T)}} = \sqrt{1 + (aT)^2},$$

the expression for the length can be rewritten as follows:

$$L = \frac{\sqrt{e^{2al_0} + \gamma^2 - 1} - \gamma}{a} \approx \frac{l_0}{\gamma}\left(1 + (1 + U^2)\,\frac{al_0}{2} + \dots\right), \tag{6.18}$$

where the approximate equality was obtained by expanding in a series in al_0 (it can be verified by carrying γ/a to the right side and squaring). This approximation is true for $al_0 \ll 1$ or, restoring the fundamental speed, $a \ll a_0 = c^2/l_0$. For a one metre rod, we have $a_0 \sim 10^{16}\,g$, where $g = 9.8\,m/c^2$, which is a very high value. If the distance between the ships is one light year (p. 87), then $a_0 \sim g$ and the deviation from the Lorentz formula $L = l_0/\gamma$ becomes noticeable.

Therefore, if the product al_0 is small, the ratio of the ruler's length in S to its length in S_0 corresponds to the Lorentz contraction:

$$L = l_0\sqrt{1 - U^2}.$$

In the general case, the contraction of the "ruler" differs from the Lorentz contraction. However, when the ships stop accelerating, the contraction will be exactly the same as the Lorentz contraction (see the "stopping paradox", p. 163).

To obtain *large* values of l_0, non-inertial observers use radar measurements rather than rulers to measure velocities in their immediate neighbourhood. According to (6.10), l_0 differs from the initial distance x_0 between the ships and has different values for the observers on the first and the second ships. The proper length of the ruler considered above is equal to the radar distance measured by the observer on the first ship.

▷* Any observer in a non-inertial reference frame can use radiolocation to measure distances. Let us consider such an experiment, conducted by the observer on the second ship in the direction of the third ship moving ahead of him. If the coordinates of these ships were x_0 and x_1 in S_0 at the initial instant of time, their trajectories $F(T)$ and $G(T)$ will change with time as:

$$1 + a\,F(T) = \sqrt{(1 + ax_0)^2 + (aT)^2}, \qquad 1 + a\,G(T) = \sqrt{(1 + ax_1)^2 + (aT)^2}.$$

The signal is sent in the laboratory frame S_0 at T_1, reflects from the third ship at T, and returns to the second ship at T_2:

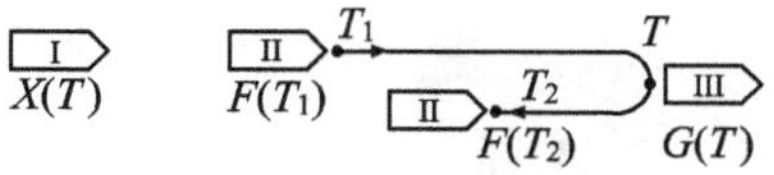

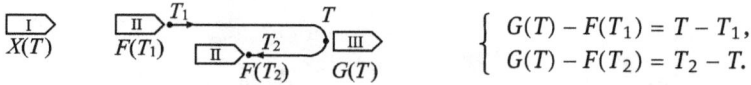

$$\begin{cases} G(T) - F(T_1) = T - T_1, \\ G(T) - F(T_2) = T_2 - T. \end{cases}$$

From these equations it follows that

$$(1 + aF(T_1) - aT_1)(1 + aF(T_2) + aT_2) = (1 + a\,G(T) - aT)(1 + a\,G(T) + aT).$$

Substituting the times $aT_i = (1 + ax_0)\,\mathrm{sh}(at_i'/(1 + ax_0))$ into the left-hand side,

$$1 + aF(T) \pm aT = (1 + ax_0)\ \exp(\pm at'/(1 + ax_0)).$$

and the explicit form of the trajectory $G = G(T)$ into $(1+aG)^2 - (aT)^2$ on the right-hand side, we obtain:

$$l_1' = \frac{t_2' - t_1'}{2} = \frac{1 + ax_0}{a}\ \ln\left(\frac{1 + ax_1}{1 + ax_0}\right) = e^{al_0}\,(l_1 - l_0). \tag{6.19}$$

Here l_0 and l_1 are the radar distances to the second and the third ship *from the viewpoint* of the observer on the first ship located at the origin of the non-inertial frame.

If these calculations for periodic signals (p. 142) are repeated in the situation when the signals are sent from the third ship to the second ship, the following relationship is obtained:

$$t_2' = l_1' + \frac{1 + ax_0}{1 + ax_1}\, t_1'' = l_1' + t_1''\, e^{-a\,(l_1 - l_0)} = l_1' + t_1''\, e^{-a_2 l_1'}, \tag{6.20}$$

where l_1' is the distance (6.19) between the ships and $a_2 = a/(1 + ax_0)$.

Besides the fact that a common time is missing from non-inertial frames, the distances between their points are measured by ***specific observers***. If the time flow is different for different observers, radar distances will be different for them. For example, the distance to the first ship measured from the second ship is $l_0' = l_0\, e^{al_0}$, which is e^{al_0} times larger than the same distance measured from the first ship.

6.3 Coordinate transformations

Let us consider the coordinate and time transformations of some event observed in the inertial frame $S_0 : \{T,\ X\}$ and the non-inertial frame $S : \{t,\ x\}$. We assume that $\{t,\ x\}$ are measurements obtained by the observer at the point $x = 0$ of the frame S (the first ship). For example, let the event be remote, having occurred next to the second ship at a distance

$$l_0 = \ln(1 + ax_0)/a$$

determined by radar measurements. This distance can be taken as *the coordinate of the event* $x = l_0$ in S. Suppose that at time t' when the event happened (by the clock of the second ship), a light signal was sent to the first ship:

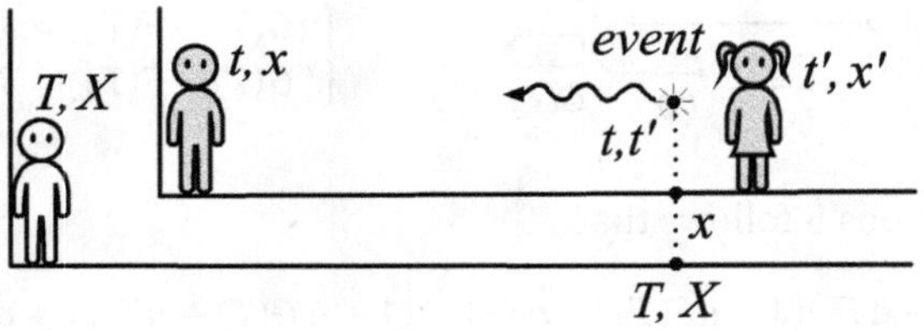

Taking into account the time for the signal to travel (6.17), the time of the event is $t = t'e^{-ax}$ by the clock of the first ship. On the other hand, the time t' of the second ship is related to the time T of the event in the inertial frame (clock T in the neighbourhood of the event) as follows:

$$T = \frac{1 + ax_0}{a}\,\mathrm{sh}\left(\frac{at'}{1 + ax_0}\right) = \frac{e^{ax}}{a}\,\mathrm{sh}(at),$$

where to obtain the second inequality we have used $t' = t\,e^{ax}$ and $1 + ax_0 = e^{ax}$. The X- coordinate of the event in S_0 coincides with the coordinate of the second ship (6.9):

$$aX = F(T) = (1 + ax_0)\,\mathrm{ch}\left(\frac{at'}{1 + ax_0}\right) - 1 = \mathrm{ch}(at)\,e^{ax} - 1.$$

As a result, the transformation between the inertial and the non-inertial frames can be written as:

$$aT = \mathrm{sh}(at)\,e^{ax}, \qquad 1 + aX = \mathrm{ch}(at)\,e^{ax}. \tag{6.21}$$

We can also use [18] another parameterization of the coordinate $e^{ax} = 1 + a\tilde{x}$:

$$aT = \mathrm{sh}(at)\,(1 + a\tilde{x}), \qquad 1 + aX = \mathrm{ch}(at)\,(1 + a\tilde{x}). \tag{6.22}$$

Then the coordinate $\tilde{x}$ of the second ship is its position in the laboratory frame at the starting time.

⊳ The transformations (6.21) have a somewhat different meaning than those between two inertial frames. Observers in an inertial frame have a *common* synchronized time, and the Lorentz transformations are the same for all such observers. Therefore, each inertial frame is usually supposed to have a single observer "spread" throughout the space. In a non-inertial frame, each observer has his own time. He can draw conclusions on remote events only from the information received from the observer who records the event exactly at the point where it occurred. Therefore

! the transformations (6.21) refer to the *specific* non-inertial observer and to an *arbitrary* inertial observer.

The relationships (6.21) can be reversed:

$$t = \frac{1}{2a} \ln\left[\frac{1 + aX + aT}{1 + aX - aT}\right], \qquad x = \frac{1}{2a} \ln[(1 + aX)^2 - (aT)^2]. \tag{6.23}$$

As an exercise (< H_{28}), it is worth finding the velocity of the point $X = 0$ (origin of frame S_0) in the frame S and confirming that it coincides with the velocity of the point $x = 0$ relative to S_0.

The direct (6.21) and inverse (6.23) transformations have different functional forms. This reflects the ***inequivalence*** of the inertial and the non-inertial observer.

Consider the interval (p. 22) between the events in the coordinates of the non-inertial observer. Substituting (6.21) into

$$ds^2 = dT^2 - dX^2,$$

we have

$$ds^2 = e^{2ax}(dt^2 - dx^2). \tag{6.24}$$

The propagation of light corresponds to a zero interval,

$$ds = 0.$$

If the light travels along the x-axis, its trajectory in the coordinates (t, x) is a linear function. This is exactly what is implied in the definition of the radar distance $x = l_0$. Using the coordinate $\tilde{x}$ (6.22), we obtain the ***Møller interval***:

$$ds^2 = (1 + a\tilde{x})^2\, dt^2 - d\tilde{x}^2. \tag{6.25}$$

The difference between the intervals (6.24) and (6.25) is due to the different "numbering" of the points in the non-inertial frame. In both cases, the time t is the observer's proper time at the origin of the non-inertial frame. The coordinate x has the meaning of physical distance between the origin and the event, in contrast to the coordinate $\tilde{x}$ which is "less physical" but also provides a unique characteristic of the event.

▷ The transformations (6.23) between inertial and non-inertial frames make sense only when

$$(1 + aX)^2 > (aT)^2. \tag{6.26}$$

This inequality does not depend on the specific parameterization used for the coordinates of the event. For example, from (6.22) we have for the Møller coordinate $\tilde{x}$:

$$(1 + a\tilde{x})^2 = (1 + aX)^2 - (aT)^2 > 0.$$

Thus, not all events registered in the inertial frame can participate in the transformation (6.23), but only those whose coordinates and times which satisfy the inequality (6.26). This restriction has a simple physical interpretation.

First of all, note that it follows from the second relation (6.23) that the events in the direction of travel (coordinates $x > 0$) correspond to the following region in the frame S_0 for the observer at $x = 0$:

$$1 + aX > \sqrt{1 + (aT)^2}.$$

The positive coordinate $X > 0$ of the event can be arbitrarily large. As for the events in the opposite direction ($x < 0$), they are seen only if

$$X > T - \frac{1}{a}.$$

In fact, the event will be visible only if the flash of light occurring at $\{T, X\}$ "catches up" with the observer on the first ship at some finite time T_1. This happens when the equation:

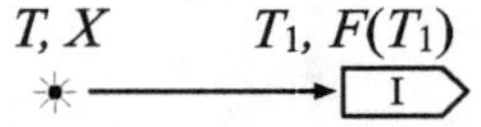

$$\frac{1}{a}\left(\sqrt{1 + (aT_1)^2} - 1\right) - X = T_1 - T$$

is solvable with respect to the arrival time T_1. Leaving the square root on the left-hand side and squaring both sides, we obtain:

$$a\,(T_1 - T) = \frac{1 - (1 + aX)^2 + (aT)^2}{2\,(1 + aX - aT)} = \frac{1 - e^{ax}}{2\,(1 + aX - aT)},$$

where we have substituted in (6.21) to obtain the second equality. The event happens behind the ship, so $x < 0$ and the numerator is positive. The time taken for the signal to travel $T_1 - T$ becomes infinite at $X = T - 1/a$ for the observers in both the inertial and the non-inertial reference frames. The signals from any event to the left of the coordinate $T - 1/a$ of the inertial frame will never reach the observer on the first ship (the origin of the non-inertial frame). Note that it is this observer that the transformations (6.21), (6.23) are written for, and their singularity leads to the inequality (6.26).

Therefore, if the motion is uniformly accelerated, the observers can "run away" from events, because they constantly increase their velocity (even though it always remains smaller than unity). After the ship with its observer starts, the observer sees the emission frequency of the light sources on the ships of the squadron behind them gradually becoming more and more red in a "wave-like" manner (p. 142) (as the "image" of their start reaches him). The farther away the ships are, the more pronounced the reddening is. The observer will never know about the existence of the visibility boundary

$$X = T - \frac{1}{a},$$

since the reddening wave will reach it only at $t = \infty$.

Define the *event horizon* as a line (surface) in the event space $S_0 : \{T, X\}$ which bounds the area of those events that can be seen by the ***specific*** observer in the frame $S : \{t, x\}$.

To describe the event horizon, two reference frames are required: the frame S, and a frame S_0 that encompasses a wider set of events than S.

The shaded area in the left figure below shows all of the events seen by the observer at $x = 0$, in the coordinates of the inertial observer $\{T, X\}$. Lines with a 45 degrees slope ($c = 1$) are the trajectories of light signals. Those which cross the ship's trajectory (the bold curved line) will never be received by the observer. Obviously, not all light signals sent from behind the ship will intersect its trajectory:

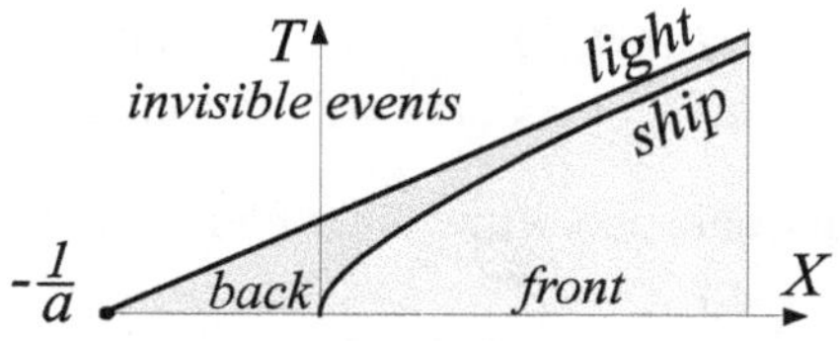

In the plane $\{T, X\}$ we can draw a grid of lines corresponding to constant values of t (simultaneous events in S) and to constant values of x (the second figure above). Their equations are:

$$aT = \text{th}(at)\,(1 + aX), \qquad (1 + aX)^2 - (aT)^2 = e^{2ax}. \tag{6.27}$$

From (6.23) it follows that the extreme visible point

$$X = T - \frac{1}{a}$$

behind the ship has coordinate $x = -\infty$ in S. We do not draw a curvilinear coordinate grid at $T < 0$, since non-inertial frames are such only for $T > 0$ (the acceleration began at $T = t = 0$).

6.4 Acceleration from a non-zero velocity *

Let us find time and coordinate transformations for the situation that a uniformly accelerating frame accelerates from some initial velocity U_0. We introduce three reference frames: the "rest" frame $S_0 : \{T, X\}$; the inertial frame $S_0' : \{T', X'\}$ moving uniformly with velocity U_0 relative to S_0; and the frame $S : \{t, x\}$, which, with respect to S_0', starts moving from rest with uniform acceleration as described above. We have two successive coordinate and time transformations:

$$\begin{cases} aT' = \text{sh}(at)\, e^{ax}, \\ aX' = \text{ch}(at)\, e^{ax} - 1, \end{cases} \qquad \begin{cases} T = \gamma_0\,(T' + U_0\,X'), \\ X = \gamma_0\,(X' + U_0\,T'), \end{cases}$$

where $\gamma_0 = 1/\sqrt{1 - U_0^2}$, and the origins of all of the frames $X = X' = x = 0$ coincide at $T = T' = t = 0$. Substituting the first pair of transformations into the second, we obtain

$$\begin{cases} a\,T &= [\mathrm{sh}(at)\,\mathrm{ch}(\alpha_0) + \mathrm{ch}(at)\,\mathrm{sh}(\alpha_0)]\,e^{ax} - \mathrm{sh}\,\alpha_0, \\ a\,X &= [\mathrm{sh}(at)\,\mathrm{sh}(\alpha_0) + \mathrm{ch}(at)\,\mathrm{ch}(\alpha_0)]\,e^{ax} - \mathrm{ch}\,\alpha_0, \end{cases}$$

where a "rapidity" parameter was introduced in place of the velocity:

$$\alpha_0 = \mathrm{ath}(U_0), \qquad \gamma_0 = \mathrm{ch}(\alpha_0), \qquad U_0\gamma_0 = \mathrm{sh}(\alpha_0). \tag{6.28}$$

Using standard angle-sum identities for the hyperbolic functions, we can rewrite the transformations in a more compact form:

$$\begin{cases} a\,T &= \mathrm{sh}(at + \alpha_0)\,e^{ax} - \mathrm{sh}\,\alpha_0, \\ a\,X &= \mathrm{ch}(at + \alpha_0)\,e^{ax} - \mathrm{ch}\,\alpha_0. \end{cases} \tag{6.29}$$

As an exercise (< H_{29}), it would be worthwhile to find a non-relativistic approximation of (6.29).

We can also obtain transformations between non-inertial frames, one of which (with coordinates $\{t, x\}$) had zero velocity relative to the laboratory frame at the initial time, and the second of which (with coordinates $\{t', x'\}$) had a non-zero velocity:

$$\begin{cases} \mathrm{sh}(at)\,e^{ax} &= \mathrm{sh}(at' + \alpha_0)\,e^{ax'} - \mathrm{sh}\,\alpha_0, \\ \mathrm{ch}(at)\,e^{ax} &= \mathrm{ch}(at' + \alpha_0)\,e^{ax'} - \mathrm{ch}\,\alpha_0. \end{cases} \tag{6.30}$$

As is easily verified, these transformations are in the form of ordinary Lorentz transformations when $a \to 0$. For $a \neq 0$ they form a group (p. 10), i.e., a composition of such transformations with different parameters α_0 has the same functional dependence on the coordinates and time as (6.30).

⊳ It is easy to verify that the transformations (6.29) again lead to the interval (6.24). Therefore, all uniformly accelerating frames have the same interval regardless of their initial velocity U_0. In the general case, by definition (to be clarified on p. 214):

! reference frames are *equivalent* if their intervals have the same functional form.

This definition of "equivalence" encompasses the whole set of inertial frames, since the Lorentz transformations leave the interval

$$ds^2 = dT^2 - dX^2$$

form-invariant (i.e, its functional form does not change). A uniformly accelerating frame is not equivalent to any inertial frame, since the intervals between the events in these frames are different.

Note that the definition of "equivalence" implies that the functional dependence between the coefficients defining the interval ds is invariant. The numerical value of the interval ds between two infinitely close events is the same in all frames, whether they are equivalent or not.

▷ Consider also translation transformations inside a uniformly accelerating frame. The Møller coordinates (6.22) between the second ($\tilde{x}'$) and the first ship ($\tilde{x}$) have the form:

$$\tilde{x} = \tilde{x}' + x_0, \qquad t = \frac{t'}{1 + ax_0}. \tag{6.31}$$

Substituting these into the interval (6.25), we obtain

$$ds^2 = (1 + a_2\tilde{x}')^2\, dt' - d\tilde{x}'^2,$$

where

$$a_2 = \frac{a}{1 + ax_0}.$$

This is quite a predictable result. The proper acceleration of the second ship is a_2. Therefore, the interval of the observer on the second ship in coordinates $\{t', x'\}$ should depend on a_2, exactly as the interval (6.25) in the coordinates $\{t, x\}$ of the first ship depends on a.

Note again that transformations in non-inertial frames are written for specific observers rather than for the reference frame as a whole. In particular, the transformations between the inertial frame and the observer on the second ship have the form:

$$a_2 T = \operatorname{sh}(a_2 t')\, e^{a_2 x'}, \qquad a_2 X = \operatorname{ch}(a_2 t')\, e^{a_2 x'} - \frac{a_2}{a}, \tag{6.32}$$

where x_0 in $a_2 = a/(1 + ax_0)$ is a *constant* giving the position of the observer at the instant of acceleration, $\{t', x'\}$ is the time and the coordinate of the event registered by *this observer*, and $e^{a_2 x'} = 1 + a_2\tilde{x}'$.

6.5 Rotating frames

We now consider two spaceships revolving with the same angular velocity ω along an orbit of radius R in the laboratory frame $S_0 : \{T, X\}$. We assume that the angular distance between them is $\Phi < \pi$. The ships are moving with velocity

$$V = \omega R < 1,$$

therefore the time dilation on the ships with respect to the laboratory (inertial) frame is the same, and their clocks show the same time:

$$t = T\sqrt{1 - V^2}, \qquad t' = \tau + T\sqrt{1 - V^2}, \tag{6.33}$$

where t is the time of the first ship, and t' is the time of the second ship travelling ahead of the first ship in the direction of the revolution. The time T of the laboratory frame S_0 in each formula is as measured by *different* synchronized clocks located along the trajectory of the ships. Clock synchronization on the ships requires certain measurements to specify the parameter τ. Obviously,

$$\tau = \tau(\Phi)$$

is a function of an angle. If counted from the first ship, $\tau(0) = 0$.

Let the observer on the first ship use radiolocation to measure the distance to the second ship. We shall first assume that the light signal travels along the circle. Such motion can be achieved by means of a set of mirrors arranged along the circle and revolving with the same angular velocity ω. Alternatively, a circular light guide can be used. In the laboratory frame S_0 the signal is emitted at T_1 and reaches the second ship at T. For the time $T - T_1$, the ship advances through the angle $\omega(T - T_1)$ along the circle. Therefore, the light travels the angular distance $\Phi + \omega(T - T_1)$. The signal is then reflected and returns to the first ship at the time T_2. This backward signal covers a smaller angular distance $\Phi - \omega(T_2 - T)$:

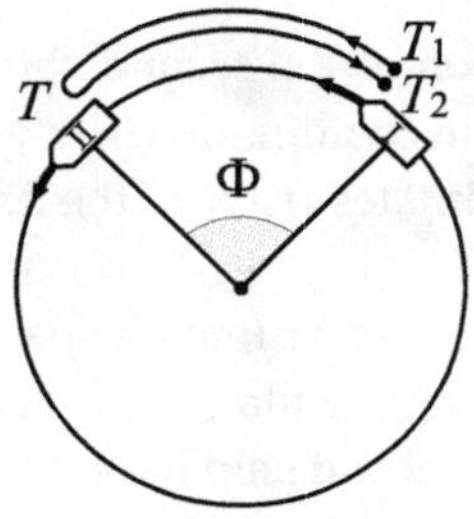

$$\begin{cases} R(\Phi + \omega(T - T_1)) = T - T_1, \\ R(\Phi - \omega(T_2 - T)) = T_2 - T, \end{cases}$$

$$T - T_1 = \frac{R\Phi}{1 - \omega R}, \qquad T_2 - T = \frac{R\Phi}{1 + \omega R}.$$

As before, we assume the length of the signal's path (in the laboratory frame) to be equal to the duration of its travel ($c = 1$).

As a result of adding and subtracting the last two formulae, we have:

$$\frac{T_2 - T_1}{2} = \frac{R\Phi}{1 - V^2} \tag{6.34}$$

and

$$T = \frac{T_1 + T_2}{2} + \frac{\omega R^2 \Phi}{1 - V^2}. \tag{6.35}$$

Similar relationships will be obtained from the same radar experiment by the observer on the second ship. However, in all formulae the replacement

$$\omega \to -\omega$$

should be made. The same replacement is necessary if the first ship measures the distance over a longer pathway

$$\Phi \mapsto 2\pi - \Phi$$

(against the revolution).

Consider the time of the first ship using the first relation (6.33). This gives the following radar distance between the ships:

$$l = \frac{t_2 - t_1}{2} = \frac{R\Phi}{\sqrt{1 - V^2}}. \tag{6.36}$$

The distance l is one Lorentz factor $1/\sqrt{1-V^2}$ longer ($V = \omega R$ is the linear velocity of the ships) than the length $R\,\Phi$ of the arc between ships in the laboratory frame. This result does not depend on the specific ship used by the observer to measure the distance (which does not change after $\omega \mapsto -\omega$ substitution).

By substituting the times (6.33) into the second relation (6.34), we obtain the time at which the signal was reflected (as per the clock of the second ship):

$$t' = \frac{t_1 + t_2}{2} + V\,l + \tau. \tag{6.37}$$

Now we can find the function $\tau = \tau(\Phi)$. There are various ways to synchronize clocks on the ships. Consider first the following option. Let there be a flash of light in the centre of the orbit. In the laboratory frame, it will reach each ship at the same time. It seems natural to use this instant as the time origin on the ships by setting $\tau = 0$. However, this does not work. Due to the relativity of simultaneity, information about the flash will reach the ships *non-simultaneously* (from "their point of view"). Indeed, imagine that R approaches infinity, and that ω approaches zero so that the velocity $V = \omega R$ remains smaller than unity. In this case, the ships will move almost in a straight line. If the angular distance Φ between them is small, the reference frame associated with the ships does not differ from the co-moving inertial frame travelling with velocity V relative to the laboratory frame. In this frame, information about the flash will reach the second ship later than the first ship (p. 52).

▷ Another, more physical method (from the viewpoint of the co-moving inertial frame), is to select τ such that

$$t' = \frac{t_1 + t_2}{2}. \tag{6.38}$$

This rule is absolutely natural for inertial frames. It implies that the speed of light is equal in all directions, which follows from the isotropy of space (remember, p. 4, that synchronization is not necessarily done using light; any uniformly moving particle can be used to achieve this).

Standards of length and time of non-inertial observers coincide with those used by the inertial observer in the co-moving frame. Therefore, (*in a neighbourhood of the inertial observer*) the speed of light is the same in all directions. Accordingly, we can take the rule (6.38) for a non-inertial frame.

Using (6.38), we have from equation (6.37):

$$\tau = \tau(\Phi) = -V\,l. \tag{6.39}$$

With this choice of τ the times needed for the signal to go back and forth will be the same by the measurements of *both* clocks:

$$t' - t_1 = t_2 - t' = l.$$

This *effective speed of light* equals unity and does not depend on the direction of the signal. The adjective "effective" is used to emphasize that there are two clocks used by two different observers to measure the velocity in one direction.

Whatever $\tau(\Phi)$ function is chosen, the rate of time flow is the same on both ships. Indeed, say the first ship sends signals to the second ship with period Δt_1. From (6.36) and (6.37) it follows that

$$t' - t_1 = l(1 + V) + \tau = const. \tag{6.40}$$

Therefore, the time interval between the signals received on the second ship will have the same value

$$\Delta t' = \Delta t_1 .$$

The situation is similar when periodic signals are sent from the second ship. Since the source and the receiver on the rotating ring are at rest relative to each other, they will register no changes in the frequency of the signal (remember that the effect of frequency change takes place in rigid uniformly accelerating frames even for devices at rest relative to each other).

Thus, the choice (6.39) gives quite natural physical results. However, the whole picture is not so simple, and despite the clocks having the same rate,

---

! global synchronization between all clocks travelling around a circle is impossible.

The function $\tau(\Phi)$ in (6.39) is linear with respect to the angle Φ (see (6.36)). The points

$$\Phi = 0 \quad \text{and} \quad \Phi = 2\pi$$

correspond to the same ship (the first one). Any attempt to synchronize the clocks with each other by sending a signal along the whole circle will lead to $\tau \neq 0$ although, by definition, $\tau(0) = 0$ is chosen on the first ship.

In particular, if the first ship synchronizes with the second ship by a longer path

$$\Phi \mapsto 2\pi - \Phi,$$

then

$$t' = \frac{t_1 + t_2}{2} - V\,l + \tau, \tag{6.41}$$

and another value of the initial countdown τ is to be chosen. In this case, the synchronization procedure is symmetrical, and if it is conducted by the observer on the second ship by sending a signal along the short path to the first ship, they will again get (6.39), and if they choose the longer path, they will get

$$\tau = V\,l .$$

As a result, the observer on the first ship can unambiguously synchronize his clock with all ships that are no farther than

$$|\Phi| < \pi$$

from him. All ships can repeat this procedure successively with each other. However, when such a "synchronization chain" reaches a ship at a distance greater than π, direct synchronization along the shortest path will show de-synchronization between it and the first ship.

For example, for two diametrically opposite clocks ($\Phi = \pi$) the radar distance is equal in the direction of revolution and against it. However, the synchronization procedure of two such clocks will depend on the direction of the signal. The difference between these two values τ is:

$$\Delta\tau = 2Vl = \frac{2\pi VR}{\sqrt{1-V^2}}.$$

This property of rotating reference frames looks very unusual. But this "unusualness" is not much stranger than other "usual" relativistic effects such as the relativity of simultaneity or time dilation.

▷ We can try to synchronize clocks by carrying a copy of the observer's clock "slowly" to other points of the reference frame. Let us consider how this procedure would look. Say the observer sends two copies of their clock in different directions along the circle. Their velocities are chosen such that the clocks come back at the same time. Consider this motion from the laboratory frame, where the angular velocity of the clock moving in the direction of rotation is $\omega + \Delta\omega_1$, and the velocity of the clock moving in the opposite direction is $\omega - \Delta\omega_2$. Since the $\Delta\omega_i$ are assumed small, the latter will make n revolutions around the circle, plus some additional angle Φ_0, by the time it meets the base clock after a time T:

$$\omega T = 2\pi n + \Phi_0. \tag{6.42}$$

The clock moving in the direction of rotation makes one additional revolution, while the second clock is saved a revolution:

$$(\omega + \Delta\omega_1)\, T = 2\pi(n+1) + \Phi_0, \qquad (\omega - \Delta\omega_2)\, T = 2\pi(n-1) + \Phi_0.$$

Substituting in the expression for ωT from (6.42), we obtain that

$$\Delta\omega_1 = \Delta\omega_2 = \Delta\omega \qquad \textit{and} \qquad T = 2\pi/\Delta\omega.$$

The proper times of the moving clocks τ_1 and τ_2 are:

$$\tau_{1,2} = T\sqrt{1-(\omega \pm \Delta\omega)^2 R^2} \approx T\sqrt{1-(\omega R)^2 \mp 2\omega\,\Delta\omega\, R^2},$$

where in the approximate equality we have expanded the square up to first order in $\Delta\omega$. Taking $1-(\omega R)^2$ outside the root sign and expanding the latter in a series, we obtain:

$$\tau_{1,2} \approx T\sqrt{1-V^2} \mp \frac{2\pi V R}{\sqrt{1-V^2}},$$

where $T = 2\pi/\Delta\omega$ has been substituted into the second term. The first term is the proper time of the base clock. Therefore, the clock moving along the rotation will run behind the base clock, whereas the clock moving with smaller velocity (in the laboratory frame) will be fast. The difference between the times of these two copies is

$$\Delta\tau = \tau_2 - \tau_1 \approx \frac{4\pi V R}{\sqrt{1 - V^2}}, \tag{6.43}$$

which ***does not depend*** on $\Delta\omega$. Even though the linear velocities of the copies are different in the laboratory frame $(\omega \pm \Delta\omega)R$, they will have the same magnitude (up to first order in $\Delta\omega\, R$) in the inertial co-moving frame for the base clock, which is natural for an observer in the frame of the base clock if he expects the sent copies to come back simultaneously.

In the laboratory frame, the clocks will first meet at $T = \pi/\Delta\omega$ at the point diametrically opposite to the base clock. At this moment their times will be different. In general, unambiguous synchronization is impossible for rotating frames even if the clocks are carried slowly. It can only be done if the clocks are placed along a non-closed arc. In this case, copies carried slowly will give the same result as a light signal with a subsequent recalculation of time according to (6.38).

▷ In relation to the synchronization problem, consider an experiment where an observer sends two light signals from his ship in opposite directions around the circle. Let us find the propagation times for each signal in the laboratory frame S_0. The signal sent in the direction of the revolution returns to the ship after time T_2. It covers the distance $2\pi R$ plus the arc $R\omega\, T_2$ passed by the ship for the time T_2. The second light signal sent towards the ship covers a smaller distance in the time T_1:

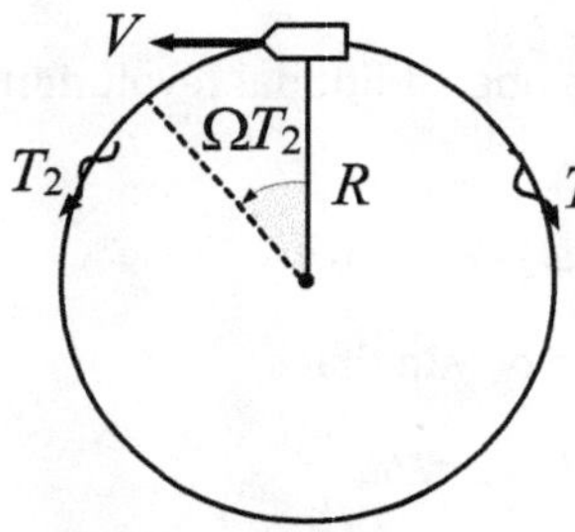

$$\begin{cases} R\,(2\pi + \omega\, T_2) = T_2, \\ R\,(2\pi - \omega\, T_1) = T_1, \end{cases}$$

$$\Delta T = T_2 - T_1 = \frac{4\pi V R}{1 - V^2}.$$

According to the ship's clock, the difference between these times ,

$$\Delta t = \Delta T \sqrt{1 - V^2},$$

is

$$\Delta t = \frac{4\pi V R}{\sqrt{1 - V^2}}, \tag{6.44}$$

which coincides with (6.43). This effect was discovered in 1913 by Georges Sagnac by means of a rotating ring interferometer. What he measured was the interference of signals due to the shift of the relative phases of light waves rather than the difference between the times. The accuracy of the experiment allowed only the principal approximation

$$\Delta t \approx \Delta T \approx 4\,S\,\omega$$

of (6.44), where

$$S = \pi R^2$$

is the area bounded by the trajectory of the signals in the laboratory frame. Remember that the root in the denominator (6.44) is related to the effect of time dilation when the clock is moving with velocity ωR. This dilation is of the second order of smallness with respect to the velocity.

6.6 Rotations with different radii

Until now we have considered only those points of the rotating frame which are equidistant from the centre. Let two spaceships in the laboratory frame now revolve with the same angular velocity ω but at different distances R_1 and R_2 from the centre. Their velocities are

$$V_1 = \omega R_1 \quad \text{and} \quad V_2 = \omega R_2,$$

and the relationship between their clocks and those in the laboratory frame has the form:

$$t = T\sqrt{1-(\omega R_1)^2}, \qquad t' = \tau + T\sqrt{1-(\omega R_2)^2}. \tag{6.45}$$

Suppose the observer on the first ship, with orbital radius R_1, conducts a radar measurement. We now suppose that the propagation of the light signal is rectilinear (in the laboratory frame).

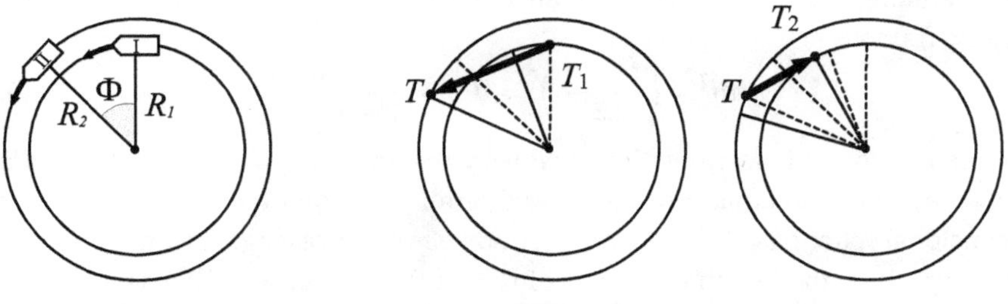

For the time $T - T_1$ required for the signal to travel in one direction, the second ship advances by an angular distance $\omega\,(T - T_1)$. Therefore, the path length of the signal is found from the cosine theorem with angle $\Phi + \omega\,(T - T_1)$, where Φ is the angle between the straight lines drawn from the centre to the ships. After being reflected, the signal starts moving back to the first ship. To calculate the length of the return path of the signal over the time $T_2 - T$, the angle $\Phi - \omega\,(T_2 - T)$ is substituted into the cosine theorem. This shows the squared distances passed by the light are related to the squared times of its travel as follows:

$$\begin{cases} R_1^2 + R_2^2 - 2R_1R_2\cos\left[\Phi + \omega\,(T - T_1)\right] = (T - T_1)^2, \\ \\ R_1^2 + R_2^2 - 2R_1R_2\cos\left[\Phi - \omega\,(T_2 - T)\right] = (T_2 - T)^2. \end{cases} \tag{6.46}$$

Consider the solutions of these transcendental equations with respect to time in the following form:

$$T - T_1 = g(\omega), \qquad T_2 - T = g(-\omega),$$

where $g(\omega)$ is a function of the angular velocity ω (as well as the radii R_1, R_2, and the angle Φ, which will be omitted in our notation).

Combining these equations and passing to the time of the first ship (6.45), we have:

$$l = \frac{t_2 - t_1}{2} = \frac{1}{2}\sqrt{1 - (\omega R_1)^2}\,\left[g(\omega) + g(-\omega)\right]. \tag{6.47}$$

The right-hand side does not depend on time. Therefore, the radar experiment will lead the observer to the conclusion that the distance between the ships is constant. Therefore, a set of observers revolving around the centre with constant angular velocity ω relative to S_0 constitute a rigid non-inertial frame S.

The formula for the radar distance (6.47) is simpler if the ships are next to each other. We exchange the angle Φ for an infinitely small $d\Phi$ and assume that $R_1 = R$ and $R_2 = R + dR$. Expanding the cosine $\cos(\alpha) \approx 1 - \alpha^2/2$ in (6.46) and solving the resulting quadratic equation, we easily obtain:

$$T - T_1 = \frac{R^2\omega\, d\Phi + \sqrt{\left[1 - (\omega R)^2\right] dR^2 + R^2 d\Phi^2}}{1 - (\omega R)^2}.$$

The expression for $T_2 - T$ is the same, with the exception that ω has been replaced by $-\omega$. Therefore, the squared radar distance between two infinitely close points of the rotating frame is

$$dl^2 = \left(\frac{t_2 - t_1}{2}\right)^2 = dR^2 + \frac{R^2 d\Phi^2}{1 - (\omega R)^2}. \tag{6.48}$$

This expression differs from the Euclidean distance in polar coordinates $dR^2 + R^2\, d\Phi^2$ by the denominator in the second term. Therefore, the geometry of a three-dimensional space based on the radiolocation *definition* of length is not the Euclidean geometry. Thus, the length of a circle of radius R with centre on the axis of rotation ($dR = 0$, $0 \leqslant \Phi < 2\pi$) is $2\pi R/\sqrt{1 - \omega^2 R^2}$, which is greater than the Euclidean value $2\pi R$.

Ships at different distances from the centre will experience different degrees of time dilation. If the second ship sends a signal at t_1', the first will receive it at t_2:

$$\frac{t_2}{\sqrt{1 - (\omega R_1)^2}} = \frac{t_1' - \tau}{\sqrt{1 - (\omega R_2)^2}} + g(-\omega).$$

If the signals are sent with a period $\Delta t'_1$, the first ship will receive them with period

$$\Delta t_2 = \Delta t'_1 \sqrt{\frac{1-(\omega R_1)^2}{1-(\omega R_2)^2}}.$$

The observers can synchronize their clocks, but later on the equality $t' = (t_1 + t_2)/2$ will fail.

6.7 Non-rigid uniformly accelerating frames *

As is well known, we have to distinguish between geometric and coordinate quantities in ordinary three-dimensional Euclidean space. Points in space are geometric entities that can be "numbered" with various kinds of coordinates (Cartesian $\{x, y, z\}$, polar $\{r, \phi, z\}$, etc.). The distance between two points or the angle between two lines are also geometric concepts. Geometric properties do not depend on the choice of coordinate system.

When describing physics in terms of space-time (p. 168), the situation is absolutely similar. Points in this space (events) are numbered by means of an arbitrarily chosen quadruplet of numbers

$$\{x^0,\ x^1,\ x^2,\ x^3\}$$

where the number x^0 does not necessarily mean the physical time of the event in this reference frame.

In this chapter, we consider non-inertial reference frames and their relation to inertial frames. Physical time T and Cartesian coordinates X, Y, Z will still be used to number events in the inertial frame. The quadruplet of numbers

$$\{T,\ X,\ Y,\ Z\}$$

forms what are called ***Lorentz coordinates***. The interval between two infinitely close events in these coordinates is

$$ds^2 = dT^2 - dX^2 - dY^2 - dZ^2. \tag{6.49}$$

In the non-inertial reference frame (NIRF), the same events are enumerated with four numbers $\{x^0,\ x^1,\ x^2,\ x^3\}$. Assume that $\{x^1, x^2, x^3\}$ are the coordinates of a *fixed* point in the three-dimensional space of the non-inertial reference frame. That points can be fixed does not mean that the reference frame is rigid. It is like a frame of reference formed by a cloud of bees flying out of their hive. Each bee is enumerated with a triplet of numbers $\{x^1, x^2, x^3\}$. In themselves, these numbers may not have a simple geometric meaning, but what is important is that they define an *unchanging* number in the frame of the bees and unambiguously characterize each bee (or point in the three-dimensional space of the non-inertial reference frame). There can be a lot of observers in space moving with different velocities relative to each other. Any subset within them can be called a reference frame. With such a common point, the choice of observers associated with this frame is quite arbitrary. Hence:

! a *reference frame* is a set of observers which continuously fill space and can measure time and distance in their immediate neighbourhood.

▷ It is convenient to define a non-inertial frame of reference by ***specifying the trajectories*** of all of its points relative to the inertial (laboratory) frame. Consider as an example a ***non-rigid*** uniformly accelerated reference frame such that all of its points have the same trajectory relative to the laboratory frame. The corresponding coordinate transformation can be written as:

$$X = x + \frac{1}{a}\left[\sqrt{1 + (aT)^2} - 1\right], \tag{6.50}$$

where x is the coordinate (number) of a fixed point in the NIRF, and $\{T, X\}$ are the time and the coordinate of this point in the laboratory frame. Note that the parameter x in (6.50) is a constant found from the initial conditions

$$X(0) = x.$$

The value of the constant determines the fixed point of the NIRF. When considering all points of the NIRF, we already assume x to be variable and (6.50) becomes a coordinate transformation between two reference frames. The values $\{T, X\}$ have a clear physical meaning, whereas that of the coordinate x is still to be discovered. To do this, we have to relate the number x with certain measurement procedures conducted by a ***specific*** observer. For example, this observer can be at the origin $x = 0$ of the NIRF. It is important to remember that such an observer is not "distributed" over the whole space, but is at a fixed point in the NIRF. They measure time, distance, and velocity in its immediate neighbourhood. Information about remote events occurring at points with coordinate $x \neq 0$, can only be obtained using signals (e.g., light signals).

To proceed, we will need the explicit form of the interval in the coordinates of the non-inertial observer. Therefore, besides the coordinate transformation (6.50), we will need the relationship between the times of the events. As with spatial coordinates, we can use quite arbitrary numbers t to enumerate the time of the event. Assume that they are ordered so that smaller values of t correspond to earlier events. For example, we can use to this end the proper time t of the clock moving along the trajectory (6.50) and arriving in the vicinity of the event at time t:

$$aT = \operatorname{sh}(at). \tag{6.51}$$

This is the relationship between the time t that passes for the observer after they start accelerating (p. 139) and the time T shown by synchronized clocks distributed along the trajectory of the motion in the inertial frame.

▷ Hence, the transformation from the Lorentz coordinates of the inertial frame $\{T, X, Y, Z\}$ to the coordinates of a non-rigid uniformly accelerating frame $\{t, x, y, z\}$ has the form:

$$T = \frac{1}{a}\,\mathrm{sh}(at), \qquad X = x + \frac{1}{a}\,[\mathrm{ch}(at) - 1]\,, \qquad Y = y, \qquad Z = z. \tag{6.52}$$

Substituting these formulas into the interval (6.49), we obtain [19]:

$$ds^2 = dt^2 - 2\,\mathrm{sh}(at)\,dt dx - dx^2 - dy^2 - dz^2. \tag{6.53}$$

In the inertial reference frame, the interval of physical time dT measured by *one* clock is equal to the interval ds between events happening at one point in space in the frame, i.e. $dX = dY = dZ = 0$. Similarly, the *proper physical time* $d\tau_0$ of the given point in an arbitrary reference frame is the interval ds between events happening at this point $dx = dy = dz = 0$. For the metric (6.53), the physical time $d\tau_0$ coincides with the coordinate time dt, since we chose the proper time t of the observer (6.51) as the time t used in (6.52).

Note that $d\tau_0$ is a "tick" of a clock at a fixed point (x, y, z). This clock is used by the observer *at this* point, and it is exactly this observer who directly measures the ticks. But how can an observer, e.g. at the origin of the frame, find out the time of this clock (and, consequently, the time of the event in its neighbourhood)? Only by receiving light (or some other) signal from the remote clock and making corrections for the time of its travel. In both inertial and non-inertial frames, the events associated with the light propagation have zero interval $ds = 0 < C_{10}$. For the metric (6.53), this condition leads to a differential equation describing the trajectory of a light pulse. Assuming the light travels parallel to the x-axis (past the observers in the NIRF with $y, z = const$), we have

$$\left(\frac{dx}{dt}\right)^2 + 2\,\mathrm{sh}(at)\,\frac{dx}{dt} - 1 = 0.$$

Completing the square, we easily obtain

$$\frac{dx}{dt} = \pm\,\mathrm{ch}(at) - \mathrm{sh}(at) = \pm e^{\mp at}. \tag{6.54}$$

The solution of this differential equation has the form:

$$x(t) = const - \frac{e^{\mp at}}{a}, \tag{6.55}$$

where *const* is a constant of integration, and the signs correspond to the direction of the signal (minus means that the coordinate x grows $(dx/dt > 0)$, and plus means that it diminishes).

▷ Let us use coordinates $\{t, x\}$ to consider a radar experiment carried out by an observer at a point with coordinate $x = 0$. At the moment t_1, he sends a light pulse that reaches a point with coordinate $x > 0$ at time t, which is then reflected and comes back at time t_2. To find the constant in (6.55), we choose the initial condition $x(t_1) = 0$ for the direction away from the observer, and $x(t_2) = 0$ for the reverse direction:

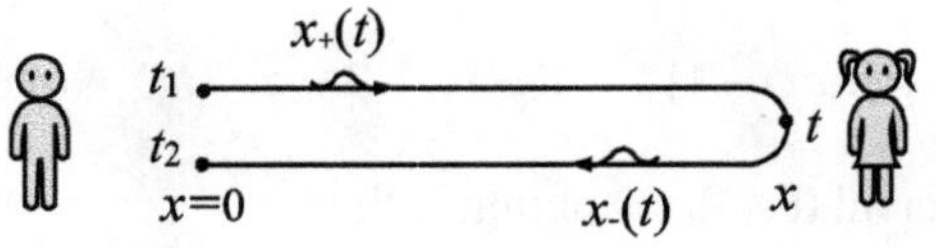

$$x_+(t) = \frac{1}{a}\left(e^{-at_1} - e^{-at}\right),$$

$$x_-(t) = \frac{1}{a}\left(e^{+at_2} - e^{+at}\right).$$

At the point of reflection

$$x_+(t) = x_-(t) = x,$$

so

$$e^{-at_1} - e^{-at} = ax, \qquad e^{at_2} - e^{at} = ax, \tag{6.56}$$

Rearranging to get rid of t, we can find the radar distance:

$$l = \frac{t_2 - t_1}{2} = \frac{1}{2a}\ln\left[1 + ax\,\frac{2\,\mathrm{ch}(at_1) - ax}{1 - ax\,e^{at_1}}\right] \approx x\,\mathrm{ch}(at_1), \tag{6.57}$$

where the approximate equality is written for small $ax \ll 1$. Knowing t_1 and t_2, the observer at the origin of the frame can determine the coordinate of the point x at which the signal is reflected. It follows from (6.57) that the radar distance to a fixed point of such a uniformly accelerating frame changes with time (depends on t_1). Therefore, the frame (6.53) is called a *non-rigid uniformly accelerating frame*. The physics in such a frame differ from those in a rigid uniformly accelerating frame as considered in the first section.

We can find the frequency of a signal received from a remote source in a similar way. Differentiating the second equation in (6.56) and rearranging to clear the term e^{at}, we have

$$dt_2 = dt\,(1 - ax\,e^{-at_2}).$$

The times dt and dt_2 are the periods with which the signals are sent and received according to the *different* clocks of observers at points $x > 0$ and $x = 0$. The ratio of these periods depends on the x coordinate and varies with the time t_2 at which the signal is received.

Note that the function $x(t)$ in (6.55) implies that there are observers in the NIRF along the trajectory of the light pulse. Each of them has their own clock to measure the proper time t, counted from the zero origin of the start of the acceleration. However, the x coordinate is, in general, not a physical distance, so *the coordinate speed of light* (6.54) is not unity.

6.8 The stopping paradox and the twin paradox

"*The stopping paradox.*" In Chapter 3 (p. 55) we considered a squadron of spaceships S'_0 moving with velocity U_0 relative to the inertial frame $S_0 : \{T, X\}$. Due to the relativity of simultaneity, the observers from each reference frame register different times on the clocks moving relative to them:

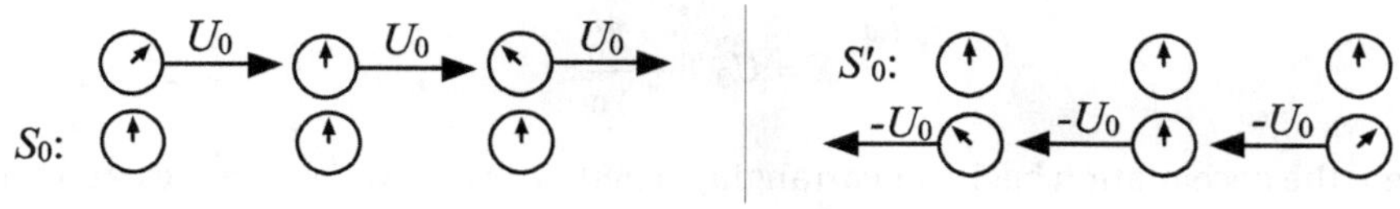

If the squadron stops, the simultaneous stop of the ships (as seen from S'_0) will not be such for observers in S_0. Let the ships keep the distance between them constant during their deceleration. For example, when the velocity of the central ship is zero relative to S_0, it should be the same for other ships which are ***at rest relative to it.*** As a result, non-simultaneous (in S_0) deceleration strangely ends with a simultaneous stop for all observers, who are now at rest in both of the frames S and S_0.

We can calculate the deceleration of these two ships. Suppose that they simultaneously begin to slow down at time $T' = 0$ by the clock of frame $S'_0 : \{T', X'\}$. The first (rear) ship has coordinate $X' = 0$ in S'_0, and the second (the next one) has coordinate $X' = X'_0$. The Lorentz transformations are

$$T = \gamma_0 (T' + U_0 X'), \qquad X = \gamma_0 (X' + U_0 T').$$

Assuming $T' = 0$ and $X' = X'_0$, we find that the second ship in the frame S_0 will start decelerating at the time

$$T = U_0 \gamma_0 X'_0,$$

having the coordinate $X = \gamma_0 X'_0$ (the second figure below). At $T = 0$ (when the first ship starts decelerating in S_0) the second ship has coordinate $X = X'_0/\gamma_0$:

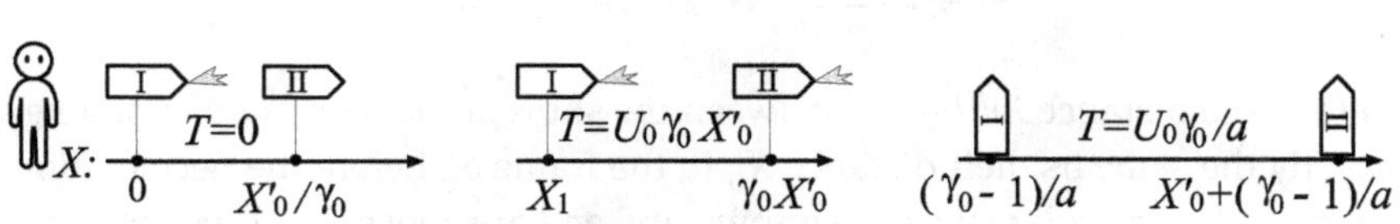

Suppose that the ships of the squadron form a rigid uniformly accelerating (or decelerating, in this case) frame $S : \{t, x\}$ with the origin on the first ship. Before the deceleration, the radar distance between the ships is X'_0. As soon as the acceleration begins, the distance becomes, p.141,

$$x_0 = -\ln(1 - aX'_0)/a,$$

(the acceleration is directed along the x axis).

Consider the transformations between the inertial and the non-inertial observers (6.29). During deceleration, we replace $a \mapsto -a$:

$$\begin{cases} aX &= -\operatorname{ch}(\alpha_0 - at)\, e^{-ax} + \gamma_0, \\ aT &= -\operatorname{sh}(\alpha_0 - at)\, e^{-ax} + \gamma_0 U_0, \end{cases} \tag{6.58}$$

where $\alpha_0 = \operatorname{ath}(U_0)$. Until the time $T = U_0\gamma_0 X_0'$ the point (the ship) moves uniformly:

$$X = U_0\, T + \frac{X_0'}{\gamma_0}.$$

Then, the deceleration begins. Rearranging (6.58) to clear t, we obtain the expression for the trajectory of a fixed point with coordinate $x = x_0 = -\ln(1 - aX_0')/a$:

$$aX = \gamma_0 - \sqrt{(1 - aX_0')^2 + (U_0\gamma_0\, - a\,T)^2}, \qquad T > U_0\gamma_0 X_0'. \tag{6.59}$$

The velocity of this point varies with time as follows:

$$U = \frac{dX}{dT} = (U_0\gamma_0 - aT)\left[(1 - aX_0')^2 + (U_0\gamma_0\, - a\,T)^2\right]^{-1/2}. \tag{6.60}$$

We see that, ***independently*** of its position, the point X_0' have zero velocity at $T = U_0\gamma_0/a$ (in this case, deceleration takes place for

$$U_0\gamma_0\, aX_0' < aT < U_0\gamma_0,$$

so the minus sign was chosen in (6.59)). Starting with a non-simultaneous deceleration from the point of view of the frame S_0, the ships of the squadron will move with ***different*** velocities and, as a result, they will stop at the same time. This relativity of rigidity of a reference frame is what underlies the stopping paradox.

For $T = U_0\gamma_0/a$ the coordinates of the first ($X_0' = 0$) and the second ($X_0' > 0$) ships (which have stopped) in the frame S_0 will be:

$$X^{(I)} = \frac{\gamma_0 - 1}{a}, \qquad X^{(II)} = X_0' + \frac{\gamma_0 - 1}{a}. \tag{6.61}$$

Therefore, the distance $X^{(II)} - X^{(I)}$ between the ships at the moment of stopping will be exactly the same as their distance X_0' in the frame S_0' before the deceleration. To observers in S_0, the ruler whose beginning and end are above the ships appears contracted in the direction of motion before the deceleration (Lorentz contraction). During the deceleration, this ruler gradually stretches, and at the moment of stopping, the Lorentz contraction of the ruler finally disappears for observers in S_0. The situation with accelerating ships is similar. In this case, a ruler which is "rigid" (in the sense of its proper length) relative to the laboratory frame gradually shrinks (in quite a complex way, p. 144). After the engines are switched off, it turns out to be γ times shorter, in exact accordance with the Lorentz formula. After stopping, this contraction again disappears, as a result of hard braking.

▷ The *twin paradox* was discussed in Chapter 3, (p. 55), and, as we saw, there is nothing paradoxical about it. Moreover, inertial frames alone are enough to understand the reason for this apparent paradox. The key to understanding the problem is the relativity of simultaneity, which must be taken into account, together with the effect of time dilation. The travelling brother sees all clocks in the reference frame of his stay-at-home brother going slowly. However, due to the relativity of simultaneity, the time on clocks placed along its trajectory "jumps" to the future. Therefore, although the clock of the stay-at-home brother is slower, it gets further and further ahead of the traveller's clock.

Non-inertial frames make it possible to carry out calculations from the point of view of each brother during the stages of accelerated motion. In the general case, a clock moving in a laboratory reference frame with velocity $\mathbf{U}(T)$ has the following accumulated time:

$$\tau = \int_0^{T_0} ds = \int_0^{T_0} \sqrt{1 - \mathbf{U}^2(T)}\, dT. \tag{6.62}$$

This is the time shown by the traveller's clock when, from the viewpoint of the stay-at-home brother, a period T_0 has passed.

To find the time on the stay-at-home brother's clock in the non-inertial frame, we have to determine how its coordinate $x(t)$ changes in this frame. Knowing the function $x(t)$, we can calculate the proper time of his clock:

$$\tau_0 = \int_0^{t_0} ds = \int_0^{t_0} \frac{ds(t)}{dt}\, dt. \tag{6.63}$$

Note that the formulas (6.62) and (6.63) give the proper times of different clocks (the traveller's and the stay-at-home brother's, respectively). Obviously, after calculating τ and τ_0, the relationship between the coordinate times T_0 and t_0 should be found.

▷ As an example, consider the following calculations for a rigid uniformly accelerating frame of reference. Assume that the traveller synchronizes a time origin $T = t = 0$ with his stay-at-home brother when passing him with velocity U_0. He then starts decelerating until he comes to a complete halt in the stay-at-home brother's frame S_0. Remaining in the same inertial frame, the brothers can unequivocally compare the times of their clocks and find out whose clock is slow.

Substituting the velocity (6.60) into the relationship (6.62), for the traveller with coordinate $x = x_0 = X_0' = 0$ we have

$$\tau = \int_0^{T_0} \frac{dT}{\sqrt{1 + (U_0\gamma_0 - aT)^2}} = \frac{1}{a}\left[\operatorname{ash}(U_0\gamma_0) - \operatorname{ash}(U_0\gamma_0 - aT_0)\right].$$

The traveller's stop $U(T_0) = 0$ happens at $T_0 = U_0\gamma_0/a$ by the stay-at-home brother's clock (see the "stopping paradox").

At this instant, the traveller's clock shows the time τ:

$$a\tau = \operatorname{ash}(U_0 y_0)$$

or

$$\frac{1}{a}\operatorname{sh}(a\tau) = \frac{U_0 y_0}{a} = T_0. \tag{6.64}$$

Naturally, this result gives the same time as the one calculated for the clock accelerated from a state of rest to the velocity U_0 (see p. 87).

▷ Let us now find the proper time of the stay-at-home brother in the non-inertial frame. Its trajectory is obtained from (6.58), with $X = 0$:

$$y_0\, e^{ax} = \operatorname{ch}(\alpha_0 - at), \qquad \frac{dx}{dt} = -\operatorname{th}(\alpha_0 - at).$$

Therefore, the interval of the stay-at-home brother's proper time is

$$d\tau_0 = \sqrt{e^{-2ax}(dt^2 - dx^2)} = \frac{y_0\, dt}{\operatorname{ch}^2(\alpha_0 - at)}.$$

Integrating from zero to t_0, we have

$$\tau_0 = \frac{y_0}{a}\left[\operatorname{th}(\alpha_0) - \operatorname{th}(\alpha_0 - at_0)\right].$$

The stay-at-home brother stops relative to the traveller when his velocity dx/dt becomes zero. This happens when

$$t_0 = \frac{\alpha_0}{a}.$$

At this moment:

$$\tau_0 = \frac{y_0 \operatorname{th}(\alpha_0)}{a} = \frac{U_0 y_0}{a} = T_0. \tag{6.65}$$

Naturally, we again get the value T_0, which is the time that elapses before the stay-at-home brother and the traveller cease moving relative to each other in S_0.

Thus, the relationship (6.64) is true from the viewpoint of any observer, and gives the absolute time dilation for the clock moving with variable velocity, as compared to the clock at rest. Similarly, we can consider the classical version of the flight (acceleration from a state of rest, uniform motion and deceleration, and then a return journey in the same sequence). However, these calculations add nothing new to the effect.

7 Covariant formalism

This chapter is concerned with the covariant formulation used to describe physical quantities. The formulation is based on the concept of four-dimensional space-time. Any event with coordinates $\{t, x, y, z\}$ is a point in such a space, where distance is specified by an invariant interval (p. 22). The concepts of four-vectors and four-vector tensors will be used to make many relationships in the theory of relativity appear very elegant and simple. To do this, however, we will have to learn how to work confidently with expressions which sometimes contain a lot of indices. This is not a difficult thing to do, it is enough just to examine each formula carefully and rewrite it, if necessary, several times on a piece of parchment with a camel hair brush.

https://doi.org/10.1515/9783110515886-007

7.1 Four-dimensional space-time

Let us consider the rotation of the Cartesian coordinate axes (x, y) through an angle ϕ. The coordinates of the *same* point in the initial and in the rotated Cartesian frames are related as follows:

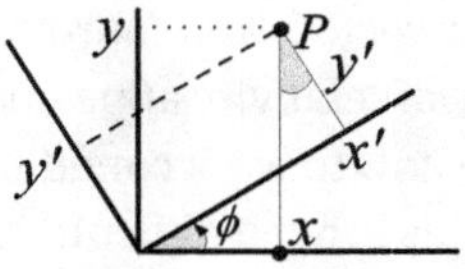

$$\begin{cases} x' = x\cos\phi + y\sin\phi \\ y' = y\cos\phi - x\sin\phi. \end{cases} \tag{7.1}$$

These transformations are obtained from elementary geometrical considerations. The dashed angles of the two similar right-angled triangles in the figure are equal to ϕ. Therefore, a vertical line with length y dropped from the point P consists of two segments: the hypotenuse $y'/\cos\phi$ of the upper triangle and the side $x\,\mathrm{tg}\,\phi$ of the lower. Their sum gives the second transformation equation (7.1). Similarly, x' (the projection of the point P onto the x'-axis) consists of the side $y'\,\mathrm{tg}\,\phi$ and the hypotenuse $x/\cos\phi$, which gives the first equation. Rotating the axes does not change distances. For example, for the distance from the origin: $x'^2 + y'^2 = x^2 + y^2$.

▷ The Lorentz transformations have a similar linear form:

$$\begin{cases} x' = x\gamma - t v\gamma = x\,\mathrm{ch}\,\alpha - t\,\mathrm{sh}\,\alpha \\ t' = t\gamma - x v\gamma = t\,\mathrm{ch}\,\alpha - x\,\mathrm{sh}\,\alpha, \end{cases}$$

where we used the hyperbolic cosine

$$\mathrm{ch}\,\alpha = \gamma = \frac{1}{\sqrt{1-v^2}}$$

and the hyperbolic sine $\mathrm{sh}\,\alpha = v\gamma$, so that $v = \mathrm{th}\,\alpha$. Due to their properties (p. 355), we have:

$$\gamma^2 - (v\gamma)^2 = \mathrm{ch}^2\,\alpha - \mathrm{sh}^2\,\alpha = 1.$$

Given that $\imath\,\mathrm{sh}(\alpha) = \sin(\imath\alpha)$ and $\mathrm{ch}(\alpha) = \cos(\imath\,\alpha)$, we can use the imaginary unit $\imath$ to make the Lorentz transformations appear formally the same as rotations in ordinary space:

$$\begin{cases} x' = x\cos(\imath\alpha) + \imath t\sin(\imath\alpha) \\ \imath t' = \imath t\cos(\imath\alpha) - x\sin(\imath\alpha). \end{cases}$$

If $(x, \imath t)$ are taken as Cartesian coordinates, these transformations define rotations of the coordinate axes through an *imaginary* angle $\phi = \imath\alpha$. Also, the squared distance from the origin

$$(x)^2 + (\imath t)^2 = x^2 - t^2,$$

remains constant and coincides, apart from a sign, with the interval between two events at the points $(0, 0)$ and (x, t) (p. 22). A space with such unusual properties is called a *pseudo-Euclidean space*.

▷ In the general approach to the concept of "space", abstract sets of points are considered. They can be "numbered" with sets of n numbers $x_1, \ldots, x_n$ called ***coordinates.*** The number n is called the ***dimension of the space.*** However, such an abstract set of points is not enough to define the concept of the *proximity* of two points. Coordinates by themselves do not provide such a measure of proximity, since the points can be numbered arbitrarily. Therefore, the first step in converting a set of points into a space is to define the *distance* between two infinitely close points. In three-dimensional Euclidean space this distance has the form:

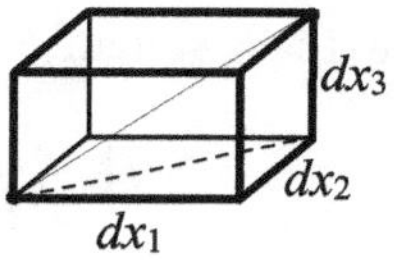

$$dl^2 = dx_1^2 + dx_2^2 + dx_3^2.$$

The usual geometrical interpretation of this uses Pythagoras' theorem to express the length of the diagonal of a parallelepiped with sides dx_1, dx_2, dx_3. The distance between two points being zero,

$$dl^2 = 0,$$

means that they coincide. Defining distances between infinitely close points is convenient when considering spaces with variable curvatures. In the Euclidean distance, only plus signs appear before the squared coordinate differentials. Therefore, it is said to have the *signature* $(+, +, +)$.

If the distance is defined in a different way, it can create a different space, whose properties will differ from those of Euclidean space. Consider a four-dimensional space of points, or *events*, numbered with coordinates t, x, y, z. Let us *define* the infinitesimal distance ds as follows:

$$ds^2 = dt^2 - dx^2 - dy^2 - dz^2 = dt^2 - d\mathbf{r}^2. \tag{7.2}$$

Its signature is $(+, -, -, -)$. In other respects, this distance is similar to the Euclidean distance in Cartesian coordinates; therefore, the space is called *pseudo*-Euclidean space. The proximity of two points is quite peculiar in this space. For example, having zero distance $ds = 0$ does not mean that the points coincide (see p. 22). Also, ds^2 can be negative, even though it is denoted with a square.

In early books on the theory of relativity, the imaginary unit was commonly used to make the pseudo-Euclidean distance look like the Euclidean distance:

$$-ds^2 = (\imath t)^2 + (d\mathbf{r})^2.$$

This trick was used to "sweep under the carpet" the pseudo-Euclidean nature of the space; however, it is not reasonable and is rarely used nowadays.

7.2 Space-time on paper

Coordinate axes and other geometrical objects in two-dimensional Euclidean space are easily represented on a plane, which is a physical representation of the space. Things are not so simple for pseudo-Euclidean space, even in two dimensions $\{t, x\}$. However, we can start by drawing two orthogonal axes. Let the vertical axis represent the time t of the event, and the horizontal axis represent its coordinate x. Uniform rectilinear motion

$$x = ut$$

with velocity u will be represented with a straight line (the first figure):

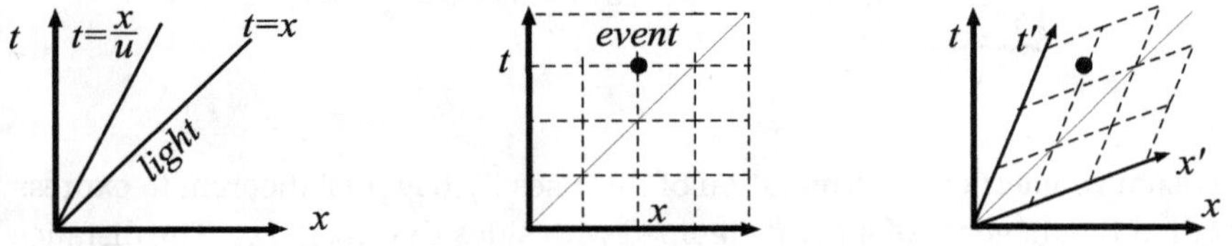

The median (middle line $x = t$) between the coordinate axes correspond to the propagation of a light signal. All other trajectories coming out of $x = 0$, $t = 0$ are closer to the axis t than the median, since physical velocities are always smaller than unity (the fundamental velocity $c = 1$).

The coordinate and the time of some event (the point in the second figure) are defined similarly to the Euclidean Cartesian coordinates using a coordinate grid. This is drawn as an array lines parallel to each coordinate axis (the dashed lines in the figure). This means that the x-axis is given by the equation $t = 0$, and the t-axis is given by the equation $x = 0$. The grid is formed by the horizontal lines $t = const$ and the vertical lines $x = const$.

Now we can use the same plane to draw the t'- and x'-axes of a reference frame S' moving with velocity v relative to S. If their origins coincide ($x' = x = 0$) at $t' = t = 0$, the trajectory of the clock at the origin ($x' = 0$) of S' is the line $x = vt$, or the coordinate axis t'. Similarly, the line $t' = \gamma(t - vx) = 0$ corresponds to the coordinate axis x' and has the equation $t = vx$. The angle between this axis and the x-axis is the same as the angle between the t'-axis and the t-axis (the third figure).

The spatial x'-axis and the time axis t' of the frame S', *when represented on the plane* (t, x) of the frame S, appear flattened along the median $t = x$ (the trajectory of the light signal). The same trajectory will be a median in S'. The coordinate grid of S' is also flattened. Its time lines are parallel to the x'-axis and are given be the equations $t' = const$, or

$$t = vx + \frac{t'}{\gamma}.$$

Similarly, for the spatial lines of the grid, we have $x' = const$.

It is important to remember that an event (a *point* in space-time) exists all by itself and does not depend on any choice of coordinate system (or, more precisely, reference frame). Its *description* (coordinates), however, will be different in different systems. This is the same as saying that the Euclidean plane rotates, and thus the coordinates (x, y) change, but the point in space to which they refer remains the same.

Coordinate grids in different reference frames make it easy to obtain the coordinates of events. For example, two simultaneous events in S should lie on the horizontal straight line $t = const$. Simultaneous events in S' lie on the lines $t' = const$, which make an angle with the line $t = const$. This is a geometric illustration of the relativity of simultaneity.

To introduce a geometry on the (t, x)-plane, let us define the distance

$$s^2 = (t_2 - t_1)^2 - (x_2 - x_1)^2$$

between two points as the interval between the events they represent (p. 22). This distance appears unusual when we try to apply our "Euclidean intuition" to it. As an example, consider the Euclidean and the pseudo-Euclidean unit circles, the sets of points at a unit distance from the centre (the coordinate origin in the figures below):

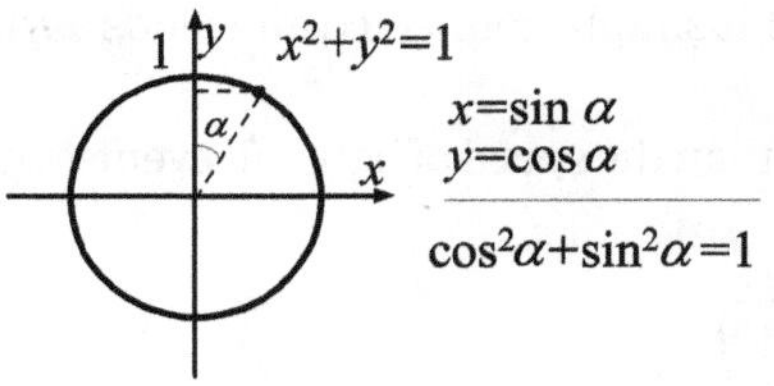

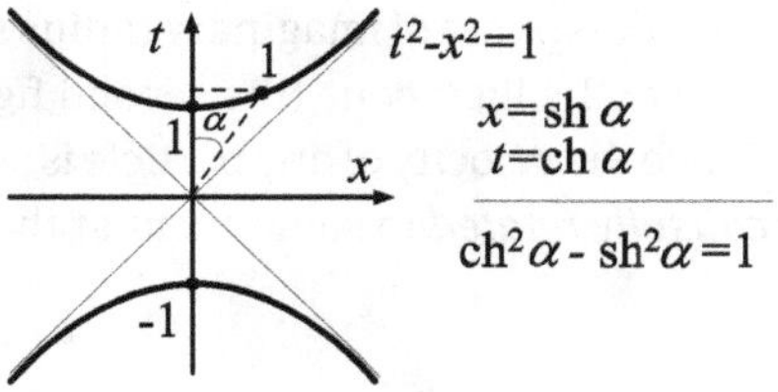

The unit circle $s = 1$ in pseudo-Euclidean space is a hyperbola

$$t^2 - x^2 = 1.$$

At great distances from the coordinate origin the hyperbola tends to the trajectories $t = \pm x$ of the light signal emitted from the point $t = x = 0$, in both directions parallel to the x-axis.

An ordinary circle in the Euclidean plane defines the trigonometric functions sine and cosine. Similarly, a circle in pseudo-Euclidean space (a "pseudo-circle") defines the hyperbolic sine $\mathrm{sh}\,\alpha$ and the hyperbolic cosine $\mathrm{ch}\,\alpha$, which are projections of the points of the pseudo-circle onto the x- and t-axes. In this case, the length of the hypotenuse $\sqrt{t^2 - x^2}$ of a right-angled triangle is *shorter* than its leg t. Instead of the Euclidean Pythagorean theorem $a^2 + b^2 = c^2$ we now have the pseudo-Euclidean theorem

$$a^2 - b^2 = c^2.$$

In general, pseudo-Euclidean space appears somewhat awkward when depicted on the *Euclidean* plane; so the best physical model for it is space-time itself rather than a sheet of paper.

▷ So far we have considered two-dimensional pseudo-Euclidean space. In fact, we want to consider the four-dimensional case, but at most three dimensions (t, x, y) can be depicted on a sheet of paper. In this case, the light signals, given by $t^2 = x^2 + y^2$, form a *light cone*, and the pseudo-circle becomes a *pseudo-sphere* or, in terms of the Euclidean space, a hyperboloid (the first figure):

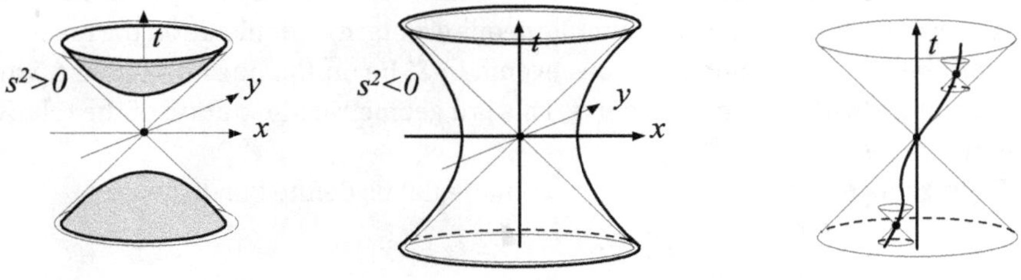

While the squared distance is always positive in Euclidean space, it can be negative in pseudo-Euclidean space. Therefore, there are two unit spheres:

$$t^2 - x^2 - y^2 = 1 \quad \text{and} \quad t^2 - x^2 - y^2 = -1.$$

The pseudo-sphere of imaginary radius $s^2 = -1$ is simply-connected and tends asymptotically to the light cone (the second figure).

Since the velocity of any particle is smaller than the speed of light, all events which are *causally related* to some event at the point

$$t = x = y = 0$$

are inside the light cone. Causality means that one event can effect another event with some agent moving between them with a velocity not exceeding the fundamental velocity.

If a particle passes the point $x = y = 0$ at $t = 0$, its trajectory $x(t)$ will always remain inside the light cone, and the tangent line to this trajectory at any point will always be inside a local cone with apex at this point (the third figure).

Because of the properties of the pseudo-Euclidean distance, the length element

$$d\tau = \sqrt{dt^2 - dx^2 - dy^2} = dt\sqrt{1 - \mathbf{u}^2} < dt$$

will always be smaller than the length of the trajectory of a particle at rest at the point $x = y = 0$ (along the t-axis), even though this distance looks longer in the Euclidean plane. Therefore, we should be very careful when applying Euclidean intuition to pseudo-Euclidean space-time charts.

The events inside the upper part of the cone are called the *absolute future*, and those inside the lower part are called the *absolute past* relative to $t = x = y = 0$. The events outside of the cone are called the *absolute elsewhere*. The word "absolute" means that these properties cannot be changed by Lorentz transformations. For example, if $t > 0$ and $t^2 - x^2 > 0$, then $t' > 0$ in any reference frame.

▷ Let us consider the geometric interpretations of some kinematic effects. Even though the pseudo-Euclidean plane is a somewhat difficult thing to imagine, it becomes a little more clear with this geometric analysis of the theory of relativity.

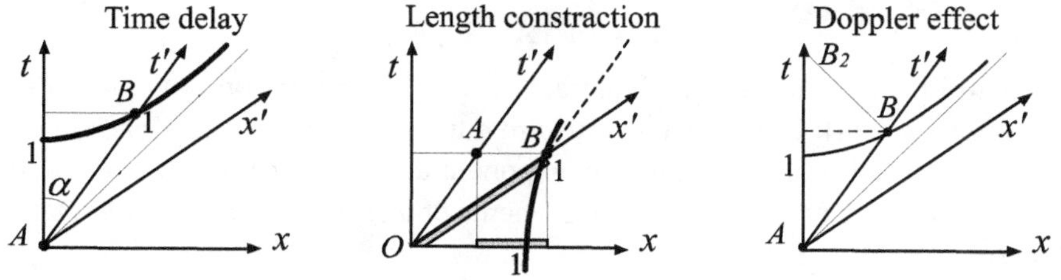

▷ Suppose that two "ticks" are registered by a clock at rest in the frame S' at the point $x' = 0$. The first event A happens at the origin $t' = x' = 0$, and the second event B lies on the t'-axis (the first figure above). If the second "tick" happens at time $t' = 1$, the same event has coordinate $t = t' \operatorname{ch} \alpha$ (hyperbolic projection). By definition,

$$\operatorname{ch} \alpha = \frac{1}{\sqrt{1 - v^2}},$$

and we obtain relativistic time dilation. The fact that $t' < t$ can be seen by drawing a pseudosphere which cuts off the same time units (segments of unit length) on the t'- and t-axes. As is seen in the figure, the point with $t = 1$ lies lower than the projection of the event $t' = 1,\ x' = 0$ (point B in the figure).

▷ Length contraction can be analysed in a similar way (the second figure above). Suppose that a rod of unit length is at rest in S' (represented by the line segment OB). Its left end is at the coordinate origin and moves along a trajectory coinciding with the t'-axis. The trajectory of the right-hand end (the dashed line) is parallel to it. The length of the rod is measured in S by simultaneously recording the coordinates of its beginning and end. The simultaneous events A and B happen at the same time in S but at different times in S'. Projecting the points A and B onto the x-axis gives the rod's length in S. As is seen, this projection is shorter than the length unit cut off by the pseudo-circle (now its radius is -1).

▷ The third figure illustrates the Doppler effect. The light source at the origin in S' emits light signals so that the first signal is emitted at $t' = t = 0$ and the second signal is emitted $t' = 1$ later. It will arrive at the origin $x = 0$ of S when the light trajectory (the thin line BB_2) intersects the t-axis. This time is longer than both the length unit on the axis $t = 1$ cut off by a single pseudo-sphere and the projection of the event B of emitting the signal (the dashed line).

As can be seen, there are some caveats when representing pseudo-Euclidean space on the Euclidean plane. It is to be kept in mind that real space-time geometry is not Euclidean (see also p. 316).

7.3 Four-vectors

Many equations in the theory of relativity can be written in an elegant covariant form. Let

$$A^\alpha = \{A^0, A^1, A^2, A^3\}$$

be a quadruple of numbers which are called the components of a *four-vector* (four-dimensional vector). These components are numbered with superscripts, which are not to be confused with the notation for exponentiation. We assume that for observers in two inertial frames S and S' the components of a vector A^α and A'^α are related as follows:

$$A'^0 = \frac{A^0 - vA^1}{\sqrt{1-v^2}}, \qquad A'^1 = \frac{A^1 - vA^0}{\sqrt{1-v^2}}, \qquad A'^2 = A^2, \qquad A'^3 = A^3. \tag{7.3}$$

The zeroth component of a four-vector is called the ***timelike component***, and three other components are called ***spacelike components***. We will write spacelike vectors as ordinary three-dimensional vectors (three-vectors): $\mathbf{A} = \{A^1, A^2, A^3\}$. Sometimes they are written as Cartesian projections $\{A_x, A_y, A_z\}$, but are actually assumed to be components with superscripts.

As can easily be seen, the time and the coordinates form a four-vector which is denoted according to its spatial components:

$$x^\alpha = \{t, \mathbf{r}\} = \{t, x, y, z\},$$

and (7.3) are Lorentz transformations (1.13), p. 16. Such four-vectors can be defined for a wide variety of physical quantities.

By analogy with the vector Lorentz transformations (1.18), p. 17, we can write a more general transformation for four-vectors using $\gamma = 1/\sqrt{1-\mathbf{v}^2}$ and $\Gamma = (\gamma - 1)/\mathbf{v}^2$:

$$A'^0 = \gamma\,(A^0 - \mathbf{v}\mathbf{A}), \qquad \mathbf{A}' = \mathbf{A} - \gamma\mathbf{v}A^0 + \Gamma\,\mathbf{v}\,(\mathbf{v}\mathbf{A}). \tag{7.4}$$

The inverse transformation is obtained by replacing $\mathbf{v} \mapsto -\mathbf{v}$.

In addition to the components of four-vectors, we can also define quadruples of quantities with subscripts $A_\alpha = \{A_0, A_1, A_2, A_3\}$, which we call *covector* components. We assume that they are related to the vector components as follows:

$$A_\alpha = \{A_0, A_1, A_2, A_3\} = \{A^0, -A^1, -A^2, -A^3\} = \{A^0, -\mathbf{A}\}.$$

In other words, four-vectors and four-covectors have the same timelike components, while their spacelike components have opposite signs. Components with upper indices are also called *contravariant* components and those with lower indices are called *covariant* components.

Covector components are used to define a number which is *invariant* (the same) in different inertial frames:

$$\mathrm{A}^2 \;=\; A_\alpha A^\alpha \;=\; A_0A^0 + A_1A^1 + A_2A^2 + A_3A^3 \;=\; (A^0)^2 - \mathbf{A}^2 \;=\; \text{inv}. \tag{7.5}$$

We denote this invariant by A^2, where the raised two designates a power (square) rather than an index. To avoid confusion with the indices of the vector components, we will denote four-vectors using bold rather than italic typeface (similarly, bold typeface is used for three-vectors). Repeated upper and lower indices *α imply* summation from 0 to 3, so the summation symbol *is not used.*

Let us verify the invariance of the quantity A^2. To do this, we can write it down in the frame S' in terms of dashed quantities and substitute in the transformation (7.3) to find its value in the frame S. Omitting the unchanged components A^2, A^3, we obtain:

$$\mathrm{A}^2 = (A'^0)^2 - (A'^1)^2 = \frac{(A^0 - vA^1)^2}{1 - v^2} - \frac{(A^1 - vA^0)^2}{1 - v^2} = (A^0)^2 - (A^1)^2.$$

For the four-vector $x^\alpha = \{t, \mathbf{r}\}$ the expression is invariant: $\mathrm{x}^2 = t^2 - \mathbf{r}^2$. The equation $\mathrm{x}^2 = 0$ describes the propagation of a spherical wave as a result of a flash of light emitted at the origin of the coordinate system. The light front of this signal is spherical from the viewpoint of any observer.

Let us define the *scalar product* of two four-vectors A and B:

$$\mathrm{A} \cdot \mathrm{B} = A_\alpha B^\alpha = A^\alpha B_\alpha = A^0 B^0 - \mathbf{AB} = \text{inv}. \tag{7.6}$$

The dot used to denote the product can be omitted, i.e. $\mathrm{A} \cdot \mathrm{B} \equiv \mathrm{AB}$. It is easy to verify that the scalar product is also invariant: $\mathrm{A} \cdot \mathrm{B} = inv$. The invariant

$$\mathrm{A}^2 = \mathrm{A} \cdot \mathrm{A}$$

will be called the *square of the four-vector.*

The square of a three-vector and the scalar product of two three-vectors do not depend on the orientation of the coordinate system. Therefore, they are *invariant* with respect to rotations of a Cartesian system in three-dimensional space. As we saw above, the Lorentz transformations (7.3) can be interpreted as rotations in a pseudo-Euclidean four-dimensional space-time with distance defined by the interval:

$$ds^2 = (dt)^2 - (d\mathbf{r})^2.$$

A four-vector, its square, scalar product and the distance ds^2 are *geometric objects* of four-dimensional space which are not dependent on the orientation of its coordinate axes (the reference frame).

7.4 The metric tensor and bases

We can define the matrix *of the metric tensor*, denoted with the help of two lower or two upper indices, to be:

$$g_{\alpha\beta} = g^{\alpha\beta} = \begin{pmatrix} 1 & 0 & 0 & 0 \\ 0 & -1 & 0 & 0 \\ 0 & 0 & -1 & 0 \\ 0 & 0 & 0 & -1 \end{pmatrix}, \equiv \mathrm{diag}(1, -1, -1, -1). \tag{7.7}$$

Since all of the non-zero elements lie on the diagonal, the matrix can also be denoted using the abbreviation on the right, which can be compared with the concept of "*signature*".

We assume that the metric tensor looks the same in all inertial frames carrying Cartesian coordinates $\{x, y, z\}$. It is easy to verify that:

$$A_\alpha = g_{\alpha\beta}A^\beta, \qquad A^\alpha = g^{\alpha\beta}A_\beta.$$

Indeed, we can write in explicit form the summation rule for repeated (lower and upper) indices. For example,

$$A_1 = g_{1\beta}A^\beta = g_{10}A^0 + g_{11}A^1 + g_{12}A^2 + g_{13}A^3 = g_{11}A^1 = -A^1,$$

where we used that the off-diagonal entries of the metric tensor are zero and that $g_{11} = -1$.

The name "metric tensor" derives from the fact that $g_{\alpha\beta}$ can be used to write the squared distance (*the metric*) in four-dimensional pseudo-Euclidean space:

$$ds^2 = g_{\alpha\beta}\, dx^\alpha dx^\beta = (dx^0)^2 - (dx^1)^2 - (dx^2)^2 - (dx^3)^2 = dt^2 - d\mathbf{r}^2.$$

Similarly, the scalar product of two four-vectors is defined as:

$$\mathrm{A} \cdot \mathrm{B} = g_{\alpha\beta}\, A^\alpha B^\beta = g^{\alpha\beta}\, A_\alpha B_\beta = A^0 B^0 - \mathbf{AB}.$$

The contraction of the tensors $g^{\alpha\beta}$ and $g_{\alpha\beta}$ gives the *Kronecker symbol*:

$$g^{\alpha\gamma} g_{\gamma\beta} = \delta^\alpha_\beta = \begin{cases} 1, & \alpha = \beta, \\ 0, & \alpha \neq \beta \end{cases}, \tag{7.8}$$

which is zero for different indices α and β, and unity for indices which coincide. The same expression can be written as a product of two matrices (7.7) to give the identity matrix.

The metric tensor may seem superfluous, but in arbitrary curvilinear (non-Cartesian) coordinates, the coefficients $g_{\alpha\beta}$ depend on the coordinates, and can be non-diagonal. In this case, the metric tensor is a key concept for describing the geometry of the space.

▷ By analogy with ordinary vector analysis, we can use four four-vectors $\mathrm{e}_0, \mathrm{e}_1, \mathrm{e}_2, \mathrm{e}_3$ to introduce a *basis* for four-dimensional space. The indices are the numbers of the vectors rather than their components (bold typeface). We will, in addition, introduce four more vectors $\mathrm{e}^0, \mathrm{e}^1, \mathrm{e}^2, \mathrm{e}^3$, which we will call the *reciprocal basis*. By definition, the scalar products of the basis four-vectors are:

$$\mathrm{e}^\alpha \cdot \mathrm{e}_\beta = \delta^\alpha_\beta, \qquad \mathrm{e}_\alpha \cdot \mathrm{e}_\beta = g_{\alpha\beta}, \qquad \mathrm{e}^\alpha \cdot \mathrm{e}^\beta = g^{\alpha\beta}. \tag{7.9}$$

A four-vector can be decomposed over the original or the reciprocal basis vectors as:

$$\mathrm{A} = A^\alpha\, \mathrm{e}_\alpha = A_\alpha\, \mathrm{e}^\alpha.$$

In the first case, the expansion coefficients are the components of the four-vector (*contra-variant* components), and in the second case they are the components of the covector (*covariant* components). More precisely, the same four-vector can be decomposed over two different bases (the "initial" e_α and the reciprocal e^α), which give the components with upper and lower indices (p. 347).

Remember that a vector in three-dimensional space can be decomposed over three basis vectors

$$\mathbf{a} = a_x\,\mathbf{i} + a_y\,\mathbf{j} + a_z\,\mathbf{k}.$$

It is important to understand that the vector $\mathbf{a}$ is a geometric object ("directed arrow"). It does not depend on the choice of coordinate axes, although its components (projections) on the Cartesian axes $\{a_x, a_y, a_z\}$ obviously change when the coordinate system is rotated (the basis changes).

The situation is the same in four-dimensional space-time. Any four-vector is a physical object. For example, observed from *any* inertial frame, "x" is the same event. This four-vector can be decomposed over different bases. The bases can correspond to different inertial frames. The expansion coefficients are ordinary numbers with upper or lower indices, with dashes or without them.

In view of (7.9), the scalar product of two four-vectors can be written as:

$$\mathrm{A}\cdot\mathrm{B} = (A^\alpha \mathrm{e}_\alpha)\cdot(B^\beta \mathrm{e}_\beta) = (\mathrm{e}_\alpha\cdot\mathrm{e}_\beta)\,A^\alpha B^\beta = g_{\alpha\beta}\,A^\alpha B^\beta.$$

The products of other vector expansions over basis vectors are written similarly. For example:

$$\mathrm{A}\cdot\mathrm{B} = (A_\alpha \mathrm{e}^\alpha)\cdot(B^\beta \mathrm{e}_\beta) = (\mathrm{e}^\alpha\cdot\mathrm{e}_\beta)\,A_\alpha B^\beta = \delta^\alpha_\beta\,A_\alpha B^\beta = A_\alpha B^\alpha.$$

Note that the first equation indicates that the indices α and β in the expansions of A and B take different values, since these are different sums.

7.5 Examples of four-vectors

We now consider several examples of four-vectors. Let us assume that a particle has velocity $\mathbf{u} = d\mathbf{r}/dt$. The interval between two successive positions of the particle separated by an infinitely small time interval dt is:

$$ds^2 = dt^2 - d\mathbf{r}^2 = dt^2\,(1 - \mathbf{u}^2).$$

The square root of this expression is called the *proper time* of the particle:

$$ds = dt\,\sqrt{1 - \mathbf{u}^2}. \tag{7.10}$$

The proper time is an invariant of the Lorentz transformations. The time interval dt and the particle velocity $\mathbf{u}$ will be different for each observer, but the combination ds will be the same.

Since the four-coordinates $x^\alpha = \{t, \mathbf{r}\}$ transform as a four-vector, and the proper time ds (the interval *along the trajectory*) is invariant, the following *four-velocity* vector can be defined:

$$U^\alpha = \frac{dx^\alpha}{ds} = \{U^0, \mathbf{U}\} = \left\{\frac{1}{\sqrt{1-\mathbf{u}^2}}, \frac{\mathbf{u}}{\sqrt{1-\mathbf{u}^2}}\right\}. \tag{7.11}$$

The components $\{U^0, \mathbf{U}\}$ of the four-velocity transform in the same way as the components of any four-vector (7.3). Thus, for the spatial components, we have:

$$\frac{u'_x}{\sqrt{1-\mathbf{u}'^2}} = \frac{u_x - v}{\sqrt{1-v^2}\,\sqrt{1-\mathbf{u}^2}}, \qquad \frac{u'_y}{\sqrt{1-\mathbf{u}'^2}} = \frac{u_y}{\sqrt{1-\mathbf{u}^2}},$$

and u'_z transforms similarly to u'_y. The zeroth component transforms as:

$$\frac{1}{\sqrt{1-\mathbf{u}'^2}} = \frac{1 - vu_x}{\sqrt{1-v^2}\,\sqrt{1-\mathbf{u}^2}}.$$

Using this relation to replace the term $\sqrt{1-\mathbf{u}'^2}$ in the transformations for the spatial components of the velocity, we obtain the law for transforming velocities (1.20), p. 19, between these two inertial frames:

$$u'_x = \frac{u_x - v}{1 - vu_x}, \qquad u'_y = \frac{u_y\sqrt{1-v^2}}{1 - vu_x}.$$

Note that the square of the four-velocity is unity:

$$\mathrm{U}^2 = U_\alpha U^\alpha = (U^0)^2 - \mathbf{U}^2 = \frac{1}{1-\mathbf{u}^2} - \frac{\mathbf{u}^2}{1-\mathbf{u}^2} = 1.$$

This also follows directly from the definition of the four-velocity and the interval $\mathrm{U}^2 = dx_\alpha\, dx^\alpha/ds^2 = 1$.

▷ Similarly, the four-acceleration is defined as the derivative of four-velocity:

$$A^\alpha = \frac{dU^\alpha}{ds} = \frac{1}{\sqrt{1-\mathbf{u}^2}}\,\frac{d}{dt}\left\{\frac{1}{\sqrt{1-\mathbf{u}^2}}, \frac{\mathbf{u}}{\sqrt{1-\mathbf{u}^2}}\right\}.$$

Introducing the usual three-dimensional acceleration $\mathbf{a} = d\mathbf{u}/dt$ and differentiating component-wise we obtain:

$$A^\alpha = \frac{dU^\alpha}{ds} = \left\{\frac{\mathbf{ua}}{(1-\mathbf{u}^2)^2}, \frac{\mathbf{a} + [\mathbf{u}\times[\mathbf{u}\times\mathbf{a}]]}{(1-\mathbf{u}^2)^2}\right\}. \tag{7.12}$$

Using (7.4), we can write the transformation laws for acceleration, even though this is more easily acheived by differentiating the vector transformation for the velocity directly. We can verify that the scalar product of the four-velocity and the four-acceleration is zero: $\mathrm{U}\cdot\mathrm{A} = 0$. This relation can also be obtained by differentiating $\mathrm{U}^2 = 1$ with respect to s:

$$\frac{dU^2}{ds} = \frac{d(U^\alpha U_\alpha)}{ds} = \frac{dU^\alpha}{ds}U_\alpha + U^\alpha\frac{dU_\alpha}{ds} = A^\alpha U_\alpha + U^\alpha A_\alpha = 2\,A^\alpha U_\alpha = 0,$$

where we took the derivative of the product and used that $A^\alpha U_\alpha = A_\alpha U^\alpha$ (7.6).

▷ The equations for the Doppler and aberration effects can also be written using covariant notation. To do this, we define the *wave four-vector*:

$$k^\alpha = \{\omega, \mathbf{k}\} = \{\omega, \omega\,\mathbf{n}\}, \tag{7.13}$$

where $\omega = 2\pi\nu$ is the *angular frequency* (ν is the ordinary frequency) and $\mathbf{k}$ is the wave vector directed along the wave propagation and equal to $|\mathbf{k}| = 2\pi/\lambda$, where λ is the wavelength. Since $\nu\lambda = c = 1$, we introduce a unit vector $\mathbf{n}^2 = 1$ in the second equation and express λ in terms of ν. Note that

$$\mathrm{k}^2 = k_\alpha k^\alpha = \omega^2(1 - \mathbf{n}^2) = 0.$$

The transformation (7.4) of the zeroth component gives the Doppler effect (p. 59):

$$\frac{\omega}{\omega'} = \frac{\sqrt{1-\mathbf{v}^2}}{1-\mathbf{vn}}.$$

This expression differs from (3.6), p. 61, by the minus sign in the denominator. This happens because of the different meanings of the unit vector $\mathbf{n}$. In (3.6) $\mathbf{n}$ was the direction to the source (*from* the observer), while in the above transformation it is the direction *to* the observer. This is the reason for the change of sign. The aberration effect (p. 77) is obtained from the transformations of the spatial components k^α.

7.6 Matrix transformations

Let us consider three inertial reference frames S, S', and S'', and let their observers measure the coordinates and the time of some event $\{t, x, y\}$, $\{t', x', y'\}$, and $\{t'', x'', y''\}$, respectively. As usual, let S' move to the right relative to S with velocity v along the x-axis, the coordinate axes of these two frames being parallel. Let S'' move relative to S' with velocity v' upwards along the y'-axis. We can write down the Lorentz transformations between the pairs of reference frames:

$$\left\{\begin{array}{lcl} t' & = & \gamma\, t - v\gamma\, x \\ x' & = & \gamma\, x - v\gamma\, t \\ y' & = & y, \end{array}\right. \qquad \left\{\begin{array}{lcl} t'' & = & \gamma'\, t' - v'\gamma'\, y' \\ x'' & = & x' \\ y'' & = & \gamma'\, y' - v'\gamma'\, t'. \end{array}\right.$$

Substituting the first transformation into the second one, we obtain the relationship between the coordinates and the time for observers in S and S'' (p. 70). This substitution can be made in the matrix form using the following transformations:

$$\begin{pmatrix} t' \\ x' \\ y' \end{pmatrix} = \begin{pmatrix} \gamma & -v\gamma & 0 \\ -v\gamma & \gamma & 0 \\ 0 & 0 & 1 \end{pmatrix} \cdot \begin{pmatrix} t \\ x \\ y \end{pmatrix}, \qquad \begin{pmatrix} t'' \\ x'' \\ y'' \end{pmatrix} = \begin{pmatrix} \gamma' & 0 & -v'\gamma' \\ 0 & 1 & 0 \\ -v'\gamma & 0 & \gamma' \end{pmatrix} \cdot \begin{pmatrix} t' \\ x' \\ y' \end{pmatrix}.$$

The *composition of transformations* is now reduced to matrix multiplication:

$$\begin{pmatrix} t'' \\ x'' \\ y'' \end{pmatrix} = \begin{pmatrix} \gamma' & 0 & -v'\gamma' \\ 0 & 1 & 0 \\ -v'\gamma' & 0 & \gamma' \end{pmatrix} \cdot \begin{pmatrix} \gamma & -v\gamma & 0 \\ -v\gamma & \gamma & 0 \\ 0 & 0 & 1 \end{pmatrix} \cdot \begin{pmatrix} t \\ x \\ y \end{pmatrix}.$$

Matrix multiplication is associative, so we can first multiply the square matrices representing the two transformations, and then multiply the resulting matrix by a column vector containing the coordinates and the time:

$$\begin{pmatrix} t'' \\ x'' \\ y'' \end{pmatrix} = \begin{pmatrix} \gamma\gamma' & -v\gamma\gamma' & -v'\gamma' \\ -v\gamma & \gamma & 0 \\ -v'\gamma'\gamma & vv'\gamma\gamma' & \gamma' \end{pmatrix} \cdot \begin{pmatrix} t \\ x \\ y \end{pmatrix}.$$

The resulting matrix above is obtained by the "crowbar" rule (p. 336): the sum of the row elements (the crowbar) of the first matrix multiplied by the elements of the column ("wall") of the second matrix is written in the place at which the hole is "breached". For example, for the upper left element of the matrix we have:

$$\gamma' \cdot \gamma + 0 \cdot (-v\gamma) + (-v'\gamma') \cdot 0 = \gamma\gamma',$$

the others are obtained similarly.

Note that even though both of the original matrices were symmetric (about the diagonal), the matrix of the resulting transformation is *asymmetric*.

We defined the components of a vector in four-dimensional space to be a quadruple

$$A^\alpha = (A^0, A^1, A^2, A^3)$$

of numbers which change together with the frame of reference according to the Lorentz transformations. In the general case, these transformations can be written as follows:

$$A'^\alpha = \Lambda^\alpha{}_\beta A^\beta = \Lambda^\alpha{}_0 A^0 + \Lambda^\alpha{}_1 A^1 + \Lambda^\alpha{}_2 A^2 + \Lambda^\alpha{}_3 A^3. \tag{7.14}$$

If the relative velocity of the reference frames has an arbitrary direction, from the vector transformations (7.4) we obtain the following matrix:

$$\Lambda^\alpha{}_\beta = \begin{pmatrix} \Lambda^0{}_0 & \Lambda^0{}_1 & \Lambda^0{}_2 & \Lambda^0{}_3 \\ \Lambda^1{}_0 & \Lambda^1{}_1 & \Lambda^1{}_2 & \Lambda^1{}_3 \\ \Lambda^2{}_0 & \Lambda^2{}_1 & \Lambda^2{}_2 & \Lambda^2{}_3 \\ \Lambda^3{}_0 & \Lambda^3{}_1 & \Lambda^3{}_2 & \Lambda^3{}_3 \end{pmatrix} = \begin{pmatrix} \gamma & -v_x\gamma & -v_y\gamma & -v_z\gamma \\ -v_x\gamma & 1+\Gamma v_x^2 & \Gamma v_x v_y & \Gamma v_x v_z \\ -v_y\gamma & \Gamma v_y v_x & 1+\Gamma v_y^2 & \Gamma v_y v_z \\ -v_z\gamma & \Gamma v_z v_x & \Gamma v_z v_y & 1+\Gamma v_z^2 \end{pmatrix}.$$

The matrix will look even more complicated if the coordinate axes of the reference frames also rotate relative to each other. Note that the indices α and β in the matrix $\Lambda^\alpha{}_\beta$ not only have different heights, but they are also shifted horizontally relative to each other. The first index α enumerates the rows of the matrix, and the second β enumerates its columns. The situation is similar with indices for ordinary matrices, but here they are on the same level (usually both below). For the scalar product of vectors $A^\alpha B_\alpha$ the covariant summation rule is used for repeated indices placed at different levels. We will use the same rule for matrices as well. Therefore, the order of the indices in $\Lambda^\alpha{}_\beta$ is important in both directions.

A sequence of two Lorentz transformations with different matrices

$$A''^\alpha = \Lambda'^\alpha_{\ \mu} A'^\mu, \qquad A'^\mu = \Lambda^\mu{}_\beta A^\beta, \qquad A''^\alpha = \Lambda''^\alpha_{\ \beta} A^\beta$$

can be substituted into each other, so that:

$$\Lambda''^\alpha_{\ \beta} = \Lambda'^\alpha_{\ \mu} \Lambda^\mu{}_\beta, \qquad \Lambda'' = \Lambda'\Lambda.$$

On the left-hand side, there is a matrix product written in the explicit index form. The summation over μ is according to the "crowbar" rule, since the lower right index μ of the first matrix counts the elements of the row with index α, and the upper left index μ of the second matrix counts the elements of the column with index β. On the right, the same expression is written in matrix form without indices.

7.7 Properties of the Lorentz matrix

We can define another matrix form for the Lorentz transformations by changing the "height level of the indices" via contraction with the metric tensor:

$$\Lambda_\alpha{}^\beta = g_{\alpha\mu} \Lambda^\mu{}_\nu g^{\nu\beta}. \tag{7.15}$$

The subscript α in the matrix $\Lambda_\alpha{}^\beta$ is on the left and numbers the rows, while the index β numbers the columns. Despite the letter Λ being the same, the matrices $\Lambda_\alpha{}^\beta$ and $\Lambda^\mu{}_\nu$ are different, so it is necessary to pay attention to the order and the height level of the indices. Furthermore, in matrix (non-index) expressions we denote the matrix $\Lambda^\mu{}_\nu$ by Λ, and mark the matrix $\Lambda_\mu{}^\nu$ with a tilde $\tilde{\Lambda}$.

To obtain the explicit form of the matrix $\tilde{\Lambda}$, we multiply Λ on the left and on the right by the matrix of the metric tensor $\mathbf{g}$, i.e., $\tilde{\Lambda} = \mathbf{g}\Lambda\mathbf{g}$. Thus, for motion along the x-axis in two dimensions we have:

$$\tilde{\Lambda} = \begin{pmatrix} 1 & 0 & 0 \\ 0 & -1 & 0 \\ 0 & 0 & -1 \end{pmatrix} \begin{pmatrix} \gamma & -v\gamma & 0 \\ -v\gamma & \gamma & 0 \\ 0 & 0 & 1 \end{pmatrix} \begin{pmatrix} 1 & 0 & 0 \\ 0 & -1 & 0 \\ 0 & 0 & -1 \end{pmatrix} = \begin{pmatrix} \gamma & v\gamma & 0 \\ v\gamma & \gamma & 0 \\ 0 & 0 & 1 \end{pmatrix}.$$

In inertial frames with Cartesian coordinates, metric tensors with lower and upper indices are diagonal and are *the same*. As for other matrices, their first index enumerates the rows, and the second enumerates the columns. In contrast to the matrices $\tilde{\Lambda}$ and Λ, both indices of the matrix of the metric tensor $\mathbf{g}$ are at the same height level (above or below).

Using matrix $\tilde{\Lambda}$, we can write the Lorentz transformation of a covector:

$$A'_\alpha = g_{\alpha\mu} A'^\mu = g_{\alpha\mu} \Lambda^\mu{}_\nu A^\nu = g_{\alpha\mu} \Lambda^\mu{}_\nu g^{\nu\beta} A_\beta,$$

so

$$A'_\alpha = \Lambda_\alpha{}^\beta A_\beta. \tag{7.16}$$

Note that the summation index β, which enumerates the columns of the matrix $\tilde{\Lambda}$, "approaches" the four-vector A_β. Such detailed calculations are rarely carried out with metric tensors. More often, for two summation indices, the relative height level is simply changed:

$$A_\alpha B^\alpha = A^\alpha B_\alpha, \qquad g_{\alpha\beta} A^\alpha B^\beta = g^{\alpha\beta} A_\alpha B_\beta.$$

Even if the index in the equation is not a summation index, it can be raised or lowered simultaneously on the right- or left-hand side side of the expression:

$$g_{\alpha\beta} A^\beta = A_\alpha \quad \Leftrightarrow \quad g^{\alpha\beta} A_\beta = A^\alpha,$$

$$A'^\alpha = \Lambda^\alpha{}_\beta A^\beta \quad \Leftrightarrow \quad A'_\alpha = \Lambda_\alpha{}^\beta A_\beta,$$

which significantly accelerates the derivation of the corresponding relationships.

▷ We now consider one important property of the Lorentz transformation matrices. By definition, the scalar product of two four-vectors is the same in all reference frames:

$$\mathrm{A} \cdot \mathrm{B} = g_{\alpha\beta} A'^\alpha B'^\beta = \underline{g_{\alpha\beta} \Lambda^\alpha{}_\mu \Lambda^\beta{}_\nu} A^\mu B^\nu = \underline{g_{\mu\nu}} A^\mu B^\nu = inv.$$

These relationships are true (due to the arbitrariness of the vectors) if the Lorentz matrix satisfies the equation

$$g_{\alpha\beta} \Lambda^\alpha{}_\mu \Lambda^\beta{}_\nu = g_{\mu\nu}, \tag{7.17}$$

which is called the ***orthogonality condition***. Taking into account (7.8), and having contracted this relationship with $g^{\gamma\mu}$ over the index μ, we obtain:

$$g^{\gamma\mu} g_{\alpha\beta} \Lambda^\alpha{}_\mu \Lambda^\beta{}_\nu = \Lambda_\beta{}^\gamma \Lambda^\beta{}_\nu = (\tilde{\Lambda}^T)^\gamma{}_\beta \Lambda^\beta{}_\nu = \delta^\gamma_\nu. \tag{7.18}$$

The word "***contraction***" means multiplication of one index expression by a second and summation of this product with respect to some indices. In the first equation of (7.18), the contraction with the metric tensors gives the matrix $\tilde{\Lambda}$, and the second equation contains the transposed matrix denoted with a ^{T}: $(\tilde{\Lambda}^T)^\gamma{}_\beta = \tilde{\Lambda}_\beta{}^\gamma$. The transpose is obtained from the original matrix $\tilde{\Lambda}_\beta{}^\gamma$ by switching its rows and columns. This is done to obtain a matrix product given by summation over the index β. We must emphasize that

$$(\tilde{\Lambda}^T)^\alpha{}_\beta \neq \Lambda^\alpha{}_\beta.$$

Denote now the Kronecker symbol δ^γ_ν using the identity matrix **1**. Then the orthogonality condition (7.18) can be written without indices:

$$\tilde{\Lambda}^T \Lambda = \mathbf{1}.$$

The determinant of the product of two matrices is equal to the product of their determinants. Also, the transposition does not change the determinant. Therefore:

$$\det(\tilde{\Lambda}^T\Lambda) = \det(\mathbf{g}^T\Lambda^T\mathbf{g}^T\Lambda) = (\det\mathbf{g})^2\,(\det\Lambda)^2 = 1.$$

It follows rom the definition (7.7) that $\det\mathbf{g} = -1$, so $(\det\Lambda)^2 = 1$. Direct calculation of the determinant of the matrix Λ from the beginning of the section shows that the positive root should be extracted:

$$\det\Lambda = \det\tilde{\Lambda} = 1.$$

The determinant of $\tilde{\Lambda}$ is obtained from the definition $\tilde{\Lambda} = \mathbf{g}\Lambda\mathbf{g}$. Hence, the matrices of the Lorentz transformations for both vectors (Λ) and covectors ($\tilde{\Lambda}$) have unit determinants and are orthogonal.

▷ Since the height levels of indices can be changed, the orthogonality condition (7.17) can also be rewritten in equivalent forms:

$$\Lambda^{\alpha}{}_{\mu}\,\Lambda_{\alpha\nu} = g_{\mu\nu}, \qquad \Lambda^{\alpha}{}_{\mu}\,\Lambda_{\alpha}{}^{\nu} = \delta^{\nu}_{\mu}.$$

Note once again that for the *different* matrices

$$\Lambda^{\alpha}{}_{\beta}, \qquad \Lambda_{\alpha\beta}, \qquad \Lambda_{\alpha}{}^{\beta},$$

the same letter is used, because they are obtained from the initial $\Lambda^{\alpha}{}_{\beta}$ by a fixed procedure (contraction with a metric tensor). Under this agreement, the Kronecker symbol δ^{α}_{β} can also be denoted by g^{α}_{β}, since the condition (7.8), p. 176, is equivalent to increasing or decreasing the index of one of the metric tensors. Due to the symmetry of the Kronecker symbol, there is in this case no need to pay attention to the order of the indices in the horizontal direction:

$$\delta^{\alpha}{}_{\beta} = \delta_{\beta}{}^{\alpha} = \delta^{\alpha}_{\beta}.$$

Sometimes, the indices are not shifted horizontally, but are written one under another even for Lorentz matrices Λ^{α}_{β} which are not symmetric. It is assumed that the superscript numbers the rows, and the subscript numbers the columns. We will not use this convention, and will always pay attention to the horizontal ordering of indices for asymmetric matrices.

▷ In the orthogonality condition (7.17), the Lorentz matrices are contracted with respect to the first index. Although these matrices are not in general symmetric, an analogous orthogonality condition holds when the contraction is done over the second indices. Indeed, we can perform a contraction of the Lorentz transformation

$$A'^{\mu} = \Lambda^{\mu}{}_{\alpha}\,A^{\alpha}$$

with the matrix $\Lambda_{\mu}{}^{\beta}$. Using (7.17), we easily obtain the expression for the inverse transformation:

$$A^{\beta} = A'^{\mu}\,\Lambda_{\mu}{}^{\beta}. \tag{7.19}$$

The same condition can also be written for covector components by simultaneously changing the height levels of the indices:

$$A_\beta = A'_\mu \Lambda^\mu{}_\beta. \tag{7.20}$$

We now substitute (7.19) into

$$g_{\alpha\beta} A^\alpha B^\beta = g_{\mu\nu} A'^\mu B'^\nu,$$

to replace the primed quantities. Repeating the arguments used for (7.17), we obtain another form of the orthogonality condition:

$$g_{\alpha\beta} \Lambda_\mu{}^\alpha \Lambda_\nu{}^\beta = g_{\mu\nu}, \tag{7.21}$$

where the Lorentz matrices are contracted along their second indices.

▷ When a three-dimensional coordinate system rotates, the basis vectors directed along the axes also rotate (change). The basis vectors of a four-dimensional space change in exactly the same way as when the reference frame changes (rotates in the four-space). Such transformations are associated with "inverse" matrix multiplication (of, say, a row vector by a matrix):

$$\mathrm{e}_\alpha = \mathrm{e}'_\beta \Lambda^\beta{}_\alpha, \qquad \mathrm{e}^\alpha = \mathrm{e}'^\beta \Lambda_\beta{}^\alpha, \tag{7.22}$$

where the tilde sign is omitted for the matrix $\Lambda_\beta{}^\alpha$. Note that the index of a basis vector is its number, and does not refer to a component, so there are four-vectors

$$\mathrm{e}_0, \;\; \mathrm{e}_1, \;\; \mathrm{e}_2, \;\; \mathrm{e}_3$$

on the left-hand side of the transform, and on its right-hand side (to the left of the matrix). To obtain th transformations (7.22), the vector must be decomposed over two bases corresponding to two inertial frames,

$$\mathrm{A} = A^\alpha \mathrm{e}_\alpha = A'^\alpha \mathrm{e}'_\alpha,$$

and the transformation law for the vector components must be used:

$$A'^\alpha = \Lambda^\alpha{}_\beta A^\beta.$$

We suggest doing this as an exercise. Remember that a four-vector is a physical object that has different projections on different bases.

Contracting (7.22) with Lorentz matrices and taking into account the orthogonality conditions, we obtain direct transformations for the basis vectors:

$$\mathrm{e}'_\alpha = \Lambda_\alpha{}^\beta \mathrm{e}_\beta, \qquad \mathrm{e}'^\alpha = \Lambda^\alpha{}_\beta \mathrm{e}^\beta. \tag{7.23}$$

For example, for the second relation in (7.22) we have:

$$\mathrm{e}^\alpha \Lambda^\mu{}_\alpha = \mathrm{e}'^\beta \Lambda_\beta{}^\alpha \Lambda^\mu{}_\alpha = \mathrm{e}'^\beta \delta^\mu_\beta = \mathrm{e}'^\mu.$$

Note that the basis vectors e_α in (7.23) are transformed like covariant components A_α, and the vectors of the reciprocal basis e^α are transformed like contravariant components A^α.

Note also the relationship

$$e'^\beta \cdot e_\alpha = \Lambda^\beta{}_\alpha,$$

which is easily obtained from (7.22) by multiplying it by the corresponding basis vector and using the conditions (7.9). Therefore, the Lorentz matrix that connects two inertial frames can be obtained if basis four-vectors are known in each of these frames. Equivalently, a set of four linearly independent four-vectors determines a unique reference frame.

7.8 Four-tensors

Assume there are two four-vectors A and B. Their components can be used to make four different products with two indices: $A^\alpha B^\beta$, $A_\alpha B_\beta$, $A^\alpha B_\beta$, $A_\alpha B^\beta$. These products are transformed by means of two Lorentz matrices (7.14), (7.16):

$$A'^\alpha B'^\beta = \Lambda^\alpha{}_\mu \Lambda^\beta{}_\nu A^\mu B^\nu, \qquad A'_\alpha B'_\beta = \Lambda_\alpha{}^\mu \Lambda_\beta{}^\nu A_\mu B_\nu,$$

etc., (summing over repeated indices). A quantity with $n+m$ indices, n at the top and m at the bottom, which transforms as a product of n vector and m covector components, will be called a ***tensor*** of type (n, m). The number $n+m$ is called the ***order*** of the tensor. For example, for a tensor $T_{\alpha\beta}{}^\gamma$ of type $(1, 2)$:

$$T'_{\alpha\beta}{}^\gamma = \Lambda^\gamma{}_\sigma \Lambda_\alpha{}^\mu \Lambda_\beta{}^\nu T_{\mu\nu}{}^\sigma.$$

Since the orthogonality condition is fulfilled for Lorentz matrices, the metric coefficients $g_{\alpha\beta}$, $g^{\alpha\beta}$ are tensors:

$$g'^{\alpha\beta} = \Lambda^\alpha{}_\mu \Lambda^\beta{}_\nu g^{\mu\nu} = g^{\alpha\beta}.$$

Note that the components $g^{\alpha\beta}$ or $g_{\alpha\beta}$ do not change only in Cartesian coordinates on a flat space (see Chapter 9). Similarly, due to orthogonality, the Kronecker symbol δ^α_β is also a tensor:

$$\delta'^\alpha_\beta = \Lambda^\alpha{}_\mu \Lambda_\beta{}^\nu \delta^\mu_\nu = \Lambda^\alpha{}_\mu \Lambda_\beta{}^\mu = \delta^\alpha_\beta.$$

Using $g^{\alpha\beta}$ or $g_{\alpha\beta}$, we can define a ***new*** tensor for any given tensor by raising or lowering its indices. Although the resulting tensor will be a different tensor, it is commonly denoted with the same letter. In this case, it is necessary to pay attention to the horizontal ordering of the indices:

$$T_{\alpha\beta\gamma} = g_{\gamma\mu} T_{\alpha\beta}{}^\mu, \qquad T_\alpha{}^\beta{}_\gamma = g^{\beta\mu} T_{\alpha\mu\gamma}.$$

All inertial frames are equivalent. Therefore, the equations describing a physical law should have the same form in all frames. This is why such equations are written in tensor notation. For example, as we shall see in the next volume, the motion of a particle in an electromagnetic field is described by the equation

$$m\,\frac{dU^\alpha}{ds} = q\,F^{\alpha\beta}\,U_\beta,$$

where m and q are the invariant mass and charge of the particle and ds is the invariant interval. This equation includes tensor quantities such as the four-velocity vector of the particle U^α and the tensor of the electromagnetic field $F^{\alpha\beta}$. Since they are transformed with the help of the same matrix, the equation will appear exactly the same in any other reference frame.

▷ A vector has four components, a tensor of second order (with two indices) sixteen components. The number of components in a tensor increases rapidly with the number of indices. A tensor with n indices has 4^n components. Therefore, tensors which have some symmetry with respect to the permutation of their indices are of particular importance. Tensors of type (0, 2) can be *symmetric* or *antisymmetric*:

$$S_{\alpha\beta} = S_{\beta\alpha}, \qquad A_{\alpha\beta} = -A_{\beta\alpha}.$$

The metric tensor $g_{\alpha\beta}$ is symmetric. As an exercise (⋖ H30), we suggest finding the Lorentz transformations for an arbitrary tensor $T^{\alpha\beta}$ and for the case when it is antisymmetric.

The contraction of a symmetric tensor against an antisymmetric tensor is zero:

$$S_{\alpha\beta}\,A^{\alpha\beta} = -S_{\beta\alpha}\,A^{\beta\alpha} = -S_{\alpha\beta}\,A^{\alpha\beta} = 0.$$

In the first equation, the indices are rearranged using symmetry, and in the second equation they are renamed ($\alpha \mapsto \beta$, $\beta \mapsto \alpha$), because they are summation indices. An expression that is equal to itself with opposite sign can only be zero.

Absolutely antisymmetric tensors are of special importance in the theory of relativity. The word "absolute" means that they change their sign under permutation of *any* two of their indices. Their properties will be discussed in the next chapter in more detail.

By the definition of a tensor, the components of a four-vector with upper indices A^α are tensors of type (1,0), and the corresponding covector (the components of the vector with lower indices A_α) is a tensor of type (0, 1).

A quantity that has the same value in all reference frames is called a *scalar* and is a zero-rank tensor. Note that having the same value does not mean having the same functional form. As an example, let $\varphi(t, \mathbf{r})$ be a *scalar function*. This means that each point of space and time is associated with some number φ. This number is "attached" to the point and, *by definition*, does not depend on the choice of reference frame, i.e.

it is the same for all observers. Nevertheless, the form of the function will be different for different observers. For example, if

$$\varphi = t + x$$

in one frame, it will be

$$\varphi' = \gamma(1 + v)(t' + x')$$

in another frame. But the values of the functions for a given event (a point in space-time) will be the same $\varphi' = \varphi$. Therefore, the prime is usually not used for scalar functions. The outputs of vector-valued functions $A^{\alpha}(t, \mathbf{r})$ or tensor-valued functions $T^{\alpha\beta}(t, \mathbf{r})$ are also "attached" to a specific point of space-time. However, when turning to another frame, their values change, since they are multiplied by corresponding matrices Λ.

8 Dynamics in covariant notation

In this chapter we continue our consideration of covariant formalism. We will now apply it to dynamical problems. Energy and momentum are the components of a momentum four-vector whose square is invariant and is equal to the mass of the particle. Similarly, a force applied to a small ("test") particle can be written in covariant notation. When describing particle collisions and decay, conservation laws can be studied much more easily with the help of covariant notation. We will introduce invariants s, t, u which are widely used in the physics of elementary particles, and consider a graphical representation of the permitted energy ranges of various four-particle interactions. We will also consider antisymmetric tensors and covariant notation for angular momentum and classical spin.

https://doi.org/10.1515/9783110515886-008

8.1 Momentum and force

Any physical quantity can be expressed in terms of four-vectors. Once such a four-vector is written, the transformation of a physical quantity between two inertial reference frames is easily found from the relations (7.4), p. 174.

For example, multiplication of the velocity four-vector (7.11), p. 178, by the mass gives the momentum four-vector of a particle (or simply four-momentum):

$$p^\alpha = m\,U^\alpha = m\,\frac{dx^\alpha}{ds} = \left\{\frac{m}{\sqrt{1-\mathbf{u}^2}},\ \frac{m\mathbf{u}}{\sqrt{1-\mathbf{u}^2}}\right\} = \{E,\ \mathbf{p}\}, \tag{8.1}$$

where we utilised the fact that the invariant interval along the particle trajectory is

$$ds = dt\,\sqrt{1-\mathbf{u}^2},$$

and $dx^\alpha = \{dt,\ d\mathbf{r}\}$. The mass of the particle is invariant and its square coincides with the square of the four-momentum:

$$\mathrm{p}^2 = p^\alpha p_\alpha = E^2 - \mathbf{p}^2 = m^2.$$

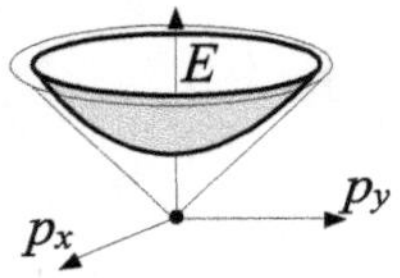

The same expression immediately follows from the relation

$$\mathrm{U}^2 = 1$$

for the four-velocity. The equation $\mathrm{p}^2 = m^2$ describes a hyperboloid in the four-dimensional momentum space with axes E, p_x, p_y, p_z. For "ordinary" particles we have

$$m^2 > 0, \qquad E > 0,$$

so their energy and momentum are on the upper cup of the hyperboloid, which is called the *mass shell*. Zero mass particles "lie" on the cone.

▷ Consider the vector transformation between two frames for energy and momentum (see also p. 101):

$$E' = \gamma\,(E - \mathbf{v}\mathbf{p}), \qquad \mathbf{p}' = \mathbf{p} - \gamma\mathbf{v}E + \Gamma\,\mathbf{v}\,(\mathbf{v}\mathbf{p}). \tag{8.2}$$

The photon is a massless particle, so we can define a wave four-vector in accordance with Planck's law:

$$p^\alpha = \hbar\,k^\alpha, \qquad k^\alpha = \{\omega,\ \mathbf{k}\} = \{\omega,\ \omega\,\mathbf{n}\}, \tag{8.3}$$

where $\mathbf{n}$ is a unit vector in the direction of the photon motion, and $\omega = 2\pi\nu$ is its angular frequency. Using these relationships and the transformation law for four-momenta (8.2), we can again find the equations describing the Doppler effect and aberration (p. 178).

▷ The four-force vector is obtained by differentiating the four-momentum along the invariant interval:

$$f^\alpha = \frac{dp^\alpha}{ds}. \tag{8.4}$$

Since the interval s (proper time, p. 178) along the particle trajectory is $ds = \sqrt{1-\mathbf{u}^2}\,dt$, we have

$$f^\alpha = \frac{dp^\alpha/dt}{\sqrt{1-\mathbf{u}^2}} = \left\{\frac{\mathbf{uF}}{\sqrt{1-\mathbf{u}^2}}, \frac{\mathbf{F}}{\sqrt{1-\mathbf{u}^2}}\right\},$$

where we have substituted in the expressions (p. 118)

$$\frac{dE}{dt} = \mathbf{uF}, \qquad \frac{d\mathbf{p}}{dt} = \mathbf{F}.$$

The values f^α are the components of a four-vector (which is transformed according to (7.4), p. 174), since p^α is a four-vector and ds is an invariant of the transformation.

▷ The scalar product of a four-momentum and a four-force is zero:

$$\mathrm{p}\cdot\mathrm{f} = p_\alpha f^\alpha = p^0 f^0 - \mathbf{pf} = \frac{E\,(\mathbf{uF}) - \mathbf{pF}}{\sqrt{1-\mathbf{u}^2}} = 0,$$

where $\mathbf{p} = \mathbf{u}E$ has been substituted into the last equation. This relationship can also be proved by differentiating the mass. Since this is constant, we have:

$$0 = \frac{dm^2}{ds} = \frac{d(p^\alpha p_\alpha)}{ds} = \frac{dp^\alpha}{ds}p_\alpha + p^\alpha\frac{dp_\alpha}{ds} = 2\,p_\alpha\frac{dp^\alpha}{ds} = 2\,p_\alpha f^\alpha.$$

The derivative of the square $p^\alpha p_\alpha$ is calculated as the derivative of a product. Then, the following property of the scalar product of any two vectors is used: $A^\alpha B_\alpha = A_\alpha B^\alpha$ (p. 175). Note that similar reasoning was used to prove the relationship $\mathrm{U}\cdot\mathrm{A} = 0$.

▷ The square of the four-force is invariant, so the quantity

$$\mathrm{f}^2 = \frac{(\mathbf{uF})^2 - \mathbf{F}^2}{1-\mathbf{u}^2} = inv$$

has the same meaning for all inertial observers.

▷ Using the definition of four-acceleration (p. 178),

$$A^\alpha = \frac{dU^\alpha}{ds},$$

the expression for the four-force can be written in the quasi-Newtonian form

$$f^\alpha = m\,A^\alpha. \tag{8.5}$$

Naturally, this relation is written for four-vectors and is not equivalent to the Newtonian law $\mathbf{F} = m\mathbf{a}$ for ordinary three-vectors.

8.2 Conservation laws

Covariant notation can be used to describe interactions between particles in a very compact way. For two particles with four-momenta

$$\mathrm{p}_1 = \{E_1, \mathbf{p}_1\} \quad \text{and} \quad \mathrm{p}_2 = \{E_2, \mathbf{p}_2\}$$

and masses m_1 and m_2, the following relations are true:

$$\mathrm{p}_1^2 = m_1^2, \qquad \mathrm{p}_2^2 = m_2^2, \qquad \mathrm{p}_1\,\mathrm{p}_2 = E_1E_2 - \mathbf{p}_1\mathbf{p}_2. \tag{8.6}$$

When a squared four-momentum occurs, it can always be immediately replaced with the squared particle mass. The latter relationship is the general definition of the scalar product of four-vectors (p. 175). The squares of four-vectors, or their scalar products, are invariant, so they can be written in any reference frame. The value obtained coincides numerically with the value of this invariant in any other frame. If some equation connects these two invariants, we can write its left-hand side in one coordinate system and its right-hand side in another system. As a result, the relationship between the values measured by observers in different frames can be obtained.

▷ Consider again the reaction in which a particle with four-momentum p decays into two particles with four-momenta p_1 and p_2 (p. 106). In this case, the law of conservation of energy and momentum has the form:

$$\mathrm{p} = \mathrm{p}_1 + \mathrm{p}_2. \tag{8.7}$$

For the zeroth components of the four-vectors this equation gives the law of conservation of energy, and for the space-like components it gives the law of conservation of momentum.

Carrying the four-momentum p_1 to the left and squaring the expression, we have

$$(\mathrm{p} - \mathrm{p}_1)^2 = m^2 + m_1^2 - 2\,\mathrm{p}\,\mathrm{p}_1 = \mathrm{p}_2^2 = m_2^2,$$

where the square is expanded according to the usual algebraic formula.

Now we can write this expression in a specific frame of reference. Let the initial particle be at rest $\mathrm{p} = \{m, \mathbf{0}\}$. In this case, the scalar product is

$$\mathrm{p}\,\mathrm{p}_1 = mE_1 - \mathbf{0}\,\mathbf{p}_1 = mE_1,$$

and

$$m^2 + m_1^2 - 2mE_1 = m_2^2.$$

As a result, the energy E_1 depends on the particle masses:

$$E_1 = \frac{m^2 + m_1^2 - m_2^2}{2m}. \tag{8.8}$$

The energy E_2 is found similarly. To do this, we have to carry the four-momentum p_2 to the left-hand side of the conservation law (or interchange the indices).

▷ Consider now the reaction of an *elastic collision* between two particles with four-momenta p_1 and p_2. Their masses do not change after the collision, and their four-momenta become p'_1 and p'_2. In this case, the law of conservation of energy and momentum has the form:

$$\mathrm{p}_1 + \mathrm{p}_2 = \mathrm{p}'_1 + \mathrm{p}'_2. \tag{8.9}$$

Let us remove the four-momentum of one of the final particles by carrying p'_1 to the left and squaring the whole expression:

$$(\mathrm{p}_1 + \mathrm{p}_2 - \mathrm{p}'_1)^2 = \mathrm{p}'^2_2 = m_2^2.$$

After some standard algebraic operations, we arrive at

$$m_1^2 + m_2^2 + m_1^2 + 2\,\mathrm{p}_1\,\mathrm{p}_2 - 2\,\mathrm{p}_1\,\mathrm{p}'_1 - 2\,\mathrm{p}_2\,\mathrm{p}'_1 = m_2^2.$$

Let the *laboratory* frame be such that the second particle is stationary in it, $\mathrm{p}_2 = \{m_2, \mathbf{0}\}$. Then,

$$m_1^2 = \mathrm{p}_1\,\mathrm{p}'_1 + \mathrm{p}_2\,(\mathrm{p}'_1 - \mathrm{p}_1) = E_1E'_1 - \mathbf{p}_1\mathbf{p}'_1 + m_2\,(E'_1 - E_1).$$

We can now express the scattering angle $\mathbf{p}_1\mathbf{p}'_1 = p_1p'_1\,\cos\theta$ in terms of the energy of the incident particle $E_1 = \sqrt{p_1^2 + m_1^2}$ and its energy after the collision $E'_1 = \sqrt{p'^2_1 + m_1^2}$.

▷ Let us now find the relationship between the momentum $|\tilde{\mathbf{p}}_1| = |\tilde{\mathbf{p}}_2| = \tilde{p}$, the scattering angle χ in the *centre of mass frame*, and the energies of the particles in the laboratory frame. In this setting we can write the conservation law as $\mathrm{p}'_1 - \mathrm{p}_1 = \mathrm{p}_2 - \mathrm{p}'_2$, and multiply it by p_2:

$$\mathrm{p}_2\,(\mathrm{p}'_1 - \mathrm{p}_1) = \mathrm{p}_2\,(\mathrm{p}_2 - \mathrm{p}'_2).$$

This equality is invariant. We can write down its left-hand part in the laboratory frame $\mathrm{p}_2 = \{m_2, \mathbf{0}\}$, and the right-hand part in the centre of mass frame $\mathrm{p}_2 = \{\tilde{E}_2,\ \tilde{\mathbf{p}}_2\}$:

$$m_2\,(E'_1 - E_1) = m_2^2 - (\tilde{E}_2\tilde{E}'_2 - \tilde{\mathbf{p}}_2\tilde{\mathbf{p}}'_2).$$

The energies of the particles and the magnitudes of the momenta do not change (p. 108) in the centre of mass frame:

$$\tilde{E}_2 = \tilde{E}'_2, \qquad |\tilde{\mathbf{p}}_1| = |\tilde{\mathbf{p}}_2| = \tilde{p}.$$

Therefore, given that $\tilde{E}_2^2 - \tilde{\mathbf{p}}_2^2 = m_2^2$, we obtain the relation

$$E'_1 - E_1 = -\frac{\tilde{p}^{\,2}}{m_2}\,(1 - \cos\chi), \tag{8.10}$$

which we already found above using the transformation laws for energy and momentum between two reference frames (p. 111).

8.3 The invariants *s*, *t*, and *u*

If a collision between two particles changes both their velocities and their masses, this interaction is called a two-particle *inelastic collision*. Some examples of such collisions are:

$$\begin{array}{ccccccc} p & + & \gamma & \mapsto & p & + & \pi^0, \\ \pi^+ & + & \pi^- & \mapsto & \bar{p} & + & p, \end{array}$$

where p, $\bar{p}$ are a proton and an antiproton, γ is a photon, π^0, and $\pi^\pm$ signifies a neutral and a charged pion. Inelastic two-particle collisions take place in those reactions where a particle decays into three other particles. For example:

$$\begin{array}{ccl} \mu^- & \mapsto & e^- + \bar{\nu}_e + \nu_\mu, \\ K^- & \mapsto & \pi^0 + e^- + \bar{\nu}_e, \end{array}$$

where μ is a muon, K^- is a kaon, and ν_e and ν_μ are an electron and a muon neutrino respectively.

▷ What the above scattering and decay reactions have in common is that there are four particles participating in them. For the sake of uniformity, we can represent them in the form of a *four-tail diagram* (the first figure below) where all four-momenta are directed towards the centre. In the scattering reaction $1+2 \mapsto 3+4$ we have to change the sign of the four-momenta for particles 3 and 4: $\mathrm{p}_3 \mapsto -\mathrm{p}_3$, $\mathrm{p}_4 \mapsto -\mathrm{p}_4$. For the decay reaction $1 \mapsto 2+3+4$, particles 2, 3 and 4 change the signs of their four-momenta. We can also assume that $\mathrm{p} = \{E,\ \mathbf{p}\}$ for the initial particles, and $\mathrm{p} = \{-E,\ -\mathbf{p}\}$ for the resulting particles.

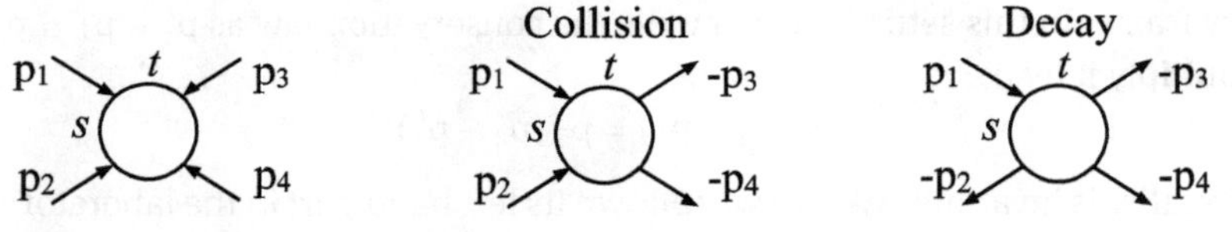

Under this agreement, the conservation law will have the same form for both the scattering and the decay reactions:

$$\mathrm{p}_1 + \mathrm{p}_2 + \mathrm{p}_3 + \mathrm{p}_4 = 0. \tag{8.11}$$

▷ To describe such reactions, we can introduce the following *invariants*:

$$\begin{array}{ccccc} s & = & (\mathrm{p}_1 + \mathrm{p}_2)^2 & = & (\mathrm{p}_3 + \mathrm{p}_4)^2, \\ t & = & (\mathrm{p}_1 + \mathrm{p}_3)^2 & = & (\mathrm{p}_2 + \mathrm{p}_4)^2, \\ u & = & (\mathrm{p}_1 + \mathrm{p}_4)^2 & = & (\mathrm{p}_2 + \mathrm{p}_3)^2. \end{array} \tag{8.12}$$

This is widely accepted notation, and s, t are not to be confused with the interval and the time. In the second equations of each definition, we used the law of conservation of four-momentum (8.11).

▷ The sum of these three invariants is the sum of the squared particle masses:

$$s + t + u = m_1^2 + m_2^2 + m_3^2 + m_4^2. \tag{8.13}$$

To prove this, multiply the conservation law by p_1:

$$\mathrm{p}_1 (\mathrm{p}_1 + \mathrm{p}_2 + \mathrm{p}_3 + \mathrm{p}_4) = m_1^2 + \mathrm{p}_1\,\mathrm{p}_2 + \mathrm{p}_1\,\mathrm{p}_3 + \mathrm{p}_1\,\mathrm{p}_4 = 0.$$

On the other hand, expanding the squares in the definitions of s, t, u, gives us:

$$s + t + u = (m_1^2 + m_2^2 + 2\,\mathrm{p}_1\,\mathrm{p}_2) + (m_1^2 + m_3^2 + 2\,\mathrm{p}_1\,\mathrm{p}_3) + (m_1^2 + m_4^2 + 2\,\mathrm{p}_1\,\mathrm{p}_4).$$

Using these two relationships, we easily obtain (8.13). Therefore, s, t and u are not independent. For example, if s and t are assumed independent, the invariant u will be expressed in terms of them.

▷ Let us now consider two-particle scattering in more detail. This is usually studied in one of two reference frames: in the *centre of mass* frame or in the ***laboratory frame***. In the first case, the particles collide with each other so that their total three-momentum is zero before and after the interaction. In the laboratory frame, the particles of the first type collide with the resting particles of the second type, which are referred to as the *target*.

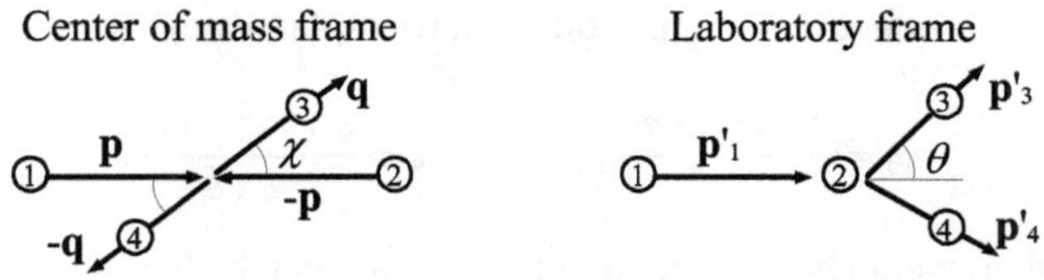

We will denote those quantities relating to the laboratory frame with primes, and leave quantities unprimed in the centre of mass frame.

In the centre of mass frame

$$\mathbf{p}_1 + \mathbf{p}_2 = \mathbf{p}_3 + \mathbf{p}_4 = 0,$$

it is convenient to introduce two momenta

$$\mathbf{p} = \mathbf{p}_1 = -\mathbf{p}_2, \qquad \mathbf{q} = \mathbf{p}_3 = -\mathbf{p}_4.$$

Since the masses of the particles change, the law of conservation of energy,

$$\sqrt{\mathbf{p}^2 + m_1^2} + \sqrt{\mathbf{p}^2 + m_2^2} = \sqrt{\mathbf{q}^2 + m_3^2} + \sqrt{\mathbf{q}^2 + m_4^2}$$

does not imply equality of the magnitudes of the momenta before and after the collision, and in the general case $|\mathbf{p}| \neq |\mathbf{q}|$. The article energies also change. If $m_3 = m_1$ and $m_4 = m_2$, we have $|\mathbf{p}| = |\mathbf{q}|$.

Such an interaction is possible only if the total energy of the initial particles in the centre of mass frame is greater than the sum of masses of the resulting particles, $E_1 + E_2 > m_3 + m_4$. This inequality of energies is called the *reaction threshold.*

▷ The invariant s is the square of the total energy in the centre of mass frame. Indeed, if $\mathbf{p}_1 + \mathbf{p}_2 = 0$, then

$$s = (E_1 + E_2)^2 = (E_3 + E_4)^2 .$$

Using s, we can express the particle energies in the centre of mass frame before and after the collision. To do this, we calculate the following invariant:

$$\mathrm{p}_1\,(\mathrm{p}_1 + \mathrm{p}_2) = E_1(E_1 + E_2) = E_1\sqrt{s}.$$

On the other hand, after opening the parentheses,

$$\mathrm{p}_1\,(\mathrm{p}_1 + \mathrm{p}_2) = m_1^2 + \mathrm{p}_1\,\mathrm{p}_2 = E_1\sqrt{s},$$

and rearranging for $\mathrm{p}_1\,\mathrm{p}_2$, we can substitute into the definition of the invariant:

$$s = (\mathrm{p}_1 + \mathrm{p}_2)^2 = m_1^2 + m_2^2 + 2\,\mathrm{p}_1\,\mathrm{p}_2 = m_2^2 - m_1^2 + 2E_1\sqrt{s}.$$

As a result, we obtain the energy E_1 (and similarly E_2):

$$E_1 = \frac{s + m_1^2 - m_2^2}{2\sqrt{s}}, \qquad E_2 = \frac{s + m_2^2 - m_1^2}{2\sqrt{s}}. \tag{8.14}$$

To find the particle energies after the collision, we replace indices $1 \mapsto 3$, $2 \mapsto 4$. As a result:

$$E_3 = \frac{s + m_3^2 - m_4^2}{2\sqrt{s}}, \qquad E_4 = \frac{s + m_4^2 - m_3^2}{2\sqrt{s}}. \tag{8.15}$$

▷ The squared momenta before and after the reaction can be found from the standard relationships $\mathbf{p}^2 = E_1^2 - m_1^2$ and $\mathbf{q}^2 = E_3^2 - m_3^2$, whence

$$\mathbf{p}^2 = \frac{\lambda(s,\, m_1^2,\, m_2^2)}{4s}, \qquad \mathbf{q}^2 = \frac{\lambda(s,\, m_3^2,\, m_4^2)}{4s}, \tag{8.16}$$

where we have introduced the so-called *triangle function*:

$$\lambda(x,\, y,\, z) = x^2 + y^2 + z^2 - 2\,xy - 2\,xz - 2\,yz. \tag{8.17}$$

▷ The invariant t is related to the scattering angle in the centre of mass frame. Since particle 3 is a resulting particle, we have to replace $\mathrm{p}_3 \mapsto -\mathrm{p}_3$:

$$t = (\mathrm{p}_1 - \mathrm{p}_3)^2 = m_1^2 + m_3^2 - 2\,\mathrm{p}_1\,\mathrm{p}_3 = m_1^2 + m_3^2 - 2\,E_1E_3 + 2\,|\mathbf{p}||\mathbf{q}|\,\cos\chi.$$

Using the above relationships, we have ($\lessdot$ H_{31}):

$$\cos\chi = \frac{s\,(t-u) + (m_1^2 - m_2^2)\,(m_3^2 - m_4^2)}{\sqrt{\lambda(s,\, m_1^2,\, m_2^2)\,\lambda(s,\, m_3^2,\, m_4^2)}}, \tag{8.18}$$

where (8.13) was used.

▷ Similarly, we can express the quantities s, t and u in terms of the invariants in the *laboratory frame*, where the second particle is assumed to be at rest $\mathrm{p}_2 = \{m_2, \mathbf{0}\}$. Squaring the definitions of the invariants containing p_2, $s = (\mathrm{p}_1 + \mathrm{p}_2)^2$,

$$t = (\mathrm{p}_2 - \mathrm{p}_4)^2 \quad \text{and} \quad u = (\mathrm{p}_2 - \mathrm{p}_3)^2,$$

we obtain:

$$E_1' = \frac{s - m_1^2 - m_2^2}{2\,m_2}, \qquad E_3' = \frac{m_2^2 + m_3^2 - u}{2\,m_2}, \qquad E_4' = \frac{m_2^2 + m_4^2 - t}{2\,m_2}.$$

Remember that u can be expressed in terms of s, t and the sum of the squared particle masses (8.13).

▷ The squared momenta in the laboratory frame can be found from the relationships between the energy, the momentum, and the mass $\mathbf{p}_1'^2 = E_1'^2 - m_1^2$, etc. These are given by

$$\mathbf{p}_1'^2 = \frac{\lambda(s,\ m_1^2,\ m_2^2)}{4\,m_2^2}, \qquad \mathbf{p}_3'^2 = \frac{\lambda(u,\ m_3^2,\ m_2^2)}{4\,m_2^2}, \qquad \mathbf{p}_4'^2 = \frac{\lambda(t,\ m_4^2,\ m_2^2)}{4\,m_2^2}.$$

▷ The exit angle of the third particle relative to the momentum of the first particle is expressed in terms of the invariant t:

$$t = (\mathrm{p}_1 - \mathrm{p}_3)^2 = m_1^2 + m_3^2 - 2\,\mathrm{p}_1\,\mathrm{p}_3 = m_1^2 + m_3^2 - 2\,E_1'E_3' + 2\,|\mathbf{p}_1'||\mathbf{p}_3'|\,\cos\theta.$$

Substituting in the energies and the squares of the momenta, we have ($\lessdot$ H_{32}):

$$\cos\theta = \frac{(s - m_1^2 - m_2^2)\,(m_2^2 + m_3^2 - u) + 2\,m_2^2\,(t - m_1^2 - m_3^2)}{\sqrt{\lambda(s,\ m_1^2,\ m_2^2)\,\lambda(u,\ m_2^2,\ m_3^2)}}. \tag{8.19}$$

▷ Since s and t are invariants (have the same value in the laboratory frame and in the centre of mass frame), they can be used to find the relationships between the energies, the momenta, and the angles in the two different frames. For example, we can do this by expressing s in terms of the energy E_1':

$$s = m_1^2 + m_2^2 + 2\,\mathrm{p}_1\,\mathrm{p}_2 = m_1^2 + m_2^2 + 2\,m_2 E_1'.$$

Since the variable $s = (E_1 + E_2)^2$ is positive in the centre of mass frame, it will naturally also be positive in the laboratory frame. As an example, find the relationship between the energies of the first particle in these two reference frames:

$$E_1 = \frac{m_1^2 + m_2 E_1'}{\sqrt{m_1^2 + m_2^2 + 2\,m_2 E_1'}}.$$

Similarly, t can be expressed in terms of the scattering angle θ of the third particle. It is then easy find the relation between scattering angles θ and χ in both frames.

8.4 Dalitz plots *

Consider the annihilation reaction of pions and the consequent production of a proton and an antiproton

$$\pi^+ + \pi^- \mapsto p + \bar{p}.$$

This is an *inelastic* scattering reaction, since not only the momenta but also the "type" of the initial particles (pions) changes when they turn into protons. In this case, the initial masses $m_1 = m_2 = m = 135$ MeV differ from those of the resulting particles $m_3 = m_4 = M = 938$ MeV. Therefore, the cosine of the scattering angle in the centre of mass frame (8.18) is

$$\cos\chi = \frac{s + 2\,(t - m^2 - M^2)}{\sqrt{(s - 4m^2)\,(s - 4M^2)}}. \tag{8.20}$$

The reaction threshold is

$$s \geqslant (m_3 + m_4)^2 = 4M^2.$$

However, there is an additional restriction due to the fact that

$$\cos^2\chi \leqslant 1.$$

This inequality has the following form:

$$t\,(t + s - 2\,(M^2 + m^2)) + (M^2 - m^2)^2 \leqslant 0. \tag{8.21}$$

Thus, we have an area on the (s, t)-plane which is permitted by the energy called the *Dalitz plot*:

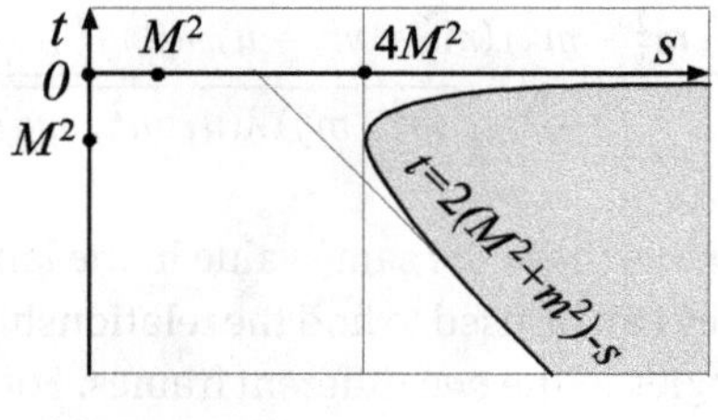

In this diagram, two independent invariants s and t are plotted along orthogonal Cartesian axes, and the region of permitted energies is coloured grey. The reaction threshold $s \geqslant 4M^2$ is represented by a thin vertical line, the permitted area being to the right of it. This leads to $t < 0$. The boundary of the permitted area is obtained by placing an equality sign in the relation (8.21):

$$t + s - 2\,(M^2 + m^2) + \frac{(M^2 - m^2)^2}{t} = 0.$$

For $t \to \infty$ this equation tends to a straight line

$$t = 2\,(M^2 + m^2) - s.$$

The second asymptote corresponds to the limit $s \to \infty$, $t \to 0$.

▷ Consider next the elastic scattering of a pion by a proton:

$$\pi^+ + p \mapsto \pi^+ + p.$$

In this case, no new particles are produced; rather, the initial particles simply change their momenta. This is why the reaction is called elastic. The particle masses are

$$m_1 = m_3 = m, \qquad m_2 = m_4 = M,$$

and the reaction threshold is

$$s \geqslant (M + m)^2.$$

The cosine of the scattering angle is

$$\cos\chi = \frac{s^2 + 2\,s\,(t - m^2 - M^2) + (M^2 - m^2)^2}{(s - (M + m)^2)\,(s - (M - m)^2)}.$$

Constraints on the energy are most easily obtained from the two inequalities

$$\cos\chi \leqslant 1 \quad \text{and} \quad \cos\chi \geqslant -1.$$

The first of them gives

$$s\,t \leqslant 0,$$

and since $s > 0$, we have $t \leqslant 0$. The second inequality gives a relation similar to (8.21), but with s and t interchanged:

$$s\,(s + t - 2\,(M^2 + m^2)) + (M^2 - m^2)^2 \geqslant 0. \tag{8.22}$$

We can now draw the permitted area for the energy in the Dalitz plot:

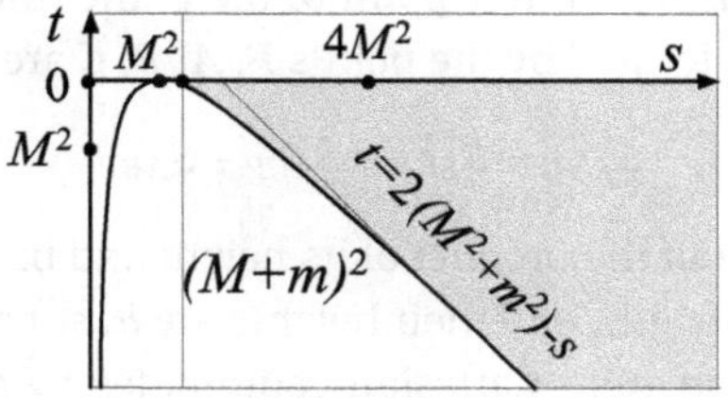

Note that this boundary energy line is obtained from the reaction line

$$\pi^+ + \pi^- \mapsto \bar{p} + p$$

by interchanging the invariants s and t. This is due to the fact that these two reactions are actually a single four-tail. If the second and the third particles are interchanged in $\pi^+ + \pi^- \mapsto \bar{p} + p$, we obtain the reaction $\pi^+ + p \mapsto \pi^+ + p$ (as the proton and the antiproton have equal masses). Obviously, the invariants $s = (p_1 + p_2)$ and $t = (p_1 + p_3)$ then change places. These reactions are said to occur in different *channels* of the same four-tail diagram. The first reaction takes place in the s-channel, and the second takes place in the t-channel. Accordingly, the line of the boundary area is obtained by rotating the figure through 90 degrees.

8.5 Mandelstam diagrams *

Besides Dalitz plots, *Mandelstam diagrams* are also used to represent all three invariants (s, t, u) simultaneously. Such a diagram is an oblique coordinate system with axes s and t drawn at an angle of 60 degrees. However, instead of the axes we will draw the levels $s = 1$, $s = 2$, ..., parallel to the axis $s = 0$ and separated from it by 1, 2, ... The same is then done for the levels of t and u. The lines $s = 0$, $t = 0$, and $u = 0$ form a regular triangle ABC of height

$$h = s + t + u = m_1^2 + m_2^2 + m_3^2 + m_4^2.$$

Positive values of the invariants are plotted from the zero line in the direction of the triangle, and negative values are plotted in the opposite direction:

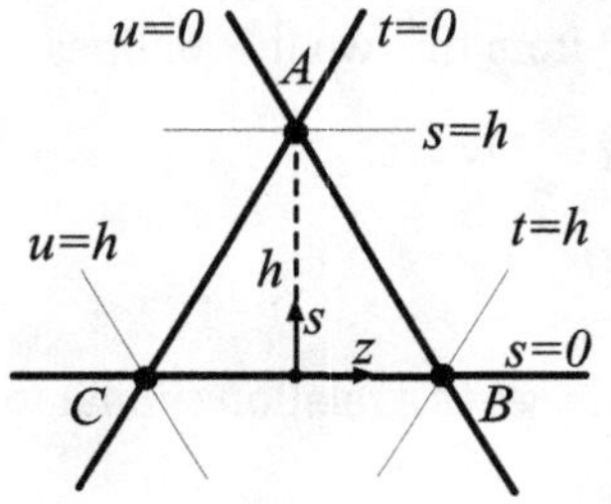

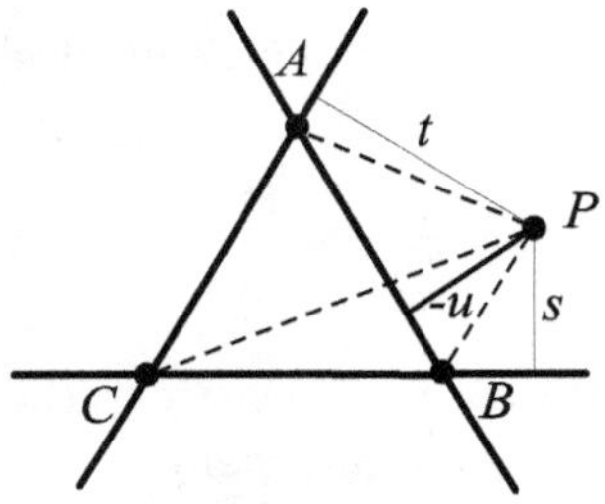

In the first figure, bold lines indicate zero-levels and thin lines indicate the levels $s = h$, $t = h$, $u = h$. At the point A, the invariants take the values $s = h$, $t = u = 0$. Their sum is h. This property is satisfied for any point of the plane (the second figure). Indeed, the areas of the triangles formed by the points P, A, B, C are related as follows:

$$S_{ABC} = S_{BCP} + S_{ACP} - S_{ABP}.$$

The area of a triangle is half the product of its height and base. The bases of all of the triangles have the same length, and their heights are h, s, t and $-u$, so $h = s + t + u$.

We can move to rectangular Cartesian coordinates by drawing a vertical axis s along the height h orthogonal to the line $s = 0$, which will be the horizontal z-axis (see the first figure). The rectangular coordinates (z, s) are related to the invariant t as

$$t = \frac{h + \sqrt{3}\,z - s}{2}. \tag{8.23}$$

The third invariant is

$$u = \frac{h - \sqrt{3}\,z - s}{2}, \tag{8.24}$$

and as a result, the sum of all of these invariants is h.

▷ Consider again the annihilation reaction of two pions followed by the production of a proton-antiproton pair:

$$\pi^+ + \pi^- \mapsto \bar{p} + p.$$

The allowed area for this reaction is determined by the threshold inequality $s \leqslant 4M^2$ and by the relation (8.21). In the latter, we can express t in terms of z and s:

$$3\,z^2 - (s - h)^2 + 4\,(M^2 - m^2)^2 \leqslant 0, \tag{8.25}$$

where $h = 2\,(M^2 + m^2)$. This area is shown in the figure below,

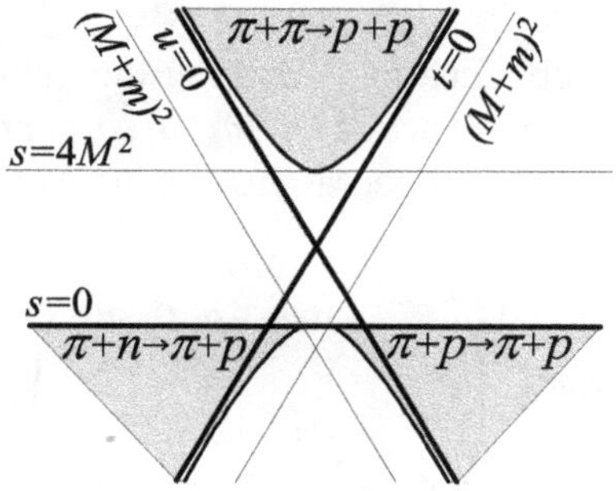

and corresponds to the parabola in the upper part of the figure with asymptotic lines $u = 0$ and $t = 0$. The reaction

$$\pi^+ + p \mapsto \pi^+ + p,$$

and the similar reaction for the antiparticles can be illustrated in the same figure, assuming that the pions are numbered 1 and 2, and the protons are numbered 3 and 4, as in the previous reaction. Then we have to interchange s and t in the inequalities (8.22)

$$t \leqslant 0, \qquad s \geqslant (M + m)^2,$$

to obtain (8.25) with the inequality reversed, and

$$s \leqslant 0, \qquad t \geqslant (M + m)^2.$$

This area is shown in the lower right corner of the figure. The third possible channel is shown in the left corner; it corresponds to the inelastic reaction of a neutral pion scattered by a neutron to form a proton and a charged pion:

$$\pi^0 + n \mapsto \pi^- + p.$$

In fact, this reaction is not completely symmetric to the above two reactions, since the particle masses are different. However, this is a small difference. In fact, a neutron is about 0.1% heavier than a proton, which makes 1% of the pion's mass. Similarly, a neutral pion π^0 is only 1% lighter than its charged counterparts $\pi^\pm$.

Thus, all three reactions are obtained by interchanging the particles in the reaction and renaming the invariant variables s, t, and u.

8.6 Antisymmetric tensors

Antisymmetric tensors are of great importance in relativistic physics. An antisymmetric second-order tensor $A^{\alpha\beta}$ has six independent non-zero components. If two indices coincide, then the corresponding entry is zero (due to antisymmetry). The non-zero components are

$$A^{01},\quad A^{02},\quad A^{03},\quad A^{12},\quad A^{13},\quad A^{23}.$$

All other components are obtained by permuting the indices. For example, $A^{10} = -A^{01}$, etc. Two three-dimensional vectors $\mathbf{a} = \{a_x, a_y, a_z\}$ and $\mathbf{b} = \{b_x, b_y, b_z\}$ have six components, so they can be used to represent an antisymmetric second-order tensor in the form of the following table:

$$A^{\alpha\beta} = \begin{pmatrix} 0 & -a_x & -a_y & -a_z \\ a_x & 0 & b_z & -b_y \\ a_y & -b_z & 0 & b_x \\ a_z & b_y & -b_x & 0 \end{pmatrix}. \tag{8.26}$$

In abbreviated form, this table is written as $A^{\alpha\beta} = (\mathbf{a}, \mathbf{b})$. An antisymmetric tensor with lower indices is obtained by contracting with the metric tensor: $A_{\alpha\beta} = g_{\alpha\mu}\, g_{\beta\nu}\, A^{\mu\nu}$. As a result:

$$A_{\alpha\beta} = \begin{pmatrix} 0 & a_x & a_y & a_z \\ -a_x & 0 & b_z & -b_y \\ -a_y & -b_z & 0 & b_x \\ -a_z & b_y & -b_x & 0 \end{pmatrix}. \tag{8.27}$$

As an example, since the metric tensor $g_{\mu\nu} = \mathrm{diag}(1, -1, -1, -1)$ is diagonal, we have:

$$\begin{aligned} A_{01} &= g_{0\mu}\, g_{1\nu}\, A^{\mu\nu} = g_{00}\, g_{11}\, A^{01} = -A^{01}, \\ A_{12} &= g_{1\mu}\, g_{2\nu}\, A^{\mu\nu} = g_{11}\, g_{22}\, A^{12} = +A^{12}. \end{aligned}$$

Thus, components with an index zero change their signs, and those without a zero remain the same. This can be written in the following form: $A_{\alpha\beta} = (-\mathbf{a}, \mathbf{b})$.

Similarly, one of the indices can be lowered without the other. Such a tensor is written $A^{\alpha\mu}\, g_{\mu\beta} = A^{\alpha}{}_{\beta}$, and found by multiplying the matrices:

$$\begin{pmatrix} 0 & -a_x & -a_y & -a_z \\ a_x & 0 & b_z & -b_y \\ a_y & -b_z & 0 & b_x \\ a_z & b_y & -b_x & 0 \end{pmatrix} \begin{pmatrix} 1 & 0 & 0 & 0 \\ 0 & -1 & 0 & 0 \\ 0 & 0 & -1 & 0 \\ 0 & 0 & 0 & -1 \end{pmatrix} = \begin{pmatrix} 0 & a_x & a_y & a_z \\ a_x & 0 & -b_z & b_y \\ a_y & b_z & 0 & -b_x \\ a_z & -b_y & b_x & 0 \end{pmatrix}.$$

In this case, the table is not antisymmetric any more: $A^{\alpha}{}_{\beta} \neq -A^{\beta}{}_{\alpha}$.

▷ The relationship between the components of an antisymmetric tensor measured in two inertial frames can be obtained from the general tensor transformation rule:

$$A'^{\alpha\beta} = \Lambda^{\alpha}{}_{\mu}\, \Lambda^{\beta}{}_{\nu}\, A^{\mu\nu} = \Lambda^{\alpha}{}_{\mu}\, A^{\mu\nu} (\Lambda^{T})_{\nu}{}^{\beta} = (\Lambda\, \mathbf{A}\, \Lambda^{T})^{\alpha\beta},$$

where we made use of the transpose matrix of the Lorentz transformation

$$(\Lambda^T)_\nu^{\ \beta} = \Lambda^\beta_{\ \nu}.$$

When obtaining the components $A'^{\alpha\beta}$ in S', if $\Lambda^\beta_{\ \nu}$ is symmetric, then we have to multiply $A^{\alpha\beta}$ from the left and from the right by the matrix Λ:

$$\begin{pmatrix} \gamma & -\gamma v & 0 & 0 \\ -\gamma v & \gamma & 0 & 0 \\ 0 & 0 & 1 & 0 \\ 0 & 0 & 0 & 1 \end{pmatrix} \cdot \begin{pmatrix} 0 & -a_x & -a_y & -a_z \\ a_x & 0 & b_z & -b_y \\ a_y & -b_z & 0 & b_x \\ a_z & b_y & -b_x & 0 \end{pmatrix} \cdot \begin{pmatrix} \gamma & -\gamma v & 0 & 0 \\ -\gamma v & \gamma & 0 & 0 \\ 0 & 0 & 1 & 0 \\ 0 & 0 & 0 & 1 \end{pmatrix}.$$

Multiplying the matrices, we obtain the following transformations:

$$\begin{aligned} a'_x &= a_x, & a'_y &= \gamma\,(a_y + v\,b_z), & a'_z &= \gamma\,(a_z - v\,b_y), \\ b'_x &= b_x, & b'_y &= \gamma\,(b_y - v\,a_z), & b'_z &= \gamma\,(b_z + v\,a_y). \end{aligned} \tag{8.28}$$

▷ Using these transformations we can verify that the following combinations of vectors:

$$\mathbf{a}^2 - \mathbf{b}^2 = inv, \qquad \mathbf{ab} = inv, \tag{8.29}$$

are invariant (that is, they have the same value in all inertial frames). In particular, if the vectors **a** and **b** are orthogonal in one reference frame, they will remain so in any other frame.

▷ ***An absolutely antisymmetric tensor*** is a tensor of any order (with an arbitrary number of indices) which changes its sign when any *two* of its indices are interchanged. Obviously, a second-order antisymmetric tensor is an absolutely antisymmetric tensor.

Such a tensor $A^{\alpha\beta\gamma}$ of order three has only four independent non-zero components: A^{012}, A^{013}, A^{023}, A^{123}. The other components are obtained by permuting the indices:

$$A^{120} = -A^{102} = A^{012} = -A^{021}.$$

Therefore, as we will see below, $A^{\alpha\beta\gamma}$ can be expressed in terms of the four components of a four-vector.

An absolutely antisymmetric tensor $A^{\alpha\beta\gamma\delta}$ of order four has only one non-trivial component A^{0123}. Absolutely antisymmetric tensors of order five and above in four-dimensional space-time are zero. Every component will have at least one pair of indices which are the same.

▷ The absolutely antisymmetric ***Levi-Civita symbol*** is an important concept in index mathematics. In a two-dimensional space, the symbol is introduced with two indices, $\varepsilon_{\alpha\beta}$, so that $\varepsilon_{12} = 1$ and interchanging the indices reults in a change of sign. In a three-dimensional space, the Levi-Civita symbol has three indices, and $\varepsilon_{123} = 1$. Using this, we can write a vector product $[\mathbf{a}\times\mathbf{b}]_i = \varepsilon_{ijk}\, a_j b_k$ (p. 341):

$$[\mathbf{a}\times\mathbf{b}]_3 = \varepsilon_{312}\, a_1\, b_2 + \varepsilon_{321}\, a_2\, b_1 + \ldots = a_1 b_2 - a_2 b_1,$$

etc. The terms denoted by the dots are zero, since Levi-Civita symbols with equal indices are zero.

Similarly, a four-dimensional Levi-Civita symbol $\varepsilon_{\alpha\beta\gamma\delta}$ can be defined. Its components are obtained from the value $\varepsilon_{0123} = 1$. For example:

$$\varepsilon_{2103} = -\varepsilon_{1203} = \varepsilon_{1023} = -\varepsilon_{0123} = -1,$$

$$\varepsilon_{0103} = \varepsilon_{1111} = 0.$$

If two indices have the same value, the symbol is zero, and it changes its sign when *any* two symbols are interchanged. Remember that the indices of the coordinates of an event in a four-dimensional space are commonly numbered starting from zero.

▷ With the help of Levi-Civita symbols, matrix determinants can be written. For example, for a 2 × 2 matrix,

$$\det|a_{ij}| = \varepsilon_{ij}\, a_{1i}\, a_{2j} = \varepsilon_{12}\, a_{11}\, a_{22} + \varepsilon_{21}\, a_{12}\, a_{21} = a_{11}\, a_{22} - a_{12}\, a_{21}.$$

Similarly, for the determinants of 3 × 3 and 4 × 4 matrices:

$$\det|a_{ij}| = \varepsilon_{ijk}\, a_{1i}\, a_{2j}\, a_{3k}, \qquad \det|a^{\alpha\beta}| = \varepsilon_{\mu\nu\sigma\tau}\, a^{0\mu}\, a^{1\nu}\, a^{2\sigma}\, a^{3\tau},$$

where the matrix elements in the products are written in ascending ordering of their first indices. The second indices are used for the contraction with the Levi-Civita symbol.

▷ We can prove that a four-dimensional Levi-Civita symbol *is a tensor* with respect to Lorentz transformations. As an example, consider the transformation

$$\varepsilon'_{0123} = \tilde{\Lambda}_0{}^{\alpha}\tilde{\Lambda}_1{}^{\beta}\tilde{\Lambda}_2{}^{\gamma}\tilde{\Lambda}_3{}^{\delta}\, \varepsilon_{\alpha\beta\gamma\delta} = \det\tilde{\Lambda} = 1,$$

where the definition of the determinant and the orthogonality of the Lorentz transformations were used (p. 182). Thus, if $\varepsilon_{0123} = 1$ in one inertial frame, it will be the same in any other frame. Permutation of the indices does not change the result. If any two indices coincide, e.g. ε'_{0023}, the contraction with the Lorentz matrices will be zero, since the symmetric tensor $\tilde{\Lambda}_0{}^{\alpha}\tilde{\Lambda}_0{}^{\beta}$ gets contracted with the antisymmetric tensor $\varepsilon_{\alpha\beta\gamma\delta}$ with respect to the indices α and β.

▷ The Levi-Civita symbol can be used to create new tensors. Define, as an example, the following antisymmetric tensor:

$$^{*}A_{\alpha\beta} = \frac{1}{2}\, \varepsilon_{\alpha\beta\mu\nu}\, A^{\mu\nu}. \tag{8.30}$$

Tensors with an asterisk $^*A$ are called *dual* (or *conjugate*) to a tensor contracted with $\varepsilon_{\alpha\beta\mu\nu}$. We can find out the components of the dual tensor. If we write in explicit form the sum over the repeated indices, omitting identical indices corresponding to the zeros of the Levi-Civita symbol, we obtain:

$$^*A_{01} = \frac{1}{2}\,(\varepsilon_{0123}\,A^{23} + \varepsilon_{0132}\,A^{32}) = \varepsilon_{0123}\,A^{23} = A^{23} = b_x,$$

$$^*A_{12} = \frac{1}{2}\,(\varepsilon_{1203}\,A^{03} + \varepsilon_{1230}\,A^{30}) = \varepsilon_{0123}\,A^{03} = A^{03} = -a_z,$$

where $\varepsilon_{0132}\,A^{32} = \varepsilon_{0123}\,A^{23}$, since both quantities are antisymmetric. In the second equation, the index 0 in the Levi-Civita symbol should be moved to the beginning:

$$\varepsilon_{1203} = -\varepsilon_{1023} = \varepsilon_{0123}.$$

Note that the factor 1/2 in the definition of the dual tensor is related to the number of possible permutations of two indices (2! = 2). Similarly, for the other components we obtain:

$$^*A_{\alpha\beta} = \begin{pmatrix} 0 & b_x & b_y & b_z \\ -b_x & 0 & -a_z & a_y \\ -b_y & a_z & 0 & -a_x \\ -b_z & -a_y & a_x & 0 \end{pmatrix}, \qquad ^*A^{\alpha\beta} = \begin{pmatrix} 0 & -b_x & -b_y & -b_z \\ b_x & 0 & -a_z & a_y \\ b_y & a_z & 0 & -a_x \\ b_z & -a_y & a_x & 0 \end{pmatrix}.$$

The vectors **a** and **b** in the dual tensor change places as compared with the initial tensor. The tables for the conjugate tensors can be written in abbreviated form as follows:

$$^*A_{\alpha\beta} = (-\mathbf{b}, -\mathbf{a}), \qquad ^*A^{\alpha\beta} = (\mathbf{b}, -\mathbf{a}).$$

⊳ The invariants obtained above can be written in explicit covariant form:

$$A_{\alpha\beta}\,A^{\alpha\beta} = 2\,(\mathbf{b}^2 - \mathbf{a}^2), \qquad ^*A_{\alpha\beta}\,A^{\alpha\beta} = -4\,\mathbf{ab}. \tag{8.31}$$

The duals of third- and the fourth-order tensors are defined similarly:

$$^*A_\alpha = \frac{1}{6}\,\varepsilon_{\mu\nu\sigma\alpha}\,A^{\mu\nu\sigma}, \qquad ^*A_{\alpha\beta\gamma} = \varepsilon_{\mu\alpha\beta\gamma}\,A^{\mu}.$$

Conjugation usually implies contraction with respect to the first indices of the Levi-Civita symbol. For a tensor with two indices $^*A_{\alpha\beta}$ it makes no difference whether one takes the first or last two, since $\varepsilon_{\alpha\beta\mu\nu} = \varepsilon_{\mu\nu\alpha\beta}$ (< H_{33}).

8.7 Angular momentum

A good physical example of an antisymmetric tensor is the *angular four-momentum*:

$$L^{\alpha\beta} = x^\alpha\,p^\beta - x^\beta\,p^\alpha, \tag{8.32}$$

where $x^\alpha = \{t, \mathbf{r}\}$ is a particle's position at time t, and $p^\alpha = \{E, \mathbf{p}\}$ are its energy and momentum. We can write its components in the explicit form $L^{\alpha\beta}$. By definition:

$$L^{\alpha\beta} = -L^{\beta\alpha}.$$

Its diagonal elements are zero: $L^{00} = L^{11} = L^{22} = L^{33} = 0$. For the space-like components we have:

$$\begin{aligned} L^{23} &= x^2 p^3 - x^3 p^2 &= y\,p_z - z\,p_y &= L_x, \\ L^{31} &= x^3 p^1 - x^1 p^3 &= z\,p_x - x\,p_z &= L_y, \\ L^{12} &= x^1 p^2 - x^2 p^1 &= x\,p_y - y\,p_x &= L_z. \end{aligned}$$

These are the components of the angular three-momentum vector, which is equal to the vector product of the radius vector and the momentum:

$$\mathbf{L} = \mathbf{r} \times \mathbf{p} = \begin{vmatrix} \mathbf{i} & \mathbf{j} & \mathbf{k} \\ x & y & z \\ p_x & p_y & p_z \end{vmatrix}. \tag{8.33}$$

The components of the angular momentum tensor with one index zero are written similarly:

$$\begin{aligned} L^{10} &= x^1 p^0 - x^0 p^1 &= x\,E - t\,p_x &= G_x, \\ L^{20} &= x^2 p^0 - x^0 p^2 &= y\,E - t\,p_y &= G_y, \\ L^{30} &= x^3 p^0 - x^0 p^3 &= z\,E - t\,p_z &= G_z. \end{aligned}$$

These components can also be represented as a three-vector ($\mathbf{p} = \mathbf{u}\,E$):

$$\mathbf{G} = E\,\mathbf{r} - t\,\mathbf{p} = (\mathbf{r} - \mathbf{u}\,t)\,E. \tag{8.34}$$

Taking into account the antisymmetry property, the angular momentum four-tensor can be written in matrix form:

$$L^{\alpha\beta} = \begin{pmatrix} 0 & -G_x & -G_y & -G_z \\ G_x & 0 & L_z & -L_y \\ G_y & -L_z & 0 & L_x \\ G_z & L_y & -L_x & 0 \end{pmatrix}. \tag{8.35}$$

As before (p. 202), we write $L^{\alpha\beta} = (\mathbf{G},\ \mathbf{L})$.

▷ The relationship between the components of the angular momentum measured in two inertial frames can be obtained from the formulas for an arbitrary antisymmetric tensor (8.28) with $\mathbf{a} = \mathbf{G}$ and $\mathbf{b} = \mathbf{L}$:

$$\begin{aligned} G'_x &= G_x, \quad & G'_y &= \gamma\,(G_y + v\,L_z), \quad & G'_z &= \gamma\,(G_z - v\,L_y), \\ L'_x &= L_x, \quad & L'_y &= \gamma\,(L_y - v\,G_z), \quad & L'_z &= \gamma\,(L_z + v\,G_y). \end{aligned} \tag{8.36}$$

We can also write the transformations in vector form. In this case, it is easier to substitute the vector Lorentz transformations for coordinates (1.18), p. 17, and momentum (4.13), p. 101 into $\mathbf{L} = \mathbf{r} \times \mathbf{p}$:

$$\mathbf{L}' = \mathbf{r}' \times \mathbf{p}' = \{\mathbf{r} - \gamma\, \mathbf{v}\, t + \Gamma\, \mathbf{v}\,(\mathbf{vr})\} \times \{\mathbf{p} - \gamma\, \mathbf{v}\, E + \Gamma\, \mathbf{v}\,(\mathbf{vp})\}.$$

Multiplying the expressions inside the brackets and using $t\,\mathbf{p} = E\,\mathbf{r} - \mathbf{G}$ to rewrite t, we have:

$$\mathbf{L}' = \mathbf{L} + \gamma\,[\mathbf{v} \times \mathbf{G}] + \Gamma\left\{(\mathbf{rv})[\mathbf{v} \times \mathbf{p}] + (\mathbf{pv})[\mathbf{r} \times \mathbf{v}]\right\}.$$

We now multiply the expression in brackets by an *arbitrary* vector $\mathbf{a}$, which will then be “reduced” (that is, we assume $\mathbf{a} = \{1, 0, 0\}$, etc.):

$$\mathbf{a}\,\{\ldots\} = [\mathbf{a} \times \mathbf{v}]\,[(\mathbf{rv})\,\mathbf{p} - (\mathbf{pv})\,\mathbf{r}] = -[\mathbf{a} \times \mathbf{v}]\,[\mathbf{v} \times \mathbf{L}] = -\mathbf{a}\,[\mathbf{v} \times [\mathbf{v} \times \mathbf{L}]].$$

Omitting $\mathbf{a}$ and expanding the vector triple product, we obtain:

$$\mathbf{L}' = \gamma\,(\mathbf{L} + \mathbf{v} \times \mathbf{G}) - \Gamma\,\mathbf{v}\,(\mathbf{vL}), \tag{8.37}$$

$$\mathbf{G}' = \gamma\,(\mathbf{G} - \mathbf{v} \times \mathbf{L}) - \Gamma\,\mathbf{v}\,(\mathbf{vG}), \tag{8.38}$$

where the transformations for the vector $\mathbf{G}$ are found similarly to those for $\mathbf{L}$. If $\mathbf{v}$ is directed along the x-axis, we again have (8.36).

▷ Note that it follows from the definition of the vectors that $\mathbf{G}\,\mathbf{L} = 0$. Obviously, the Lorentz transformations preserve this orthogonality. Another quantity, $\mathbf{a}^2 - \mathbf{b}^2$, which is invariant for any antisymmetric tensor and a relative speed $\mathbf{v}$ pointing in any direction (< H_{34}), gives the relation:

$$\mathbf{G}^2 - \mathbf{L}^2 = (\mathbf{p}^2 - \mathbf{E}^2)\,(t^2 - \mathbf{r}^2) + (E\,t - \mathbf{rp})^2.$$

This expression can be written in explicit covariant form using four-vectors:

$$\mathbf{G}^2 - \mathbf{L}^2 = -\mathrm{p}^2\,\mathrm{x}^2 + (\mathrm{p\,x})^2 = (\mathrm{p\,x})^2 - m^2\,\mathrm{x}^2,$$

where block letters stand for the four-momentum vectors “p” and for the points “x” of space-time, and $m^2 = \mathrm{p}^2$ is the squared particle mass. Due to the invariance of the four-product, the difference between the squared vectors $\mathbf{G}$ and $\mathbf{L}$ is also invariant.

▷ The components of tensor $L^{\alpha\beta}$ are invariant only if the force acting on the particle has a certain form. The conditions to make the space-like part of the tensor

$$\mathbf{L} = \mathbf{r} \times \mathbf{p}$$

invariant were considered when we discussed the concept of force (p. 126). Now we can find out when the vector $\mathbf{G}$ is an integral of motion. Let its time derivative be zero:

$$\frac{d\mathbf{G}}{dt} = \frac{d(E\,\mathbf{r} - t\,\mathbf{p})}{dt} = (\mathbf{uF})\,\mathbf{r} + E\,\mathbf{u} - \mathbf{p} - t\,\mathbf{F} = (\mathbf{uF})\,\mathbf{r} - t\,\mathbf{F} = 0.$$

Here, we substituted in the force $\mathbf{F} = d\mathbf{p}/dt$ and used that

$$\frac{dE}{dt} = \mathbf{uF} \quad \text{and} \quad \mathbf{p} = E\mathbf{u}.$$

Since $\mathbf{r}$ is the coordinate of the particle, and t is the current time, independent of the particle, the expression on the right above is zero only if the particle is free: $\mathbf{F} = 0$. In this case, it moves along the trajectory

$$\mathbf{r} = \mathbf{r}_0 + \mathbf{u}\,t,$$

where $\mathbf{r}_0$ is a constant vector, and $\mathbf{G} = E\,\mathbf{r}_0$ is conserved, since the energy of a free particle is conserved.

▷ Consider a set of interacting particles and suppose that $\mathbf{G}$ is the sum of the expressions $E_i\,\mathbf{r}_i - t\,\mathbf{p}_i$ taken over all particles. In this case, $\mathbf{G}$ turns out to be an integral of motion *if* the force $\mathbf{F}_i$ acting on the i-th particle as a result of it interacting with the others satisfies the equations:

$$\sum_i \mathbf{F}_i = 0, \qquad \sum_i (\mathbf{u}_i\mathbf{F}_i)\,\mathbf{r}_i = 0. \tag{8.39}$$

The first relationship is Newton's third law and is satisfied, for example, if the force is the sum of the paired interactions between the i-th particle and all of the other particles, the values of which are proportional to the distances between the particles. The second condition is more difficult to satisfy.

In general, the theory of relativity only describes the dynamics of a "test particle" placed in an external stationary (constant) force field. If the interaction of "equivalent" particles is to be considered, the simple concept of the force depending on the particles' positions (and, possibly, their velocities) does not work any more. Consider as an example the interaction of two fast-moving particles. Since no impact can propagate faster than the fundamental speed, there will always be a delay. Each particle is "perceived" by the other particle as being in some place other than its "real" position. As a result, the dynamical problem becomes too complicated. Some simplification can be achieved by using the concept of a force field. This field obeys certain equations and is, to some extent, an independent entity. We will consider these issues in more detail in the next volume when discussing electromagnetic interactions.

▷ Consider the physical meaning of the vector $\mathbf{G}$, *assuming* that it is conserved. Since the total energy is also conserved, $\sum E = const$ (where we sum over all particles), we see that

$$\frac{\mathbf{G}}{\sum E} = \frac{\sum E\mathbf{r}}{\sum E} - \frac{\sum \mathbf{p}}{\sum E}\,t = \mathbf{R} - \mathbf{V}\,t = const,$$

is also conserved. We used here the radius vector of the system's *centre of energy* $\mathbf{R}$ and its "total speed" $\mathbf{V}$:

$$\mathbf{R} = \frac{\sum E\mathbf{r}}{\sum E} \approx \frac{\sum m\mathbf{r}}{\sum m}, \qquad \mathbf{V} = \frac{\sum \mathbf{p}}{\sum E} \approx \frac{\sum m\mathbf{u}}{\sum m}. \tag{8.40}$$

The approximate equalities are written in the non-relativistic limit where the centre of energy coincides with the *centre of mass*. The constancy of the vector **G** results in the system's centre of energy moving uniformly along a straight line:

$$\mathbf{R} = \mathbf{V}\,t + const.$$

Note that **R** was calculated using the energies of motion

$$E = \frac{m}{\sqrt{1 - \mathbf{u}^2}}$$

rather than the total energy of the particle, which takes its interactions into account. Therefore, it would be more precise to call **R** the *centre of motion-energy*.

▷ If a frame S' is assigned to the particle, the energy of the particle is equal to its mass, and its momentum is zero (the particle is at rest). Therefore, $\mathbf{G}' = m\,\mathbf{r}'$, $\mathbf{L}' = 0$. We can write transformations inverse to (8.37) by interchanging the primed and the unprimed values ($\mathbf{v} \mapsto -\mathbf{v}$):

$$\mathbf{L} = \gamma\,\mathbf{L}' - \Gamma\,\mathbf{v}\,(\mathbf{v}\mathbf{L}') - \gamma\,[\mathbf{v} \times \mathbf{G}'] = -m\,\gamma\,[\mathbf{v} \times \mathbf{r}'] = m\,\gamma\,[\mathbf{r} \times \mathbf{v}].$$

In the last equation we changed $\mathbf{r}'$ for $\mathbf{r}$, since in the vector Lorentz transformations these vectors differ only in terms which are proportional to the relative velocity vector $\mathbf{v}$, which then vanish when taking the vector product. Since $m\,\gamma\,\mathbf{v}$ is the particle momentum, we again obtain the angular momentum

$$\mathbf{L} = \mathbf{r} \times \mathbf{p}.$$

▷ In the section concerning force (p. 127) we considered a modified angular momentum $\mathbf{L} = E\,[\mathbf{r} \times \mathbf{p}]$ which is conserved if the force depends on the particle velocity in a certain way. A similar angular momentum can also be represented in covariant tensor form. However, this requires a third-order tensor:

$$L^{\alpha\beta\gamma} = (x^\alpha p^\beta - x^\beta p^\alpha)\,p^\gamma. \tag{8.41}$$

Since $p^0 = E$, the space-like components $L^{\alpha\beta 0}$ will give the modified angular momentum.

8.8 Spin

If we consider a system of particles (for example, a rotating gyroscope) rather than a single point particle, it is convenient to introduce the four-vector of *spin* in addition to the angular momentum tensor.

This can be interpreted as the proper angular momentum in the frame where the centre of energy is at rest.

We can total physical quantities to characterize the system of particles:

$$\mathcal{E} = \sum E, \quad \mathbf{P} = \sum \mathbf{p}, \quad \mathbf{L} = \sum [\mathbf{r} \times \mathbf{p}], \quad \mathbf{R} = \frac{\sum E\mathbf{r}}{\sum E},$$

where we omitted the indices numbering the particles. Therefore, $\mathcal{E}$ is the total energy of motion, $\mathbf{P}$ is the total momentum, $\mathbf{L}$ is the total angular momentum, and $\mathbf{R}$ is the system's centre of energy. We define the four-vector of *spin*: by

$$S_\nu = \frac{1}{2}\, \varepsilon_{\nu\alpha\beta\gamma}\, L^{\alpha\beta}\, U^\gamma, \tag{8.42}$$

where $\varepsilon_{\nu\alpha\beta\gamma}$ is the Levi-Civita symbol, and U^α is the total four-velocity of the system of particles. This is found from the total four-momentum $P^\alpha = \{\mathcal{E}, \mathbf{P}\}$:

$$U^\alpha = \frac{P^\alpha}{M} = \{U^0, \mathrm{U}\} = \frac{\{1, \mathbf{u}\}}{\sqrt{1-\mathbf{u}^2}},$$

where $\mathbf{u}$ is a three-dimensional vector, the "total velocity", and $M = \sqrt{\mathcal{E}^2 - \mathbf{P}^2}$ is the mass of the system of particles (*not taking into account* the energy of their interaction).

▷ The physical meaning of classical spin becomes clear if its definition (8.42) is written in three-dimensional notation $S^\alpha = \{S^0, \mathbf{S}\}$. We have

$$S_0 = \varepsilon_{0231}\, L^{23}\, U^1 + \varepsilon_{0312}\, L^{31}\, U^2 + \varepsilon_{0123}\, L^{12}\, U^3 = L^1 U^1 + L^2 U^2 + L^3 U^3,$$

where only the terms with different indices are shown (since the Levi-Civita symbol is zero for other indices). The factor 1/2 disappears, since the sums with $L^{\alpha\beta}$ contain two summands with rearranged indices: $\varepsilon_{0231}\, L^{23} + \varepsilon_{0321}\, L^{32} = 2\, \varepsilon_{0231}\, L^{23}$ (due to the antisymmetry of both tensors). The values of the Levi-Civita symbol are calculated by rearranging indices and using $\varepsilon_{0123} = 1$ to obtain $\varepsilon_{0231} = -\varepsilon_{0213} = \varepsilon_{0123} = 1$, etc. The sums for the four-vector's space-like components are rewritten similarly. For example:

$$S_1 = \varepsilon_{1023}\, L^{02}\, U^3 + \varepsilon_{1032}\, L^{03}\, U^2 + \varepsilon_{1230}\, L^{23}\, U^0 = G^2\, U^3 - G^3\, U^2 - L^1\, U^0.$$

The space-like components of the four-spin S^α with upper indices form a three-vector $\mathbf{S} = \{S^1, S^2, S^3\} = \{S_1, -S_2, -S_3\}$.

As a result, the time-like and space-like components of the spin four-vector are:

$$S^0 = \mathbf{LU}, \qquad \mathbf{S} = \mathbf{L}\, U^0 - \mathbf{G} \times \mathbf{U}. \tag{8.43}$$

We can express the total (over all particles) vector $\mathbf{G}$ in terms of the energy and the momentum:

$$\mathbf{G} = \mathcal{E}\,\mathbf{R} - t\,\mathbf{P}$$

(8.34), or the four-velocity: $U^\alpha = \{U^0, \mathrm{U}\} = \{\mathcal{E}, \mathbf{P}\}/M$:

$$S^0 = \frac{\mathbf{LP}}{M} = \frac{\mathbf{SP}}{\mathcal{E}} = \mathbf{uS}, \qquad \mathbf{S} = \frac{\mathcal{E}}{M}\,(\mathbf{L} - \mathbf{R} \times \mathbf{P}). \tag{8.44}$$

Thus, the three-vector of spin $\mathbf{S}$ is proportional to the difference between the total angular momentum $\mathbf{L}$ and the angular momentum of the system as a whole $\mathbf{R} \times \mathbf{P}$. This difference represents the ***proper angular momentum*** (the total angular momentum minus the angular momentum of the system as a whole). The factor $\mathcal{E}/M$ makes $\mathbf{S}$ a space-like component of the four-vector. The zeroth component of the four-spin is fully determined by the three-vector of spin and the system's "integral velocity"

$$\mathbf{u} = \frac{\mathbf{P}}{\mathcal{E}}.$$

▷ Due to the antisymmetry of the Levi-Civita symbol, the product of the four-spin and the four-speed is zero in any reference frame,

$$\mathrm{S}\,\mathrm{U} = S_\alpha U^\alpha = 0. \tag{8.45}$$

Therefore, in a system at rest $U^\alpha = \{1, \mathbf{0}\}$ (or $\mathbf{P} = \mathbf{0}$), the spin has only vector components $S^\alpha = \{0, \mathbf{S}\}$. In such systems, the three-dimensional spin vector coincides with the angular momentum of the system $\mathbf{S} = \mathbf{L}$. For a point particle

$$\mathbf{L} = \mathbf{R} \times \mathbf{P},$$

so its classical spin is always zero. In the case of a system of particles, the spin is, generally speaking, not equal to zero.

▷ The concept of spin was introduced in quantum mechanics as a dynamical variable (in addition to coordinates and momentum) to describe a ***point*** electron. Therefore, spin is sometimes believed to be a purely quantum physical quantity. However, it is not. Quantum mechanics operates just as well with the spin of a proton or a nucleus, which are certainly not point objects. In this case, the spin should have a corresponding classical prototype. The situation is similar with quantum energy, momentum, or angular momentum. Of course, some ingenuity is required to introduce non-quantized spin for a point particle, but for a non-point object such as a rotating gyroscope, the concept of non-quantized spin appears to be quite a convenient characteristic of the physical system. We will return to the description of classical spin in Chapter 10 and consider it from more general perspectives in the next volume.

9 Curvilinear coordinates *

In this chapter we will continue our consideration of non-inertial frames, which we began in Chapter 6. We will now use the more formal covariant technique of curvilinear coordinates to describe arbitrary reference frames whose points move along certain trajectories relative to an inertial frame.

We will show how expressions for various physical quantities are obtained in such arbitrary coordinates. We will then discuss the concept of rigid frames in the theory of relativity and in some dynamical problems.

https://doi.org/10.1515/9783110515886-009

9.1 Arbitrary reference frames

A reference frame $S : \{t, x, y, z\}$ can be defined by specifying the law of motion for each of its points relative to the laboratory (inertial) frame $S_0 : \{T, X, Y, Z\}$. Let the coordinates $x^i = \{x, y, z\}$ uniquely determine a given point of the frame S, and let this point move with respect to the laboratory frame S_0 along the trajectory:

$$X^i(T) = F^i(T, x, y, z). \tag{9.1}$$

We assume that the $F^i(T, x, y, z)$ are smooth functions (differentiable with respect to each argument), and $X^i = \{X, Y, Z\}$.

Time t of a non-inertial laboratory frame can be defined in any way which is convenient using a function $T = T(t, x, y, z)$. We will only assume that earlier events in S correspond to smaller values of t than later events.

! This is a *coordinate* time and, generally speaking, it does not coincide with the physical time of the fixed clock at the point x^i.

At first glance, such an arbitrary definition of time looks strange. However, as we will see later, a rule can be specified for any function $T(t, x, y, z)$, with which physical time and physical length can be computed in the non-inertial frame.

Replacing the time T by the function $T(t, x, y, z)$ in the trajectory (9.1), we obtain the transformation from the frame S to the laboratory frame S_0:

$$T = T(t, x, y, z), \qquad X^i = X^i(t, x, y, z).$$

Substituting into

$$ds^2 = dT^2 - dX^2 - dY^2 - dZ^2,$$

we obtain the interval between the events in the non-inertial frame:

$$ds^2 = g_{\alpha\beta}\, dx^\alpha dx^\beta.$$

The metric coefficients $g_{\alpha\beta}$ are now functions of the time and coordinates of the event

$$x^i = \{t, x^1, x^2, x^3\},$$

and completely determine the properties of the reference frame.

In any given frame we can introduce another numbering of events (*without changing the frame*):

$$t' = t'(t, x, y, z), \qquad x'^i = x'^i(x, y, z). \tag{9.2}$$

The first transformation defines a new coordinate time, and the rest of them define another way of numbering the spatial points of the system. It is important to note that the spatial transformations do not depend on the time t (otherwise, we would be in another frame). Let us consider some examples.

▷ Suppose the points of a non-inertial frame move along the trajectory:

$$X = \frac{1}{a}\left[\sqrt{(1+ax)^2 + (aT)^2} - 1\right], \qquad Y = y, \quad Z = z,$$

where x, y, and z are fixed numbers characterizing each given point and corresponding to its position in the laboratory frame at $T = 0$. Let us consider the following transformation of time: $aT = (1 + ax)\,\mathrm{sh}(at)$. This choice is for the sake of simplicity only. Substituting aT into the trajectory of the point, we obtain the transformations between the reference frames (p. 146):

$$aT = (1+ax)\,\mathrm{sh}(at), \quad aX = (1+ax)\,\mathrm{ch}(at) - 1, \quad Y = y, \quad Z = z. \tag{9.3}$$

The differentials of these transformations are:

$$dT = (1+ax)\,\mathrm{ch}(at)\,dt + \mathrm{sh}(at)\,dx, \qquad dX = (1+ax)\,\mathrm{sh}(at)\,dt + \mathrm{ch}(at)\,dx$$

and $dY = dy$, $dZ = dz$. Substituting these into the interval between the events in the laboratory frame $ds^2 = dT^2 - dX^2 - dY^2 - dZ^2$, we obtain the same interval in the rigid uniformly accelerating non-inertial frame:

$$ds^2 = (1+ax)^2\,dt^2 - dx^2 - dy^2 - dz^2, \tag{9.4}$$

with non-zero metric coefficients $g_{00} = (1+ax)^2$ and $g_{11} = g_{22} = g_{33} = -1$. The interval for a fixed point ($dx = dy = dz = 0$) is equal to its *proper time*:

$$ds = d\tau_0 = (1+ax)\,dt.$$

We can now find the relationship between dT and dt from the transformations between the frames (for fixed coordinates $x = const$):

$$dT = (1+ax)\,\mathrm{ch}(at)\,dt = \mathrm{ch}(at)\,d\tau_0 = \frac{d\tau_0}{\sqrt{1-U^2}},$$

where in the final equation U is the velocity of the point in the laboratory frame $U = dX/dT = \mathrm{th}(at)$, also obtained from the expressions for the differentials. Thus, the proper time of the clock in the non-inertial frame is related to the laboratory time by the standard relativistic formula (3.2), p. 53.

The *numbering system* for the points of the non-inertial frame can be changed by the coordinate transformation

$$t = t', \qquad 1 + ax = e^{ax'}, \qquad y = y', \qquad z = z',$$

which gives the interval

$$ds^2 = e^{2ax'}(dt'^2 - dx'^2) - dy'^2 - dz'^2. \tag{9.5}$$

Obviously, this does not change the reference frame.

▷ To study a rotating reference frame, we write the interval of the laboratory frame in cylindrical coordinates $X = R\cos\Phi$, $Y = R\sin\Phi$:

$$ds^2 = dT^2 - dR^2 - R^2\,d\Phi^2 - dZ^2, \tag{9.6}$$

where Φ is the polar angle, and R is the distance from the axis of rotation.

We will number the points of the rotating system with three numbers (r, ϕ, z). Imagine a disc rotating in the plane $Z = 0$ with constant angular velocity ω. The trajectory of an arbitrary point in it (r, ϕ) is given by the following equations:

$$R = r, \qquad \Phi = \phi + \omega T.$$

For a given ϕ and r, the angular coordinate Φ changes with time, T, with constant angular velocity ω. For the time we choose the simplest transformation $T = t$. As a result, the transformations between the rotating and the laboratory frames have the form:

$$T = t, \qquad R = r, \qquad \Phi = \phi + \omega t, \qquad Z = z, \tag{9.7}$$

where the coordinates (t, r, ϕ, z) are called the ***Born coordinates***. Substituting the differentials of these transforms

$$dT = dt, \qquad dR = dr, \qquad d\Phi = d\phi + \omega dt, \qquad dZ = dz$$

into the interval (9.6), we have:

$$ds^2 = (1 - \omega^2 r^2)\,dt^2 - 2\omega\, r^2\,dt\,d\phi - dr^2 - r^2\,d\phi^2 - dz^2. \tag{9.8}$$

The proper time $d\tau_0 = ds$ of the fixed point ($dr = d\phi = dz = 0$) is

$$d\tau_0 = dt\sqrt{1 - \omega^2 r^2} = dT\,\sqrt{1 - \omega^2 r^2}.$$

Since ωr is the linear velocity of the point in the laboratory frame, we again obtain the standard formula for time dilation.

Zero intervals ($ds^2 = 0$) give the equations for the trajectories of light pulses. For example, if a pulse moves in a circle,

$$r = const,$$

as can be achieved by means of mirrors or a light guide, the trajectory has the form:

$$\frac{d\phi}{dt} = \pm\frac{1}{r} - \omega.$$

From this it follows that the angular coordinate of the light pulse is linear with respect to the coordinate time t.

▷ In the general case, the transformation from the non-inertial frame $x^\mu = (t,\ x,\ y,\ z)$ to the inertial frame $X^\mu = (T,\ X,\ Y,\ Z)$ has the form

$$X^\mu = X^\mu(x^0, x^1, x^2, x^3), \tag{9.9}$$

where $\mu = 0, \ldots, 3$. Expanding the differential $dX^\mu = \partial_\nu X^\mu \, dx^\nu$, we obtain

$$dT = \partial_0 X^0 \, dt + \partial_i X^0 \, dx^i, \qquad dX^k = \partial_0 X^k \, dt + \partial_i X^k \, dx^i, \tag{9.10}$$

where $\partial_i = \partial/\partial x^i$ and we are summing over repeated indices. By substituting dT, dX^k into the interval $ds^2 = dT^2 - dX^k \, dX^k$, we obtain

$$ds^2 = g_{\mu\nu} \, dx^\mu dx^\nu, \tag{9.11}$$

where the coefficients of the metric tensor are (summing over k from 1 to 3):

$$\begin{aligned} g_{00} &= \partial_0 X^0 \, \partial_0 X^0 - \partial_0 X^k \, \partial_0 X^k, \\ g_{0i} &= \partial_0 X^0 \, \partial_i X^0 - \partial_0 X^k \, \partial_i X^k, \\ g_{ij} &= \partial_i X^0 \, \partial_j X^0 - \partial_i X^k \, \partial_j X^k. \end{aligned} \tag{9.12}$$

From (9.10), it follows that the velocity components of a fixed point in the non-inertial frame ($dx^k = 0$) are

$$U^k = \frac{dX^k}{dT} = \frac{\partial_0 X^k}{\partial_0 X^0}. \tag{9.13}$$

Therefore, the coefficient g_{00} can be rewritten as

$$g_{00} = (\partial_0 X^0)^2 \, (1 - \mathbf{U}^2), \tag{9.14}$$

where $\mathbf{U}^2 = (U^1)^2 + (U^2)^2 + (U^3)^2$. Let us find the proper time of the clock by using the interval ds with $dx^i = 0$:

$$d\tau_0 = \sqrt{g_{00}} \, dt = \sqrt{1 - \mathbf{U}^2} \, \partial_0 X^0 \, dt = \sqrt{1 - \mathbf{U}^2} \, dT,$$

where have used (9.10) with $dx^i = 0$ to identify dT in the final equality. Thus, we again obtain relativistic time dilation.

Sometimes it is more convenient to use the following special transformation:

$$T = t, \qquad \mathbf{X} = \mathbf{R}(t, \, x^1, x^2, x^3), \tag{9.15}$$

where the function $\mathbf{R}(t, \, x^i)$ defines the trajectory $\mathbf{X}(T) = \mathbf{R}(T, \, x^i)$ of a given point ($x^i = const$) in the non-inertial frame relative to the laboratory frame. The corresponding interval is

$$ds^2 = (1 - \mathbf{U}^2) \, dt^2 - 2(\mathbf{U} \, \partial_i \mathbf{R}) \, dt \, dx^i - (\partial_i \mathbf{R} \, \partial_j \mathbf{R}) \, dx^i \, dx^j, \tag{9.16}$$

where $\mathbf{U} = \partial_0 \mathbf{R} = \partial \mathbf{R}/\partial t = \partial \mathbf{R}/\partial T$ is the velocity of the point in the non-inertial frame, as seen in the laboratory frame.

9.2 Linearly accelerating frames *

We now consider another class of non-inertial reference frames (NIRF) for which the interval between events takes on a simple form. First, let the NIRF move along the X-axis of the laboratory frame with varying velocity. Consider the following linear transformations with respect to the coordinate x [18]:

$$T = \gamma(t)\, v(t)\, x + \int_0^t \gamma(\tau)\, d\tau, \qquad X = \gamma(t)\, x + \int_0^t \gamma(\tau)\, v(\tau)\, d\tau, \tag{9.17}$$

where $v = v(t)$ is an arbitrary function of time, $\gamma = \gamma(t) = 1/\sqrt{1-v^2}$, $Y = y$ and $Z = z$. If the velocity v is constant, then (9.17) gives the Lorentz transformations. In general we have

$$dT = \gamma\,(dt + v\,dx) + \gamma^3\,\dot{v}\,x\,dt, \qquad dX = \gamma\,(dx + v\,dt) + \gamma^3\; v\dot{v}\,x\,dt,$$

where we used that

$$\frac{d\gamma}{dt} = \gamma^3 v\dot{v}, \qquad \frac{d(\gamma v)}{dt} = \gamma^3\,\dot{v},$$

and the dot denotes differentiation with respect to t. Substituting these differentials into the interval of the laboratory frame $ds^2 = dT^2 - dX^2 - dY^2 - dZ^2$, we have:

$$ds^2 = (1 + \gamma^2\dot{v}x)^2\, dt^2 - dx^2 - dy^2 - dz^2. \tag{9.18}$$

Only g_{00} has a non-trivial value in such a frame. For $\gamma^2\,\dot{v} = a = const$ the interval corresponds to a rigid uniformly accelerating frame (9.3) such that

$$v(t) = \mathrm{th}(at), \qquad \gamma(t) = \mathrm{ch}(at). \tag{9.19}$$

In general, an arbitrary fixed point ($dx = 0$) of the frame with coordinate x moves relative to the laboratory frame with velocity

$$U(T) = \frac{dX}{dT} = v(t),$$

where we should use the Lorentz time T of the laboratory frame rather than the coordinate time t in the function $v(t)$. The time T is found from the first transformation of (9.17) and, generally speaking, ***depends*** on x. A rigid uniformly accelerating frame (9.3) is distinguished from the broader class of linearly accelerating non-inertial reference frames (9.17) by the fact that the spatial part of its interval $dx^2 + dy^2 + dz^2$ is Euclidean, and that its proper time is constant (g_{00} depends on the coordinate x, but does not depend on t).

⊳ The transformations (9.17) can be obtained by considering an inertial frame (IRF) which is co-moving with respect to the NIRF. At a given instant of time, such a frame has the same velocity relative to the laboratory frame as an observer at the origin of the NIRF. Two observers in the NIRF and in the IRF are at rest relative to each other and have the same time and length standards.

Let the origin of the NIRF have velocity $U(T)$ in the laboratory frame $S_0 : (T, X)$. At the time T_0, its coordinate is

$$X_0 = \int_0^{T_0} U(T)\, dT = \int_0^{\tau} \gamma(\tau)\, U(\tau)\, d\tau,$$

where we have used the proper time of the origin

$$d\tau = \sqrt{1 - U^2(T)}\, dT,$$

or $dT = \gamma(\tau)\, d\tau$, in the second equation, and, as usual,

$$\gamma = \frac{1}{\sqrt{1 - U^2}}.$$

The time T_0 can also be expressed as an integral over the proper time:

$$T_0 = \int_0^{T_0} dT = \int_0^{\tau} \gamma(\tau)\, d\tau.$$

If the clock of the co-moving frame $S_0' : \{T', X'\}$ shows the time $T' = 0$ at the point $X' = 0$ at the time instant T_0, the transformation between S_0' and S_0 can be written as:

$$\begin{aligned} T &= T_0 + \gamma\,(T' + UX'), \\ X &= X_0 + \gamma\,(X' + UT'). \end{aligned}$$

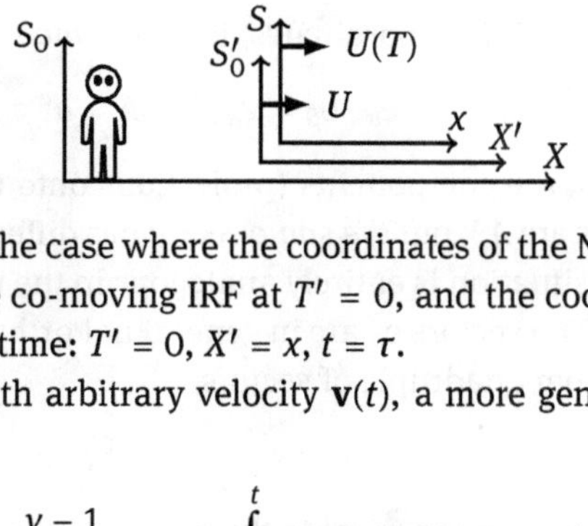

The transformations (9.17) are obtained in the case where the coordinates of the NIRF points coincide with the coordinates of the co-moving IRF at $T' = 0$, and the coordinate time in the NIRF is equal to its proper time: $T' = 0$, $X' = x$, $t = \tau$.

If the origin of the NIRF is moving with arbitrary velocity $\mathbf{v}(t)$, a more general transformation can be written:

$$T = \gamma\,\mathbf{vr} + \int_0^t \gamma(\tau)\, d\tau, \qquad \mathbf{X} = \mathbf{r} + \frac{\gamma - 1}{v^2}\,(\mathbf{vr})\,\mathbf{v} + \int_0^t \gamma(\tau)\,\mathbf{v}(\tau)\, d\tau. \tag{9.20}$$

It would be a good exercise to find its interval in the coordinates $(t, \mathbf{r})$ (< H_{35}).

9.3 Curvilinear bases

In inertial reference frames, it is convenient to use *Lorentz coordinates*

$$X^\alpha = \{T, \mathbf{R}\} = \{T, X, Y, Z\},$$

where T is physical time, and $\mathbf{R} = \{X,\ Y\ Z\}$ are the Cartesian coordinates of the event. The interval between two infinitely close events in Lorentz coordinates is

$$ds^2 = dT^2 - d\mathbf{R}^2 = \eta_{\alpha\beta}\, dX^\alpha dX^\beta,$$

where $\eta_{\alpha\beta} = \mathrm{diag}(1,-1,-1,-1)$ is the diagonal metric tensor. The coordinates and the time of the same event in a non-inertial reference frame will be called *curvilinear coordinates* and denoted using small typeface, x^α. In general, the interval between the same events in these coordinates is expressed in terms of the non-diagonal metric tensor $g_{\alpha\beta}$:

$$ds^2 = g_{\alpha\beta}\, dx^\alpha dx^\beta.$$

Remember that a four-vector A (block-letter typeface) is a physical object. It does not depend on the coordinate system (it is an "arrow" in four-space). However, its components A^α depend on the specific coordinate description of the space (reference frame). Suppose that the points of four-space are numbered using Lorentz coordinates. Let us introduce four four-vectors, forming a *basis*, at each of its points:

$$\mathrm{n}_0,\quad \mathrm{n}_1,\quad \mathrm{n}_2,\quad \mathrm{n}_3$$

(the index represents the number of the vector). The products of these vectors give the coefficients of the metric tensor $\eta_{\alpha\beta}$, which can be used to decompose any four-vector (p. 176):

$$\mathrm{n}_\alpha \cdot \mathrm{n}_\beta = \eta_{\alpha\beta}, \qquad \mathrm{A} = A^\alpha\, \mathrm{n}_\alpha, \qquad A_\alpha = \eta_{\alpha\beta}\, A^\beta = \mathrm{A}\cdot \mathrm{n}_\alpha.$$

In curvilinear coordinates, the *curvilinear basis* is defined by:

$$\mathrm{e}_\alpha \cdot \mathrm{e}_\beta = g_{\alpha\beta}, \qquad \mathrm{A} = a^\alpha\, \mathrm{e}_\alpha, \qquad a_\alpha = g_{\alpha\beta}\, a^\beta = \mathrm{A}\cdot \mathrm{e}_\alpha.$$

Thus, the components (projections onto the basis) of a four-vector A in the Lorentz basis are A^α, but *the same* vector has different components a^α in the curvilinear basis. The situation is entirely analogous in the usual three-dimensional space [1].

The vectors e_α are in general not orthogonal. Therefore, it is convenient to define another quadruple of vectors

$$\mathrm{e}^0,\ \mathrm{e}^1,\ \mathrm{e}^2,\ \mathrm{e}^3,$$

which are orthogonal to them:

$$\mathrm{e}^\alpha \cdot \mathrm{e}_\beta = \delta^\alpha_\beta, \qquad \mathrm{e}^\alpha \cdot \mathrm{e}^\beta = g^{\alpha\beta}, \qquad \mathrm{e}^\alpha = g^{\alpha\beta}\, \mathrm{e}_\beta, \qquad \mathrm{A} = a_\alpha \mathrm{e}^\alpha.$$

These vectors form a *reciprocal basis*, whose expansion coefficients a_α are covariant components of the four-vector A.

▷ Consider an infinitely small displacement $d\mathrm{X}$ in the space of events, represented by a four-vector which connects two close events. It can be decomposed in terms of any system of basis vectors, and its components are the coordinate differentials:

$$d\mathrm{X} = \mathrm{n}_\beta\, dX^\beta = \mathrm{e}_\alpha\, dx^\alpha. \tag{9.21}$$

The square of this vector is the interval between the events:

$$ds^2 = d\mathrm{X} \cdot d\mathrm{X} = (\mathrm{e}_\alpha \cdot \mathrm{e}_\beta)\, dx^\alpha dx^\beta = g_{\alpha\beta}\, dx^\alpha dx^\beta.$$

Fixing all but one of the coordinates x^α in (9.21), we obtain the relationship

$$\mathrm{e}_\alpha = \frac{\partial X^\beta}{\partial x^\alpha}\, \mathrm{n}_\beta \equiv \partial_\alpha X^\beta\, \mathrm{n}_\beta.$$

Thus, for a rigid uniformly accelerating frame $aT = \mathrm{sh}(at)\, e^{ax}$, $aX = \mathrm{ch}(at)\, e^{ax} - 1$, we have:

$$\begin{cases} \mathrm{e}_t \equiv \mathrm{e}_0 = \frac{\partial T}{\partial t}\, \mathrm{n}_0 + \frac{\partial X}{\partial t}\, \mathrm{n}_1 = \big[\mathrm{n}_0\, \mathrm{ch}(at) + \mathrm{n}_1\, \mathrm{sh}(at)\big]\, e^{ax}, \\ \mathrm{e}_x \equiv \mathrm{e}_1 = \frac{\partial T}{\partial x}\, \mathrm{n}_0 + \frac{\partial X}{\partial x}\, \mathrm{n}_1 = \big[\mathrm{n}_0\, \mathrm{sh}(at) + \mathrm{n}_1\, \mathrm{ch}(at)\big]\, e^{ax}. \end{cases} \tag{9.22}$$

From $\mathrm{n}_0^2 = -\mathrm{n}_1^2 = 1$ and $\mathrm{n}_0 \cdot \mathrm{n}_1 = 0$ it follows that $\mathrm{e}_t^2 = -\mathrm{e}_x^2 = e^{2ax}$ and $\mathrm{e}_t \cdot \mathrm{e}_x = 0$, which gives the following interval: $ds^2 = e^{2ax}\,(dt^2 - dx^2)$.

These basis vectors can be interpreted visually on the pseudo-Euclidean plane. To do this, imagine a coordinate grid on the plane such that a given coordinate is constant along each line (p. 149). In Lorentz coordinates $\{T,\ X\}$, the grid is rectangular. On the same axes $\{T,\ X\}$, the grid lines $t = const$ and $x = const$ will be curved:

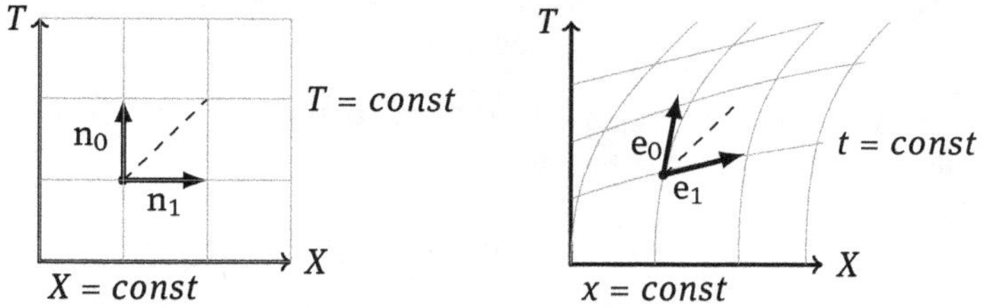

The basis vectors are tangential to the corresponding coordinate lines, as follows from the definition $d\mathrm{X} = \mathrm{e}_\alpha\, dx^\alpha$ (e_0 is proportional to $d\mathrm{X}$ when x^0 changes, etc.). They depend on the particular point in four-space and have different orientations and magnitudes at different points (in contrast to n_0, n_1). The vectors e_0 and e_1 do not appear orthogonal, because pseudo-Euclidean space cannot be accurately represented on the page, which is Euclidean (p. 170). The components of two orthogonal vectors $\mathrm{a} \cdot \mathrm{b} = 0$ are related as follows: $a^0 b^0 = a^1 b^1$. Therefore, the median between two orthogonal vectors is at 45° to the T-axis, i.e., it is the trajectory of a light signal (the dashed line in the figures).

▷ Let us now consider the transformations of various quantities under a change of curvilinear coordinates. Consider two reference frames, and decompose the displacement vector dX in the space of events over their bases:

$$d\mathrm{X} = \mathrm{e}_\alpha\, dx^\alpha = \mathrm{e}'_\alpha\, dx'^\alpha.$$

From this equation we obtain the following relationship between the bases:

$$\mathrm{e}'_\alpha = \frac{\partial x^\beta}{\partial x'^\alpha}\, \mathrm{e}_\beta, \qquad \mathrm{e}_\alpha = \frac{\partial x'^\beta}{\partial x^\alpha}\, \mathrm{e}'_\beta. \tag{9.23}$$

By the chain rule, we have

$$\frac{\partial x'^\alpha}{\partial x'^\beta} = \delta^\alpha_\beta = \frac{\partial x'^\alpha}{\partial x^\gamma}\frac{\partial x^\gamma}{\partial x'^\beta}, \qquad \frac{\partial x^\alpha}{\partial x^\beta} = \delta^\alpha_\beta = \frac{\partial x^\alpha}{\partial x'^\gamma}\frac{\partial x'^\gamma}{\partial x^\beta}, \tag{9.24}$$

where δ^α_β is the Kronecker symbol, and we are summing over γ. Therefore, the matrix $\partial x^\alpha/\partial x'^\beta$ is the inverse of $\partial x'^\alpha/\partial x^\beta$.

The vectors of the reciprocal basis are transformed as

$$\mathrm{e}'^\alpha = \frac{\partial x'^\alpha}{\partial x^\beta}\,\mathrm{e}^\beta, \qquad \mathrm{e}^\alpha = \frac{\partial x^\alpha}{\partial x'^\beta}\,\mathrm{e}'^\beta. \tag{9.25}$$

These relations correspond to the definition $\mathrm{e}^\alpha \cdot \mathrm{e}_\beta = \delta^\alpha_\beta$ and the conditions (9.24) for the transformation matrices. Multiplying them by e_γ and e'_γ respectively, we obtain the following representation for the transformation matrices:

$$\mathrm{e}'^\alpha \cdot \mathrm{e}_\beta = \frac{\partial x'^\alpha}{\partial x^\beta}, \qquad \mathrm{e}^\alpha \cdot \mathrm{e}'_\beta = \frac{\partial x^\alpha}{\partial x'^\beta}. \tag{9.26}$$

An arbitrary four-vector A is a physical object, and can be decomposed over each basis:

$$\mathrm{A} = \mathrm{e}_\alpha A^\alpha = \mathrm{e}'_\alpha A'^\alpha.$$

Multiply both sides by e'^β or e^β. Using (9.26), we then obtain the transformation law for the contravariant components of the vector:

$$A'^\alpha = \frac{\partial x'^\alpha}{\partial x^\beta}\,A^\beta, \qquad A^\alpha = \frac{\partial x^\alpha}{\partial x'^\beta}\,A'^\beta. \tag{9.27}$$

The transformations for the covariant components can be found similarly:

$$A'_\alpha = \frac{\partial x^\beta}{\partial x'^\alpha}\,A_\beta, \qquad A_\alpha = \frac{\partial x'^\beta}{\partial x^\alpha}\,A'_\beta. \tag{9.28}$$

They contain the inverses of the matrices used in the transformations of the contravariant components.

To memorize the expressions (9.27) and (9.28), note that the contravariant components of the vector A^α are transformed like the differentials of the coordinates dx^α, while the covariant components A_α transform like partial derivative ∂_α. The relationships (9.23) represent the linear decomposition of a basis *vector* over the vectors of another basis. The similar relations (9.27) and (9.28) are decompositions of vector *components* rather, than of the vectors themselves.

As expected, the product of two four-vectors is invariant with respect to the transformations (9.27):

$$\mathrm{A}\cdot\mathrm{B} = A'^\alpha B'_\alpha = \frac{\partial x'^\alpha}{\partial x^\beta}\frac{\partial x^\gamma}{\partial x'^\alpha}\,A^\beta B_\gamma = \delta^\gamma_\beta A^\beta B_\gamma = A^\beta B_\beta.$$

Naturally, all of these relations are also true for Lorentz coordinates in different inertial frames. These frames are related by linear Lorentz transforms:

$$x'^{\alpha} = \Lambda^{\alpha}_{\ \beta}\, x^{\beta},$$

where the factors $\Lambda^{\alpha}_{\ \beta}$ depend on the relative speed and the angles of rotation of the Cartesian axes (p. 180), but do not depend on the coordinates. Therefore,

$$\frac{\partial x'^{\alpha}}{\partial x^{\beta}} = \Lambda^{\alpha}_{\ \beta}. \tag{9.29}$$

Rules for transforming tensors between different frames are found similarly to the transformations of four-vector components. A tensor can have an arbitrary number of upper and lower indices (p. 185). For each of these, it is transformed either as the components of a contravariant or a covariant vector. For example:

$$F'^{\mu\nu} = \frac{\partial x'^{\mu}}{\partial x^{\alpha}} \frac{\partial x'^{\nu}}{\partial x^{\beta}} F^{\alpha\beta}.$$

In a similar way, we can write the transformations for the other kinds of second-order tensor,

$$F_{\mu\nu} \sim A_{\mu} B_{\nu}, \qquad F^{\mu}_{\ \nu} \sim A^{\mu} B_{\nu},$$

and for higher-order tensors, where the tilde means "transforms as". In particular, the metric tensor is a tensor quantity ($\mathrm{A} \cdot \mathrm{B} = g_{\alpha\beta} A^{\alpha} B^{\beta}$ is invariant). For example, transformations from Lorentz coordinates X^{α} to curvilinear co-ordinates x^{α} have the form:

$$g_{\alpha\beta} = \frac{\partial X^{\mu}}{\partial x^{\alpha}} \frac{\partial X^{\nu}}{\partial x^{\beta}} \eta_{\mu\nu}. \tag{9.30}$$

The same relationship for the components was obtained on p. 217.

9.4 Physical length and time

Using the metric tensor $g_{\alpha\beta}$, we can write the general expression for an infinitesimal distance in space-time:

$$ds^2 = g_{\alpha\beta}\, dx^{\alpha} dx^{\beta} = g_{00}\, dt^2 + 2g_{0i}\, dt\, dx^i + g_{ij}\, dx^i dx^j,$$

where the Greek indices α, β vary from 0 to 3, the Latin symbols i, j vary from 1 to 3,

$$dx^{\alpha} = \{dt, dx^i\},$$

and $g_{0i} = g_{i0}$. Let us complete the square in this expression for those terms which depend on the differential of the coordinate time:

$$ds^2 = \left(\sqrt{g_{00}}\, dt + \frac{g_{0i}}{\sqrt{g_{00}}}\, dx^i\right)^2 - \left(\frac{g_{0i} g_{0j}}{g_{00}} - g_{ij}\right) dx^i dx^j. \tag{9.31}$$

This interval has the form of a distance in a pseudo-Euclidean space in Lorentz coordinates:

$$ds^2 = \delta\tau^2 - \delta l^2,$$

where the infinitesimals

$$\delta\tau = \sqrt{g_{00}}\, dt + \frac{g_{0i}}{\sqrt{g_{00}}}\, dx^i, \qquad \delta l^2 = \left(\frac{g_{0i}\, g_{0j}}{g_{00}} - g_{ij}\right) dx^i dx^j \tag{9.32}$$

are called the ***physical time*** and the square of the ***physical distance***. The latter is conveniently written using a three-dimensional tensor,

$$\gamma_{ij} = \frac{g_{0i}\, g_{0j}}{g_{00}} - g_{ij}, \qquad \delta l^2 = \gamma_{ij}\, dx^i dx^j. \tag{9.33}$$

Here we have used the symbol "δ" rather than "d" to denote infinitesimals. In fact, we will see below that in general, the quantities $\delta\tau$ and δl are not differentials.

Besides $\delta\tau$ and δl, there is the ***proper time*** $\delta\tau_0$ of the clock at any ***fixed*** point of space. It is equal to the interval ds between events for which $dx^1 = dx^2 = dx^3 = 0$ or $\delta l = 0$:

$$\delta\tau_0 = \sqrt{g_{00}}\, dt. \tag{9.34}$$

Obviously,

$$\delta\tau = \delta\tau_0 \quad \text{for} \quad dx^k = 0.$$

As before, we will take the proper time to be the physical time of the given clock in the non-inertial frame. The clock moves with variable velocity relative to the inertial frame, and its time slows down according to the standard relativistic formula.

▷ To clarify the meaning of δl, consider radar measurements performed by an observer at point $A : (x^1, x^2, x^3)$ to measure the distance to a nearby point $B : (x^1 + dx^1, x^2 + dx^2, x^3 + dx^3)$. The interval between the events is zero $ds = 0$ for a moving light pulse. Therefore, from (9.31), we have:

$$\sqrt{g_{00}}\, dt_\pm = -\frac{g_{0i}}{\sqrt{g_{00}}}\, dx^i \pm \sqrt{\delta l^2}. \tag{9.35}$$

The case $dt_+ > 0$ corresponds to motion in the direction of coordinate increase, and $dt_- < 0$ corresponds to the opposite direction. The ***coordinate speed of light***, e.g. along the coordinate x^1, is ($dx^1 > 0$, $dx^2 = dx^3 = 0$):

$$\frac{dx^1}{dt} = \frac{g_{00}}{\pm\sqrt{g_{01}^2 - g_{00}g_{11}} - g_{01}}. \tag{9.36}$$

Its sign is the same as that of the square root term in the denominator (verify this for the rotating reference frame, p. 216).

Let the signal be reflected from the point B at time t (by the clock of the observer in A). Then, it started from A at $t_1 = t - dt_+ < t$ and will come back at $t_2 = t - dt_- > t$. Half of the proper time interval $\sqrt{g_{00}}\,(t_2 - t_1)/2$ between these events is:

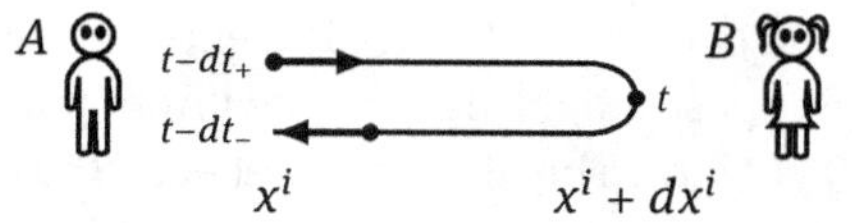

$$\frac{\sqrt{g_{00}}\,(dt_+ - dt_-)}{2} = \sqrt{\delta l^2}.$$

Thus, the physical distance δl is the ***radar distance*** measured by the observer at the point (x^1, x^2, x^3).

▷ The meaning of δl can be clarified by the following reasoning. Suppose that there is a co-moving IRF in the vicinity of some fixed point of the NIRF at some given instant of time. *This* point of the NIRF has zero velocity $U^k = 0$ relative to the IRF, and from (9.13) it follows that $\partial_0 X^k = 0$. This derivative is zero for the given values $\{t,\ x^i\}$, and we obtain for them the following form for the metric tensor (9.12):

$$g_{00} = (\partial_0 X^0)^2, \qquad g_{0i} = \partial_0 X^0\, \partial_i X^0, \qquad g_{ij} = \partial_i X^0\, \partial_j X^0 - \partial_i X^k\, \partial_j X^k,$$

whence:

$$\gamma_{ij} = \partial_i X^k\, \partial_j X^k.$$

On the other hand, the distance in the IRF is

$$\delta l^2 = dX^2 + dY^2 + dZ^2 = dX^k\, dX^k = \partial_i X^k\, \partial_j X^k\, dx^i dx^j = \gamma_{ij}\, dx^i dx^j.$$

Equivalent rulers give the same distance if used by a *given* observer in the NIRF and in *his* corresponding co-moving IRF.

▷ The physical time $\delta\tau$ for $dx^i \neq 0$ corresponds to the difference between the proper times of two infinitely close but *differently* synchronized clocks. Indeed, consider the standard synchronization procedure. Let the signal be sent at $t_1 = t - dt_+$ from point $A = (x^1,\ x^2,\ x^3)$ so that it reaches point $B = (x^1 + dx^1,\ x^2 + dx^2,\ x^3 + dx^3)$ at the instant of time t and arrives back at A at $t_2 = t - dt_-$. All of these times are coordinate times. They are related to the clocks' proper times τ_A and τ_B at the points A and B. By definition, the clocks are synchronized if the following relation is true:

$$\tau_B(t) = \frac{\tau_A(t_1) + \tau_A(t_2)}{2}. \tag{9.37}$$

For an infinitesimal change of the coordinate time, we have from (9.34) the following relation for the physical proper time of the clock at the point A:

$$\tau_A(t + dt) = \tau_A(t) + \delta\tau_A = \tau_A(t) + \sqrt{g_{00}}\, dt.$$

Using the values $dt_\pm$ found from the radar measurements (9.35), we find from (9.37) that $\tau_A(t) - \sqrt{g_{00}}\,(dt_+ + dt_0)/2$, or

$$\tau_B(t) = \tau_A(t) + \frac{g_{0i}\, dx^i}{\sqrt{g_{00}}}. \tag{9.38}$$

This relationship specifies the rule for synchronizing two infinitely close clocks; this rule is a function of the coordinate time t of the observer at the point A.

Their clocks being synchronized, we can consider the time difference between the events that occur at the different points. Suppose a light signal is sent from point A at the coordinate time t and reaches point B at $t + dt$. The difference between the proper times of the clocks at the points B and A is equal to the physical time found in (9.32):

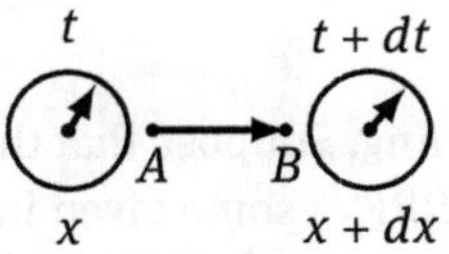

$$\delta\tau = \tau_B(t + dt) - \tau_A(t) = \sqrt{g_{00}}\, dt + \frac{g_{0i}}{\sqrt{g_{00}}}\, dx^i.$$

Two synchronized clocks can be used to determine the speed of light as it travels from A to B. It is equal to

$$\frac{\delta l}{\delta\tau} = 1,$$

where we have used the fact that

$$ds^2 = \delta\tau^2 - \delta l^2 = 0$$

for light. The *physical speed of light* just defined is always unity (or "c", if the fundamental speed constant is used). However, the *coordinate velocity* dx^i/dt of the light pulse (9.36) is, generally speaking, not unity, and can be arbitrarily large.

▷ Using the procedure (9.38) we can define a function $\tau(t,\ x^k)$ which takes the coordinates of the event measured by the observer A and calculates from it the current value of the proper time of another observer close to the event. For example, in the above equation (9.38), we have $\tau_A(t) = \tau(t, x^i)$ and $\tau_B(t) = \tau(t, x^i + dx^i)$. When can we define such a function?

The differential of a function $\tau = \tau(t,\ x^i)$ of several variables is

$$d\tau = \frac{\partial\tau}{\partial t}\, dt + \frac{\partial\tau}{\partial x^i}\, dx^i.$$

Therefore, $\delta\tau$ (9.32) is a differential if

$$\frac{\partial\tau}{\partial t} = \sqrt{g_{00}}, \qquad \frac{\partial\tau}{\partial x^i} = \frac{g_{0i}}{\sqrt{g_{00}}}. \tag{9.39}$$

Partial derivatives must be commutative. We differentiate $\partial\tau/\partial t$ in (9.39) with respect to x^i, and $\partial\tau/\partial x^i$ with respect to t, and equate the results. Repeating the same procedure for the derivatives with respect to x^i and x^j then yields

$$\frac{\partial\sqrt{g_{00}}}{\partial x^i} = \frac{\partial}{\partial t}\frac{g_{0i}}{\sqrt{g_{00}}}, \qquad \frac{\partial}{\partial x^i}\frac{g_{0j}}{\sqrt{g_{00}}} = \frac{\partial}{\partial x^j}\frac{g_{0i}}{\sqrt{g_{00}}}. \tag{9.40}$$

If the conditions (9.40) are not satisfied, the physical time $\delta\tau$ is not a total differential. In this case, the integral between two events P_1 and P_2 in four-space

$$\tau = \int_{P_1}^{P_2} \delta\tau = \int_{P_1}^{P_2} [\sqrt{g_{00}}\, dt + \frac{g_{0i}}{\sqrt{g_{00}}}\, dx^i]$$

depends on the path of integration. For example, in a rigid uniformly accelerating frame (9.4), p. 215, we have the following curvilinear integral:

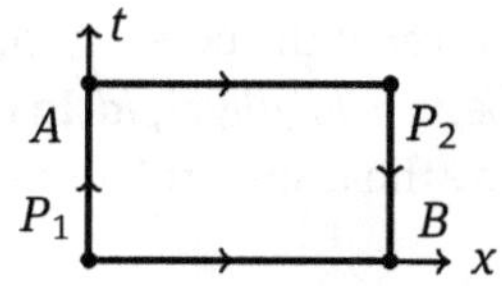

$$\tau = \int_{P_1}^{P_2} [e^{ax}\,dt + 0 \cdot dx].$$

If the integral is calculated between $P_1 = (0,\ 0)$ and $P_2 = (t,\ x)$ along the line $P_1\,A\,P_2$, we obtain $\tau = t$. The integral along the line $P_1\,B\,P_2$ gives another value,

$$\tau = e^{ax}\,t.$$

Therefore, clocks with different x coordinates cannot be synchronized. Physically, this is due to the fact that time flows differently in a uniformly accelerating frame. The initial synchronization of the clocks on the spaceships when they are in the spaceports "collapses" with time and the relation (9.38) cannot always be satisfied.

Thus, one should integrate (9.32) to obtain the clock time $\tau(t,\ x^i)$, synchronized across the entire space. This can only be done if $\delta\tau$ is a differential.

▷ Periodicity conditions for the coordinates x^i can impose additional restrictions on the possibility of synchronizing clocks. Recall the expression for physical time in the rotating frame (9.8):

$$\delta\tau = \sqrt{1-(\omega r)^2}\,dt - \frac{\omega r^2\,d\phi}{\sqrt{1-(\omega r)^2}}.$$

For clocks (belonging to observers) located at an equal distance from the centre ($r =$ *const*), this expression is a differential:

$$\tau = \sqrt{1-(\omega r)^2}\,t - \frac{\omega r^2\phi}{\sqrt{1-(\omega r)^2}}.$$

We obtained such an expression for the physical time after choosing the clock synchronization condition (6.39), p. 153 (there, physical time was denoted by t, and T was the time of the inertial frame, which coincides, in Born coordinates, with the coordinate time t of this section).

Even though $\delta\tau$ is a total differential, the function $\tau = \tau(t,\ \phi)$ is not single-valued, because the points $\phi = 0$ and $\phi = 2\pi$ are equivalent. Therefore, we cannot synchronize all clocks moving in a circle. If r is not supposed to be constant, then $\delta\tau$ is not a total differential. The reason is familiar from the situation in a rigid uniformly accelerating frame: the flow of proper time is different for points located at different distances from the centre.

In the rotating frame, the physical length is given by

$$\delta l^2 = \frac{\omega^2 r^4\,d\phi^2}{1-\omega^2 r^2} + dr^2 + r^2\,d\phi^2 + dz^2 = dr^2 + \frac{r^2\,d\phi^2}{1-\omega^2 r^2} + dz^2. \tag{9.41}$$

This result was obtained on p. 158 when we considered a radar experiment conducted by observers on ships revolving about a common centre.

The physical length (9.41) does not depend on time. This means that, when measuring the radar distance to a point next to him, the observer at the point $(r,\ \phi,\ z)$ will get the same value at any instant. A frame with this property is ***locally rigid***. In general, the frame of reference obtained from an inertial frame by the transformations

$$T = t, \quad \mathbf{X} = \mathbf{R}(t,\ x^1,\ x^2,\ x^3), \quad \mathbf{V} = \partial_0 \mathbf{R}$$

is locally rigid if the tensor

$$\gamma_{ij} = \frac{(\mathbf{V}\partial_i\mathbf{R})\,(\mathbf{V}\partial_j\mathbf{R})}{1-\mathbf{V}^2} + \partial_i\mathbf{R}\,\partial_j\mathbf{R}, \tag{9.42}$$

does not depend on time. If the frame moves by translation and is accelerating (9.17), it is locally rigid for any function $v(t)$.

▷ The impossibility of synchronizing clocks is typical for non-inertial frames. As we know, there are no such problems in inertial frames. Consider as an example how the Lorentz transformations are derived from Galilean coordinate transformations [19]:

$$T = t, \qquad X = x + vt. \tag{9.43}$$

The quantities T and X will be interpreted as the physical time and the coordinate of an event in $S_0 : \{T,\ X\}$. The points of $S : \{t,\ x\}$ move with velocity v relative to S_0. This is used in the transformation $X = x + vt$ and, therefore, it follows (as $t = T$) that the trajectory of a point in the inertial frame S_0 with coordinate x has the form $X = x + vT$. Such a transformation is an example of the more general one in (9.15).

The transformations (9.43) are quite arbitrary, and the numbers $\{t,\ x\}$ are therefore coordinate values. To find their relation to physical quantities, consider the interval

$$ds^2 = dT^2 - dX^2$$

in the new coordinates:

$$ds^2 = (1-v^2)\,dt^2 - 2v\,dt\,dx - dx^2. \tag{9.44}$$

From this, using general formulas or completing the square in dt, we obtain

$$\delta\tau = \sqrt{1-v^2}\,dt - \frac{v\,dx}{\sqrt{1-v^2}}, \qquad \delta l = \frac{dx}{\sqrt{1-v^2}}.$$

The coefficients of the differentials dt and dx are constant in the expression for $\delta\tau$. Therefore, the conditions for the possibility of clock synchronization (9.40) are satisfied, and $\delta\tau$ and δl can be integrated:

$$\tau = t\,\sqrt{1-v^2} - \frac{vx}{\sqrt{1-v^2}}, \qquad l = \frac{x}{\sqrt{1-v^2}}.$$

Expressing $\{t,\ x\}$ in terms of $\{\tau,\ l\}$ and substituting into (9.43), we obtain the Lorentz transformations:

$$T = \frac{\tau + vl}{\sqrt{1 - v^2}}, \qquad X = \frac{x + v\tau}{\sqrt{1 - v^2}},$$

which connect physical quantities rather than coordinate quantities. The coordinate speed of light is obtained from the condition $ds = 0$:

$$\frac{dx}{dt} = \pm 1 - v.$$

For motion in the negative x-direction (the minus sign), the magnitude of the velocity is greater than unity. Of course, the physical speed of light $\delta l/\delta\tau$ is still unity.

9.5 Invariance of physical quantities *

Transferring to a new coordinate time and another enumeration of the points in three-space

$$x'^0 = x'^0(x^0,\ x^1, x^2, x^3), \qquad x'^i = x'^i(x^1,\ x^2,\ x^3) \tag{9.45}$$

does not change the reference frame, and should therefore leave physical time and length unchanged. Let us demonstrate that this is the case. First, we write the interval of physical time (9.32) in the following form:

$$\delta\tau = \frac{d\mathrm{x} \cdot \mathrm{e}_0}{|\mathrm{e}_0|}. \tag{9.46}$$

Indeed, since $d\mathrm{x} = dx^\alpha\, \mathrm{e}_\alpha$ and $\mathrm{e}_\alpha \mathrm{e}_\beta = g_{\alpha\beta}$, we have

$$d\mathrm{x} \cdot \mathrm{e}_0 = dx^\alpha\, \mathrm{e}_\alpha \cdot \mathrm{e}_0 = g_{0\alpha}\, dx^\alpha = g_{00}\, dt + g_{0i}\, dx^i.$$

Dividing this by $|\mathrm{e}_0| = \sqrt{\mathrm{e}_0 \cdot \mathrm{e}_0} = \sqrt{g_{00}}$, we obtain $d\tau$.

Since it follows from (9.23) that

$$\mathrm{e}_0 = \frac{\partial x'^\beta}{\partial x^0}\, \mathrm{e}'_\beta = \frac{\partial x'^0}{\partial x^0}\, \mathrm{e}'_0,$$

the transformation (9.45) does not change the orientation of the basis vector e_0.

Nevertheless, the basis vectors do change with the coordinates. However, if the transformation does not change the reference frame, then it does not change the unit vector $\mathrm{e}_0/|\mathrm{e}_0|$ either. Therefore, $\delta\tau$ (9.46) is invariant under the transformations (9.45). Accordingly, due to the invariance of the interval $ds^2 = \delta\tau^2 - \delta l^2$, the physical length δl is invariant, in contrast to its tensor γ_{ij}. By writing the transformation law for the metric tensor

$$g_{\mu\nu} = \partial_\mu x'^\alpha \partial_\nu x'^\beta\, g'_{\alpha\beta}$$

and separating the time-like and space-like components (neglecting the terms with $\partial_0 x'^i = 0$), we easily obtain (< H36):

$$\gamma_{ij} = \partial_i x'^p \partial_j x'^q \gamma'_{pq}, \tag{9.47}$$

where γ'_{ij} is expressed in terms of $g'_{\mu\nu}$ in the same way as γ_{ij} is expressed in terms of $g_{\mu\nu}$. Therefore, γ_{ij} is a tensor with respect to spatial transformations (9.45).

The unit vector $\mathbf{e}_0/|\mathbf{e}_0|$ can be interpreted as the velocity four-vector of a given point of the non-inertial frame. Indeed, if we write the transforms in curvilinear coordinates (p. 220) rather than in the Lorentz coordinates, we obtain:

$$\mathbf{u} = \frac{\mathbf{e}_0}{|\mathbf{e}_0|} = \frac{\partial_0 X^\beta \, \mathbf{n}_\beta}{\sqrt{\mathbf{e}_0 \mathbf{e}_0}} = \frac{\partial_0 X^0 \mathbf{n}_0 + \partial_0 X^i \mathbf{n}_i}{\sqrt{g_{00}}} = \frac{n_0}{\sqrt{1-\mathbf{U}^2}} + \frac{U^i n_i}{\sqrt{1-\mathbf{U}^2}},$$

where the relations (9.13), (9.14), p. 217, were used to obtain the last equality.

▷ Let us consider some properties of the quantities γ_{ij} (9.33) which determine the physical length δl in the non-inertial frame.

▷ The matrix γ_{ij} is the inverse of the spatial components of the metric tensor with upper indices g^{ij}:

$$\gamma_{jk}\,\gamma^{ki} = \delta^i_j, \qquad \gamma^{ij} = -g^{ij}. \tag{9.48}$$

Indeed, writing the orthogonality condition for metric tensors with upper and lower indices $g^{\alpha\gamma}\,g_{\gamma\beta} = \delta^\alpha_\beta$ with $\alpha = i$, $\beta = 0$, we find that

$$g^{i0}\,g_{00} + g^{ik}\,g_{k0} = 0 \quad \Rightarrow \quad g^{i0} = -\frac{g^{ik}\,g_{k0}}{g_{00}}. \tag{9.49}$$

If $\alpha = i$, $\beta = j$, then substituting back in this expression for g^{i0} yields

$$\delta^i_j = g^{i0}\,g_{0j} + g^{ik}\,g_{kj} = -g^{ik}\left(\frac{g_{k0}g_{0j}}{g_{00}} - g_{kj}\right) = -g^{ik}\gamma_{kj}.$$

▷ Sometimes it is convenient to use the three-vector $\boldsymbol{\gamma}$, whose components with upper and lower indices are defined to be

$$\gamma_i = -\frac{g_{0i}}{g_{00}}, \qquad \gamma^i = -g^{0i}. \tag{9.50}$$

According to (9.49) and (9.48), these components are related to each other by means of a three-dimensional metric γ_{ij}, so that $\gamma_i = \gamma_{ij}\gamma^j$ and $\gamma^i = \gamma^{ij}\gamma_j$. Within this reference frame, coordinate transformations of the form $x'^0 = x^0$, $x'^i = x'^i(x^1, x^2, x^3)$ do not change the contraction:

$$\boldsymbol{\gamma}^2 = \gamma_i\gamma^i = \gamma_{ij}\,\gamma^i\gamma^j = \gamma^{ij}\,\gamma_i\gamma_j = \frac{1}{g_{00}} - g^{00}, \tag{9.51}$$

where the last equality follows from the orthogonality condition

$$g^{\alpha\gamma}\,g_{\gamma\beta} = \delta^\alpha_\beta$$

with $\alpha = \beta = 0$.

▷ We can also verify the relation

$$g = -g_{00}\,\gamma, \tag{9.52}$$

where $g = \det(g_{\alpha\beta})$ and $\gamma = \det(\gamma_{ij})$, i.e., on the left-hand side we have the determinant of the 4×4 matrix with entries $g_{\alpha\beta}$, and on the right-hand side the determinant of the 3×3 matrix with entries γ_{ij}. Their ratio is $-g_{00}$.

▷ Note that the condition of coordinate ***admissibility*** is $g_{00} > 0$, as follows from the fact that the root expression $\sqrt{g_{00}}$ in the equation for the proper time (9.34) should be positive. Distances should also be positive, so the form $\gamma_{ij}\,dx^i dx^j > 0$ is positive definite, which means that $\gamma = \det(\gamma_{ij}) > 0$, and due to (9.52), we thus have $g = \det(g_{\alpha\beta}) < 0$.

9.6 Rigid reference frames

Rigidity (no matter how defined), like many other effects in the theory of relativity, is a relative concept. This is easily illustrated with the example of the non-rigid uniformly accelerating frame from p. 160. Its points move along trajectories

$$X = x + \frac{1}{a}\left[\sqrt{1 + (aT)^2} - 1\right], \tag{9.53}$$

relative to the laboratory frame. In this case, points with different x-coordinates have the same proper acceleration. The distance between them in the laboratory frame remains constant ($\Delta X = \Delta x$ does not depend on the time T). However, the radar distance between two points varies with time (p. 162). Therefore, observers within this frame do not consider it rigid.

Conversely, the distance between any two observers in a rigid uniformly accelerating frame (p. 138) is constant. However, in the laboratory frame, these observers move with different velocities along trajectories

$$X = \frac{1}{a}\left[\sqrt{(1 + ax)^2 + (aT)^2} - 1\right]. \tag{9.54}$$

This "rigid" uniformly accelerating frame does not appear as such to stationary observers (ΔX decreases with time, and the distance between the points in the non-inertial frame diminishes).

! From now on, when referring to a reference frame as being rigid, we will assume that it is such for observers associated with the frame. Similarly to the proper time, we will call this the *proper rigidity* of the frame.

Note also that we always assume ***kinematic rigidity*** of frames and the bodies associated with them. This means that deformation effects due to forces at work inside of a "solid body" are beyond the scope of our consideration. A rigid frame is assumed to be a set of points, the distance between which remains in some sense constant when the frame moves. We can imagine that there is an observer with a clock and a ruler who is attached to each point. An observer in the co-moving inertial frame has equivalent time and length standards. In general, different co-moving inertial frames must be considered at different points of the non-inertial frame.

There are three possible definitions of proper rigidity:

!

I. *Co-moving rigidity*: all points of the reference frame have zero velocity in the inertial frame co-moving relative to one of its points.

II. *Local rigidity*: the tensor

$$\gamma_{ij} = -g_{ij} + \frac{g_{0i} g_{0j}}{g_{00}}$$

defining the element of infinitesimal physical length does not depend on time.

III. *Global rigidity*: the radar distance between any two points of the reference frame is constant.

These three definitions are not equivalent to each other. This may seem most unexpected for the last two definitions. The procedure that gives γ_{ij} is based on measurements of the radar distance between two infinitesimally close points. Nevertheless, it turns out that this constancy, generally speaking, does not ensure global rigidity of the reference frame. That is, infinitesimal radar distances can be constant while the distance between remote points of the frame changes with time. We will demonstrate this below using the example of a non-inertial frame moving with an arbitrary translational velocity.

All three of these criteria are satisfied in a rigid uniformly accelerating frame. The constancy of radar distances was used in Chapter 6 to find the trajectories of such a frame. Its metric in Møller coordinates

$$ds^2 = (1 + ax)^2 \, dt^2 - dx^2 - dy^2$$

gives the Euclidean physical length

$$\delta l^2 = dx^2 + dy^2 ,$$

which does not depend on time. Therefore, the local rigidity criterion is also satisfied. As will be shown below, this frame is also rigid in the co-moving sense. In this respect, it is distinguished from all other non-inertial reference frames.

A rotating frame is rigid both in the local (9.41) and in the global sense (p. 158). However, it does not satisfy the criterion of co-moving rigidity. It is possible for the inertial frame to move with such a velocity that, in it, only one point of the rotating disc will have zero velocity. Obviously, this is true both in the theory of relativity and in classical mechanics.

▷ Using the criterion of *co-moving rigidity*, we can obtain the law of motion for the points of a rigid uniformly accelerating reference frame. Let an arbitrary point x of the frame move along the trajectory

$$X = x + \frac{1}{a_x}\left[\sqrt{1 + (a_x T)^2} - 1\right], \tag{9.55}$$

where a_x is a constant dependending on the initial position of the point $x = X(0)$. We can remove the root in the trajectory by rewriting it in the following form:

$$(1 - a_x x)^2 + 2a_x (1 - a_x x) X = 1 + a_x^2 (T^2 - X^2). \tag{9.56}$$

Substituting in the Lorentz transformations between the laboratory frame S_0 and the inertial frame $S'_0 : \{T', X'\}$ moving with constant velocity U_0 with respect to S_0, we obtain

$$\begin{cases} T &= \gamma_0 (T' + U_0 X') \\ X &= \gamma_0 (X' + U_0 T'), \end{cases}$$

where

$$\gamma_0 = \frac{1}{\sqrt{1 - U_0^2}}.$$

The right-hand side of (9.56) is invariant, therefore

$$(1 - a_x x)^2 + 2a_x (1 - a_x x)\gamma_0 (X' + U_0 T') = 1 + a_x^2 (T'^2 - X'^2).$$

For a given point ($x = const$), we differentiate the left- and right-hand sides with respect to T' and assume that

$$U' = \frac{dX'}{dT'} = 0$$

(the point x has zero velocity in the co-moving inertial frame). This gives

$$T' = \frac{1 - a_x x}{a_x}\gamma_0 U_0. \tag{9.57}$$

This time should be the same for any coordinate x (every point in S is at rest in S'_0). This is possible only if the factor in front of $\gamma_0 U_0$ in (9.57) does not depend on x:

$$a_x = \frac{a}{1 + ax}, \tag{9.58}$$

where $a = const$ is the proper acceleration of the point $x = 0$. Substituting a_x into (9.55), we obtain (9.54). Such a system of points satisfies the criterion of co-moving rigidity and constitutes a rigid uniformly accelerating frame (p. 141).

▷ Let us now consider a frame moving with an arbitrary velocity along the X-axis (p. 218). Its interval in space-time has the form $ds^2 = [1 + w(t)\,x]^2\, dt^2 - dx^2$ and corresponds to the Euclidean physical length. Therefore, this frame is *locally rigid*.

However, we must test whether it satisfies the criterion of *global rigidity*. Setting the interval to zero, $ds^2 = 0$, gives the following differential equation:

$$\frac{dx}{dt} = \pm(1 + w\,x). \tag{9.59}$$

Let a light signal be sent from the frame's origin $x = 0$ at the time instant t_1 and reflected from a point with coordinate $x > 0$. It then returns at t_2. Suppose the signal moves in the positive x-direction (we take the plus sign). Since $x(t_1) = 0$, we have from (9.59), for $t = t_1$:

$$\left.\frac{dx}{dt}\right|_{t=t_1} = 1, \qquad \left.\frac{d^2x}{dt^2}\right|_{t=t_1} = \left(\frac{dw}{dt}\,x + \frac{dx}{dt}\,w\right)_{t=t_1} = w(t_1),$$

where the second derivative is obtained by differentiating (9.59). Therefore, the trajectory of the receding signal has the form:

$$x_+(t) \approx (t - t_1) + w(t_1)\,\frac{(t - t_1)^2}{2} + \ldots$$

Similarly, we can find the trajectory of the signal as it approaches the origin $x(t_2) = 0$, which corresponds to the minus sign in (9.59):

$$x_-(t) \approx (t_2 - t) + w(t_2)\,\frac{(t_2 - t)^2}{2} + \ldots$$

These two trajectories coincide after the reflection: $x_+(t) = x_-(t) = x$. By solving the quadratic equations with respect to $t - t_1 > 0$ and $t_2 - t > 0$ and summing the solutions, we obtain

$$t_2 - t_1 \approx \frac{\sqrt{1 + 2\,w_1\,x} - 1}{w_1} + \frac{\sqrt{1 + 2\,w_2\,x} - 1}{w_2},$$

where $w_1 = w(t_1)$ and $w_2 = w(t_2)$. For a small time interval $t_2 - t_1$, the coordinate of the reflection point is of the same order of smallness. Therefore, expanding the root up to second order in x, we have

$$l = \frac{t_2 - t_1}{2} \approx x - \frac{w(t_1)}{2}\,x^2, \tag{9.60}$$

where we have approximated the term $(w_1 + w_2)\,x^2$ by $2w_1\,x^2$, which is accurate up to the second order of smallness. Since the proper time coincides with the coordinate time for the observer at the origin $x = 0$, the expression obtained is the radar distance to the point with the coordinate x.

This distance is constant up to the first order of smallness, as is seen from the constancy of the physical length

$$\delta l^2 = dx^2 + dy^2.$$

However, the next approximation with respect to x depends on the time at which the signal is sent, unless $w(t)$ is constant. Therefore, constancy of the infinitesimal radar distance

$$\delta l^2 = \gamma_{ij}\,dx^i dx^j$$

does not, generally speaking, imply constancy of the final radar distance between two points (local rigidity does not ensure global rigidity). This property of non-inertial frames is closely related to another feature. In Møller coordinates, the radar distance in a rigid uniformly accelerating frame is given by

$$l = \ln(1 + ax)/a.$$

At the same time, $\delta l^2 = dx^2 + dy^2$, and at first sight it seems that we should have $l = x$ for motion along the x-axis.

The reason for these apparent contradictions lies in the way the physical length

$$\delta l^2 = \gamma_{ij}\, dx^i dx^j$$

is measured. It is obtained by an observer who measures the propagation *time* of a light signal travelling to and from an infinitely close point. Summing a number of small elements δl along some curve implies that numerous observers are aligned along this curve, and that each of them obtains their own value δl. However, time flow is, generally speaking, different for different observers in a non-inertial frame. Therefore, a sum of measurements of radar distances made with clocks located at different points differs from a single measurement of the same distance performed by one observer with a single clock.

▷ The rotating reference frame is a good illustration of this statement. Length can be measured along any line of light propagation. Clocks have the same rate for observers equidistant from the centre. Let the signal move along a circle ($r = const$) from the point $\phi = 0$ to the point $\phi > 0$ and back. Setting the interval (9.8) to zero gives the equation

$$\frac{d\phi}{dt} = \pm\frac{1}{r} - \omega.$$

Repeating the reasoning of the previous page, we have

$$l = \sqrt{g_{00}}\,\frac{t_2 - t_1}{2} = \frac{r\phi}{\sqrt{1 - (\omega r)^2}},$$

where the factor $\sqrt{g_{00}}$ is introduced to obtain the physical time. We get the same distance by integrating the expression for the physical length (9.41), p. 227, in a rotating frame with $r = const$.

The situation is different when the light signal travels along the radius

$$\phi = const.$$

In this case, its equation of motion

$$\frac{dr}{dt} = \pm\sqrt{1 - (\omega r)^2}$$

gives the following radar distance for an observer located at the centre of rotation:

$$l = \frac{1}{\omega}\arcsin(\omega r).$$

This expression differs from the result $l = r$ obtained by integrating (9.41) along the line $\phi = const$. Remember that points at different distances r from the axis of the rotating frame have different velocities ωr and experience different dilations of their proper time.

▷ Thus, if the rate of time flow is constant along the trajectory of the light signal, then identical results are obtained from a single radar measurement of the distance and by summing measurements of infinitesimal distances. A varying rate of time flow along the trajectory will lead to different measurement results.

In this regard, let us note one more point. Deviation of the physical length

$$\delta l^2 = \gamma_{ij}\, dx^i dx^j$$

from the Euclidean expression is usually interpreted as evidence of the non-Euclidean nature of three-space in non-inertial frames. It is worth mentioning that this non-Euclidean nature differs substantially from that of ordinary curved spaces. Geometry has no time. The length of a line is the sum of the lengths of its infinitesimal elements. However, both of these statements fail in non-inertial frames. Therefore, considerations of the geometric properties of space, e.g. with the metric (9.41), are somewhat formal. For example, the physical length in a rigid uniformly accelerating frame is Euclidean:

$$\delta l^2 = dx^2 + dy^2,$$

as is obtained from the analysis of how light propagates over infinitesimal distances. However, even in this situation, the same light signal moves along curved lines rather than along straight lines when propagating over finite distances.

Behind the geometric properties of the metric γ_{ij} we should see a lot of observers using different clocks to measure radar distances in their immediate vicinities. Based on γ_{ij}, the geometry of the three-space of a non-inertial system combines such infinitesimal local measurements.

9.7 Born rigidity *

Historically, Max Born was the first (in 1909) to introduce the concept of rigidity in the theory of relativity [7]. He considered a body whose points are uniquely characterized (enumerated) by three coordinates $x^i = \{x, y, z\}$, and which moves along the trajectory $X^\alpha = X^\alpha(\tau, x^i)$ in the laboratory frame, where τ is the proper time of the clock attached to the point. In classical mechanics, a body is considered rigid if the distance between any two of its points as measured at a given point in time never changes thereafter (the first figure below). This definition is not relativistically invariant and will be violated in any other frame of reference (since simultaneity is relative). Therefore, Born demanded that a rigid body should have invariant infinitesimal distances in the four-space of the hyperplane orthogonal to the trajectories of any two neighbouring points (the second figure below):

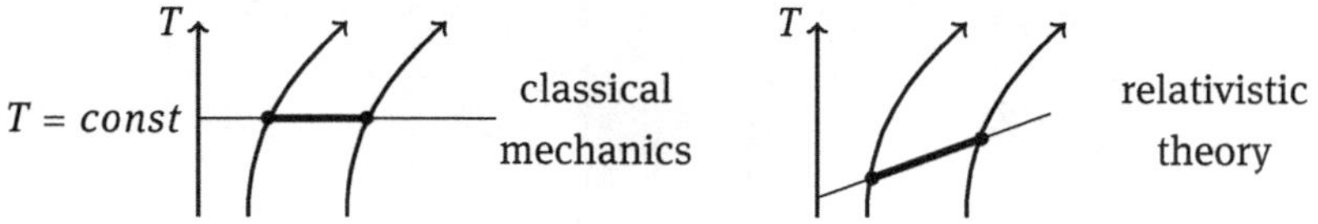

Note that four-vectors which are orthogonal ($a^0b^0 = a^1b^1$) are represented by non-orthogonal lines in the Euclidean plane.

Following Born, we will refer to rigid bodies rather than to rigid non-inertial frames. We can write the trajectory of an arbitrary point of such a body with respect to the laboratory frame $S_0 : \{T, X, Y, Z\}$:

$$X^\alpha = X^\alpha(\tau, x^1, x^2, x^3), \tag{9.61}$$

where the x^i are the coordinates, which uniquely determine a fixed point of the body, τ is its proper time, and, as usual, $X^\alpha = \{T, \mathbf{X}\}$. Below, we will use index-free notation, where block-letter typeface denotes four-vectors:

$$\mathrm{A} \cdot \mathrm{B} = A^\alpha B_\alpha,$$

etc.

The interval along the trajectory of the point coincides with the change of its proper time:

$$d\tau^2 = dX^\alpha dX_\alpha = (\partial_0 X^\alpha)(\partial_0 X_\alpha)\, d\tau^2 \equiv (\partial_0 \mathrm{X})^2\, d\tau^2,$$

where we substituted in the differentials $dX^\alpha = \partial_0 X^\alpha\, d\tau$, written for constant coordinates x^i and $\partial_0 = \partial/\partial\tau$. Therefore:

$$(\partial_0 \mathrm{X})^2 = 1. \tag{9.62}$$

Note that this compact equation is in fact a short form of the equation $(\partial_0 T)^2 - (\partial_0 \mathbf{X})^2 = 1$.

Consider two neighbouring points with coordinates x^i and $x^i + dx^i$. The positions of the first and the second point correspond to the instances τ and $\tau + d\tau$ of their proper times, respectively. The distance between these points in Minkowski space is given by the four-vector

$$d\mathrm{X} = \partial_0 \mathrm{X}\, d\tau + \partial_i \mathrm{X}\, dx^i, \tag{9.63}$$

where $\partial_i = \partial/\partial x^i$. This is the usual differential of a function of four variables τ, x^1, x^2, x^3. To find the value $d\tau$, we require that the vector $d\mathrm{X}$ be orthogonal to the four-vector $\partial_0\mathrm{X}$ tangential to the trajectory with respect to varying time:

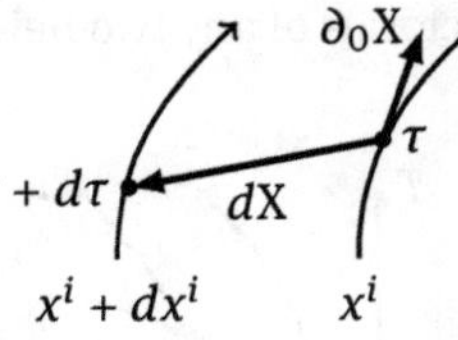

$$dX \cdot \partial_0 X = 0. \tag{9.64}$$

According to Born's definition, a body such as this is considered *rigid* if the length of the vector dX does not change with time:

$$(dX)^2 = const. \tag{9.65}$$

From the orthogonality relationship, (9.63), and (9.62), we have

$$d\tau = -(\partial_0 X \cdot \partial_i X)\, dx^i.$$

Substituting this value into the squared distance (9.63) between the points

$$(dX)^2 = d\tau^2 + 2(\partial_0 X \cdot \partial_i X)\, d\tau dx^i + (\partial_i X \cdot \partial_j X)\, dx^i dx^j,$$

we have:

$$(dX)^2 = \{\partial_i X \cdot \partial_j X - (\partial_0 X \cdot \partial_i X)(\partial_0 X \cdot \partial_j X)\}\, dx^i dx^j.$$

The first term in brackets is g_{ij}, and the second is the product $g_{0i}g_{0j}$. Indeed, the interval is

$$ds^2 = (dX)^2 = (\partial_\alpha X \cdot \partial_\beta X)\, dx^\alpha dx^\beta = g_{\alpha\beta}\, dx^\alpha dx^\beta,$$

and as a result, the metric coefficients $g_{\alpha\beta}$ of the non-inertial frame associated with the body are $\partial_\alpha X \cdot \partial_\beta X$. In this case, since τ is the proper time in the transformations (9.61),

$$g_{00} = (\partial_0 X)^2 = 1,$$

see also p. 217.

Thus, the Born rigidity criterion is equivalent to the constancy of the tensor γ_{ij} specifying the physical length. Note that Born formulated his criterion of rigidity for a special class of transforms where the coordinate time t is the proper time τ of the point in the non-inertial reference frame. If condition (9.62) is neglected, the general expression (9.33) is obtained.

9.8 Motion of particles and light

In contrast to inertial frames, space is not simultaneously isotropic and homogeneous in non-inertial frames. Therefore, the motion of a free particle is no longer uniform and rectilinear. According to classical mechanics, particles are subject to *inertial forces*, which curve their trajectories. Knowing the metric coefficients $g_{\alpha\beta}$, we can write down the differential ***geodesic equation*** [1]. Its solutions are the trajectories of motion of free particles in a non-inertial frame. However, these trajectories can also be found using relatively elementary methods.

Let us write the trajectory of a free particle in Lorentz coordinates in the laboratory frame in the (X, Y)-plane:

$$X = X_0 + V_{0x} T, \qquad Y = Y_0 + V_{0y} T,$$

where (V_{0x}, V_{0y}) are constant velocity components, and (X_0, Y_0) is the initial position of the particle. To find its trajectory in a rigid uniformly accelerating frame, for example, we have to use the transformations (9.3):

$$aT = (1 + ax)\,\mathrm{sh}(at), \qquad 1 + aX = (1 + ax)\,\mathrm{ch}(at), \qquad Y = y.$$

Substituting these into the trajectory, we have

$$1 + ax = \frac{1 + aX_0}{\mathrm{ch}(at) - V_{0x}\,\mathrm{sh}(at)}, \qquad a\,(y - Y_0) = V_{0y}\,\frac{(1 + aX_0)\,\mathrm{sh}(at)}{\mathrm{ch}(at) - V_{0x}\,\mathrm{sh}(at)}.$$

Now we have only to express the constants X_0, Y_0 and V_{0x}, V_{0y} in terms of the initial conditions. Taking $t = 0$, we find that $x_0 = X_0$ and $y_0 = Y_0$. Differentiating the left- and right-hand sides with respect to time at $t = 0$, we obtain:

$$V_{0x} = \frac{v_{0x}}{1 + ax_0}, \qquad V_{0y} = \frac{v_{0y}}{1 + ax_0},$$

where (v_{0x}, v_{0y}) is the initial velocity of the particle in the non-inertial frame. Substituting these constants into the trajectory, we obtain:

$$1 + ax = \frac{(1 + ax_0)^2}{(1 + ax_0)\,\mathrm{ch}(at) - v_{0x}\,\mathrm{sh}(at)}, \tag{9.66}$$

$$a\,(y - y_0) = \frac{v_{0y}\,(1 + ax_0)\,\mathrm{sh}(at)}{(1 + ax_0)\,\mathrm{ch}(at) - v_{0x}\,\mathrm{sh}(at)}, \tag{9.67}$$

which is the solution of the problem.

Suppose that at time $t = 0$ the particle is at the origin of the frame, $x_0 = y_0 = 0$, and moves with velocity v_{0y} parallel to the y-axis, while $v_{0x} = 0$. In this case, the expressions for the trajectory are simplified:

$$x(t) = \frac{1}{a}\left[\frac{1}{\mathrm{ch}(a\,t)} - 1\right], \qquad y(t) = \frac{v_{0y}}{a}\,\mathrm{th}(at). \tag{9.68}$$

Taking derivatives in t, we obtain the coordinate velocities:

$$v_x(t) = -\frac{\mathrm{sh}(at)}{\mathrm{ch}^2(at)}, \qquad v_y(t) = \frac{v_{0y}}{\mathrm{ch}^2(at)}. \tag{9.69}$$

The physical time of the NIRF observer with coordinate x for the interval (9.4) is $d\tau_0 = (1 + ax)\,dt$, and the physical length is $dl^2 = dx^2 + dy^2$. Therefore, the components of the ***physical velocity*** $dl/d\tau_0$ are $v_i/(1+ax)$. Due to different rates of time flow, the physical velocity is non-linear with respect to time along the y-axis. The absolute value of the physical velocity along the x-axis tends to unity:

$$v_x/(1+ax) = -\,\mathrm{th}(at).$$

For $v_{0y} = 1$ the square of physical velocity $(v_x^2 + v_y^2)/(1+ax)^2$ is always unity (the fundamental speed "c"). The proper time of the particle is

$$\tau = \int_0^t ds = \int_0^t \sqrt{(1+ax(t))^2 - v_x^2(t) - v_y^2(t)}\, dt = \frac{\sqrt{1-v_{0y}^2}}{a}\ \mathrm{th}(a\,t).$$

For the infinite coordinate time $t = \infty$, the particle goes beyond the event horizon $x_0 = -1/a$ (p. 149) of the observer at the coordinate origin. However, the proper time of the particle at this "instant" remains finite.

Finally, we can supress the dependency of the trajectory on time:

$$y = \frac{v_{0y}}{a}\sqrt{1-(1+ax)^2}.$$

For small ax, this describes a parabola (or the trajectory of a free particle in a uniformly accelerating frame in classical mechanics).

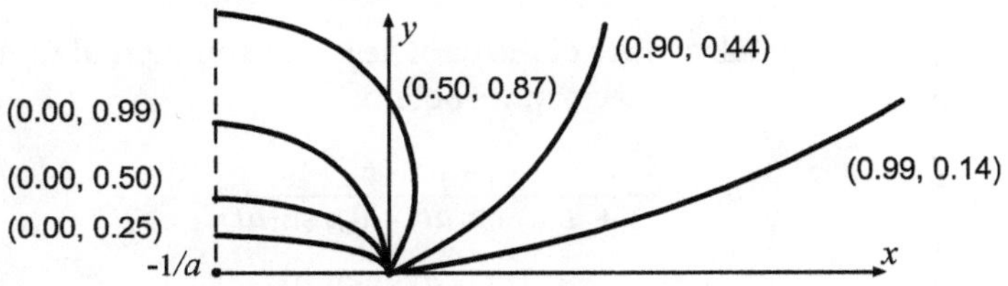

The figure above shows some example trajectories starting from the point $x_0 = y_0 = 0$. The parentheses next to the lines contain the components of the initial velocity (v_{0x}, v_{0y}).

▷ As another example, let us find the trajectory of a free particle in a rotating frame of reference (p. 216). In the laboratory frame, the trajectory of a particle that passes the point with Cartesian coordinates (X_0, Y_0) at $T = 0$ has the form

$$X = X_0 + V_0\cos(\alpha)\,T, \qquad Y = Y_0 + V_0\sin(\alpha)\,T,$$

where α is the angle between the direction of the motion and the X-axis. The magnitude of the velocity vector satisfies $V_0 < 1$, and it is unity for a light pulse ($V_0 = c = 1$). Let us write the trajectory in polar coordinates

$$X = R\cos\Phi, \quad Y = R\sin\Phi,$$

and turn to the curvilinear coordinates of the rotating frame (p. 216) in the plane $Z = z = 0$:

$$T = t, \quad R = r, \quad \Phi = \phi + \omega t.$$

After some simple calculations (< H_{37}), we arrive at

$$\begin{cases} x &= r_0 \cos(\phi_0 - \omega t) \quad + \quad V_0 t \cos(\alpha - \omega t), \\ y &= r_0 \sin(\phi_0 - \omega t) \quad + \quad V_0 t \sin(\alpha - \omega t), \end{cases} \tag{9.70}$$

where r_0 and ϕ_0 give the initial position of the particle in the coordinates of the non-inertial frame, $x = r\cos\phi$, and $y = r\sin\phi$. We can also write the dependence of time on the distance from the centre of rotation:

$$r^2 = r_0^2 + 2r_0\, V_0\, t\, \cos(\phi_0 - \alpha) + V_0^2 t^2, \tag{9.71}$$

and constant velocity functions (the dot denotes differentiation with respect to t):

$$\dot{r}^2 + r^2(\dot{\phi} + \omega)^2 = V_0^2, \qquad r^2\,(\dot{\phi} + \omega) = V_0 r_0 \sin(\alpha - \phi_0). \tag{9.72}$$

The figures below show some light trajectories ($V_0 = 1$) in the rotating frame:

In the first figure, the rays of light leave the point $r_0 = 1/(2\omega)$ in different directions. In the second figure, by contrast, the rays converge to this point. Finally, the third figure shows the trajectories when the observers at the bullet points exchange light signals with the observers on the axis of rotation.

⊳ Qualitatively, the curving of light trajectories can be understood by analogy with the Coriolis inertial force of classical mechanics. This force depends on the velocity of a body in a rotating frame, and for a frame rotating anticlockwise, it is directed along the axis of rotation and to the right of the velocity (in vector notation, the acceleration has the form $2[\mathbf{v}\times\boldsymbol{\omega}]$). Similarly, the trajectory of a light pulse is always curved to the right of its velocity vector.

Due to this key feature of rotating frames, light rays travel along different paths between the observers during radar measurements, so the trajectory of a light ray depends on its direction. The third figure above shows this using a "flower" of light rays illustrating the divergence of the trajectories. The arrows depict the directions of the pulses.

In contrast to rotating frames, in rigid uniformly accelerating frames the trajectory of light in radar experiments is the same in both the “forward” and “backward” directions, like a volleyball in a gravitational field moving along the same parabola, whichever team set it in motion. The trajectory of such a light pulse can be found from equations (9.66) and (9.67), where we have to set

$$v_{0x} = (1 + ax_0)\cos\alpha,$$

$$v_{0y} = (1 + ax_0)\sin\alpha.$$

Then, the physical speed of light will always be unity.

The fact that light rays curve has the effect that in a non-inertial frame, the surrounding world appears absolutely differently as in an inertial frame. An observer sees an object in the direction from which a light "ray" was emitted by the object. If the "ray" moves along a curved trajectory, the object will be seen to be in some place other than its "actual" position. In the second picture above each line represents the sequence of points that the observer sees as a straight line.

Uniformly accelerating frames are also associated with visual distortion of the surrounding world. For example, observers on the y-axis of such a frame will be seen by their colleague at the origin of the same frame as being located in front of them, rather than above or below. The farther the observers are from the origin, the greater will be the angle between them and the y-axis. Qualitatively, this can easily be imagined by drawing a parabola between two points on the y-axis. For a frame moving with uniform acceleration along the x-axis, the y-axis plays the role of the Earth's surface. A stone thrown from one point to another will travel along a parabola starting and ending at an angle to the surface. The behaviour of light rays is similar.

9.9 Dynamics in non-inertial frames

Let us return to the general solution for the trajectory of a free particle in a rigid uniformly accelerating frame (9.66), (9.67). By differentiating these expressions with respect to time, we can easily find the components of the particle's coordinate velocity:

$$v_x = \frac{(1+ax)^2}{(1+ax_0)^2}\left[v_{0x}\,\mathrm{ch}(at) - (1+ax_0)\,\mathrm{sh}(at)\right], \qquad v_y = v_{0y}\frac{(1+ax)^2}{(1+ax_0)^2}.$$

Squaring these expressions and replacing $\mathrm{ch}^2(at)$ by $1 + \mathrm{sh}^2(at)$ and $\mathrm{sh}^2(at)$ by $\mathrm{ch}^2(at) - 1$, we obtain the square of the physical velocity:

$$\frac{\delta l^2}{\delta\tau_0^2} = \frac{v_x^2 + v_y^2}{(1+ax)^2} = 1 - \frac{(1+ax)^2}{(1+ax_0)^4}\left[(1+ax_0)^2 - v_{0x}^2 - v_{0y}^2\right].$$

This relationship can be rewritten in the form

$$\frac{1+ax}{\sqrt{1-\mathbf{v}^2/(1+ax)^2}} = \frac{1+ax_0}{\sqrt{1-\mathbf{v}_0^2/(1+ax_0)^2}},$$

where $\mathbf{v}^2 = v_x^2 + v_y^2$, and the initial value $\mathbf{v}_0^2$ is defined analogously. The right-hand side of this equation is constant. Therefore, when a particle moves, the following quantity is conserved (is an integral of motion):

$$E = m\,\frac{1 + ax}{\sqrt{1 - \mathbf{v}^2/(1 + ax)^2}} = const. \tag{9.73}$$

This we call the *energy of a particle* with mass m.

Expanding the root in a Taylor series in the non-relativistic, low-velocity limit, we have:

$$E \approx m\,(1 + ax) + \frac{m\mathbf{v}^2/2}{1 + ax} \approx m + \frac{m\mathbf{v}^2}{2} + m a\,x,$$

where we have neglected the acceleration in the denominator of the second approximate equality, assuming that $\mathbf{v}^2 ax \ll 1$. A uniformly accelerating non-inertial reference frame is equivalent to the homogeneous gravitational field $g = a$ (directed opposite to the x-axis) of classical mechanics. Therefore, the term $m a\,x$ corresponds to the potential energy, and the expression for E is the total energy of the particle (kinetic plus potential), including the resting energy $E_0 = m$.

Thus, the conserved quantity (9.73) is the ***total energy*** of the relativistic particle in the non-inertial frame. This energy takes into account the force of inertia acting on the particle, so it depends not only on the particle's velocity, but also on its position.

Multiplying the numerator and the denominator in the expression for the energy by $1 + ax$, we obtain

$$E = \frac{m\,(1 + ax)^2}{\sqrt{(1 + ax)^2 - \mathbf{v}^2}}.$$

Since the interval along the particle's trajectory in a rigid uniformly accelerating frame is

$$ds^2 = (1 + ax)^2\,dt - dx^2 - dy^2 = \left[(1 + ax)^2 - \mathbf{v}^2\right] dt^2$$

and given that $g_{00} = (1 + ax)^2$, the energy can be rewritten as

$$E = m\,g_{00}\,\frac{dt}{ds}. \tag{9.74}$$

Introducing the four-vector $dx^\alpha = \{dt,\ d\mathbf{x}\}$, we can define the four-speed and four-momentum of the particle, proceeding similarly as for an inertial frame:

$$u^\alpha = \frac{dx^\alpha}{ds}, \qquad p^\alpha = m\,u^\alpha. \tag{9.75}$$

Using the metric tensor, lower the index by defining

$$p_\alpha = g_{\alpha\beta}\,p^\beta. \tag{9.76}$$

As $ds^2 = g_{\alpha\beta}\,dx^\alpha dx^\beta$, the square of the velocity four-vector is unity, and the square of the four-momentum is, as usual, the square of the invariant particle mass:

$$m^2 = p_\alpha p^\alpha = g_{\alpha\beta}\,p^\alpha p^\beta. \tag{9.77}$$

For a rigid uniformly accelerating frame we have $g_{0i} = 0$ and $g_{ij} = -\delta_{ij}$. Therefore, the conserved total energy E coincides with the zero component p_0 of the contravariant four-vector p_α:

$$p_0 = g_{0\alpha}\, p^\alpha = g_{00}\, m\,\frac{dt}{ds} + g_{0i}\, m\,\frac{dx^i}{ds} = m\, g_{00}\,\frac{dt}{ds}.$$

Note that in a non-inertial frame, the coefficients of the metric tensor depend on the coordinates, in contrast to their representation in Lorentz coordinates in an inertial frame. Therefore, p_α is a different function of the coordinates than p^α, and it is $p_0 = E$ that is conserved in a rigid non-inertial frame, rather than p^0.

Since, generally speaking, the particle's velocity changes its direction under the influence of inertial forces, the three-dimensional momentum is not conserved. This is true for both p_i and for p^i. Only the total energy is conserved.

▷ Let us make some generalizations. Define the *square of the physical velocity* of a particle:

$$\tilde{\mathbf{v}}^2 = \frac{\delta l^2}{\delta\tau^2} = \frac{\gamma_{ij}\, dx^i dx^j}{\left(\sqrt{g_{00}}\, dt + g_{0i}\, dx^i/\sqrt{g_{00}}\right)^2}, \tag{9.78}$$

where δl is physical length and $\delta\tau$ physical time. Dividing the numerator and the denominator by dt^2, we obtain the relationship between the square of the physical velocity (marked by the tilde) and the components $v^i = dx^i/dt$ of the coordinate velocity:

$$\tilde{\mathbf{v}}^2 = \frac{\gamma_{ij}\, v^i v^j}{g_{00}\,(1 - \gamma_i\, v^i)^2}, \tag{9.79}$$

where we used the shorthand

$$\gamma_i = -\frac{g_{0i}}{g_{00}}.$$

We will call the following quantities the *components of the physical velocity* (again with tildes):

$$\tilde{v}^i = \frac{v^i/\sqrt{g_{00}}}{1 - \gamma_i\, v^i}, \tag{9.80}$$

so that

$$\tilde{\mathbf{v}}^2 = \gamma_{ij}\,\tilde{v}^i\tilde{v}^j = \tilde{v}_i\tilde{v}^i, \qquad \tilde{v}_i = \gamma_{ij}\,\tilde{v}^j.$$

Note also the relation

$$\frac{1}{1 - \gamma_i v^i} = 1 + \gamma_i\tilde{v}^i\,\sqrt{g_{00}},$$

which is obtained by contracting the definition (9.80) with γ_i.

The proper time of the particle (the interval along its trajectory) can be written as

$$ds = \left[\delta\tau^2 - \delta l^2\right]^{1/2} = \sqrt{1 - \tilde{\mathbf{v}}^2}\,\delta\tau,$$

or, substituting in the physical time interval:

$$ds = \sqrt{1 - \tilde{\mathbf{v}}^2}\,\left(\sqrt{g_{00}}\, dt + g_{0i}\, dx^i/\sqrt{g_{00}}\right) = \sqrt{1 - \tilde{\mathbf{v}}^2}\,(1 - \gamma_i v^i)\,\sqrt{g_{00}}\, dt.$$

Therefore, the components of the coordinate four-velocity $u^\alpha = dx^\alpha/ds$ are

$$u^0 = \frac{\gamma_i \tilde{v}^i + 1/\sqrt{g_{00}}}{\sqrt{1-\tilde{\mathbf{v}}^2}}, \qquad u^i = \frac{\tilde{v}^i}{\sqrt{1-\tilde{\mathbf{v}}^2}}. \tag{9.81}$$

Note once more that $\tilde{\mathbf{v}}$ is the physical velocity rather than the coordinate velocity.

▷ Now we can find the total energy of the particle:

$$E = m\, g_{0\alpha} u^\alpha = m\, g_{00}\, (u^0 - \gamma_i u^i).$$

Substituting in the components of the four-velocity (9.81), we finally have

$$E = \frac{m\,\sqrt{g_{00}}}{\sqrt{1-\tilde{\mathbf{v}}^2}}. \tag{9.82}$$

As will be shown in the next volume, if the metric coefficients $g_{\alpha\beta}$ are not time-dependent in a non-inertial frame, the expression (9.82) is an integral of motion. In other words, the total energy of a particle is always conserved in the stationary case.

▷ Thus, a uniformly rotating frame is stationary. Let us demonstrate that the energy of a particle (9.82) is constant in such a frame. For the interval (p. 216)

$$ds^2 = (1-\omega^2 r^2)\, dt^2 - 2\omega\, r^2\, dt\, d\phi - dr^2 - r^2\, d\phi^2 - dz^2,$$

the physical time and the squared physical length are:

$$\delta\tau = \sqrt{1-(\omega r)^2}\, dt - \frac{\omega r^2\, d\phi}{\sqrt{1-(\omega r)^2}}, \qquad \delta l^2 = dr^2 + \frac{r^2\, d\phi^2}{1-(\omega r)^2}.$$

Therefore, the squared physical velocity (9.78) is

$$\tilde{\mathbf{v}}^2 = \frac{(1-\omega^2 r^2)\, \dot{r}^2 + r^2 \dot{\phi}^2}{(1-\omega^2 r^2 - \omega r^2 \dot{\phi})^2},$$

where the point denotes differentiation with respect to time t. Accordingly, the particle energy is

$$E = m\, \frac{1-\omega r^2\, (\omega+\dot{\phi})}{\sqrt{1-\dot{r}^2 - r^2\, (\omega+\dot{\phi})^2}}. \tag{9.83}$$

Due to the relationships (9.72), p. 241, the numerator and the denominator in the expression for the energy are constant.

Note a simple particular solution that gives constant energy:

$$\dot{\phi} = -\omega, \qquad \dot{r} = const.$$

In the laboratory frame, a trajectory satisfy these equations corresponds to motion along a straight line passing through the centre of rotation. In the coordinates of the non-inertial frame this motion looks like an unwinding spiral.

9.10 Bell's paradox and the Ehrenfest paradox

Bell's paradox concerns non-rigid uniformly accelerating frames (p. 162). Suppose a string is stretched between two spaceships. These ships start accelerating simultaneously and with *equal* speed relative to the resting (laboratory) frame S_0. The length L of the strings is constant in this frame. In the non-inertial frame associated with the ships, the interval between two events has the form (p. 161):

$$ds^2 = dt^2 - 2\,\mathrm{sh}(at)\,dt\,dx - dx^2 - dy^2.$$

The corresponding element of physical length depends on time:

$$\delta l^2 = \mathrm{ch}^2(at)\,dx^2 - dy^2.$$

If the observer in the non-inertial frame holds a "rigid" ruler normal to the acceleration, its length does not change ($\delta l = dy$). The same ruler oriented along the acceleration will stretch over time according to

$$\delta l = \mathrm{ch}(at)\,dx.$$

Having directed one ruler along the acceleration and another perpendicular to it, the non-inertial observer will notice a discrepancy between their lengths. Similarly, the final radar distance varies with time (6.57).

Thus, the string has constant length in one frame and an increasing length in the other. This begs the question, will the string be break at some later time? To make the situation even more unusual, imagine that the trajectory of each point of the string is controlled from the laboratory frame, so that it "gently" accelerates without any jerks or tension (as seen by the laboratory observers). In this case, however, the *simultaneous* acceleration of the string's points will no longer look so "gentle" or simultaneous in any other inertial frame.

Generally speaking, the problem of breaking a string goes beyond kinematics, and a consistent approach to this problem would require building a model of the string. If the string is attached to the ships only, the changing speed of its ends (the ships) should be transmitted along the string in a certain way which causes some strain in it. In kinematics, it is usually assumed that if the radar distance between two points is constant then there is no additional tension in a string stretched between them (thus, the effects of inertial forces which are weak at small accelerations are ignored).

If the radar distance between the points increases, a mechanical tension should arise between them. From this point of view, the string in the non-rigid reference frame from Bell's thought experiment should break.

At the same time, it must not be forgotten that Lorentz contraction in inertial frames is a purely kinematic effect. When the length of a uniformly moving rod is measured, the observer in the frame of the rod "disagrees" with the measuring procedure used by the stationary observer (p. 65).

Therefore, it would be wrong to claim that the rod undergoes mechanical compression. For example, let a rod move with constant velocity relative to some fixed observers, and be oriented transversely to the direction of its velocity. If the rod now turns slowly so as to align with its velocity vector, the stationary observers will register its contraction. However, the rod remains in the inertial frame (with isotropic space) and obviously does not experience any tension forces associated with this contraction. This is the effect of the relativity of measurement procedures in the two reference frames.

The situation is somewhat different in non-inertial frames. Light is used to measure the radar distance. The same electromagnetic field underlies the inner structure of material bodies (rods, strings). Therefore, if the radar distance increases, this should also affect the forces acting inside the substance.

In connection with this "paradox", it makes sense to remember that an absolutely hard ("rigid") body contradicts the principles of the theory of relativity. Rigidity is relative, and there can be no bodies with invariant dimensions for observers in different frames. One of the reasons for this effect is that the velocity of any physical motion in space is always smaller than the fundamental velocity (the speed of light). Suppose as an example that there exists an absolutely rigid rod. Then, pushing one of its ends, we would have to observe an instant reaction at its other end at some distance from the first. Such a perturbation from the push must therefore be transmitted along the rod with infinite speed. Superluminal speeds are "bad", as they lead to an imaginary factor $\gamma = 1/\sqrt{1-v^2}$ in the Lorentz transformations. Besides, they cause serious causality problems (p. 254).

▷ The *Ehrenfest paradox* concerns geometric effects in rotating frames of reference. Like many other "paradoxes", it is due to the unusual nature of the physics of high velocities, rather than an inconsistency in the theory of relativity.

Suppose there is a stationary circular gutter in the laboratory reference frame, with a ring of radius R inside it, which quickly rotates with angular velocity ω.

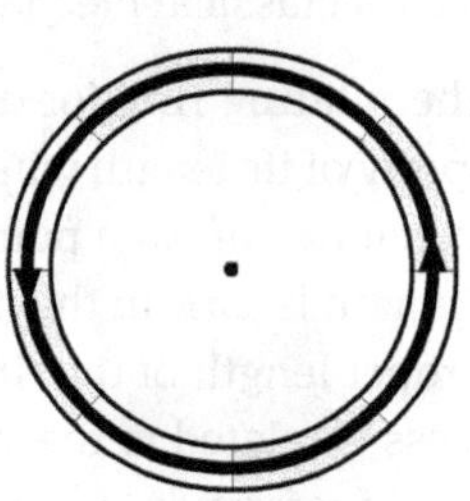

An observer at rest relative to the gutter in the laboratory frame can measure the length ΔL of a segment of the rotating ring by simultaneously (by its clock) measuring the positions of the beginning and the end of the segment.

In accordance with the effect of length contraction (p. 65), ΔL will be shorter than the segment's proper length Δl:

$$\Delta L = \Delta l\,\sqrt{1 - V^2},$$

where

$$V = \omega R$$

is the linear velocity of the point on the ring. The sum of the lengths ΔL of all of the segments coincides with the length of the resting gutter and is equal to $2\pi R$. Thus, the "proper length" of the ring l (equal to the sum of all of the measurements Δl) is greater than $2\pi R$:

$$l = \frac{2\pi R}{\sqrt{1 - V^2}}.$$

The same result is obtained by integrating the physical length (9.41) with respect to the angle Φ from 0 to 2π. This length is equal to a sum of radar measurements performed by *different* non-inertial observers. The circumference of the circle can also be measured by a single non-inertial observer by calculating the time taken by a light signal to travel around the circle. In this case, they obtain different distances depending on the direction of motion (clockwise or anticlockwise):

$$l_- = 2\pi R\,\sqrt{\frac{1 - V}{1 + V}}, \qquad l_+ = 2\pi R\,\sqrt{\frac{1 + V}{1 - V}},$$

and their average value will be equal to

$$l = \frac{l_+ + l_-}{2}$$

(see the Sagnac effect, p. 156). All this looks unusual from the position of classical physics. However, there is nothing paradoxical (or contradictory) to it. It contradicts only our intuition, which is based on classical mechanics.

▷ It is sometimes argued that the rotating ring (or disc) must somehow bend in order to produce this unusual property of its length whilst remaining locally Euclidean. Of course, this is not so. The trajectories of each point of the rotating ring are specified in the laboratory frame. Its length is $2\pi R$ in this frame and any bending is out of the question. The fact that the proper length of the ring differs from $2\pi R$ and also depends on the measurement process is related to the unusual properties of measuring procedures in non-inertial frames of reference. As we saw, three-dimensional space, understood as a set of infinitesimal radar distances, can have non-Euclidean geometry in a non-inertial frame. Such a space has negative curvature in the rotating frame of reference. However, this non-Euclidean behaviour is due to the different rates of the clocks in the non-inertial frame. Therefore, in some sense, it is of a physical rather than a geometric nature (p. 236).

▷ It is somewhat more difficult to understand the physics of a rotating disc or ring at the stage of their acceleration, where they gain the angular velocity ω. A consistent description of such physical systems would require, as with the string in Bell's paradox, an appropriate mechanical model of the substance of which the system is built. In any case, a gradually accelerating disc is not rigid. The non-inertial frame associated with it can be found, for example, by writing the following coordinate transformation:

$$T = t, \qquad R = r, \qquad \Phi = \phi + f(t)\,t,$$

where the function $f(t)$ specifies the time dependence of the angular velocity. Substituting these transformations into

$$ds^2 = dT^2 - dR^2 - R^2\,d\Phi^2,$$

we obtain the metric:

$$ds^2 = [1 - \omega^2(t)\,r^2]\,dt^2 - 2r^2\,\omega(t)\,dt\,d\phi - dr^2 - r^2\,d\phi^2,$$

where

$$\omega(t) = f + t\,\frac{df}{dt}.$$

The element of physical length (p. 224) in such a non-inertial frame is

$$\delta l^2 = dr^2 + \frac{r^2\,d\phi^2}{1 - \omega^2(t)\,r^2}.$$

If $d\omega/dt \neq 0$, then the radar distance increases along any circle. Thus, accelerated rotation can cause tension not only along the radius due to inertial forces, but also in the transverse direction. Therefore, if the reasoning in the above discussion of Bell's paradox is true, there must be breaking forces along the circles of the disc.

10 The motion of rods and gyroscopes *

This chapter concerns the motion of extended objects such as rods and rotating gyroscopes. Their proper acceleration is assumed to be small in the co-moving inertial frame. Therefore, these objects can, conditionally, be considered rigid, and treated like objects in classical mechanics. If they move at high velocity relative to the laboratory frame, their behaviour differs substantially from what we would expect from classical mechanics, due to relativistic kinematics. The relativity of simultaneity requires some corrections to be made to the conservation laws concerning total quantities of a system of particles. As a result, a rotating relativistic object undergoes "kinematic deformations" in addition to Lorentz contraction.

https://doi.org/10.1515/9783110515886-010

10.1 Rigid bodies in the theory of relativity

Classical mechanics operates with the concept of absolutely rigid (or solid) bodies. This approximation neglects the elasticity and deformation of real bodies. This assumption is quite appropriate for many materials and relatively weak impacts. However,

---

! the concept of an *absolutely rigid body* fundamentally contradicts the basic principles of the theory of relativity.

Indeed, when one end of a rigid ruler is pushed, the other end should react immediately, but all massive bodies are known to move at velocities smaller than the fundamental one $c = 1$. Besides, a real ruler ultimately consists of atoms tied up by electromagnetic forces. The speed of electromagnetic interaction cannot exceed the fundamental velocity either. Therefore, if one end of a ruler is pushed, the perturbation wave should propagate along it at a smaller speed than the fundamental one.

▷ The theory of relativity has another strong argument that prohibits greater velocities than the fundamental velocity. Suppose that the perturbation propagates along the ruler with velocity $u > 1$, and the length of the ruler is l when oriented along the x-axis in the frame S. Then, if its left-hand end is pushed to the right at time t_1, its right-hand end starts moving after the time

$$t_2 = t_1 + \frac{l}{u}.$$

Obviously, due to *causality*, in S, we always have $t_2 > t_1$. This means that the cause (the push of the ruler) precedes the effect (the shift of its right end). Consider the same process in the frame S' moving along the rod with velocity v relative to S. According to the Lorentz transformations, the events of the push and the response of the right-hand end in S' will take place at time instants

$$t'_1 = \gamma(t_1 - vx_1), \qquad t'_2 = \gamma(t_2 - vx_2).$$

After subtracting these two relations from each other and expressing the proper length of the rod as $l = x_2 - x_1$, we substitute in

$$t_2 = t_1 + \frac{l}{u}$$

to obtain

$$t'_2 - t'_1 = \gamma l(1 - vu)/u. \tag{10.1}$$

For $u > 1$ the expression in parentheses may be negative, in which case $t'_2 < t'_1$. This means that the effect in S' happens *before* the cause. First, the right end of the ruler moves, and then someone pushes its left end. This is quite an unusual idea. Therefore, causality is supposed to hold true for all observers. Invariance of causality "prohibits" rigid rulers.

Does this mean that the theory of relativity has nothing to do with rigid bodies? ?

No, it does not. Indeed, suppose there is a rod moving with small velocity and acceleration relative to some frame S'. Classical mechanics must be true for small velocities and accelerations and, therefore, the abstraction of the rigid body is quite adequate. In this case, from our everyday experience we know we have a sufficiently rigid, *real* rod. Relativistic effects will appear when the rod is observed from another inertial frame S moving at a high velocity with respect to S'. Let us consider how the concept of classical mechanics relates, in this case, to the theory of relativity.

The rod moves with non-relativistic velocity u' and acceleration a' in S', so classical mechanics is applicable. Therefore, any quantity f' characterizing the rod (its length, bending, etc.) can be represented (under certain conditions) as a series:

$$f' = f_0' + \lambda f_1' + \dots,$$

where the small parameter λ is a function of the velocity, the acceleration, and the length of the rod. The first approximation f_0' corresponds to the classical description of the rod. The other terms are neglected by observers in S' for

$$\lambda \to 0.$$

Consider the same motion from the perspective of the frame S moving at a relativistic velocity $v \sim c$ relative to S'. In this case, there are some relativistic factors, which under certain conditions are accounted for by a large, but *finite* factor $g(v)$. The corresponding quantity in S will, for example, be

$$f = (f_0' + \lambda f_1' + \dots)\, g(v).$$

If some effect (bending, etc.) is infinitely small in S', it will also be so in S:

$$\lambda g(v) \to 0.$$

These formulas are only illustrations. A more consistent consideration would require some dynamic model of the rod's "internal structure", but the general idea is that in any frame, for small accelerations, *physical deformations* of the rod can be neglected.

We will further assume that S is a laboratory frame (where observations are performed). A rod is supposed to move relative to it, with high velocity $\mathbf{v}$, but small acceleration $\mathbf{a}$. We will call this a *conditionally rigid rod*. Note that the velocity $\mathbf{v}$ is not assumed to be small (in contrast to the acceleration) and can be arbitrarily close to the fundamental velocity. We will now consider how the special case of translational motion (without rotation) in S' is described in S.

10.2 The behaviour of accelerated rods

Imagine that some force acts with no torque on the centre of mass of a conditionally rigid rod. From the perspective of classical mechanics, the latter should move in space without changing its orientation in any frame. This is not so in the theory of relativity.

Let one end of the rod coincide with the origin of the frame S' (point A in the figure below). Suppose that another "identical" rod moves with a *small*, constant velocity $\mathbf{u}'$ relative to the first one, so that both rods coincide at time $t' = 0$. Equivalently, we can assume that there is a single rod whose points *simultaneously* move with the same velocity $\mathbf{u}'$ at $t' = 0$. Such a rod will accelerate by moving parallel to its previous position. The ends of the rods at point A and the coordinate origins coincide in S at $t = 0$. However, due to the relativity of simultaneity (p. 69), at their other end, the rods will not coincide. For stationary observers, they rotate around the point A. The same changes will appear if the rod moves parallel to itself (from its own perspective) with additional velocity $\mathbf{u}'$.

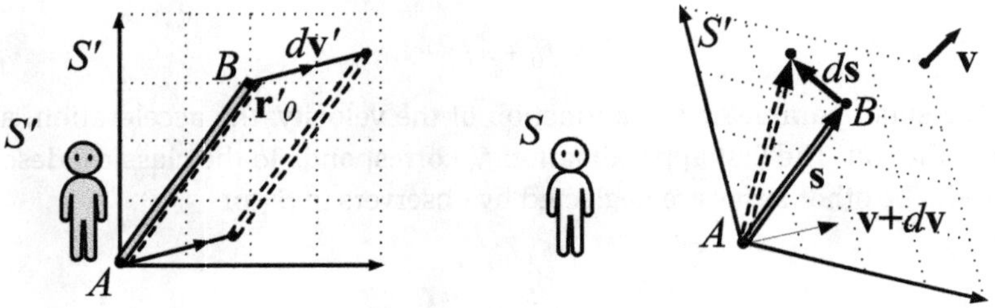

The trajectories of an arbitrary point moving with constant velocity in S' and S are

$$\mathbf{r}' = \mathbf{r}'_0 + \mathbf{u}'t',$$
$$\mathbf{r} = \mathbf{r}_0 + \mathbf{u}t.$$

We can use these equations to find the relationship between the velocities $\mathbf{u}$, $\mathbf{u}'$ and the constants $\mathbf{r}_0$, $\mathbf{r}'_0$, by first substituting into the inverse Lorentz transformations (1.18), p. 17:

$$t = \gamma\,(t' + \mathbf{v}\mathbf{r}'_0 + (\mathbf{v}\mathbf{u}')t') \tag{10.2}$$
$$\mathbf{r}_0 + \mathbf{u}t = \mathbf{r}'_0 + \mathbf{u}'t' + \mathbf{v}\gamma t' + \Gamma\mathbf{v}(\mathbf{v}\mathbf{r}'_0) + \Gamma\mathbf{v}(\mathbf{v}\mathbf{u}')t'. \tag{10.3}$$

We then substitute the expression for t from (10.2) into the left-hand side of equation (10.3) to obtain

$$\mathbf{r}_0 - \mathbf{r}'_0 + \gamma\mathbf{u}\,(\mathbf{v}\mathbf{r}'_0) - \Gamma\mathbf{v}(\mathbf{v}\mathbf{r}'_0) = [\mathbf{u}' + \gamma\mathbf{v} + \Gamma\mathbf{v}(\mathbf{v}\mathbf{u}') - \gamma\mathbf{u}\,(1 + \mathbf{v}\mathbf{u}')]t',$$

where we have grouped together terms with a common factor of t'.

This relationship is true for every t' only if its left- and right-hand sides are zero. As a result, we obtain the relationship between the velocities (1.24), p. 20 (the relativistic velocity-addition formula) and the initial positions:

$$\mathbf{r}_0 = \mathbf{r}'_0 - \gamma\,\mathbf{u}\,(\mathbf{v}\mathbf{r}'_0) + \Gamma\,\mathbf{v}\,(\mathbf{v}\mathbf{r}'_0). \tag{10.4}$$

The points A of the first and the second rod coincide with each other and with the coordinate origins ($\mathbf{r}_0 = \mathbf{r}'_0 = 0$) at the time $t = t' = 0$. Point B of the first rod has velocities $\mathbf{u} = \mathbf{v}$ and $\mathbf{u}' = 0$. Therefore, using (10.4), we obtain for it the following coordinates in S at $t = 0$:

$$\mathbf{r}_{0_1} = \mathbf{r}'_0 - \frac{\gamma}{\gamma+1}\,\mathbf{v}\,(\mathbf{v}\mathbf{r}'_0). \tag{10.5}$$

This relationship coincides with (3.7), p. 67, for $t = 0$. The point B of the second rod has velocities $\mathbf{u} = \mathbf{v} + d\mathbf{v}$ and $\mathbf{u}' = d\mathbf{v}'$. From (10.4) we obtain its position in S at $t = 0$:

$$\mathbf{r}_{0_2} = \mathbf{r}'_0 - \frac{\gamma}{\gamma+1}\,\mathbf{v}\,(\mathbf{v}\mathbf{r}'_0) - \gamma\,(\mathbf{v}\mathbf{r}'_0)\,d\mathbf{v}. \tag{10.6}$$

By subtracting equation (10.5) from (10.6), we obtain the change in the position of point B relative to point A (the displacement of the end B of the second rod with respect to the first rod) for observers in S. The value $\mathbf{r}'_0$ of the points B is the same for both rods in S' (the rods coincide at $t' = 0$).

Let $\mathbf{s}$ be the vector connecting the ends A and B of the rod. Since the radius vector of the point A is zero, we have $\mathbf{s} = \mathbf{r}_{0_1}$. After the velocity has changed in (10.6), we have

$$\mathbf{r}_{0_2} = \mathbf{s} + d\mathbf{s}.$$

Therefore,

$$d\mathbf{s} = -\gamma\,(\mathbf{v}\mathbf{s}')\,d\mathbf{v} = -\gamma^2\,(\mathbf{v}\mathbf{s})\,d\mathbf{v}, \tag{10.7}$$

where $\mathbf{s}' = \mathbf{r}'_{0_1} = \mathbf{r}'_{0_2}$ is the position of the points B of the rods in S'. Using (3.8), p. 68, we can substitute $\mathbf{v}\mathbf{s}' = \gamma\,\mathbf{v}\mathbf{s}$ (the formula for length contraction) into the second equality. By introducing the three-dimensional acceleration vector $\mathbf{a} = d\mathbf{v}/dt$, we finally find the equation [25]:

$$\frac{d\mathbf{s}}{dt} = -\gamma^2\,(\mathbf{v}\mathbf{s})\,\mathbf{a}. \tag{10.8}$$

Since the points A of both rods coincide, the time derivative of $\mathbf{s}$ in the equation (10.8) is the rate of change of the rod's orientation and length (the change in the position of B with respect to A). Point A moves independently; its velocity $\mathbf{v}(t)$ changes with time. The fundamental constant "c" is also easily restored (⋖ H_{38}) in (10.8).

▷ We can now consider several special cases of accelerated motion. The figure below shows different orientations, velocities, and accelerations of the rod. The rod moves without rotation in the first two cases.

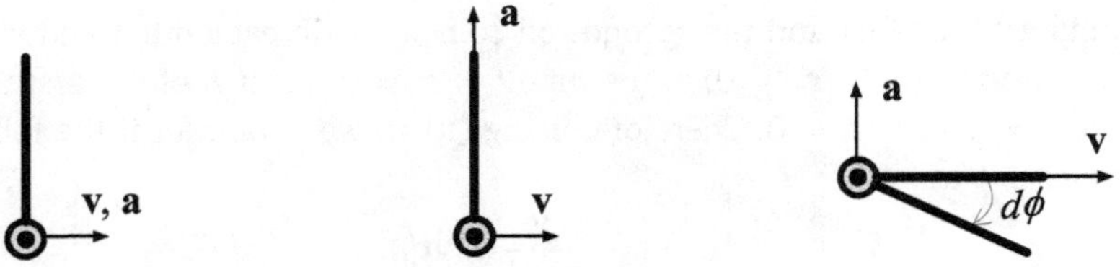

For example, if the rod is oriented across its velocity ($\mathbf{vs} = 0$), the change of the latter in magnitude (but not in direction) will not impact the rod's orientation or length. In this case, all of its points simultaneously change their velocities in both S and in S', and distances remain constant in the direction transverse to the velocity. Therefore, if an observer in S' starts pushing (accelerating) a vertical rod in the horizontal direction, it will not rotate in the laboratory frame. Nor does a vertical rod lifted *upwards* in a horizontally moving frame (the second figure) rotate (both of its ends have the same x-coordinate). However, if both ends of a horizontal rod are lifted up in S' (the third figure), their motion will not be simultaneous in S, and will cause rotation (10.8). The angle of this rotation can also be obtained (<H39) from the relativity of simultaneity.

The rotation of accelerated rods is closely related to the so-called *Thomas precession*. In 1926, Llewellyn Hilleth Thomas [22] considered the motion of a rotating ball along a curved path to explain the splitting of the lines in atomic spectra. Such a ball was proposed as a classical model of the electron. Thomas took into account the relativistic kinematic effect of the rotation of the coordinate frame, which arises when the Lorentz transformations are composed, and obtained the correct coefficients in the Hamiltonian of the interaction between the electron spin and the electromagnetic field. Today, this classical model of the electron is purely of historical significance, and the correct Hamiltonian is obtained in quantum theory from the Dirac equation.

Nevertheless, the rotation of accelerated rods and the proper angular momentum (spin) of gyroscopes are of some interest. Later, we will consider gyroscopes moving along arbitrary trajectories. In this case, the gyroscope and the rod will have different rotation dynamics, due to the different mathematical nature of the spin vector and the instantaneous form of the rod.

Equation (10.8) is true only for small accelerations, when the rod can be considered rigid. In this case, an "instantaneously inertial" (*co-moving*) reference frame is attached to the rod. Equation (10.8) was obtained under this assumption.

▷ Consider a uniformly accelerated rod whose velocity and acceleration vectors are directed along its length. Suppose that one of its ends moves with the following velocity and acceleration (see (3.38), p. 85):

$$v = \frac{wt}{\sqrt{1+(wt)^2}}, \qquad \gamma = \sqrt{1+(wt)^2}, \qquad a = \frac{w}{\gamma^3},$$

where w is some constant (the proper acceleration). The length of the rod $l = \sqrt{\mathbf{s}^2}$ satisfies the equation

$$\frac{dl}{dt} = \frac{\mathbf{s}(d\mathbf{s}/dt)}{l} = -\gamma^2 \frac{(\mathbf{sv})(\mathbf{sa})}{l} = -\gamma^2 v(t)a(t)l(t) = -\frac{w^2 t}{1+(wt)^2}.$$

Integrating this with the initial condition $l(0) = l_0$, we have

$$l(t) = \frac{l_0}{\sqrt{1+(wt)^2}} = l_0\sqrt{1-v^2(t)}. \tag{10.9}$$

This expression coincides with the instantaneous Lorentz contraction of a rod moving with velocity $v(t)$.

As we saw when discussing non-inertial frames in Chapter 6, physics looks "trickier" in an accelerated frame than in a sequence of co-moving inertial frames. In particular, if accelerated observers "keep" the same distance between each other, time will pass differently for them. The length of a rod attached to such frame changes relative to a fixed observer as follows:

$$l(t) = \frac{\sqrt{\gamma^2 + 2wl_0 + (wl_0)^2} - \gamma}{w} \approx \frac{l_0}{\gamma}\left(1 + \frac{v^2}{2} wl_0 + \dots\right),$$

where the approximate equality is written for small w. As can be seen, the ratio tends to the instantaneous Lorentz contraction (10.9) only when

$$v^2 wl_0 = v^2\gamma^3 a l_0 \ll 1.$$

This approximation is valid for a relatively small acceleration (more precisely, for

$$a \ll a_0 = \frac{c^4}{v^2\gamma^3 l_0},$$

where the speed of light c has been restored). This combination of the rod's length, velocity and acceleration should be so small that "non-inertial effects" can be ignored and equation (10.8) be used. Note that for a metre rod moving with velocity $v = 0.8\,c$ we have the rather large value $a_0 = 3 \cdot 10^{16}\ m/s^2$.

10.3 Circular motion of rods *

Let the origin of the frame S' move around a circle of radius R with velocity v constant in magnitude. If its period of revolution is T, its velocity is $v = 2\pi R/T = \omega R$, where ω is the angular frequency. The magnitude of the acceleration is $a = v^2/R = \omega v$.

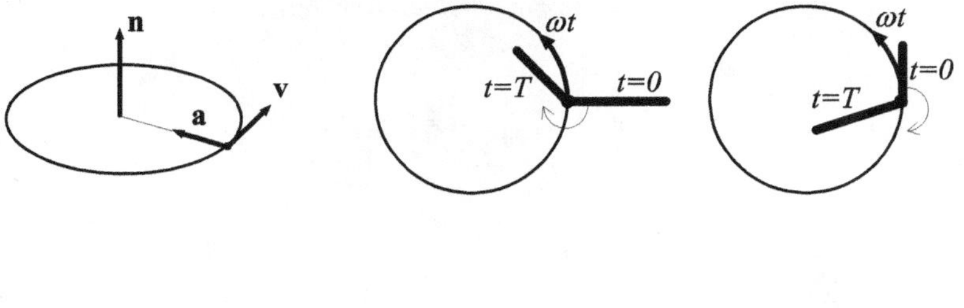

The acceleration vector is always orthogonal to the velocity vector, $\mathbf{av} = 0$. If $\mathbf{n}$ is a constant unit vector normal to the plane of the orbit, the following relations are true:

$$\mathbf{a} = \omega\,[\mathbf{n}\times\mathbf{v}], \qquad \frac{d\mathbf{a}}{dt} = \omega\,[\mathbf{n}\times\mathbf{a}] = -\omega^2\,\mathbf{v}. \tag{10.10}$$

Multiplying equation (10.8) by $\mathbf{v}$ yields

$$\frac{d\mathbf{s}}{dt} = -\gamma^2\,(\mathbf{vs})\,\mathbf{a} \qquad \Rightarrow \qquad \frac{d(\mathbf{vs})}{dt} = \mathbf{as}. \tag{10.11}$$

Differentiating with respect to time,

$$\frac{d^2(\mathbf{vs})}{dt^2} = \mathbf{a}\frac{d\mathbf{s}}{dt} + \mathbf{s}\frac{d\mathbf{a}}{dt} = -(\gamma^2 a^2 + \omega^2)\,(\mathbf{vs}),$$

and substituting in $a = \omega v$, we obtain the oscillator equation:

$$\frac{d^2(\mathbf{vs})}{dt^2} + \omega^2\gamma^2\,(\mathbf{vs}) = 0.$$

This has the following solution:

$$\mathbf{vs} = (\mathbf{vs})_0\,\cos(\omega\gamma t) + \frac{(\mathbf{as})_0}{\omega\gamma}\,\sin(\omega\gamma t), \tag{10.12}$$

where the zero index denotes the initial value of the scalar product, and the value of $d(\mathbf{vs})/dt$ for $t = 0$ is written according to (10.11).

Multiplying (10.11) by the acceleration $\mathbf{a}$ and using (10.10), we have:

$$\frac{d(\mathbf{as})}{dt} = \mathbf{s}\frac{d\mathbf{a}}{dt} - \gamma^2 a^2\,(\mathbf{vs}) = -\omega^2\gamma^2(\mathbf{vs}).$$

The right-hand side is familiar from (10.12), so the equation is easily integrated:

$$\mathbf{as} = (\mathbf{as})_0\,\cos(\omega\gamma t) - (\mathbf{vs})_0\,\omega\gamma\,\sin(\omega\gamma t). \tag{10.13}$$

Thus, the end of the rod rotates with angular velocity $\omega\gamma$ with respect to the moving basis formed by the vectors $\mathbf{v}$, $\mathbf{a}$.

Let us now find the coordinates $\mathbf{s} = \{s_x, s_y\}$ of the rod's end as functions of the coordinates of its beginning. Suppose the rod moves anti-clockwise in a circle, $\mathbf{r}(t) = R\,\{\cos(\omega t),\ \sin(\omega t)\}$. Then its velocity and acceleration components are:

$$\mathbf{v} = R\omega\,\{-\sin(\omega t),\ \cos(\omega t)\}, \qquad \mathbf{a} = -R\omega^2\,\{\cos(\omega t),\ \sin(\omega t)\}.$$

At $t = 0$, we have

$$\mathbf{v} = R\omega\,\{0, 1\}, \qquad \mathbf{a} = -R\omega^2\,\{1, 0\},$$

so

$$(\mathbf{vs})_0 = R\omega s_{y0}, \qquad (\mathbf{as})_0 = -R\omega^2 s_{x0},$$

and the solutions (10.12) and (10.13) give the following system:

$$\begin{cases} s_x \cos(\omega t) + s_y \sin(\omega t) = s_{x0} \cos(\omega\gamma t) + \ s_{y0}\, \gamma\, \sin(\omega\gamma t), \\ -s_x \sin(\omega t) + s_y \cos(\omega t) = s_{y0} \cos(\omega\gamma t) - (s_{x0}/\gamma) \sin(\omega\gamma t). \end{cases}$$

As follows from this solution, the rod has the same length at times satisfying

$$\omega t = \frac{\pi k}{\gamma}, \qquad \text{where} \qquad k = 1, 2, \ldots.$$

(< H_{40}). If γ is a rational number, the end of the rod will trace out a regular n-sided polygon, where $(\gamma - 1)/\gamma$ is equal to the irreducible fraction $2k/n$. At low velocities the rod rotates through a small angle πv^2 [25] with each revolution around the circle.

The figure below shows the trajectories of the rod's end with respect to its beginning for different velocities and initial orientations. The rod has unit length in its proper frame. The figure on the left corresponds to one revolution around the circle, its twenty-four points being separated by equal time intervals. The right-hand figure depicts the result of ten revolutions. The beginning of the rod starts moving from "three o'clock". The rotation of the rod is always opposite to the direction of its revolution around the circle.

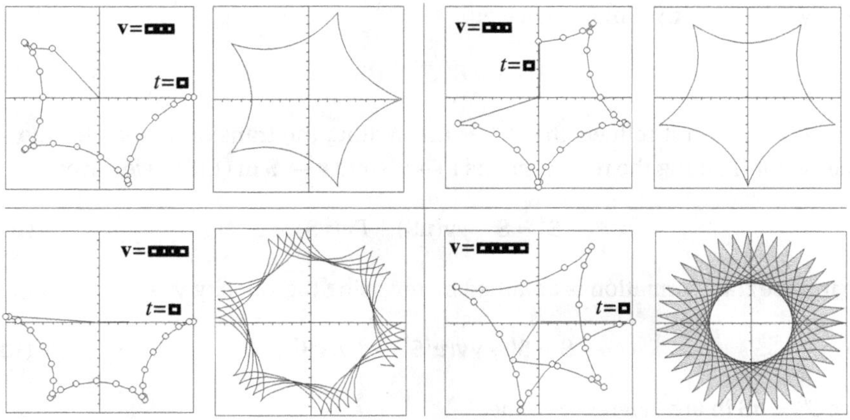

10.4 Accelerated gyroscopes

We considered above the case of an accelerated rod whose changing velocity makes it move parallel to itself from the perspective of observers in a co-moving frame. In the theory of relativity, such a rod rotates relative to observers in the stationary (laboratory) frame.

The spin (p. 209) of a rotating frame is subject to a similar effect. Suppose some force acts upon a gyroscope, changing its velocity without creating any torque. In classical mechanics, the proper angular momentum does not change (although the total momentum can change, due to the "orbital motion"). In the theory of relativity, the proper angular momentum (spin) changes (precesses) under such conditions, in the general case.

Consider three reference frames K, K', and K''. Let the velocity of K' relative K be $\mathbf{v}$, and the velocity of K'' relative to K' be $d\mathbf{v}'$. In accordance, the velocity of K'' relative to K is $\mathbf{v} + d\mathbf{v}$. These velocities are related by the velocity-addition law (1.24), p. 20:

$$d\mathbf{v}' = \frac{(\mathbf{v}+d\mathbf{v}) - \gamma\mathbf{v} + \Gamma\,\mathbf{v}\,(\mathbf{v}(\mathbf{v}+d\mathbf{v}))}{\gamma\,(1-\mathbf{v}(\mathbf{v}+d\mathbf{v}))} \approx \gamma d\mathbf{v} + \frac{\gamma^3\mathbf{v}(\mathbf{v}d\mathbf{v})}{\gamma+1}, \tag{10.14}$$

where the approximate equality is written up to the first order of smallness with respect to $d\mathbf{v}$.

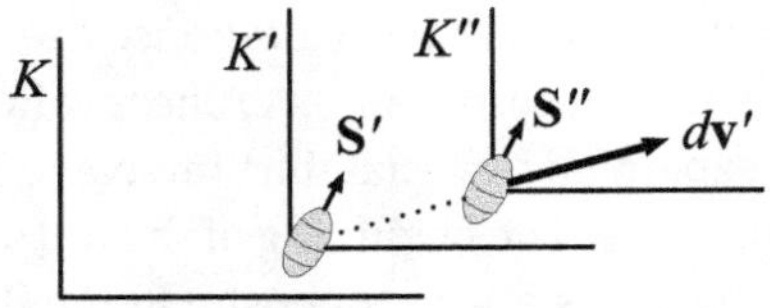

The following arguments will be true for any four-vector S^α orthogonal to the four-velocity $U^\alpha = \{\gamma_u,\ \mathbf{u}\gamma_u\}$ in four-space:

$$\mathrm{U}\cdot\mathrm{S} = U^0S^0 - \mathbf{US} = 0. \tag{10.15}$$

From this relation it follows that $S^0 = \mathbf{uS}$. Writing the transformations for the spin three-vector (making the replacements $t \mapsto \mathbf{uS}$ and $\mathbf{r} \mapsto \mathbf{S}$ in (1.18)), we have:

$$\mathbf{S}' = \mathbf{S} - \gamma\mathbf{v}(\mathbf{uS}) + \Gamma\mathbf{v}(\mathbf{vS}). \tag{10.16}$$

The inverse transformation is obtained by reversing the velocity $\mathbf{v} \mapsto -\mathbf{v}$:

$$\mathbf{S} = \mathbf{S}' + \gamma\mathbf{v}(\mathbf{u}'\mathbf{S}') + \Gamma\mathbf{v}(\mathbf{vS}'), \tag{10.17}$$

since $S^0 = \mathbf{uS}$ in any reference frame.

If the gyroscope is at rest ($\mathbf{u}' = 0$) relative to K', we have

$$\mathbf{S} = \mathbf{S}' + \Gamma\mathbf{v}(\mathbf{vS}'). \tag{10.18}$$

The inverse transformation is obtained from (10.16) after making the substitution $\mathbf{u} = \mathbf{v}$:

$$\mathbf{S}' = \mathbf{S} - \frac{\gamma}{\gamma+1}\,\mathbf{v}(\mathbf{vS}). \tag{10.19}$$

Now we can apply the transformation (10.18) to the frames K' and K''. Let there be a fixed (but rotating) gyroscope with spin $\mathbf{S}''$ in the frame K''. In this case, primes are to be added to all values, and the replacement $\mathbf{v} \mapsto d\mathbf{v}'$ should be made in the transformation (10.18). As a result, the spin remains constant in K': $\mathbf{S}' = \mathbf{S}''$, up to first order in $d\mathbf{v}'$.

Now consider an "identical" gyroscope with spin $\mathbf{S}'$ at rest in K' (the figure above). When the origins of K' and K'' coincide, the gyroscopes also "coincide", just as the rods coincided in the example above (p. 254). Therefore, we assume that the gyroscope in K'' is obtained by changing the velocity of the gyroscope in K' by adding the value $d\mathbf{v}'$. According to (10.17), the spin of the gyroscope in K'' relative to K is

$$\mathbf{S} = \mathbf{S}' + \gamma\mathbf{v}(\mathbf{S}'d\mathbf{v}') + \Gamma\mathbf{v}(\mathbf{v}\mathbf{S}'). \tag{10.20}$$

This expression gives the spin at $t+dt$, when the velocity of the gyroscope has changed by $d\mathbf{v}'$ relative to K'. Subtracting the spin at time t (10.18) from this equation, we have

$$d\mathbf{S} = \gamma\,\mathbf{v}\,(\mathbf{S}'d\mathbf{v}') = \gamma^2\,\mathbf{v}\,(\mathbf{S}d\mathbf{v}), \tag{10.21}$$

where in the second equality, we have expressed $d\mathbf{v}'$ in terms of $d\mathbf{v}$ according to (10.14), and the expression (10.19) has been substituted for $\mathbf{S}'$.

Introducing a three-dimensional acceleration vector $\mathbf{a} = d\mathbf{v}/dt$, we finally obtain:

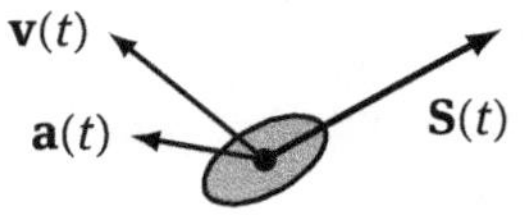

$$\frac{d\mathbf{S}}{dt} = \gamma^2\,(\mathbf{a}\mathbf{S})\,\mathbf{v}. \tag{10.22}$$

If the acceleration $\mathbf{a}$ remains orthogonal to the spin vector ($\mathbf{a}\mathbf{S} = 0$), the latter does not change during the motion. Otherwise, the spin will change as a result of the accelerated motion. This change causes both spin rotation (precession) and the change of the magnitude of the vector $\mathbf{S}$.

Note that equation (10.22) differs from equation (10.8), which describes the rotation of a "rigid" rod during curvilinear motion. Therefore, the rod and the spin will have different rotation dynamics.

▷ Equation (10.22) can be written in the covariant form

$$\frac{dS^\alpha}{d\tau} = -V^\alpha A^\beta S_\beta, \tag{10.23}$$

where $V^\alpha = \{\gamma, \mathbf{v}\gamma\}$ is the four-velocity vector, A^α is the four-acceleration vector (p. 178)

$$A^\alpha = \frac{dV^\alpha}{d\tau} = \gamma\,\frac{d}{dt}\{\gamma,\ \mathbf{v}\gamma\} = \{\gamma^4\,(\mathbf{v}\mathbf{a}),\ \gamma^2\,\mathbf{a} + \gamma^4\,\mathbf{v}\,(\mathbf{v}\mathbf{a}),\} \tag{10.24}$$

and $d\tau = \sqrt{1-\mathbf{v}^2}\,dt$ is the proper time of K'. Indeed, the contraction of the four-acceleration against the four-spin is

$$A^\beta S_\beta = A^0 S^0 - \mathbf{A}\mathbf{S} = \gamma^4\,(\mathbf{v}\mathbf{a})\,(\mathbf{v}\mathbf{S}) - \gamma^2\,(\mathbf{a}\mathbf{S}) - \gamma^4\,(\mathbf{v}\mathbf{S})\,(\mathbf{v}\mathbf{a}) = -\gamma^2\,(\mathbf{a}\mathbf{S}).$$

The differential equation (10.23) is called the ***Fermi transport equation***. It can be derived from the following considerations. Suppose that in covariant form, the change of spin can only depend on the four-velocity, the four-acceleration, and the four-spin. Then, from covariance considerations, we have

$$\frac{d\mathrm{S}}{d\tau} = \lambda \mathrm{V} + \sigma \mathrm{A} + \kappa \mathrm{S}, \tag{10.25}$$

where λ, σ and κ are some coefficients. Assume that the change of spin is zero ($d\mathbf{S}/dt = 0$), i.e., that the spin is transported parallel to its previous position in the instantaneous co-moving inertial frame where the particle is at rest ($\mathrm{V} = \{1, \mathbf{0}\}$, $\mathrm{A} = \{0, \mathbf{a}\}$, $\mathrm{S} = \{0, \mathbf{S}\}$). It is easily seen that $\sigma = \kappa = 0$ in this case. By differentiating the orthogonality condition $\mathrm{S} \cdot \mathrm{V} = 0$ for the four-spin and the four-velocity, we obtain

$$\frac{d(\mathrm{S} \cdot \mathrm{V})}{d\tau} = \frac{d\mathrm{S}}{d\tau} \cdot \mathrm{V} + \mathrm{S} \cdot \frac{d\mathrm{V}}{d\tau} = \lambda + \mathrm{S} \cdot \frac{d\mathrm{V}}{d\tau} = 0,$$

where we have made use of the fact that the square of the four-velocity is unity ($\mathrm{V}^2 = 1$). Substituting λ into (10.25), we obtain the equation (10.23).

Due to the condition $\mathrm{V} \cdot \mathrm{S} = 0$, Fermi transport does not change the square of the spin four-vector: $\mathrm{S}^2 = const$, even though the square of the spin three-vector $\mathbf{S}^2$ changes unless the spin is orthogonal to the velocity or the acceleration.

Note also the equation describing the change of the ***modified spin*** $\tilde{\mathbf{S}} = \mathbf{S}M/\mathcal{E} = \mathbf{S}/\gamma$ relative to the laboratory frame:

$$\frac{d\tilde{\mathbf{S}}}{dt} = \gamma^2 \, [\mathbf{a} \times [\mathbf{v} \times \tilde{\mathbf{S}}]. \tag{10.26}$$

The modified spin is the difference between the total angular momentum and the angular momentum of the energy centre (p. 208): $\tilde{\mathbf{S}} = \mathbf{L} - \mathbf{R} \times \mathbf{P}$.

▷ For uniformly accelerating motion (p. 257) starting from rest, integrating of the precession equation (10.22) gives the following time dependence for the longitudinal spin component relative to the velocity:

$$S_x(t) = \frac{S_{x0}}{\sqrt{1 - v^2(t)}}, \qquad S_y(t) = S_{y0}, \tag{10.27}$$

where

$$S_{x0} = S_x(0)$$

is the initial spin in the frame where the total momentum was zero. The same relation also follows from the transformation (10.18). Thus, the longitudinal spin component grows together with the gyroscope acceleration, and the transverse component remains constant. In fact, the dependence (10.27) is true not only for uniformly accelerating motion but for any form of rectilinear motion whose equation (10.22) has the following form:

$$\frac{dS_x}{S_x} = \frac{va\,dt}{1 - v^2(t)} = -\frac{1}{2}\,d[\ln(1 - v^2(t))].$$

Integrating this equation gives (10.27).

One can consider a gyroscope whose centre of energy undergoes uniform circular motion. In this case, the rotation of the rod and the precession of the spin behave in a similar manner. At small velocities v, the spin, like the rod, rotates through a small angle πv^2 [25] with each revolution.

Similarly to the spin, we can obtain the equation for the change of the total angular momentum $\mathbf{L}$ relative to the fixed (laboratory) frame. Since the projections of the vector $\mathbf{L}$ are the components of a four-tensor, its transformational properties differ from the those of the spin. Therefore, the resulting equation also differs from (10.22), and has the form:

$$\frac{d\mathbf{L}}{dt} = \gamma^2\,\mathbf{v} \times [\mathbf{L} \times \mathbf{a}].$$

If the accelerated motion is rectilinear (i.e., if the vectors $\mathbf{v}$ and $\mathbf{a}$ are parallel), the change of the angular momentum components coincides with the corresponding Lorentz transformations for the instantaneous inertial co-moving frame of the gyroscope.

Note that the initial Thomas formula for spin precession differs from equation (10.22). The reason is that Thomas considered only the Wigner rotation, which will be discussed in the following section.

10.5 Wigner rotation and Thomas precession

Suppose S, S_1, and S_2 are three inertial frames. Frame S_1 moves with velocity $\mathbf{v}_1$ relative to S, and S_2 moves with velocity $\mathbf{v}_2$ relative to S_1. The transformations between the frames are described by the Lorentz transformations (boosts)

$$\Lambda_1 = \Lambda(\mathbf{v}_1) \quad \text{and} \quad \Lambda_2 = \Lambda(\mathbf{v}_2).$$

Since the figure below is not "attached" to any specific observer, the axes of all of the frames in it are drawn parallel to each other. Remember (p. 17) that observers fix the values of the relative velocity vector $\mathbf{v}$ components to adjust their units of measurement. By doing this, each of them can build orthogonal axes in their frame, where the velocity components are

$$\mathbf{v} = \{v_x, v_y, v_z\}$$

for one observer, and the same, but with the signs reversed, for the other. At the same time, if the observer measures the positions of the "other" coordinate axes (p. 67), in general, he will find that they are neither parallel to his frame nor orthogonal to each other.

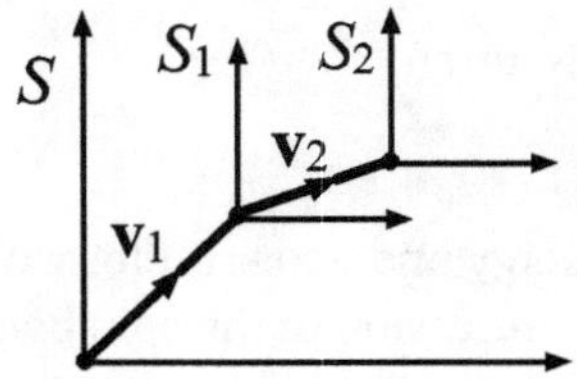

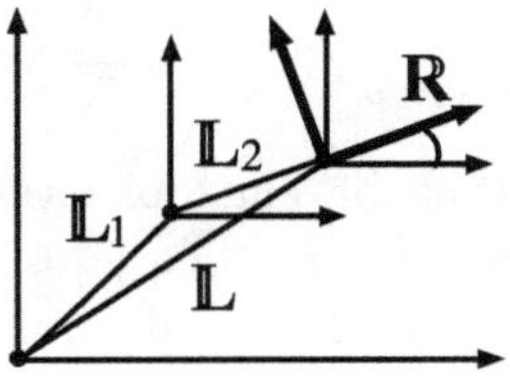

The sequence of transformations $\Lambda_2\,\Lambda_1$ adds new nuances to this picture. A composition of boosts is not necessarily a boost. This means that, regarding the adjustment procedure for the directions of the axes, transitivity is violated. If this procedure is performed by observers in S and S_1 (by fixing the value $\mathbf{v}_1$), and then a similar adjustment is made by observers in S_1 and S_2 using the component $\mathbf{v}_2$, the relative velocities of S and S_2 will not simply be equal up to a change of sign.

In section 11.4, quaternions will be used to show that the sequence of transformations Λ_1 and Λ_2 is equivalent to the pure boost Λ associated with a spatial rotation of the coordinate axes of $\mathbb{R}$. If two frames are connected by a boost, their relative velocities coincide up to a sign. When the the axes of the "moving" frame are rotated, its velocity relative to the "resting" frame will not change. The velocity of the resting frame relative to the moving frame will change, however, since the projections of the velocity onto the coordinate axes change as a result of the rotation. These considerations are another way to obtain the rotation angle, or the Wigner rotation.

▷ The velocity of S_2 relative to S is (p. 20):

$$\mathbf{v}_{20} = \frac{\mathbf{v}_2 + \gamma_1\,\mathbf{v}_1 + \Gamma_1\,\mathbf{v}_1\,(\mathbf{v}_1\mathbf{v}_2)}{\gamma_1\,(1 + \mathbf{v}_1\mathbf{v}_2)}, \tag{10.28}$$

where, as usual, $\Gamma_1 = (\gamma_1 - 1)/v_1^2$ and $\gamma_1 = 1/\sqrt{1 - v_1^2}$. Similarly, the velocity of S relative to S_2 is

$$\mathbf{v}_{02} = -\frac{\mathbf{v}_1 + \gamma_2\,\mathbf{v}_2 + \Gamma_2\,\mathbf{v}_2\,(\mathbf{v}_2\mathbf{v}_1)}{\gamma_2\,(1 + \mathbf{v}_1\mathbf{v}_2)}, \tag{10.29}$$

which we obtain via the substitution $\mathbf{v}_1 \mapsto -\mathbf{v}_2$, $\mathbf{v}_2 \mapsto -\mathbf{v}_1$. In general, $\mathbf{v}_{20}$ and $\mathbf{v}_{02}$ are not parallel to each other: $\mathbf{v}_{20} \neq -\mathbf{v}_{02}$. This is the essential difference between the velocity space of the theory of relativity and the classical rule $\mathbf{v}_{20} = \mathbf{v}_1 + \mathbf{v}_2$, which is symmetric with respect to the velocities of the successive Galilean transforms.

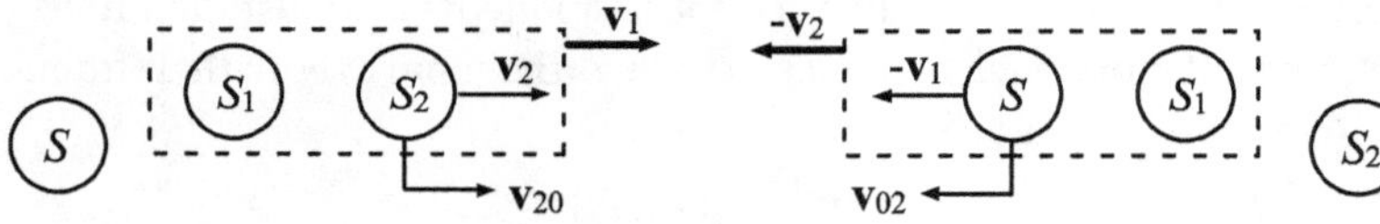

Even though the relative velocities are not parallel, their magnitudes are equal. It can be verified that

$$v^2 = \mathbf{v}_{20}^2 = \mathbf{v}_{02}^2 = \frac{(\mathbf{v}_1 + \mathbf{v}_2)^2 - [\mathbf{v}_1 \times \mathbf{v}_2]^2}{(1 + \mathbf{v}_1\mathbf{v}_2)^2}. \tag{10.30}$$

In particular, $\gamma = \gamma_1\gamma_2\,(1 + \mathbf{v}_1\mathbf{v}_2)$. Taking the vector product of (10.28) and (10.29),

$$\mathbf{n}\,\sin\phi = \frac{[\mathbf{v}_{20} \times \mathbf{v}_{02}]}{v^2},$$

we obtain the sine of the angle between the relative velocities and the unit vector $\mathbf{n}$ normal to the plane containing these vectors:

$$\mathbf{n}\,\sin\phi = -[\mathbf{v}_1 \times \mathbf{v}_2]\,\frac{\gamma_1\gamma_2 - 1 + \mathbf{v}_1\mathbf{v}_2\,(\gamma_1\Gamma_2 + \gamma_2\Gamma_1) + (\mathbf{v}_1\mathbf{v}_2)^2\Gamma_1\Gamma_2}{(\gamma^2 - 1)/(\gamma_1\gamma_2)}.$$

By writing $\mathbf{v}_1\mathbf{v}_2 = (\gamma - \gamma_1\gamma_2)/(\gamma_1\gamma_2)$, this equation can be simplified:

$$\mathbf{n}\,\sin\phi = -[\mathbf{v}_1 \times \mathbf{v}_2]\,\frac{\gamma_1\gamma_2\,(1 + \gamma + \gamma_1 + \gamma_2)}{(1 + \gamma)(1 + \gamma_1)(1 + \gamma_2)}, \tag{10.31}$$

which is the angle of the Wigner rotation that results from the composition of two pure Lorentz transformations (boosts): the velocity $\mathbf{v}_{02}$ rotates through an angle of ϕ relative to $\mathbf{v}_{20}$.

▷ Suppose a three-vector is associated with an object moving with a variable velocity. When the velocity changes, the vector shifts "parallel" to itself relative to the co-moving inertial frame, but it rotates by some angle with respect to the laboratory frame. This angle is often confused with the Wigner angle (10.31). As shown above, the "rotational dynamics" of the three-vector associated with the rod differ from those of the gyroscope due to the different transformation properties of these vectors. Therefore, Wigner rotation is related to Thomas precession, but cannot describe it by itself.

Nevertheless, let us consider, in unified notation, the descriptions of Thomas precession put forward in the books by Jackson [21] and Møller [18]. Let there be three frames of reference: S, S_1, and S_2, where S is the laboratory frame, and S_1 and S_2 are co-moving inertial frames at two consecutive moments of time.

In [21], it is assumed that S_1 and S_2 are related to S by the boosts

$$\mathrm{X}_1 = \Lambda(\mathbf{v})\,\mathrm{X}, \qquad \mathrm{X}_2 = \Lambda(\mathbf{v} + d\mathbf{v})\,\mathrm{X}, \tag{10.32}$$

where, for brevity, the "usual" matrix notation is used for four-vectors. Since

$$\Lambda^{-1}(\mathbf{v}) = \Lambda(-\mathbf{v}),$$

we have

$$\mathrm{X}_2 = \Lambda(\mathbf{v} + d\mathbf{v})\Lambda(-\mathbf{v})\,\mathrm{X}_1. \tag{10.33}$$

Using (10.31) and the relations

$$\mathbf{v}_1 = -\mathbf{v}, \qquad \mathbf{v}_2 = \mathbf{v} + d\mathbf{v},$$

we can write the infinitesimal rotation angle, up to first order in $d\mathbf{v}$ ($\lessdot$ H$_{41}$):

$$\mathbf{n}\,d\phi = \frac{\gamma^2}{\gamma + 1}\,[\mathbf{v} \times d\mathbf{v}]. \tag{10.34}$$

The transformation between S_1 and S_2 has the form:

$$t_2 = t_1 - \mathbf{r}_1 \Delta\mathbf{v}, \qquad \mathbf{r}_2 = \mathbf{r}_1 - d\phi\,[\mathbf{n} \times \mathbf{r}_1] - t_1 \Delta\mathbf{v}, \tag{10.35}$$

where

$$\Delta\mathbf{v} = \gamma\,(d\mathbf{v} + \Gamma\,(\mathbf{v}d\mathbf{v})\,\mathbf{v}).$$

Up to first order in $d\mathbf{v}$, the velocities $\mathbf{v}_{20}$ and $-\mathbf{v}_{02}$ coincide, and are equal to $\Delta\mathbf{v}$.

The resulting three-dimensional rotation

$$-d\phi\,[\mathbf{n} \times \mathbf{r}_1]$$

in (10.35) occurs between the co-moving frames S_1 and S_2. Therefore, Jackson's approach based on (10.32) concerns the rotation of co-moving frames, rather than rotation relative to the laboratory frame. Here, it is also assumed that the co-moving frames are obtained *from the laboratory frame* using boosts, and that their axes are "parallel" to S rather than to each other.

▷ Møller's reasoning [18] is closer to the approach of translating the axes of co-moving frames; here, the following sequence of transformations is considered:

$$\mathrm{X}_1 = \Lambda(\mathbf{v})\,\mathrm{X}, \quad \mathrm{X}_2 = \Lambda(d\mathbf{v}_{21})\,\mathrm{X}_1, \tag{10.36}$$

whence

$$\mathrm{X}_2 = \Lambda(d\mathbf{v}_{21})\Lambda(\mathbf{v})\,\mathrm{X}. \tag{10.37}$$

Substituting the relations

$$\mathbf{v}_1 = \mathbf{v}, \quad \mathbf{v}_2 = d\mathbf{v}_{21},$$

into (10.31) gives us, up to first order in $d\mathbf{v}_{21}$:

$$\mathbf{n}\,d\phi = -\frac{\gamma}{\gamma+1}\,[\mathbf{v} \times d\mathbf{v}_{21}]. \tag{10.38}$$

The resulting velocity (10.28), up to first order in $d\mathbf{v}_{21}$, is

$$\mathbf{v}_{20} \approx \mathbf{v} + \frac{d\mathbf{v}_{21}}{\gamma} - \frac{\mathbf{v}\,(\mathbf{v}d\mathbf{v}_{21})}{\gamma+1}. \tag{10.39}$$

The value $d\mathbf{v}_{21}$ is the velocity of S_2 relative to S_1, and represents the change of the object's velocity with respect to its previous instantaneous position in S_1. The velocity $\mathbf{v}_{20}$ is the velocity of S_2 relative to S. Therefore, Møller introduces the change of the object's velocity relative to the laboratory frame:

$$d\mathbf{v} = \mathbf{v}_{20} - \mathbf{v}.$$

Taking the vector product with $\mathbf{v}$, we can easily rewrite (10.38) as

$$\mathbf{n}\,d\phi = -\frac{\gamma^2}{\gamma+1}\,[\mathbf{v} \times d\mathbf{v}], \tag{10.40}$$

which, up to a sign, coincides with (10.34), but has a different meaning, since it gives the rotation of the frame obtained from the laboratory frame S after the Lorentz boost. In fact, there are four frames in the relationship

$$\Lambda(d\mathbf{v}_{21})\,\Lambda(\mathbf{v}) = \mathrm{R}(\mathbf{n}, d\phi)\,\Lambda(\mathbf{v} + d\mathbf{v}).$$

The left-hand side represents the successive transitions from S to S_1, and then to S_2. The right-hand side contains the Lorentz transformation $\Lambda(\mathbf{v} + d\mathbf{v})$, which describes the transformation from S to $\tilde{S}_2$ from which frame S_2 is obtained (as a result of the rotation). Therefore, the three-dimensional rotation is performed relative to $\tilde{S}_2$ rather than to the laboratory frame S.

This confusion of Wigner rotation with Thomas precession arises when the relativity of simultaneity is disregarded. The torque of a small gyroscope relates to one point, while the three-vector associated with the rod connects two points in the non-inertial frame (the beginning and the end of the rod). This is why its precession differs from the precession of the gyroscope's angular momentum. In the general case, different three-vectors moving in the laboratory frame will have different precessions, no matter how Wigner rotation is defined.

10.6 The nonlocal nature of conservation laws

In the theory of relativity, the momentum of a system of particles can be conserved in one inertial frame and not in another. The transformations (8.37), p. 207, for the angular momentum,

$$\mathbf{L} = \mathbf{r} \times \mathbf{p},$$

and the vector $\mathbf{G} = E\,\mathbf{r} - t\,\mathbf{p}$,

$$\mathbf{L} = \gamma\,(\mathbf{L}' - \mathbf{v} \times \mathbf{G}') - \Gamma\,\mathbf{v}\,(\mathbf{v}\mathbf{L}'), \tag{10.41}$$

are only first approximations [20], for composite systems. Indeed, the total angular momentum of a given reference frame is found by using synchronized clocks throughout the space. For a system of particles at different points in space, events which are simultaneous in one frame will not be simultaneous in another. To obtain the Lorentz transformations for the angular momentum of a set of particles, the transformations (10.41) must be summed over all of the particles. The quantities on the right-hand side correspond to the time t'. Their sum gives the angular momentum and the centre of energy at t'. However, if t' is fixed, the summands $\mathbf{L}$ and $\mathbf{G}$ on the left-hand side refer to different instants of time. From the Lorentz transformations (1.18), p. 17, for the k-th particle we have:

$$t_k = \mathbf{v}\mathbf{r}_k + \frac{t'}{\gamma}.$$

Since the positions of the particles $\mathbf{r}_k$ of the rotating body are different, the times t_k will also be different. Therefore, this sum will not equal the total angular momentum at time t:

$$\sum \mathbf{L}_k(\mathbf{v}\mathbf{r}_k + t'/\gamma) = \sum \gamma\,(\mathbf{L}'(t') - \mathbf{v} \times \mathbf{G}'(t')) - \Gamma \mathbf{v}\,(\mathbf{v}\mathbf{L}'(t')).$$

Similarly, the time t can be fixed on the left-hand side of the summed transformations. Then, the terms on the right-hand side will refer to different instants of time. As a result, the transformations for the total quantities appear substantially more complex than the relationship (10.41). The latter is true for a single particle, or as a first approximation with respect to the angular velocity of the rotation.

Moreover, even though the quantities $\mathbf{L}'$ and $\mathbf{G}'$ are constant in S', they change with time in S, in general. This is true not only for the angular momentum. To avoid misunderstanding, note that we are talking about the mechanical momentum, defined by

$$\mathbf{L} = \sum \mathbf{r} \times \mathbf{p}.$$

The fact that this is not conserved does not mean that we cannot introduce some quantity (which contains $\mathbf{L}$) that is constant in all inertial frames.

▷ Consider two massive balls connected by a light rod. Suppose this dumbbell rotates with angular velocity ω_0 in the (x', y')-plane of the frame S'. The centre of rotation coincides with the centre of mass and with the origin of the frame. The trajectory of one of the balls is

$$x' = r_0 \cos(\phi + \omega_0 t'), \qquad y' = r_0 \sin(\phi + \omega_0 t'), \tag{10.42}$$

where ϕ, ω_0 and r_0 are constants. The trajectory of the other ball is obtained by making either one of the replacements $\phi \mapsto \phi + \pi$ or $r_0 \mapsto -r_0$. The velocities of the balls both have magnitude $\omega_0 r_0$. The angular momentum of the dumbbell is orthogonal to the (x', y')-plane and is equal to

$$L_0 = \frac{\mu\,\omega_0 r_0^2}{\sqrt{1-(\omega_0 r_0)^2}}, \tag{10.43}$$

where $\mu = 2m$ is the total mass of the balls.

Let us find out what this dumbbell looks like for fixed observers in S when S' moves with velocity v along the x-axis. We first substitute the Lorentz transformations into the trajectory (10.42):

$$\gamma\,[x - vt] = r_0 \cos(\phi + \omega_0 \gamma\,[t - vx]\,), \qquad y = r_0 \sin(\phi + \omega_0 \gamma\,[t - vx]\,). \tag{10.44}$$

Introducing the coordinate $\tilde{x} = x - vt$ relative to the origin of S', we obtain

$$\gamma\,\tilde{x} = r_0 \cos(\phi + \omega t - \omega_0 v \gamma \tilde{x}), \qquad y = r_0 \sin(\phi + \omega t - \omega_0 v \gamma \tilde{x}), \tag{10.45}$$

where $\omega = \omega_0/\gamma$ is the angular frequency in S. The first transcendental equation gives the solution for $\tilde{x}$. The second equation gives y.

The figure below shows the positions of the balls and the rod at different times. When rotating, the rod bends, due to the relativity of simultaneity.

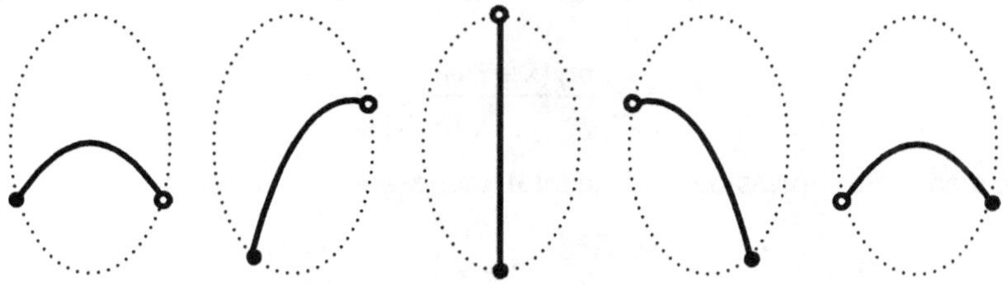

When the rod is vertical in S', both balls lie on the y'-axis. These two events have the same x'-coordinates, and are therefore also simultaneous for observers in S. At this moment, the shape of the rod is the same in both frames. The situation is different when the rod is horizontal in S' and the balls cross the x'-axis. These events will not be simultaneous in S, where the left ball will cross the x-axis before the right ball has time to do so.

▷ Differentiating (10.44) gives the velocity $\mathbf{u} = \{u_x,\ u_y\}$ of the ball:

$$u_x = \frac{v - \omega_0 y}{1 - v\omega_0 y}, \qquad u_y = \frac{\omega_0\,\tilde{x}}{1 - v\omega_0 y},$$

with the corresponding Lorentz factor

$$\gamma_u = \frac{1}{\sqrt{1 - u_x^2 - u_y^2}} = \frac{\gamma\,(1 - v\omega_0 y)}{\sqrt{1 - (r_0\omega_0)^2}},$$

where $\gamma = 1/\sqrt{1-v^2}$. Using the vector $\tilde{\mathbf{r}} = \mathbf{r} - \mathbf{v}t = \{\tilde{x}, y\}$, we can take the angular momentum relative to the instantaneous position of the origin of S' out of the total angular momentum relative to the origin of S. We will mark the former with a tilde sign,

$$\mathbf{L} = \sum m\gamma_u\,[\mathbf{r}\times\mathbf{u}] = \tilde{\mathbf{L}} + [\mathbf{v}\times\mathbf{P}]\,t, \tag{10.46}$$

where the "instantaneous" and the total momenta are:

$$\tilde{\mathbf{L}} = \sum m\gamma_u\,[\tilde{\mathbf{r}}\times\mathbf{u}], \qquad \mathbf{P} = \sum \mathbf{p}.$$

The vector $\tilde{\mathbf{L}}$ is directed along the z-axis and has length

$$\tilde{L} = \gamma L_0 - \sum \frac{m\gamma\,(\omega_0 v^2\,\gamma^2\tilde{x}^2 + v\,y)}{\sqrt{1-(r_0\omega_0)^2}}, \tag{10.47}$$

where L_0 is the angular momentum (10.43) in S'. The term γL_0 corresponds to the transformation of the momentum in accordance with (8.37), which takes no account of the relativity of simultaneity for a distributed system. Similarly, with the help of $\tilde{\mathbf{r}}$ we can write down a vector $\mathbf{G}$:

$$\mathbf{G} = \sum(E\mathbf{r} - \mathbf{p}t) = (\mathbf{v}\mathcal{E} - \mathbf{P})\,t + \sum E\,\tilde{\mathbf{r}}, \tag{10.48}$$

where $\mathcal{E} = \sum E$ is the total energy of the motion. The last term will be marked with a tilde. When divided by $\mathcal{E}$, this gives the radius vector of the centre of energy (p. 208) relative to the instantneous position of S'. The components of the total vector $\tilde{\mathbf{G}}$ have the form:

$$\tilde{\mathbf{G}} = \sum E\tilde{\mathbf{r}} = \sum \frac{m\gamma\{\tilde{x} - v\omega_0\, y\tilde{x},\ \ y - v\omega_0\, y^2\}}{\sqrt{1 - (r_0\omega_0)^2}}. \tag{10.49}$$

We can also write expressions for the total energy and momentum:

$$\mathcal{E} = \sum \frac{m\gamma(1 - v\omega_0\, y)}{\sqrt{1 - (r_0\omega_0)^2}}, \qquad \mathbf{P} = \sum \frac{m\gamma\{v - \omega_0 y,\ \omega_0\tilde{x}\}}{\sqrt{1 - (r_0\omega_0)^2}}. \tag{10.50}$$

In the case of a dumbbell, the sums in these relations contain two terms for each of the balls.

The figure below shows the time dependence of the instantaneous angular momentum $\tilde{L}$, along with the trajectories of the instantaneous centre of energy $\tilde{\mathbf{G}}/\mathcal{E}$, and the total momentum relative to the origin of S' in the (x, y)-plane. Here $v = 0.8$, $w_0 = 0.6$, and $m = r_0 = 1$. The time varies from 0 to π/ω.

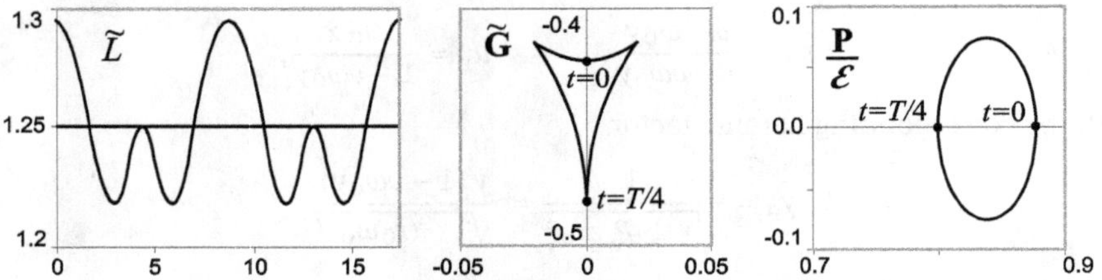

The oscillations of the angular momentum in the resting frame are due to the rapid rotation of the balls in S'. Expanding the cosine in equation (10.45) in a series with respect to $\omega_0 v y\tilde{x}$, and then expanding $\tilde{x}$ with respect to ω_0, we have:

$$y\tilde{x} \approx r_0\, c + r_0^2\omega_0 v\, s\, c + O(\omega^2 r_0^3), \qquad y \approx r_0\, s - r_0^2\, \omega_0 v\, c^2 + O(\omega^2 r_0^3).$$

where $c = \cos(\phi + \omega t)$ and $s = \sin(\phi + \omega t)$. As a result of the summation, odd powers of r_0 cancel in (10.47), since one ball has "r_0" and the other ball has "$-r_0$". Therefore, the momentum is constant in this approximation:

$$\tilde{L} \approx \gamma L_0 + O(\omega_0^2 r_0^3).$$

The correction to the result of the transformation (8.37) oscillates with frequency 2ω. The centre of energy relative to the origin of S' coincides, up to first order, with the result of the transformation (8.38):

$$\frac{\tilde{\mathbf{G}}}{\mathcal{E}} \approx r_0^2\gamma\omega_0\,\{0,\ -v\}.$$

The total energy of motion in S depends on time as follows:

$$\frac{\mathcal{E}}{\mu\gamma} \approx 1 + \frac{r_0^2\omega_0^2}{2} + r_0^2\omega_0^2 v^2 \cos^2(\omega t) + \dots$$

The total momentum is not equal to $\mathbf{v}\mathcal{E}$:

$$\frac{\mathbf{P}}{\mathcal{E}} \approx \mathbf{v} + \frac{\omega_0^2 r_0^2 v}{2\gamma^2} \left\{1 + \cos(2\omega t),\ \gamma^2 \sin(2\omega t)\right\} + \dots,$$

which makes the vector $\mathbf{G}$ not only oscillate together with the centre of energy, but also increase in length with time (10.48). All of these effects arise only at fast rotations, when the parameter ω_0 is large. For small ω_0, the transformation (10.41) is correct.

10.7 What does a rotating disc look like?

The relationships obtained for the dumbbell make it possible to find total quantities that characterize a rotating ring or a disc. Let a ring consist of a set of "dumbbells" that uniformly fill the plane of the ring in the frame S'. The figure below shows the appearance of the disc at any instant of time in S:

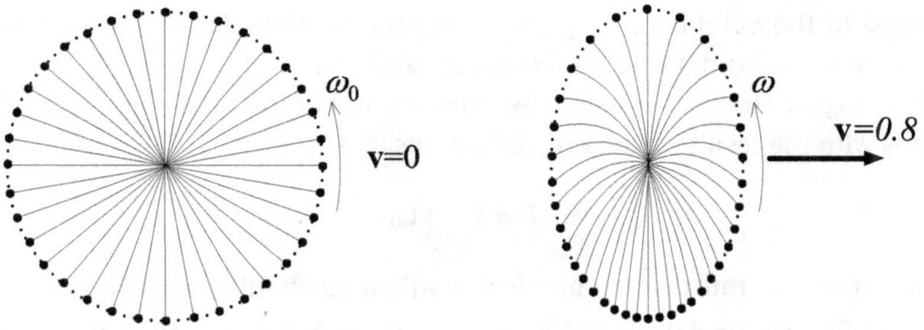

The appearance of the ring does not change with time, and there is always a crowding of masses in the lower part of the ring in the direction of the vector product $\mathbf{v}\times\mathbf{L}$, where the angular momentum $\mathbf{L}$ is orthogonal to the picture.

The equations (10.45) for $\tilde{x}$ and y are transcendental:

$$\frac{\gamma\tilde{x}}{r_0} = \cos(\phi + \omega t - \alpha\frac{\gamma\tilde{x}}{r_0}), \qquad \frac{y}{r_0} = \sin(\phi + \omega t - \alpha\frac{\gamma\tilde{x}}{r_0}),$$

where $\alpha = \omega_0 v r_0$. When summing the contributions of each of the balls, we integrate ("average") over the angle ϕ:

$$\langle f\rangle = \frac{1}{2\pi}\int_0^{2\pi} f(\phi)\, d\phi. \tag{10.51}$$

It can be shown (< H_{42}) that the averages of terms containing odd powers of $\tilde{x}$ are zero:

$$\langle \tilde{x} \rangle = \langle \tilde{x} y \rangle = 0,$$

while the non-zero values of order at most two have the form:

$$\langle y \rangle = -\frac{\omega_0 v r_0^2}{2}, \qquad \left\langle \tilde{x}^2 \right\rangle = \frac{r_0^2}{2\gamma^2}, \qquad \left\langle y^2 \right\rangle = \frac{r_0^2}{2}.$$

With these relationships, one easily finds the angular momentum and the centre of energy for a rotating ring moving in S, where all of the sums in the relationships (10.47), (10.49) and (10.50) are replaced by integrals over the angle ϕ.

The total mass of the ring must be μ, so that

$$\sum m \mapsto \frac{1}{2\pi} \int_0^{2\pi} \mu d\phi = \mu.$$

The total energy and momentum of the ring are:

$$\mathcal{E} = \frac{\mu\gamma}{\sqrt{1-(\omega_0 r_0)^2}} + \frac{\gamma L_0 v^2 \omega_0}{2}, \qquad \mathbf{P} = \mathbf{v}\mathcal{E} + \mathbf{v}\frac{L_0 \omega_0}{2\gamma}.$$

In contrast to the dumbbell, they do not depend on time. However, the relativistic relationship $\mathbf{P} = \mathbf{v}\mathcal{E}$ is not true for the total quantities.

By averaging (10.47), we reach the conclusion that the total angular momentum coincides with the result of the transformations (8.37):

$$L = \tilde{L} = \gamma L_0.$$

The angular momentum of a symmetric ring will be constant in both frames. The same is true for a disc whose axis of rotation is orthogonal to the relative velocity. Note that this result is valid only when the angular velocity of the rotation is orthogonal to the velocity $\mathbf{v}$. In general, however, the angular momentum grows linearly with time. In this case, the time-dependent term is of the second order of smallness with respect to the angular velocity of the rotation.

The vector $\tilde{\mathbf{G}}$ corresponds to the transformation (8.38), and the initial vector $\mathbf{G}$ without the tilde grows linearly with time:

$$\tilde{\mathbf{G}} = L_0 \gamma\, v \{0, -1\}, \qquad \mathbf{G} = L_0 \gamma v \{-\frac{\omega_0 t}{2\gamma^2}, -1\}.$$

This change of the vector $\mathbf{G}$ with time is due to the fact that the total momentum is not equal to the total energy multiplied by the velocity: $\mathbf{P} \neq \mathbf{v}\mathcal{E}$. As a result, although the centre of energy is constant relative to the origin of S', the vector $\mathbf{G}$ will grow linearly with time as a result of (10.48).

Nevertheless, at small angular velocities the vectors **G** and **L** can be assumed to have constant components; this is accurate up to the first order of smallness with respect to ω_0. In this approximation, the effect of the relativity of simultaneity can be ignored and the transformations (8.37) and (8.38) can also be used for the total quantities. Above, we used this approximation to find the equation for the change of angular momentum of a rotating gyroscope with respect to the laboratory frame.

11 Quaternions

In this chapter, we will be concerned with another approach to the theory of relativity. Using complex 2×2 matrices, we will define quaternions, which are completely equivalent to four-vectors. Quaternions are a convenient tool for describing rotations and compositions of Lorentz transformations. In the next volume quaternions will be used to write Maxwell's equations in a very elegant form. The development of quaternion mathematics will ultimately lead us to spinors, which encapsulate mathematical ideas more general than "ordinary" four-vectors and tensors. They also underlie the description of fundamental matter fields.

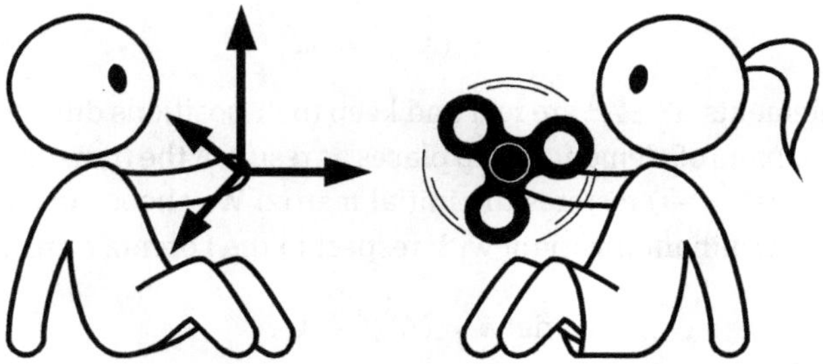

https://doi.org/10.1515/9783110515886-011

11.1 Quaternions

Four-vectors

$$A^\nu = \{A^0, A^1, A^2, A^3\} \equiv \{A^0, \mathbf{A}\}$$

constitute the main mathematical idea of the theory of relativity. So far such quadruples of numbers have been presented as columns (or lines). Four numbers can also be arranged to form a 2×2 matrix. Matrix elements can be either real or complex values. Consider the following Hermitian matrix:

$$\mathbb{A} = \begin{pmatrix} A^0 + A^3 & A^1 - \imath A^2 \\ A^1 + \imath A^2 & A^0 - A^3 \end{pmatrix}, \tag{11.1}$$

where $\imath$ is the imaginary unit ($\imath^2 = -1$), and the A^ν are four real numbers. That the matrix is *Hermitian* means that it is equal to the transpose of its complex conjugate (p. 337):

$$\mathbb{A}^+ = (\mathbb{A}^*)^{\mathrm{T}} = \mathbb{A}.$$

The diagonal elements $A^0 \pm A^3$ are real and keep their positions during transposition. Off the diagonal, pairs of elements swap places as result of the transposition, but complex conjugation ($\imath^* = -\imath$) restores the initial matrix. We choose matrix components that make the determinant invariant with respect to the Lorentz transformations:

$$\det \mathbb{A} = (A^0)^2 - \mathbf{A}^2. \tag{11.2}$$

The four-vector components change together with the reference frame. The same is true for the elements of the matrix $\mathbb{A}$. We can write this transformation in matrix form using a 2×2 matrix $\mathbb{S}$. The new matrix $\mathbb{A}'$ should remain Hermitian with respect to this transform. *Therefore* ($\lessdot$ H$_{43}$), we choose the Lorentz transformation for which:

$$\mathbb{A}' = \mathbb{S}\,\mathbb{A}\,\mathbb{S}^+, \qquad \det \mathbb{S} = 1. \tag{11.3}$$

The determinant of $\mathbb{S}$ being unity ensures the invariance of the squared four-vector:

$$\det \mathbb{A}' = \det \mathbb{S}\, \det \mathbb{A}\, \det \mathbb{S}^* = |\det \mathbb{S}|^2 \det \mathbb{A},$$

or

$$\det \mathbb{A}' = \det \mathbb{A}.$$

The matrix $\mathbb{S}$ has four elements, and each of them can be expressed in terms of the other three using the condition $\det \mathbb{S} = 1$. These three elements are, generally speaking, complex numbers, so there are $6 = 3 \cdot 2$ independent real parameters needed to define the transformation matrix $\mathbb{S}$. As we will see below, these parameters are related to the velocity components of the reference frame and to the three angles defining its orientation.

▷ A composition of two transforms (performed one after another)

$$\mathbb{A}' = \mathbb{S}_1\, \mathbb{A}\, \mathbb{S}_1^+, \qquad \mathbb{A}'' = \mathbb{S}_2\, \mathbb{A}'\, \mathbb{S}_2^+,$$

is equivalent to a single transformation:

$$\mathbb{A}'' = \mathbb{S}\, \mathbb{A}\, \mathbb{S}^+,$$

whose matrix is the product of the matrices of the two transformations:

$$\mathbb{S} = \mathbb{S}_2\, \mathbb{S}_1. \tag{11.4}$$

The order of the matrices is inverse to the order of the performed transformations.

▷ The Hermitian matrix (11.1) can be decomposed into a sum of four Hermitian 2×2 matrices:

$$\mathbb{A} = A^\mu \sigma_\mu = A^0 \sigma_0 + A^1 \sigma_1 + A^2 \sigma_2 + A^3 \sigma_3 = A^0 + \mathbf{A}\boldsymbol{\sigma}, \tag{11.5}$$

where $\sigma_0 = 1$ is the identity matrix, which will often be omitted, and

$$\sigma_1 = \begin{pmatrix} 0 & 1 \\ 1 & 0 \end{pmatrix}, \quad \sigma_2 = \begin{pmatrix} 0 & -\imath \\ \imath & 0 \end{pmatrix}, \quad \sigma_3 = \begin{pmatrix} 1 & 0 \\ 0 & -1 \end{pmatrix} \tag{11.6}$$

are the so-called *Pauli matrices*. They are combined into a vector $\boldsymbol{\sigma}$ with components $\{\sigma_1, \sigma_2, \sigma_3\}$. By direct multiplication ($\lessdot$ H$_{44}$) we can easily verify that each of tje Pauli matrices squares to the identity matrix, and that their product is the imaginary unit. Besides this, they are anti-commutative:

$$\sigma_1^2 = \sigma_2^2 = \sigma_3^2 = 1, \qquad \sigma_1 \sigma_2 \sigma_3 = \imath, \qquad \sigma_i\, \sigma_j = -\sigma_j\, \sigma_i, \quad i \neq j.$$

They can also express their own products:

$$\sigma_1\, \sigma_2 = \imath\, \sigma_3, \qquad \sigma_3\, \sigma_1 = \imath\, \sigma_2, \qquad \sigma_2\, \sigma_3 = \imath\, \sigma_1.$$

The order of the indices in the latter two equations is obtained from $\sigma_1\, \sigma_2 = \imath\, \sigma_3$ by cyclic permutation. All of these relations together are equivalent to the following single relation defining the ***algebra*** of Pauli matrices (summing over k from 1 to 3):

$$\sigma_i\, \sigma_j = \delta_{ij} + \imath\, \varepsilon_{ijk}\, \sigma_k, \tag{11.7}$$

where ε_{ijk} is the antisymmetric Levi-Civita three-tensor ($\varepsilon_{123} = 1$), and the identity matrix is omitted in the first term at the Kronecker symbol δ_{ij}. Contracting this relationship with the components of the two three-vectors **a** and **b**, we obtain ($\lessdot$ H$_{45}$):

$$(\mathbf{a}\boldsymbol{\sigma})(\mathbf{b}\boldsymbol{\sigma}) = \mathbf{ab} + \imath\, [\mathbf{a} \times \mathbf{b}]\, \boldsymbol{\sigma}. \tag{11.8}$$

We will frequently use this identity from now on.

▷ The term *quaternion* will henceforth refer to an arbitrary 2 × 2 matrix decomposed over the Pauli matrices:

$$\mathbb{Q} = Q^{\mu}\,\sigma_{\mu} = Q^0 + \mathbf{Q}\,\boldsymbol{\sigma}, \tag{11.9}$$

where Q^0 is the *scalar* part of the quaternion, and $\mathbf{Q}$ is its *vector* part. Thus, a quaternion is defined by four numbers:

$$\mathbb{Q} = \{Q^0, \mathbf{Q}\} = \{Q^0, Q^1, Q^2, Q^3\}.$$

In general, these are complex numbers (although they are real for the four-vector (11.1)). Sometimes, quaternions with complex coefficients are called biquaternions. We will not use this terminological distinction.

The product of two quaternions (matrices) is also a quaternion (matrix). Let the quaternion $\mathbb{A}$ be specified by the coefficients $\{a_0, \mathbf{a}\}$, and the quaternion $\mathbb{B}$ by the coefficients $\{b_0, \mathbf{b}\}$. Then,

$$\mathbb{A}\mathbb{B} = (a_0 + \mathbf{a}\boldsymbol{\sigma})(b_0 + \mathbf{b}\boldsymbol{\sigma}) = a_0 b_0 + a_0\,\mathbf{b}\boldsymbol{\sigma} + b_0\,\mathbf{a}\boldsymbol{\sigma} + (\mathbf{a}\boldsymbol{\sigma})(\mathbf{b}\boldsymbol{\sigma}).$$

Using (11.8), we obtain the multiplication rule for quaternions:

$$\mathbb{A}\mathbb{B} = a_0 b_0 + \mathbf{a}\mathbf{b} + (a_0\mathbf{b} + b_0\mathbf{a} + \imath\,\mathbf{a}\times\mathbf{b})\,\boldsymbol{\sigma}. \tag{11.10}$$

Multiplication of quaternions (like any matrices) is associative,

$$(\mathbb{A}\mathbb{B})\mathbb{C} = \mathbb{A}(\mathbb{B}\mathbb{C}),$$

and is, in general, non-commutative:

$$[\mathbb{A},\ \mathbb{B}] = \mathbb{A}\mathbb{B} - \mathbb{B}\mathbb{A} = 2\,\imath\,[\mathbf{a}\times\mathbf{b}]\,\boldsymbol{\sigma}. \tag{11.11}$$

The trace of any one of the Pauli matrices (the sum of its diagonal elements) is zero: $\mathrm{Tr}\,\sigma_k = 0$. The trace of the 2 × 2 identity matrix is $\mathrm{Tr}\,1 = 2$. Therefore, we can calculate the trace of a quaternion to find its scalar part:

$$Q_0 = \frac{1}{2}\,\mathrm{Tr}\,\mathbb{Q}, \qquad \mathrm{Tr}(\mathbb{A}\mathbb{B}) = \mathrm{Tr}(\mathbb{B}\mathbb{A}), \tag{11.12}$$

where the second relationship follows from (11.10). In general, it is possible to *cyclically permute* quaternions under the trace sign. For example,

$$\mathrm{Tr}(\mathbb{A}\mathbb{B}\mathbb{C}) = \mathrm{Tr}(\mathbb{C}\mathbb{B}\mathbb{A}) = \mathrm{Tr}(\mathbb{A}\mathbb{C}\mathbb{B}).$$

Indeed, since $\mathbb{A}\mathbb{B}$ is a quaternion, it follows from (11.12) that $\mathrm{Tr}((\mathbb{A}\mathbb{B})\mathbb{C}) = \mathrm{Tr}(\mathbb{C}(\mathbb{A}\mathbb{B}))$, and the parentheses can be omitted (due to the associativity of multiplication).

The *unit quaternion* is the 2 × 2 identity matrix, denoted by 1 or $\mathbb{I}$ (to emphasize that it is not a number).

▷ In addition to the matrices σ_μ, let us define four barred matrices $\bar{\sigma}_\mu$, which differ from the σ_μ in the sign of their "spatial" components:

$$\bar{\sigma}_\mu = \{1, -\boldsymbol{\sigma}\}, \qquad \sigma_\mu = \{1, \boldsymbol{\sigma}\}. \tag{11.13}$$

The *conjugate* quaternion $\bar{\mathbb{Q}}$ to $\mathbb{Q}$ will be also marked with a bar:

$$\bar{\mathbb{Q}} = Q^\mu \, \bar{\sigma}_\mu = Q^0 - \mathbf{Q}\,\boldsymbol{\sigma}. \tag{11.14}$$

Its vector part has the opposite sign to that of the initial quaternion. The product of a quaternion and its quaternionic conjugate is commutative, and is proportional to the identity matrix $\mathbb{I}$ (omitted in the equation below):

$$\bar{\mathbb{Q}}\,\mathbb{Q} = \mathbb{Q}\,\bar{\mathbb{Q}} = (Q^0)^2 - \mathbf{Q}^2 = \det \mathbb{Q} = |\mathbb{Q}|^2. \tag{11.15}$$

In particular, for the Lorentz transformations (11.3):

$$\bar{\mathbb{S}}\,\mathbb{S} = \mathbb{S}\,\bar{\mathbb{S}} = 1. \tag{11.16}$$

The determinant of the matrix $\mathbb{Q}$ is denoted by $|\mathbb{Q}|^2$ and is called the *norm of the quaternion*. The reciprocal matrix of $\mathbb{Q}$ is the matrix $\mathbb{Q}^{-1}$:

$$\mathbb{Q}^{-1} = \frac{\bar{\mathbb{Q}}}{|\mathbb{Q}|^2}, \qquad \mathbb{Q}\mathbb{Q}^{-1} = \mathbb{Q}^{-1}\mathbb{Q} = 1.$$

The norm of a product of quaternions is the product of their norms:

$$|\mathbb{A}\mathbb{B}|^2 = |\mathbb{A}|^2 \, |\mathbb{B}|^2 \tag{11.17}$$

(as follows from the corresponding property of the determinant of the matrix product). From (11.10), we have the following properties for the conjugates (< H_{46}):

$$\overline{\mathbb{A}\mathbb{B}} = \bar{\mathbb{B}}\,\bar{\mathbb{A}}, \qquad \bar{\bar{\mathbb{A}}} = \mathbb{A}. \tag{11.18}$$

Note also the relations:

$$\frac{\mathbb{A}\bar{\mathbb{B}} + \mathbb{B}\bar{\mathbb{A}}}{2} = a_0 b_0 - \mathbf{ab}, \tag{11.19}$$

$$\frac{\mathbb{A}\bar{\mathbb{B}} - \mathbb{B}\bar{\mathbb{A}}}{2} = (b_0\mathbf{a} - a_0\mathbf{b} - \imath\,\mathbf{a}\times\mathbf{b})\,\boldsymbol{\sigma}. \tag{11.20}$$

The first of them gives the covariant contraction of two four-vectors and has no vector part. The second gives a quaternion with no scalar part.

The *Hermitian conjugate* of a quaternion $\mathbb{Q} = \{Q^0, \mathbf{Q}\}$ is obtained by taking the complex conjugate of its components:

$$\mathbb{Q}^+ = \{Q_0^*, \mathbf{Q}^*\} = Q_0^* + \mathbf{Q}^*\boldsymbol{\sigma}$$

(The Pauli matrices are Hermitian: $\boldsymbol{\sigma}^+ = \boldsymbol{\sigma}$). As usual, $(\mathbb{A}\mathbb{B})^+ = \mathbb{B}^+\mathbb{A}^+$.

11.2 The algebraic approach *

Complex numbers provide a powerful tool and find use in various physical applications. The main idea in complex analysis is the imaginary unit $\imath^2 = -1$. Any complex number $z = x + \imath y$ is specified by a pair of real numbers, where x is called the real part, and y the imaginary part. Multiplication of complex numbers is associative $z_1(z_2 z_3) = (z_1 z_2) z_3$, commutative $z_1 z_2 = z_2 z_1$, and can be performed according to the usual algebraic rules, taking into account that $\imath^2 = -1$ (p. 353).

The conjugate of a complex number z is the complex number

$$\bar{z} \equiv z^* = x - \imath y,$$

with negative imaginary part. Accordingly, conjugation of the imaginary unit changes its sign: $\imath^* = -\imath$. The norm of a complex number is a non-negative real number:

$$|z|^2 = z\bar{z} = (x + iy)(x - iy) = x^2 + y^2.$$

Any complex number can be written in the trigonometric form:

$$z = |z|\,(\cos\phi + \imath \sin\phi) = |z|\, e^{\imath\phi}.$$

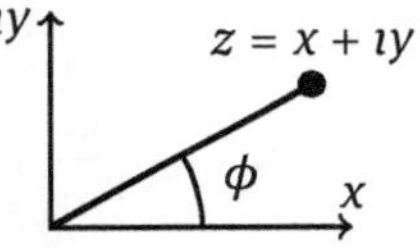

This trigonometric representation provides an elegant way of expressing the transformation of the components of a vector under a coordinate frame rotation in the (x, y)-plane. Let the components of a vector be specified by the real and imaginary parts of a complex number $z = x + \imath y$. Then we have:

$$z' = e^{-\imath\phi} z = (c_\phi - \imath s_\phi)(x + \imath y) = (x\, c_\phi + y\, s_\phi) + \imath(y\, c_\phi - x\, s_\phi).$$

Equating the real and imaginary parts of $z' = x' + \imath y'$ and $e^{-\imath\phi} z$, we can find the components of the vector $\{x', y'\}$ after the rotation (p. 168).

To obtain a tool for describing rotations in three-dimensional space, the concept of a complex number has to be generalized, by "adding dimensions" to it. In three-dimensional space, rotations can be performed about three different coordinate axes. Therefore, there are three "imaginary units", $\mathbb{I}_1$,$\mathbb{I}_2$,$\mathbb{I}_3$ which are called ***basis quaternions***. By analogy with the usual imaginary unit, it is required that

$$\mathbb{I}_1^2 = \mathbb{I}_2^2 = \mathbb{I}_3^2 = -1. \tag{11.21}$$

Let the multiplication of the eight objects

$$1,\ \mathbb{I}_1,\ \mathbb{I}_2,\ \mathbb{I}_3,\ -1,\ -\mathbb{I}_1,\ -\mathbb{I}_2,\ -\mathbb{I}_3$$

be associative, and their product be one of these objects (so that they form an associative closed algebra).

Let us find $\mathbb{I}_1\,\mathbb{I}_2$. The result cannot be equal to $\pm\mathbb{I}_1$, $\pm\mathbb{I}_2$ or ± 1. Indeed, suppose as an example that

$$\mathbb{I}_1\,\mathbb{I}_2 = \mathbb{I}_1.$$

Then, multiplying both sides by $\mathbb{I}_1$ and taking into account (11.21), we obtain $\mathbb{I}_2 = 1$. However, $\mathbb{I}_2$ and 1 are different objects. Therefore, $\mathbb{I}_1\,\mathbb{I}_2$ can only be equal to $\mathbb{I}_3$ or $-\mathbb{I}_3$. Consider the first variant:

$$\mathbb{I}_1\,\mathbb{I}_2 = \mathbb{I}_3.$$

Squaring this equation

$$\mathbb{I}_1\,\mathbb{I}_2\,\mathbb{I}_1\,\mathbb{I}_2 = \mathbb{I}_3^2 = -1,$$

multiplying it by $\mathbb{I}_2\,\mathbb{I}_1$, and using associativity and the properties (11.21), we obtain

$$\mathbb{I}_1\,\mathbb{I}_2 = -\mathbb{I}_2\,\mathbb{I}_1.$$

Thus, the multiplication of quaternions is *antisymmetric*. Using this property, we can find the products of the remaining pairs of basis quaternions:

$$\mathbb{I}_3\,\mathbb{I}_1 = (\mathbb{I}_1\,\mathbb{I}_2)\,\mathbb{I}_1 = -\mathbb{I}_2\,\mathbb{I}_1\,\mathbb{I}_1 = \mathbb{I}_2.$$

Squaring $\mathbb{I}_2$, we conclude that $\mathbb{I}_3$ and $\mathbb{I}_1$ are also anti-commutative. We can then find $\mathbb{I}_2\,\mathbb{I}_3$, etc.

As a result, the following products are obtained:

$$\mathbb{I}_1\,\mathbb{I}_2 = \mathbb{I}_3, \qquad \mathbb{I}_3\,\mathbb{I}_1 = \mathbb{I}_2, \qquad \mathbb{I}_2\,\mathbb{I}_3 = \mathbb{I}_1, \tag{11.22}$$

and $\mathbb{I}_1\mathbb{I}_2\mathbb{I}_3 = -1$. Also,

$$\mathbb{I}_i\,\mathbb{I}_j = -\mathbb{I}_j\,\mathbb{I}_i, \qquad i \neq j. \tag{11.23}$$

These relationships can be summarised using the Levi-Civita and Kronecker symbols:

$$\mathbb{I}_i\,\mathbb{I}_j = -\delta_{ij} + \varepsilon_{ijk}\,\mathbb{I}_k, \tag{11.24}$$

where we sum over k from 1 to 3. The Levi-Civita symbol is zero and the Kronecker symbol unity for equal indices i and j. Therefore, we obtain the relationships (11.21) for the squares. The Kronecker symbol is zero for unequal indices, which gives (11.22).

So far we have considered the basis quaternions as abstract objects constituting a closed algebra. As is easily seen, they can be represented using the Pauli matrices:

$$\mathbb{I}_k = -\imath\,\sigma_k.$$

The Pauli matrices are widely used in physics (especially in quantum theory). Therefore, we will henceforth work with quaternions in their matrix representation, rather than thinking of them as the abstract algebraic objects $\mathbb{I}_k$.

11.3 Rotations in three-space

Let us describe a rotation through an angle ϕ about the unit vector $\mathbf{n}$. Assume that some vector $\mathbf{r}$ is rigidly bound to a body rotating about the $\mathbf{n}$ axis (the first picture below). After the rotation, it transforms into the vector $\mathbf{r}'$ which moves around the inverted cone (the second figure):

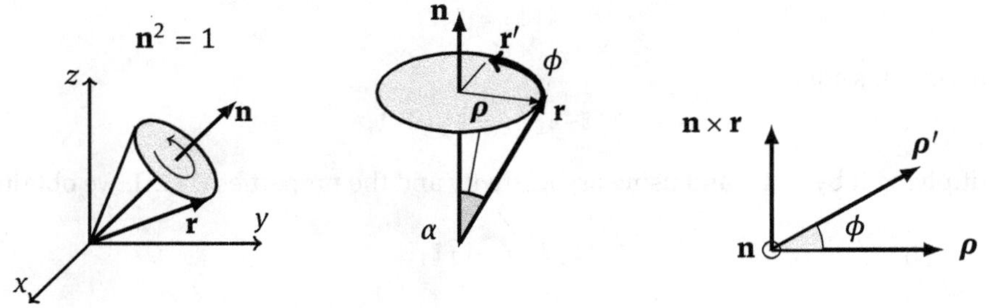

Denote the projections of $\mathbf{r}$ and $\mathbf{r}'$ onto the base of the cone by $\boldsymbol{\rho}$ and $\boldsymbol{\rho}'$ respectively. They have equal lengths, and the vector $\boldsymbol{\rho}$ transforms into $\boldsymbol{\rho}'$ when rotated through the angle ϕ. Let the vector $\mathbf{n} \times \mathbf{r}$ also lie in the base of the cone which is orthogonal to $\boldsymbol{\rho}$ (see the "top view" in the third figure). Its length is $|\mathbf{n} \times \mathbf{r}| = r \sin \alpha = |\boldsymbol{\rho}|$ (see the second figure), therefore, $\boldsymbol{\rho}'$ can be decomposed over two orthogonal vectors of equal length:

$$\boldsymbol{\rho}' = \boldsymbol{\rho} \cos \phi + [\mathbf{n} \times \mathbf{r}] \sin \phi .$$

On the other hand, the vectors $\mathbf{r}$ and $\mathbf{r}'$ can be decomposed as:

$$\mathbf{r} = \boldsymbol{\rho} + \mathbf{n}\,(\mathbf{n}\mathbf{r}), \qquad \mathbf{r}' = \boldsymbol{\rho}' + \mathbf{n}\,(\mathbf{n}\mathbf{r}'), \tag{11.25}$$

where $\mathbf{n}(\mathbf{n}\mathbf{r})$ is directed along $\mathbf{n}$ and has the same height as the cone. Given that $\mathbf{r}'\mathbf{n} = \mathbf{r}\mathbf{n}$, we finally obtain:

$$\mathbf{r}' = \boldsymbol{\rho} \cos \phi + [\mathbf{n} \times \mathbf{r}] \sin \phi + \mathbf{n}\,(\mathbf{n}\,\mathbf{r}). \tag{11.26}$$

▷ Rotations are commonly divided into ***passive*** and ***active*** rotations. In the first case, the coordinates of the same ***fixed point*** in space are compared in two coordinate systems (x, y, z) and (x', y', z') which are rotations of each other through an angle ϕ. Active rotations consider the coordinates of some vector after ***it*** has been rotated with respect to a fixed coordinate system:

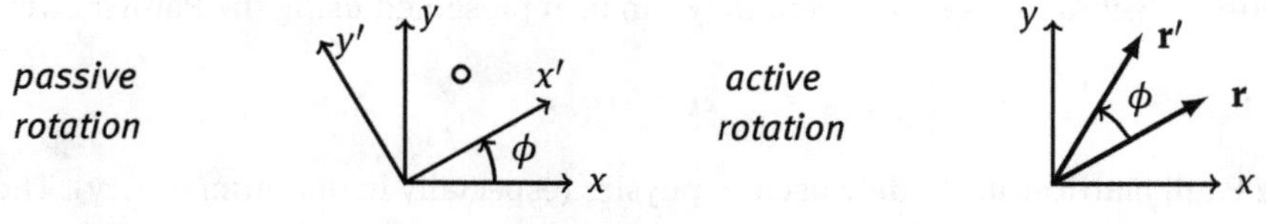

We will use the passive interpretation for spatial rotations and Lorentz transforms.

▷ The rotation of a body through an angle ϕ relative to a fixed coordinate system is equivalent to the rotation of the coordinate system through the angle "$-\phi$", the body being fixed. Therefore, substituting $\phi \mapsto -\phi$ into (11.26), changing the order of the vectors in the vector product, and then expressing $\boldsymbol{\rho}$ in terms of $\mathbf{r}$ with the help of the first relation in (11.25), we obtain

$$\mathbf{r}' = \mathbf{r}\cos\phi + \mathbf{n}(\mathbf{n}\mathbf{r})(1-\cos\phi) - [\mathbf{n}\times\mathbf{r}]\sin\phi. \tag{11.27}$$

Let the vector $\mathbf{n} = \{0, 0, 1\}$ be directed along the z-axis. Writing $\mathbf{r} = \{x, y, z\}$, and likewise for the primed quantities, we obtain from (11.27) the transformation law (p. 168) for a rotation of the coordinate system in the (x, y)-plane:

$$\begin{cases} x' = x\cos\phi + y\sin\phi, \\ y' = y\cos\phi - x\sin\phi. \end{cases}$$

▷ When describing rotations, it is convenient to use the *right-hand rule* (or the screw rule). If a screw pointing in the direction of the axis $\mathbf{n}$ is twisted through an angle ϕ, its rotation shows the direction in which the coordinate system rotates. There can be right and left coordinate systems:

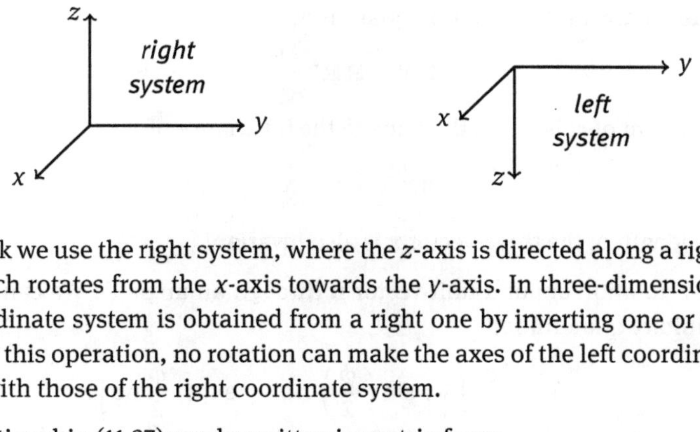

In this book we use the right system, where the z-axis is directed along a right-handed screw which rotates from the x-axis towards the y-axis. In three-dimensional space, a left coordinate system is obtained from a right one by inverting one or three of its axes. After this operation, no rotation can make the axes of the left coordinate system coincide with those of the right coordinate system.

▷ The relationship (11.27) can be written in matrix form:

$$x_i' = R_{ij}\,x_j, \qquad \mathbf{r}' = \mathbf{R}\,\mathbf{r}$$

(summing over j), or by using the Kronecker symbol δ_{ij} and the Levi-Civita tensor:

$$R_{ij} = \delta_{ij}\cos\phi + n_i n_j(1-\cos\phi) + \varepsilon_{ijk}\,n_k\sin\phi \tag{11.28}$$

(where we sum over k, and $\mathbf{r} = \{x_1, x_2, x_3\}$, $\mathbf{n} = \{n_1, n_2, n_3\}$). In general, this matrix has no symmetry, but it is *orthogonal*:

$$\mathbf{R}\mathbf{R}^T = 1,$$

as follows from the invariance of the length of the radius vector $\mathbf{r}'^2 = \mathbf{r}^2$.

11.4 Rotations and quaternions

Let us use quaternions to describe rotations of the Cartesian coordinate system. By means of the components of the radius vector $\mathbf{r} = \{x, y, z\}$ and the time t, we can define the quaternion of four-coordinates:

$$\mathbb{X} = t + \mathbf{r}\boldsymbol{\sigma}. \tag{11.29}$$

The result of its transformation is

$$\mathbb{X}' = \mathbb{R}\,\mathbb{X}\,\mathbb{R}^+, \qquad \det \mathbb{R} = 1, \tag{11.30}$$

where $\mathbb{R}$ is a rotation quaternion (a special case of the matrix $\mathbb{S}$ from the first section). The magnitude of the vector

$$\mathbf{x}^2 = x^2 + y^2 + z^2$$

does not change under a three-rotation. Nor does the time (the scalar part of the quaternion $\mathbb{X}$) change, $t' = t$. Given the properties of the trace (p. 278), we have

$$t' = \frac{1}{2}\,\mathrm{Tr}(\mathbb{R}\,\mathbb{X}\,\mathbb{R}^+) = \frac{1}{2}\,\mathrm{Tr}(\mathbb{X}\,\mathbb{R}^+\,\mathbb{R}) = \frac{1}{2}\,\mathrm{Tr}(\mathbb{X}) = t,$$

which is true for any *unitary* rotation quaternion:

$$\mathbb{R}^+\mathbb{R} = \mathbb{R}\,\mathbb{R}^+ = 1. \tag{11.31}$$

If the determinant of $\mathbb{R}$ is unity, then due to the invariance of

$$\det \mathbb{X} = t^2 - \mathbf{x}^2$$

(p. 276), the length of the three-vector $\mathbf{x}$ is also invariant.

▷ Consider a rotation about a unit vector $\mathbf{n}$ through an angle ϕ. We can define the following unitary quaternion:

$$\mathbb{R} = \cos\left(\frac{\phi}{2}\right) + \imath \sin\left(\frac{\phi}{2}\right)\mathbf{n}\boldsymbol{\sigma} = \exp\left\{\imath\,\frac{\phi}{2}\,\mathbf{n}\boldsymbol{\sigma}\right\}. \tag{11.32}$$

Its unitarity is verified by direct calculation. The second equality is obtained by expanding the exponential in a Taylor series. Thus, from (11.8), it follows that $(\mathbf{n}\boldsymbol{\sigma})^2 = 1$ for the unit vector. Therefore, $(\mathbf{n}\boldsymbol{\sigma})^3 = \mathbf{n}\boldsymbol{\sigma}$, etc. As a result:

$$\exp\left\{\imath\,\frac{\phi}{2}\,\mathbf{n}\boldsymbol{\sigma}\right\} = 1 + \imath\,\frac{\phi}{2}\,\mathbf{n}\boldsymbol{\sigma} - \frac{1}{2!}\left(\frac{\phi}{2}\right)^2 - \frac{\imath}{3!}\left(\frac{\phi}{2}\right)^3\mathbf{n}\boldsymbol{\sigma} + \frac{1}{4!}\left(\frac{\phi}{2}\right)^4 + \dots$$

Given the expansions for sin and cos, we obtain (11.32).

The quaternion $\mathbb{R}$ has real scalar part and pure imaginary vector part:

$$\bar{\mathbb{R}} = \mathbb{R}^+.$$

Therefore, due to (11.31), the norm (determinant) of this quaternion is unity (11.16).

▷ Direct calculations verify that (11.30), together with (11.32), gives equation (11.27). Suppose $c = \cos(\phi/2)$, $s = \sin(\phi/2)$. In the product $\mathbb{R}\,\mathbb{X}\,\mathbb{R}^+$ we exchange the positions of $\mathbb{R}$ and $\mathbb{X}$ using (11.11):

$$\mathbb{X}' = (\mathbb{X}\mathbb{R} + 2\,\imath^2 s\,[\mathbf{n}\times\mathbf{r}]\boldsymbol{\sigma})\,\mathbb{R}^+.$$

Removing now the parentheses and using the unitarity property (11.31),

$$\mathbb{X}' = \mathbb{X} - 2\,s\,[\mathbf{n}\times\mathbf{r}]\,\boldsymbol{\sigma}\,(c - \imath\,s\,\mathbf{n}\boldsymbol{\sigma}).$$

Multiplying the contractions of the Pauli matrices by means of the identity (11.8), we have:

$$\mathbb{X}' = \mathbb{X} - 2\,sc\,[\mathbf{n}\times\mathbf{r}]\boldsymbol{\sigma} - 2\,s^2\,[[\mathbf{n}\times\mathbf{r}]\times\mathbf{n}]\,\boldsymbol{\sigma}.$$

Expanding the vector triple product as $[\mathbf{n}\times\mathbf{r}]\times\mathbf{n} = \mathbf{r} - \mathbf{n}(\mathbf{n}\mathbf{r})$, we finally obtain:

$$t' + \mathbf{r}'\boldsymbol{\sigma} = t + ((1-2s^2)\,\mathbf{r} + 2\,s^2\,\mathbf{n}(\mathbf{n}\mathbf{r}) - 2\,sc\,[\mathbf{n}\times\mathbf{r}])\,\boldsymbol{\sigma}.$$

The change in the vector part corresponds (after using standard trigonometric identities) to the rotation of the coordinate system (11.27).

▷ Let us write the explicit form of the rotation quaternion $\mathbb{R}$:

$$\mathbb{R} = \begin{pmatrix} c_{\phi/2} + \imath n_z\,s_{\phi/2} & \imath(n_x - \imath n_y)\,s_{\phi/2} \\ \imath(n_x + \imath n_y)\,s_{\phi/2} & c_{\phi/2} - \imath n_z\,s_{\phi/2} \end{pmatrix},$$

where $s_{\phi/2} = \sin(\phi/2)$ and $c_{\phi/2} = \cos(\phi/2)$. Some direct calculations easily show that the determinant of this matrix is unity (the matrices $\mathbb{R}$ belong to the group **SU**(2) of 2×2 special unitary matrices).

Compare the transformation

$$\mathbb{X}' = \mathbb{R}\,\mathbb{X}\,\mathbb{R}^+$$

with the transformation (11.28) written in explicit matrix form:

$$\mathbf{R} = \begin{pmatrix} c_\phi + n_x^2(1-c_\phi) & n_x n_y(1-c_\phi) + n_z s_\phi & n_x n_z(1-c_\phi) - n_y s_\phi \\ n_y n_x(1-c_\phi) - n_z s_\phi & c_\phi + n_y^2(1-c_\phi) & n_y n_z(1-c_\phi) + n_x s_\phi \\ n_z n_x(1-c_\phi) + n_y s_\phi & n_z n_y(1-c_\phi) - n_x s_\phi & c_\phi + n_z^2(1-c_\phi) \end{pmatrix},$$

where $s_\phi = \sin\phi$ and $c_\phi = \cos\phi$. The 3×3 rotation matrix is more cumbersome; however, we need only multiply the matrix by a column to get the transformed vector. The 2×2 quaternionic matrices are simpler, but two matrix multiplications are required to obtain the transformed quaternion. Note also the difference in the arguments of the trigonometric functions of these matrices. They depend on the rotation angle ϕ in $\mathbf{R}$ and on $\phi/2$ in the quaternion $\mathbb{R}$.

▷ The advantages of the quaternion approach appear when composite transforms are considered (11.4). Consider the quaternions

$$\mathbb{R}_2 = c_2 + \imath s_2\, \mathbf{n}_2\boldsymbol{\sigma}, \qquad \mathbb{R}_1 = c_1 + \imath s_1\, \mathbf{n}_1\boldsymbol{\sigma},$$

where $c_i = \cos(\phi_i/2)$, etc. Their product (a composition of rotations) has the form:

$$\mathbb{R} = \mathbb{R}_2\mathbb{R}_1 = c_1c_2 - s_1s_2\,(\mathbf{n}_1\mathbf{n}_2) + \imath\,(s_1c_2\,\mathbf{n}_1 + c_1s_2\,\mathbf{n}_2 + s_1s_2\,[\mathbf{n}_1 \times \mathbf{n}_2])\,\boldsymbol{\sigma}.$$

The scalar part of the quaternion $\mathbb{R}$ is real, while its vector part is purely imaginary. Besides this, the quaternion is unitary:

$$\mathbb{R}\mathbb{R}^+ = (\mathbb{R}_2\mathbb{R}_1)(\mathbb{R}_2\mathbb{R}_1)^+ = \mathbb{R}_2\mathbb{R}_1\,\mathbb{R}_1^+\mathbb{R}_2^+ = \mathbb{R}_2\mathbb{R}_2^+ = 1.$$

Therefore, the quaternion again describes some *rotation*.

Writing the result as $\mathbb{R} = \cos(\phi/2) + \imath\sin(\phi/2)\,\mathbf{n}\boldsymbol{\sigma}$, the final angle of rotation ϕ and the axis $\mathbf{n}$ are easily expressed in terms of the angles and the axes of the initial rotations. In particular:

$$\cos\frac{\phi}{2} = \cos\frac{\phi_1+\phi_2}{2} + (1 - \mathbf{n}_1\mathbf{n}_2)\,\sin\frac{\phi_1}{2}\,\sin\frac{\phi_2}{2}. \tag{11.33}$$

It is useful to obtain this relation directly from (11.27) to fully realize the advantage quaternions have over 3 × 3 matrices when performing composite transforms.

From (11.33) it follows that the angles of sequential rotations are summed only if the axes of these rotations are parallel: $\mathbf{n}_1\mathbf{n}_2 = 1$. In the general case, the order of the rotations is important, and

$$\mathbb{R}_1\,\mathbb{R}_2 \neq \mathbb{R}_2\,\mathbb{R}_1.$$

The coefficients of the vector part of the quaternion give the final axis of rotation $\mathbf{n}$:

$$\mathbf{n}\,\sin\frac{\phi}{2} = \mathbf{n}_1\,\sin\frac{\phi_1}{2}\,\cos\frac{\phi_2}{2} + \mathbf{n}_2\,\cos\frac{\phi_1}{2}\,\sin\frac{\phi_2}{2} + [\mathbf{n}_1 \times \mathbf{n}_2]\,\sin\frac{\phi_1}{2}\,\sin\frac{\phi_2}{2},$$

where the angle of rotation ϕ is found from (11.33).

Quaternions are convenient for calculating the composition of a large number of sequential rotations:

$$\mathbb{R} = \mathbb{R}_n\,\mathbb{R}_{n-1}\ldots\mathbb{R}_1, \qquad \mathbb{X}' = \mathbb{R}\mathbb{X}\mathbb{R}^+.$$

To do this, we multiply 2 × 2 matrices rather than 3 × 3 matrices as is done in the "traditional" approach. However, we have to pay for this by making the calculation of the final transformation more complicated (three matrices must be multiplied).

▷* Rotation quaternions can be multiplied by numbers and added component-wise. The resulting quaternion will not be a rotation quaternion in general. Let us find out when this happens. Consider

$$\mathbb{R} = \alpha\,\mathbb{R}_1 + \beta\,\mathbb{R}_2, \tag{11.34}$$

where α and β are real numbers. If $\mathbb{R}_1$ and $\mathbb{R}_2$ have real scalar parts and purely imaginary vector parts, the same will be true for $\mathbb{R}$. However, this is not enough, since $\mathbb{R}$ should also have unit norm:

$$\mathbb{R}\bar{\mathbb{R}} = (\alpha\,\mathbb{R}_1 + \beta\,\mathbb{R}_2)(\alpha\,\bar{\mathbb{R}}_1 + \beta\,\bar{\mathbb{R}}_2) = (\alpha^2 + \beta^2) + \alpha\beta\,(\mathbb{R}_1\bar{\mathbb{R}}_2 + \mathbb{R}_2\bar{\mathbb{R}}_1) = 1.$$

According to (11.19), the expression $\mathbb{R}_1\bar{\mathbb{R}}_2 + \mathbb{R}_2\bar{\mathbb{R}}_1$ is proportional to the identity matrix. The coefficient of this matrix is twice the scalar part of the quaternion product $\mathbb{R}_1\bar{\mathbb{R}}_2$:

$$\frac{1}{2}\,\mathrm{Tr}(\mathbb{R}_1\bar{\mathbb{R}}_2) = \frac{1}{2}\,\mathrm{Tr}(\mathbb{R}_2\bar{\mathbb{R}}_1) = c_1c_2 + s_1s_2\,(\mathbf{n}_1\mathbf{n}_2).$$

Therefore, $\mathbb{R}$ is a rotation quaternion if:

$$\alpha^2 + \beta^2 + \mathrm{Tr}(\mathbb{R}_1\bar{\mathbb{R}}_2)\,\alpha\beta = 1. \tag{11.35}$$

⊳* Suppose that orientations $\mathbb{X}_0 = \mathbb{R}_0\mathbb{X}\,\mathbb{R}_0^+$ and $\mathbb{X}_1 = \mathbb{R}_1\mathbb{X}\,\mathbb{R}_1^+\,\mathbb{X}$ are obtained from $\mathbb{X}$ using the quaternions $\mathbb{R}_0$ and $\mathbb{R}_1$ (the above are different rotations of $\mathbb{X}$, rather than sequential rotations). The inverse of the first transform is $\mathbb{R}_0^+\,\mathbb{X}_0\mathbb{R}_0 = \mathbb{X}$. Therefore, $\mathbb{X}_1$ and $\mathbb{X}_0$ are related as follows:

$$\mathbb{X}_1 = \mathbb{Q}_1\mathbb{X}_0\,\mathbb{Q}_1^+, \qquad \mathbb{Q}_1 = \mathbb{R}_1\mathbb{R}_0^+.$$

Let us find the quaternion $\mathbb{Q}_t$ which depends on the parameter $t = [0\ldots1]$ and defines orientation

$$\mathbb{X}_t = \mathbb{Q}_t\mathbb{X}_0\mathbb{Q}_t^+$$

equal to $\mathbb{X}_0$ at $t = 0$ and to $\mathbb{X}_1$ at $t = 1$. Suppose that the rotation angle changes *uniformly*:

$$\mathbb{Q}_t = \cos(\omega t) + \imath\sin(\omega t)\,\mathbf{n}\boldsymbol{\sigma} = \cos(\omega t) + \frac{\sin(\omega t)}{\sin\omega}\,(\mathbb{R}_1\mathbb{R}_0^+ - \cos\omega),$$

where the axis of rotation $\mathbf{n}$ is expressed in terms of $\mathbb{R}_1\mathbb{R}_0^+$, so that $\mathbb{Q}_0 = 1$ and $\mathbb{Q}_1 = \mathbb{R}_1\mathbb{R}_0^+$. From the first equation it follows that

$$2\cos\omega = \mathrm{Tr}(\mathbb{Q}_1) = \mathrm{Tr}(\mathbb{R}_1\mathbb{R}_0^+).$$

From the quaternion $\mathbb{Q}_t$ we can move to the rotation $\mathbb{R}_t$ starting from the orientation $\mathbb{X}$: $\mathbb{X}_t = \mathbb{R}_t\mathbb{X}\,\mathbb{R}_t^+$. Since $\mathbb{R}_t = \mathbb{Q}_t\,\mathbb{R}_0$, we have:

$$\mathbb{R}_t = \frac{\sin(\omega(1-t))}{\sin\omega}\,\mathbb{R}_0 + \frac{\sin(\omega t)}{\sin\omega}\,\mathbb{R}_1. \tag{11.36}$$

This relationship is called the ***spherical linear interpolation*** between the orientations $\mathbb{X}_0$ and $\mathbb{X}_1$. Its coefficients satisfy (11.35).

11.5 Lorentz transformations and quaternions

We now proceed to the Lorentz transformations. Using the hyperbolic cosine and sine of the scalar parameter α and the unit vector $\mathbf{m}^2 = 1$, we can define the quaternion corresponding to a *Lorentz boost* (a Lorentz transformation without rotation):

$$\mathbb{L} = \operatorname{ch}\left(\frac{\alpha}{2}\right) - \operatorname{sh}\left(\frac{\alpha}{2}\right)\mathbf{m}\boldsymbol{\sigma} = \exp\left\{-\frac{\alpha}{2}\mathbf{m}\boldsymbol{\sigma}\right\}. \tag{11.37}$$

This representation of the exponent in terms of hyperbolic functions is verified in the same way as in the case of three-rotations (11.32).

Similarly to the rotation quaternion $\mathbb{R}$, the quaternion of the boost $\mathbb{L}$ has *unit norm*:

$$\det\mathbb{L} = \mathbb{L}\bar{\mathbb{L}} = (c - s\,\mathbf{m}\boldsymbol{\sigma})(c + s\,\mathbf{m}\boldsymbol{\sigma}) = (c^2 - s^2) = 1$$

(here $c = \operatorname{ch}(\alpha/2)$, $s = \operatorname{sh}(\alpha/2)$). Since the vector part of $\mathbb{L}$ is real, the quaternion of the boost is not unitary:

$$\mathbb{L}\,\mathbb{L}^+ = \mathbb{L}^2 = \exp(-\alpha\mathbf{m}\boldsymbol{\sigma}) = \operatorname{ch}(\alpha) - \operatorname{sh}(\alpha)\,\mathbf{m}\boldsymbol{\sigma} \neq 1.$$

As a result, transforming a quaternion of four-coordinates $\mathbb{X}$,

$$\mathbb{X}' = \mathbb{L}\mathbb{X}\mathbb{L}^+,$$

changes both its scalar (the time) and the vector (the coordinates of the event) parts. Nevertheless, the quantity

$$\det\mathbb{X} = t^2 - \mathbf{x}^2$$

is invariant. This is precisely what is required of the Lorentz transformations.

As a matrix, the quaternion of a Lorentz boost has the form:

$$\mathbb{L} = \begin{pmatrix} \operatorname{ch}(\alpha/2) - m_3\operatorname{sh}(\alpha/2) & -(m_1 - \imath m_2)\operatorname{sh}(\alpha/2) \\ -(m_1 + \imath m_2)\operatorname{sh}(\alpha/2) & \operatorname{ch}(\alpha/2) + m_3\operatorname{sh}(\alpha/2) \end{pmatrix}.$$

As is easily verified by direct calculations, $\det\mathbb{L} = 1$.

In general, for a transformation

$$\mathbb{X}' = \mathbb{S}\mathbb{X}\mathbb{S}^+,$$

the matrix $\mathbb{S}$ is only required to have unit determinant $\det\mathbb{S} = 1$. Such matrices form the special (**S**), linear (**L**) group $\mathbf{SL}(2, C)$, where the two refers to the dimension of the matrices and the C indicates that the matrix coefficients are complex numbers. The rotation quaternions $\mathbb{R}$ constitute a *subgroup* $\mathbf{SU}(2)$ of the group $\mathbf{SL}(2, C)$ (see the second volume). But, as we will see below, the matrices $\mathbb{L}$ do not constitute a group, as a product of two such matrices is not necessarily the quaternion of a boost (unless one-dimensional motion is considered).

▷ Let us write the Lorentz transformations in a more familiar form. To do this, we substitute the quaternion (11.37) into $\mathbb{X}' = \mathbb{L}\,\mathbb{X}\,\mathbb{L}^+$ with $\mathbb{X} = t + \mathbf{r}\boldsymbol{\sigma}$:

$$t' + \mathbf{r}'\boldsymbol{\sigma} = (c - s\,\mathbf{m}\boldsymbol{\sigma})\,(t + \mathbf{r}\boldsymbol{\sigma})\,(c - s\,\mathbf{m}\boldsymbol{\sigma}).$$

Multiplying out the parentheses and keeping all of factors in order (since they are matrices!), we obtain from (11.8):

$$\mathbb{X}' = (c^2 + s^2)\,t - 2\,sc\,(\mathbf{rm}) + c^2\,\mathbf{r}\boldsymbol{\sigma} - 2\,sc\,t\,\mathbf{m}\boldsymbol{\sigma} + s^2(\mathbf{m}\boldsymbol{\sigma})(\mathbf{r}\boldsymbol{\sigma})(\mathbf{m}\boldsymbol{\sigma}).$$

Using the identity

$$(\mathbf{a}\boldsymbol{\sigma})(\mathbf{b}\boldsymbol{\sigma})(\mathbf{a}\boldsymbol{\sigma}) = (2\,(\mathbf{ab})\,\mathbf{a} - \mathbf{b})\,\boldsymbol{\sigma}, \tag{11.38}$$

which follows from (11.8) after the vector triple product is expanded, we have:

$$t' + \mathbf{r}'\boldsymbol{\sigma} = (c^2 + s^2)\,t - 2\,sc\,(\mathbf{mr}) + (c^2 - s^2)\,\mathbf{r}\boldsymbol{\sigma} - 2cst\,\mathbf{m}\boldsymbol{\sigma} + 2\,s^2\,(\mathbf{mr})(\mathbf{m}\boldsymbol{\sigma}).$$

Equating the scalar and the vector parts and using some identities for hyperbolic functions (p. 355), we obtain the Lorentz transformations:

$$\begin{cases} t' &= t\,\mathrm{ch}\,\alpha - (\mathbf{mr})\,\mathrm{sh}\,\alpha, \\ \mathbf{r}' &= \mathbf{r} - t\,\mathbf{m}\;\mathrm{sh}\,\alpha + (\mathrm{ch}\,\alpha - 1)\,(\mathbf{mr})\,\mathbf{m}. \end{cases} \tag{11.39}$$

The elations (1.18), p. 17, are obtained after a change of notation:

$$\mathbf{m} = \frac{\mathbf{v}}{v}, \qquad \mathrm{ch}\,\alpha = \gamma, \qquad \mathbf{m}\,\mathrm{sh}\,\alpha = \mathbf{v}\,\gamma, \qquad v = \mathrm{th}\,\alpha, \tag{11.40}$$

where $\gamma = 1/\sqrt{1 - v^2}$, and α is the *rapidity*:

$$\alpha = \mathrm{ath}\,v = \frac{1}{2}\,\ln\frac{1+v}{1-v}. \tag{11.41}$$

Zero speed $v = 0$ corresponds to zero rapidity, and unit speed (the speed of light) to infinite rapidity.

▷ Perform two sequential Lorentz boosts $\mathbb{L}_2\,\mathbb{L}_1$ (first $\mathbb{L}_1$ and then $\mathbb{L}_2$), where

$$\mathbb{L}_2 = c_2 - s_2\,\mathbf{m}_2\boldsymbol{\sigma}, \qquad \mathbb{L}_1 = c_1 - s_1\,\mathbf{m}_1\boldsymbol{\sigma}$$

(as usual, $c_1 = \mathrm{ch}(\alpha_1/2)$, etc.). Multiplying these quaternions, we obtain the quaternion $\mathbb{S} = \mathbb{L}_2\,\mathbb{L}_1$:

$$\mathbb{S} = c_1c_2 + (\mathbf{m}_1\mathbf{m}_2)\,s_1s_2 - (\mathbf{m}_1\,s_1c_2 + \mathbf{m}_2\,c_1s_2 + \imath\,[\mathbf{m}_1 \times \mathbf{m}_2]\,s_1s_2)\,\boldsymbol{\sigma}.$$

Its vector part is complex, if $\mathbf{m}_1 \times \mathbf{m}_2 \neq 0$. Therefore, it is not a Lorentz boost (11.37), i.e., the matrices $\mathbb{L}$ do not constitute a group.

▷ Although the composition of two boosts $\mathbb{L}_2\mathbb{L}_1$ is not a boost, it can be decomposed into a composition of a boost (11.37) and a spatial rotation (11.32), e.g. as follows (first the boost, then the rotation):

$$\mathbb{L}_2\,\mathbb{L}_1 = \mathbb{R}\,\mathbb{L}. \tag{11.42}$$

Writing the quaternion of rotation $\mathbb{R}$ and the boost $\mathbb{L}$,

$$\mathbb{R} = c_\phi + \imath\, s_\phi\, \mathbf{n}\boldsymbol{\sigma}, \qquad \mathbb{L} = c_\alpha - s_\alpha\, \mathbf{m}\boldsymbol{\sigma},$$

where $c_\alpha = \mathrm{ch}(\alpha/2)$, $c_\phi = \cos(\phi/2)$ and multiplying them, we have:

$$\mathbb{S} = \mathbb{R}\,\mathbb{L} = c_\alpha c_\phi - \imath\,\mathbf{nm}\, s_\alpha s_\phi + (\imath\,\mathbf{n}\, c_\alpha s_\phi - \mathbf{m} s_\alpha c_\phi - [\mathbf{m}\times\mathbf{n}]\, s_\alpha s_\phi)\,\boldsymbol{\sigma}. \tag{11.43}$$

This quaternion coincides with $\mathbb{L}_2\,\mathbb{L}_1$ (which has real scalar part), only if the final velocity and the axis of rotation are *orthogonal* ($\mathbf{nm} = 0$). In addition, the following relations should be satisfied:

$$\begin{gathered} c_\alpha c_\phi = c_1 c_2 + (\mathbf{m}_1\mathbf{m}_2)\, s_1 s_2, \qquad \mathbf{n}\, c_\alpha s_\phi = -[\mathbf{m}_1\times\mathbf{m}_2]\, s_1 s_2, \\ \mathbf{m}\, s_\alpha c_\phi + [\mathbf{m}\times\mathbf{n}]\, s_\alpha s_\phi = \mathbf{m}_1\, s_1 c_2 + \mathbf{m}_2\, c_1 s_2. \end{gathered} \tag{11.44}$$

Now we can easily obtain the relationship between the parameters of the initial boosts and the equivalent sequence of the boost and the rotation. Let us define

$$\mathbf{m}\,\mathrm{sh}\,\alpha = \mathbf{v}\gamma,$$

and similarly for the indexed values $\mathbf{v}_1$ and $\mathbf{v}_2$. Since $\mathbf{n}$ and $\mathbf{m}$ are unit orthogonal vectors, $\mathbf{m}\times\mathbf{n}$ is also a unit vector. Squaring the third relation in (11.44), we obtain the expression for γ. Taking the vector product of this third relationship with $\mathbf{n}$ and using the result to get rid of the term $\mathbf{m}\times\mathbf{n}$, we obtain the expression for vector $\mathbf{m}$. As a result, the parameters of the total boost $\mathbb{L}$ are ($\lessdot$ H_{47}):

$$\gamma = \gamma_1\gamma_2\,(1+\mathbf{v}_1\mathbf{v}_2), \qquad \mathbf{v}\frac{\gamma}{\gamma_2} = \mathbf{v}_2 + \mathbf{v}_1\gamma_1 + \mathbf{v}_1(\mathbf{v}_1\mathbf{v}_2)\frac{\gamma_1 - 1}{v_1^2}. \tag{11.45}$$

As follows from the second relationship of (11.44), the *unit* vector along the axis of rotation is

$$\mathbf{n} = -[\mathbf{v}_1\times\mathbf{v}_2]/|\mathbf{v}_1\times\mathbf{v}_2|$$

for $s_\phi > 0$. Multiplying the first and the second relationships (11.44), we obtain the expression for the angle ($\lessdot$ H_{48}):

$$\mathbf{n}\,\sin\phi = -[\mathbf{v}_1\times\mathbf{v}_2]\,\frac{\gamma_1\gamma_2\,(1+\gamma+\gamma_1+\gamma_2)}{(1+\gamma)(1+\gamma_1)(1+\gamma_2)}. \tag{11.46}$$

The angle of rotation ϕ is called the *Wigner angle*, and the above formula for it was derived by Stapp [23] in 1956. Note that we always have $\phi < \pi/2$.

▷ The product of two boosts can also be decomposed into a sequence where, first, the coordinate system is rotated, and then the reference frame is changed as the result of a boost:

$$\mathbb{L}_2\,\mathbb{L}_1 = \mathbb{L}\,\mathbb{R}. \tag{11.47}$$

The expressions for the angle of rotation ϕ and the axis of rotation $\mathbf{n}$ (11.46) will not change in this case. Nor does the absolute value of the velocity of the final boost $\mathbb{L}$ (or the Lorentz factor γ). However, the indices 1 and 2 will swap places ($\lessdot$ H_{49}):

$$\mathbf{v}\,\frac{\gamma}{\gamma_1} = \mathbf{v}_1 + \mathbf{v}_2\,\gamma_2 + \mathbf{v}_2\,(\mathbf{v}_2\mathbf{v}_1)\,\frac{\gamma_2 - 1}{v_2^2}. \tag{11.48}$$

Therefore, the order of the boost and the rotation is important, and in general:

$$\mathbb{L}\,\mathbb{R} \neq \mathbb{R}\,\mathbb{L}, \tag{11.49}$$

since the parameters of $\mathbb{L}$ are different on the left and on the right. This property is a reflection of the non-Abelian nature of the Lorentz group (see the second volume).

Note that when multiplying quaternions of rotation

$$\mathbb{R} = \exp(\imath\,\phi\,\mathbf{n}\boldsymbol{\sigma}/2)$$

with quaternions of boost

$$\mathbb{L} = \exp(-\alpha\,\mathbf{m}\boldsymbol{\sigma}/2),$$

the arguments in the exponentials cannot be added (the matrices $\mathbf{n}\boldsymbol{\sigma}$ and $\mathbf{m}\boldsymbol{\sigma}$ do not commute in general).

▷* A quaternion $\mathbb{S} = S_0 + \mathbf{S}\,\boldsymbol{\sigma}$ with unit norm ($\mathbb{S}\bar{\mathbb{S}} = \bar{\mathbb{S}}\mathbb{S} = 1$) and complex coefficients can always be decomposed into a boost and a rotation (generally speaking, with *non-orthogonal* axes $\mathbf{nm} \neq 0$). To do this, we use (11.43):

$$S_0 = c_\alpha c_\phi - \imath\,\mathbf{nm}\,s_\alpha s_\phi, \qquad \mathbf{S} = \imath\,\mathbf{n}\,c_\alpha s_\phi - \mathbf{m}\,s_\alpha c_\phi - [\mathbf{m}\times\mathbf{n}]\,s_\alpha s_\phi,$$

and express the unit vectors $\mathbf{n}$, $\mathbf{m}$ and parameters ϕ, α in terms of the real and the imaginary parts of the quaternion $\mathbb{S}$:

$$\mathbf{n}\,\frac{s_\phi}{c_\phi} = \frac{\mathbf{S}_I}{S_{0R}}, \qquad c_\alpha = \frac{S_{0R}}{c_\phi}, \qquad \mathbf{m}\,s_\alpha = [\mathbf{S}_R\times\mathbf{n}]\,s_\phi - \mathbf{S}_R\,c_\phi - \mathbf{n}\,S_{0I}\,s_\phi,$$

where $\mathbf{S}_R$ and $\mathbf{S}_I$ are the real and imaginary parts of the vector $\mathbf{S}$, etc.

Taking an arbitrary quaternion $\mathbb{K}$ with unit norm, a Lorentzian boost quaternion can be formed as follows:

$$\mathbb{L} = \frac{\mathbb{K} + \mathbb{K}^+}{|1 + \mathbb{K}\bar{\mathbb{K}}^+|}, \qquad \mathbb{K}\bar{\mathbb{K}} = 1.$$

With this definition $\mathbb{L}$ is obviously Hermitian, and the fact that it has unit norm $\mathbb{L}\,\bar{\mathbb{L}} = 1$ can be verified by direct multiplication ($\lessdot$ H_{50}). By writing $\mathbb{K} = \mathbb{R}\,\mathbb{L}$, the quaternion $\mathbb{R} = \mathbb{K}\,\bar{\mathbb{L}}$ ($\lessdot$ H_{51}) is easily found. However, it is not a rotation quaternion, since its vector part is, in general, not purely imaginary.

11.6 Quaternions and four-vectors

By definition, any four-vector $A^\mu = \{A_0, \mathbf{A}\}$ transforms under the Lorentz transformations in the same way as a four-vector in space-time $x^\mu = \{t, \mathbf{r}\}$. Therefore, the components of a four-vector can always be associated with the quaternion

$$\mathbb{A} = A_0 + \mathbf{A}\boldsymbol{\sigma},$$

which changes under general Lorentz transformations (including boosts and rotations) as:

$$\mathbb{A}' = \mathbb{S}\,\mathbb{A}\,\mathbb{S}^+, \qquad \mathbb{S}\bar{\mathbb{S}} = 1. \tag{11.50}$$

For example, the *velocity quaternion* is

$$\mathbb{U} = \frac{d\mathbb{X}}{ds} = \gamma + \gamma\,\mathbf{u}\boldsymbol{\sigma},$$

where

$$ds = dt\sqrt{1 - \mathbf{u}^2} = dt/\gamma$$

is the proper time of the particle. The norm of the velocity quaternion (the square of the four-vector) is unity:

$$\mathbb{U}\bar{\mathbb{U}} = 1. \tag{11.51}$$

Differentiating this relationship with respect to ds and introducing the *acceleration quaternion* $\mathbb{A} = d\mathbb{U}/ds$, we obtain the quaternionic representation of the orthogonality of the four-velocity and four-acceleration:

$$\mathbb{U}\bar{\mathbb{A}} + \mathbb{A}\bar{\mathbb{U}} = 0.$$

This is a scalar relationship. In general, the invariant contraction of two quaternions $\mathbb{A}$ and $\mathbb{B}$ built from the components of the four-vectors $\mathrm{A} = \{A^0, \mathbf{A}\}$ and $\mathrm{B} = \{B^0, \mathbf{B}\}$ is the following combination (see the identity (11.19), p. 279):

$$\mathbb{A}\bar{\mathbb{B}} + \mathbb{B}\bar{\mathbb{A}} = 2\,\mathrm{A}\cdot\mathrm{B}.$$

The *momentum quaternion* (or the energy-momentum quartenion) is proportional to the velocity quaternion, with coefficient equal to the mass of the particle:

$$\mathbb{P} = m\mathbb{U} = E + \mathbf{p}\boldsymbol{\sigma}.$$

According to (11.51), the product of the momentum quaternion with its conjugate is proportional to the squared mass:

$$\mathbb{P}\bar{\mathbb{P}} = m^2.$$

Other quaternionic analogues of four-vectors are easily defined similarly to the position $\mathbb{X}$, the velocity $\mathbb{U}$, the acceleration $\mathbb{A}$, and the momentum $\mathbb{P}$.

▷ Now we can proceed to antisymmetric four-tensors. Define the *angular momentum quaternion*:

$$\mathbb{J} = \frac{1}{2}\,(\mathbb{X}\bar{\mathbb{P}} - \mathbb{P}\bar{\mathbb{X}}) = (\mathbf{G} - \imath\mathbf{L})\,\boldsymbol{\sigma}, \tag{11.52}$$

where we have introduced the two three-vectors

$$\mathbf{G} = E\mathbf{r} - t\mathbf{p} \quad \text{and} \quad \mathbf{L} = \mathbf{r}\times\mathbf{p}.$$

The six independent components of the antisymmetric tensor $J^{\mu\nu} = (\mathbf{G},\ \mathbf{L})$, where $\mathbf{G} = \{J^{10},\ J^{20},\ J^{30}\}$ and $\mathbf{L} = \{J^{23},\ J^{31},\ J^{12}\}$, are expressed in terms of these vectors (p. 205). The quaternion $\mathbb{J}$ has complex vector part and no scalar part. This is a common property of quaternions which are analogues of *antisymmetric four-tensors*:

$$\mathbb{J} = \mathbf{J}\boldsymbol{\sigma}, \qquad \mathbf{J}^* \neq \mathbf{J}.$$

After lowering the indices of the tensor $J^{\mu\nu}$, we again obtain an antisymmetric tensor $J_{\mu\nu} = (-\mathbf{G},\ \mathbf{L})$. Also, we can use the tensor $J_{\mu\nu}$ to write its *dual* antisymmetric tensor (p. 204):

$$^*J^{\mu\nu} = \frac{1}{2}\,\varepsilon^{\mu\nu\alpha\beta}J_{\alpha\beta}.$$

Its components are obtained by swapping the two three-vectors of the tensor $J_{\mu\nu}$:

$$^*J^{\mu\nu} = (\mathbf{L},\ -\mathbf{G}) \quad \text{and} \quad ^*J_{\mu\nu} = (-\mathbf{L},\ -\mathbf{G}).$$

As is easily seen, we have the following correspondence between antisymmetric tensors and quaternions:

$$J^{\mu\nu} \leftrightarrow \mathbb{J}, \qquad J_{\mu\nu} \leftrightarrow \bar{\mathbb{J}}^+, \qquad ^*J^{\mu\nu} \leftrightarrow \imath\mathbb{J}, \qquad ^*J_{\mu\nu} \leftrightarrow \imath\mathbb{J}^+. \tag{11.53}$$

There are similar rules for the relationships between the contravariant and covariant components of four-vectors and quaternions:

$$A^{\mu} \leftrightarrow \mathbb{A}, \qquad A_{\mu} \leftrightarrow \bar{\mathbb{A}}. \tag{11.54}$$

Contracting the antisymmetric tensor $J^{\mu\nu} = (\mathbf{G},\ \mathbf{L})$ against an arbitrary four-vector $A^{\mu} = \{A^0,\ \mathbf{A}\}$ gives a four-vector with components

$$J^{\mu\nu}A_{\nu} = \{\,\mathbf{GA},\ A^0\mathbf{G} + \mathbf{L}\times\mathbf{A}\,\}.$$

By direct multiplication of the quaternions we can verify that

$$\frac{1}{2}\,(\mathbb{J}\mathbb{A} + \mathbb{A}\mathbb{J}^+) = J^{\mu\nu}A_{\nu}\,\sigma_{\mu}. \tag{11.55}$$

In quaternion notation the four-vector of spin (p. 210) has the form:

$$S^{\mu} = \frac{1}{2}\,\varepsilon^{\mu\nu\alpha\beta}J_{\alpha\beta}\,U_{\nu} = {}^*J^{\mu\nu}U_{\nu} \quad \leftrightarrow \quad \mathbb{S} = \frac{\imath}{2}(\mathbb{J}\mathbb{U} - \mathbb{U}\mathbb{J}^+).$$

Thus, any relations involving four-vectors and antisymmetric four-tensors can be expressed in the language of quaternions.

▷ Let us find out how the angular momentum quaternion (and, consequently, any antisymmetric four-tensor) changes under the Lorentz transformations. Since $\mathbb{X}$ and $\mathbb{P}$ are quaternions associated with with four-vectors, we have

$$\mathbb{X}'\bar{\mathbb{P}}' = \mathbb{S}\mathbb{X}\mathbb{S}^+ \,\overline{\mathbb{S}\mathbb{P}\mathbb{S}^+} = \mathbb{S}\mathbb{X}\mathbb{S}^+ \,\bar{\mathbb{S}}^+\bar{\mathbb{P}}\bar{\mathbb{S}} = \mathbb{S}(\mathbb{X}\bar{\mathbb{P}})\bar{\mathbb{S}},$$

where we used the fact that $\mathbb{S}\bar{\mathbb{S}} = 1$ or $\mathbb{S}^+\,\bar{\mathbb{S}}^+ = 1$. A similar relation holds for $\mathbb{P}\bar{\mathbb{X}}$. Therefore,

$$\mathbb{J}' = \mathbb{S}\,\mathbb{J}\,\bar{\mathbb{S}}, \qquad \mathbb{S}\bar{\mathbb{S}} = 1. \tag{11.56}$$

This law differs from the transformation law (11.50): it has a quaternionic conjugation (with a bar) rather than a Hermitian conjugation on the right-hand side. It is not surprising that the transformation $\mathbb{J}$ differs from the four-vector transformation (11.50), as the quaternion $\mathbb{J}$ corresponds to an antisymmetric four-tensor rather than a four-vector. For three-dimensional rotations $\mathbb{S} = \mathbb{R}$ the quaternion $\mathbb{R}$ has purely imaginary vector part, and $\mathbb{R}^+ = \bar{\mathbb{R}}$. Therefore, $\mathbb{J}$, which depends on the complex vector $\mathbf{G} - \imath\mathbf{L}$, is in this case transformed like any three-vector. If

$$\mathbb{S}^+ \neq \bar{\mathbb{S}}$$

(for example, for a Lorentz boost), the relationships (11.50) and (11.56) will give different results.

▷ It is easily verified that the transformations (11.50) and (11.56) preserve the norm of the quaternion invariant. Set $\mathbf{J} = \mathbf{G} - \imath\mathbf{L}$. Then

$$\mathbb{J}\bar{\mathbb{J}} = -(\mathbf{J}\boldsymbol{\sigma})(\mathbf{J}\boldsymbol{\sigma}) = -\mathbf{J}^2 = \mathbf{L}^2 - \mathbf{G}^2 + 2\imath\,\mathbf{L}\mathbf{G} = inv$$

(the minus appears due to the conjugation). Thus, the real and the imaginary parts are invariant independently of each other:

$$\mathbf{L}^2 - \mathbf{G}^2 = inv, \qquad \mathbf{L}\mathbf{G} = inv.$$

These invariants were obtained earlier (p. 203). Interestingly, they are equivalent to the single relationship

$$\mathbf{J}^2 = inv$$

for a complex vector.

▷ Let us verify that the relationship (11.55) for multiplying an antisymmetric tensor by a four-vector gives a quaternion that transforms like a four-vector. We substitute in the transformation laws of each of the quantities:

$$\mathbb{J}'\mathbb{A}' + \mathbb{A}'\mathbb{J}'^+ = (\mathbb{S}\mathbb{J}\bar{\mathbb{S}})\,(\mathbb{S}\mathbb{A}\mathbb{S}^+) + (\mathbb{S}\mathbb{A}\mathbb{S}^+)\,(\mathbb{S}\mathbb{J}\bar{\mathbb{S}})^+ = \mathbb{S}(\mathbb{J}\mathbb{A} + \mathbb{A}\mathbb{J}^+)\mathbb{S}^+,$$

where we used the fact that the norm of the quaternion of the transformation $\mathbb{S}$ is unity. Thus, we obtain the original combination of quaternions surrounded by the quaternions $\mathbb{S}\ldots\mathbb{S}^+$, which corresponds to the transformation of a four-vector. Note that $\mathbb{J}\mathbb{A}$ is also a four-vector, but is complex, whereas the combination above is real.

▷ The relationship (11.56) applied to $\mathbb{J} = \mathbf{J}\boldsymbol{\sigma}$, where

$$\mathbf{J} = \mathbf{G} - \imath\mathbf{L},$$

gives the transformations of the vectors **G** and **L** when the inertial frame of reference is changed. We can write it explicitly:

$$\mathbf{J}'\boldsymbol{\sigma} = (c - s\mathbf{m}\boldsymbol{\sigma})\,\mathbf{J}\boldsymbol{\sigma}\,(c + s\mathbf{m}\boldsymbol{\sigma}),$$

where $c = \mathrm{ch}(\alpha/2)$, $s = \mathrm{sh}(\alpha/2)$, and we have substituted in the quaternion $\mathbb{S} = \mathbb{L}$ of the Lorentz boost (11.37). By multiplying the matrices, we obtain:

$$\mathbf{J}'\boldsymbol{\sigma} = c^2\,\mathbf{J}\boldsymbol{\sigma} - s^2(\mathbf{m}\boldsymbol{\sigma})(\mathbf{J}\boldsymbol{\sigma})(\mathbf{m}\boldsymbol{\sigma}) + sc\,[\mathbf{J}\boldsymbol{\sigma}, \mathbf{m}\boldsymbol{\sigma}].$$

Calculating the commutator $[\mathbf{J}\boldsymbol{\sigma}, \mathbf{m}\boldsymbol{\sigma}]$ with the help of (11.11), and the product of the three Pauli matrices by successive application of the identity (11.8), we have:

$$\mathbf{J}'\boldsymbol{\sigma} = (c^2 + s^2)\,\mathbf{J}\boldsymbol{\sigma} + 2\imath sc\,[\mathbf{J}\times\mathbf{m}]\,\boldsymbol{\sigma} - 2s^2(\mathbf{J}\mathbf{m})(\mathbf{m}\boldsymbol{\sigma}).$$

Passing from hyperbolic functions of the “angle” $\alpha/2$ to functions of α, we obtain the transformation for the complex vector **J**:

$$\mathbf{J}' = \mathbf{J}\,\mathrm{ch}\,\alpha - \imath\,\mathrm{sh}\,\alpha[\mathbf{m}\times\mathbf{J}] - (\mathrm{ch}\,\alpha - 1)(\mathbf{m}\mathbf{J})\mathbf{m}.$$

As in the case of the Lorentz transformations for a four-vector, a more familiar form of this transform is obtained after the rapidity α is written in terms of the velocity **v**, the factor γ, and the factor $\Gamma = (\gamma - 1)/v^2$, see p. 289:

$$\mathbf{J}' = \gamma\,(\mathbf{J} - \imath\,\mathbf{v}\times\mathbf{J}) - \Gamma\,\mathbf{v}\,(\mathbf{v}\mathbf{J}). \qquad (11.57)$$

The real part of this relation gives the transformation of the vector **G**, and the complex part gives the same for the vector **L** (see p. 207):

$$\begin{aligned}\mathbf{G}' &= \gamma\,(\mathbf{G} - \mathbf{v}\times\mathbf{L}) - \Gamma\,\mathbf{v}(\mathbf{v}\mathbf{G}),\\ \mathbf{L}' &= \gamma\,(\mathbf{L} + \mathbf{v}\times\mathbf{G}) - \Gamma\,\mathbf{v}(\mathbf{v}\mathbf{L}).\end{aligned}$$

As can be seen, the transformation (11.57) for a complex three-vector **J** looks more compact when obtained using the quaternion technique than by using the transforms for the two vector components of the antisymmetric tensor

$$J_{\alpha\beta} = x_\alpha p_\beta - x_\beta p_\alpha.$$

In general, the theory of relativity is at many points more concise when written in the language of quaternions than it is when written in conventional vector notation or using the covariant tensor apparatus. The price we pay is having to work with matrices, which ensure the non-commutative nature of the quaternions.

11.7 Equations of motion

Quaternions can be used to elegantly write and solve equations of motion. In Chapter 5 we considered a particle moving in a field of constant force and subject to a force proportional to the its velocity. Let such forces act simultaneously:

$$\frac{d\mathbf{p}}{dt} = \mathbf{F} = \mathbf{E} + \mathbf{u}\times\mathbf{B}, \qquad \frac{dE}{dt} = \mathbf{F}\mathbf{u} = \mathbf{E}\mathbf{u}. \tag{11.58}$$

In the next volume we will show that these formulas describe the Lorentz force acting on a particle of unit charge. Here **E** is an electric field and **B** is a magnetic field. Let us write equation (11.58) in quaternionic notation.

Define the four-coordinate quaternion of the particle, as well as its four-velocity quaternion:

$$\mathbb{X} = t + \mathbf{r}\boldsymbol{\sigma}, \qquad \mathbb{U} = \frac{d\mathbb{X}}{ds} = U_0 + \mathbf{U}\boldsymbol{\sigma}, \qquad \bar{\mathbb{U}}\,\mathbb{U} = \mathbb{U}\,\bar{\mathbb{U}} = 1.$$

In addition, define the quaternion of force (or the strength of the electromagnetic field), which has no scalar part and is written as

$$\mathbb{F} = (\mathbf{E} + \imath\mathbf{B})\,\boldsymbol{\sigma}. \tag{11.59}$$

It corresponds to some antisymmetric tensor. We can calculate the product

$$\mathbb{F}\,\mathbb{U} + \mathbb{U}\,\mathbb{F}^{+} \;=\; (\mathbf{E}\boldsymbol{\sigma} + \imath\mathbf{B}\boldsymbol{\sigma})\,(U_0 + \mathbf{U}\boldsymbol{\sigma}) + \;(U_0 + \mathbf{U}\boldsymbol{\sigma})(\mathbf{E}\boldsymbol{\sigma} - \imath\mathbf{B}\boldsymbol{\sigma}).$$

By multiplying out the parentheses, we obtain

$$\mathbb{F}\,\mathbb{U} + \mathbb{U}\,\mathbb{F}^{+} \;=\; 2\,\mathbf{E}\mathbf{U} + 2(U_0\mathbf{E} + \mathbf{U}\times\mathbf{B})\,\boldsymbol{\sigma}.$$

The interval (the proper time of the particle) is $ds = \sqrt{1-\mathbf{u}^2}\,dt$. Given that

$$U^{\nu} = \{1, \mathbf{u}\}/\sqrt{1-\mathbf{u}^2},$$

the quaternionic equation of motion can be written as:

$$m\,\frac{d\mathbb{U}}{ds} = \frac{1}{2}\,(\mathbb{F}\,\mathbb{U} + \mathbb{U}\,\mathbb{F}^{+}). \tag{11.60}$$

The scalar part (11.60) gives the equation for the change of energy, and the vector part does the same for the three-dimensional Lorentz force (11.58).

If there is only a magnetic field

$$\mathbb{F} = \imath\,\mathbf{B}\boldsymbol{\sigma},$$

the field strength quaternion is *anti-Hermitian*:

$$\mathbb{F}^{+} = -\mathbb{F}.$$

In this case, the commutator of the field strength and the speed $\mathbb{F}\,\mathbb{U} - \mathbb{U}\,\mathbb{F}$ appears on the right-hand side of (11.60). If its trace is zero, then the scalar part of the four-velocity is conserved: $\mathrm{Tr}(\mathbb{U}) = const.$

▷ Let us now find the general solution of equation (11.60). For this purpose, we represent the velocity quaternion in the following form:

$$\mathbb{U}(s) = e^{\mathbb{G}(s)}\,\mathbb{U}(0)\,e^{\mathbb{G}^+(s)}, \tag{11.61}$$

where $\mathbb{G}(s)$ is some quaternion equal to zero for $s = 0$ (the initial conditions are contained in $\mathbb{U}(0)$, and the Hermitian conjugation in the second exponential is assumed to preserve the hermiticity of the four-velocity). Substituting (11.61) into the equations of motion (11.60):

$$m\frac{d\mathbb{U}}{ds} = m\,\frac{d\mathbb{G}}{ds}\,\mathbb{U} + m\,\mathbb{U}\,\frac{d\mathbb{G}^+}{ds} = \frac{1}{2}\,(\mathbb{F}\,\mathbb{U} + \mathbb{U}\,\mathbb{F}^+).$$

This relation is true if the quaternion $\mathbb{G}$ satisfies the equation:

$$\frac{d\mathbb{G}}{ds} = \frac{1}{2m}\,\mathbb{F}. \tag{11.62}$$

Let the field intensities vary along the trajectory of the charge. This means that they depend on the proper time s of the charge. By integrating (11.62), we obtain the following form for the solution of the equations of motion (11.60):

$$\mathbb{U}(s) = \exp\left(\frac{1}{2m}\int_0^s \mathbb{F}(s)\,ds\right)\,\mathbb{U}(0)\,\exp\left(\frac{1}{2m}\int_0^s \mathbb{F}^+(s)\,ds\right). \tag{11.63}$$

If the external electric and magnetic fields are *constant*,

$$\mathbb{F} = (\mathbf{E} + \imath\,\mathbf{B})\,\boldsymbol{\sigma} = const,$$

the field strength quaternion $\mathbb{F}$ can be taken out of the integral sign, and the solution simplifies to:

$$\mathbb{U}(s) = \exp\left(\frac{1}{2m}\,\mathbb{F}s\right)\,\mathbb{U}(0)\,\exp\left(\frac{1}{2m}\,\mathbb{F}^+s\right). \tag{11.64}$$

The exponential factors are matrices. If there is only a constant electric field, they can be simplified, similarly to the rotation or Lorentz transformations, by expanding the exponent in a Taylor series and using the fact that $(\mathbf{e}\boldsymbol{\sigma})^2 = 1$, $(\mathbf{e}\boldsymbol{\sigma})^3 = \mathbf{e}\boldsymbol{\sigma}$, …, where $\mathbf{e} = \mathbf{E}/|\mathbf{E}|$ is a unit vector in the direction of the electric field. Defining $\omega = |\mathbf{E}|/m$, we finally obtain:

$$\exp\left(\frac{1}{2m}\,\mathbf{E}\boldsymbol{\sigma}s\right) = \mathrm{ch}(\omega s/2) + \mathbf{e}\boldsymbol{\sigma}\,\mathrm{sh}(\omega s/2). \tag{11.65}$$

Similarly, if there is only a magnetic field, we obtain:

$$\exp\left(\imath\,\frac{1}{2m}\,\mathbf{B}\boldsymbol{\sigma}s\right) = \cos(\omega s/2) + \imath\mathbf{b}\boldsymbol{\sigma}\sin(\omega s/2), \tag{11.66}$$

where $\omega = |\mathbf{B}|/m$ and $\mathbf{b} = \mathbf{B}/|\mathbf{B}|$. Multiplying these quaternions by the initial velocity quaternion $\mathbb{U}(0)$ and integrating once again, we easily obtain the expressions for the trajectory of the particle.

▷ Let a charge move in a constant electric field (the magnetic field is zero). In this case, the strength quaternion is Hermitian ($\mathbb{F}^+ = \mathbb{F}$). A simple solution is obtained if the charge is initially at rest and, consequently, the velocity quaternion is a unit quaternion $\mathbb{U}(0) = 1$ at $s = 0$. Then equation (11.64) simplifies to

$$\mathbb{U}(s) = \exp\Big(\frac{1}{m}\,\mathbb{F}s\Big) = \exp\Big(\frac{1}{m}\,\mathbf{E}\boldsymbol{\sigma}s\Big). \tag{11.67}$$

Expanding the exponent in a series, we obtain:

$$\mathbb{U}(s) = \mathrm{ch}(\omega s) + \mathbf{e}\boldsymbol{\sigma}\,\mathrm{sh}(\omega s), \tag{11.68}$$

where

$$\omega = \frac{|\mathbf{E}|}{m}, \quad \text{and } \mathbf{e} = \frac{\mathbf{E}}{|\mathbf{E}|}$$

is a unit vector in the direction of the electric field. Take the scalar and the vector parts of the quaternion in (11.68):

$$U_0(s) = \mathrm{ch}(\omega s), \qquad \mathbf{U}(s) = \mathbf{e}\ \mathrm{sh}(\omega s). \tag{11.69}$$

Obviously, this solution satisfies the general relationship for the four-speed components $U_0^2 - \mathbf{U}^2 = 1$.

Since

$$U^\mu = \{U_0, \mathbf{U}\} = \frac{dx^\mu}{ds},$$

integrating (11.69) over s gives the expressions for the laboratory time and the coordinates of the charge:

$$t = \frac{1}{\omega}\ \mathrm{sh}(\omega s), \qquad \mathbf{r}(s) = \mathbf{r}(0) + \frac{\mathbf{e}}{\omega}\,(\mathrm{ch}(\omega s) - 1). \tag{11.70}$$

Since $t = 0$ for $s = 0$, the integration constant is chosen to be zero in the first case. The constant in the expression for the radius vector is chosen to make the vector $\mathbf{r}(s)$ equal to the initial position of the charge $\mathbf{r}(0)$ for $s = 0$. Using the identity $\mathrm{ch}^2\,\alpha - \mathrm{sh}^2\,\alpha = 1$ for hyperbolic functions, we can express the hyperbolic cosine via the sine, and thus proceed from the parametric representation of the solution (11.70) to an explicit expression of the particle coordinates as functions of the laboratory time (see also p. 121):

$$\mathbf{r}(t) = \mathbf{r}(0) + \frac{\mathbf{e}}{\omega}\,\big(\sqrt{1 + (\omega t)^2} - 1\big).$$

If the initial velocity of the charge is not zero, $\mathbb{U}(0)$ is not a unit quaternion. To find the corresponding solution $\mathbb{U}(s)$, the following quaternions must be multiplied:

$$\Big\{\mathrm{ch}\Big(\frac{\omega s}{2}\Big) + \mathbf{e}\boldsymbol{\sigma}\,\mathrm{sh}\Big(\frac{\omega s}{2}\Big)\Big\}\,\{U_0(0) + \mathbf{U}(0)\boldsymbol{\sigma}\}\,\Big\{\mathrm{ch}\Big(\frac{\omega s}{2}\Big) + \mathbf{e}\boldsymbol{\sigma}\,\mathrm{sh}\Big(\frac{\omega s}{2}\Big)\Big\},$$

which, as usual, is done using identity (11.8), p. 277.

▷ We can similarly consider motion in a constant magnetic field, when the strength quaternion is anti-Hermitian ($\mathbb{F}^+ = -\mathbb{F}$). Using (11.66), we write the solution (11.64) as

$$\mathbb{U}(s) = \{\cos(\omega s/2) + \imath\mathbf{b}\boldsymbol{\sigma}\sin(\omega s/2)\}\,\mathbb{U}(0)\,\{\cos(\omega s/2) - \imath\mathbf{b}\boldsymbol{\sigma}\sin(\omega s/2)\},$$

where

$$\omega = \frac{|\mathbf{B}|}{m} \quad \text{and} \quad \mathbf{b} = \frac{\mathbf{B}}{|\mathbf{B}|}.$$

Substituting the initial velocity quaternion

$$\mathbb{U}(0) = U_0 + \mathbf{U}_0\boldsymbol{\sigma}$$

into the solution and multiplying the Pauli matrices with the help of identities (11.8), we obtain:

$$\mathbb{U}(s) = U_0 + (\mathbf{U}_0\mathbf{b})(\mathbf{b}\boldsymbol{\sigma}) + [\mathbf{b}\times[\mathbf{U}_0\times\mathbf{b}]]\,\boldsymbol{\sigma}\,\cos(\omega s) + [\mathbf{U}_0\times\mathbf{b}]\boldsymbol{\sigma}\,\sin(\omega s).$$

The only term U_0 not appearing with a Pauli matrix is constant. This means that the zero component of the four-velocity (the energy of motion of the charge) is independent of time. The vector part $\mathbb{U}(s)$ gives the solution (5.21), p. 125:

$$\mathbf{U}(s) = (\mathbf{U}_0\mathbf{b})\,\mathbf{b} + [\mathbf{b}\times[\mathbf{U}_0\times\mathbf{b}]]\,\cos(\omega s) + [\mathbf{U}_0\times\mathbf{b}]\,\sin(\omega s). \tag{11.71}$$

Note that the vector $\mathbf{u}$ on p. 125 is the "usual" velocity rather than the spatial part of the velocity four-vector considered in this section. However, since the particle energy (U_0) is constant in a magnetic field, these are the same up to a constant scaling factor. For the same reason, the interval of motion in a magnetic field is related to the three-velocity and the laboratory time by the simple relationship

$$s = \int_0^t \sqrt{1-\mathbf{u}^2(t)}\,dt = t\sqrt{1-\mathbf{u}_0^2} = t/U_0,$$

where the squared velocity of the charge remains constant, $\mathbf{u}(t) = \mathbf{u}_0$, during the motion, so that the root can be taken out of the integral sign.

Finally, integrating (11.71) over s, we obtain the coordinates of the charge as functions of the proper time $s = t/U_0$ (or the laboratory time t):

$$\mathbf{r}(s) = \mathbf{r}(0) + (\mathbf{U}_0\mathbf{b})\,\mathbf{b}\,s + \frac{1}{\omega}\,[\mathbf{b}\times[\mathbf{U}_0\times\mathbf{b}]]\,\sin(\omega s) - \frac{1}{\omega}\,[\mathbf{U}_0\times\mathbf{b}]\,(\cos(\omega s) - 1),$$

where the integration constants are chosen to make the vector $\mathbf{r}(0)$ correspond to the initial position of the charge. Geometrically, this gives a spiral wrapped around a cylinder of radius $|\mathbf{U}_0\times\mathbf{b}|/\omega$.

11.8 Covariant notation *

Let us find the relationship between the 2×2 matrix $\mathbb{S}$ of the transformation (11.3) and the 4×4 matrix of the "usual" Lorentz transformations $\Lambda^{\mu}{}_{\nu}$ for four-vectors:

$$A'^{\mu} = \Lambda^{\mu}{}_{\nu}\, A^{\nu}, \qquad A^{\beta} = A'^{\mu}\, \Lambda_{\mu}{}^{\beta}.$$

The contraction of two Lorentz matrices over their first or the second indices is a metric tensor (by the orthogonality property, p. 182):

$$g_{\mu\nu}\, \Lambda^{\mu}{}_{\alpha}\, \Lambda^{\nu}{}_{\beta} = g_{\alpha\beta}, \qquad g^{\alpha\beta}\, \Lambda^{\mu}{}_{\alpha}\, \Lambda^{\nu}{}_{\beta} = g^{\mu\nu}. \tag{11.72}$$

From the transformation law (11.3) for quaternions related to four-vector components (11.1), it follows that:

$$\mathbb{A}' = \mathbb{S}\, \mathbb{A}\, \mathbb{S}^{+} \quad \Rightarrow \quad A'^{\nu}\sigma_{\nu} = \Lambda^{\nu}{}_{\mu}\, A^{\mu}\, \sigma_{\nu} = \mathbb{S}\, A^{\mu}\sigma_{\mu}\, \mathbb{S}^{+}.$$

Since the components of the four-vector here are arbitrary, we obtain the relationship between the quaternionic matrix of the transformation $\mathbb{S}$ and the Lorentz matrix:

$$\mathbb{S}\, \sigma_{\mu}\, \mathbb{S}^{+} = \sigma_{\nu}\, \Lambda^{\nu}{}_{\mu}. \tag{11.73}$$

Contracting this expression with $\Lambda_{\alpha}{}^{\mu}$, left-multiplying it by $\bar{\mathbb{S}}$, and right-multiplying it by $\bar{\mathbb{S}}^{+}$, we have

$$\bar{\mathbb{S}}\, \mathbb{S}\, \Lambda_{\alpha}{}^{\mu} \sigma_{\mu}\, \mathbb{S}^{+} \bar{\mathbb{S}}^{+} = \bar{\mathbb{S}}\, \sigma_{\nu}\, \bar{\mathbb{S}}^{+}\, \Lambda^{\nu}{}_{\mu}\, \Lambda_{\alpha}{}^{\mu}.$$

Given that the determinant of the matrix $\mathbb{S}$ is unity, $\bar{\mathbb{S}}\mathbb{S} = 1$, see (11.15), the orthogonality (11.72) can be used to obtain another form of the relationship between these matrices:

$$\bar{\mathbb{S}}\, \sigma_{\alpha}\, \bar{\mathbb{S}}^{+} = \Lambda_{\alpha}{}^{\mu}\, \sigma_{\mu}. \tag{11.74}$$

As an exercise (< H_{52}), we suggest finding the dependence of the matrix $\Lambda^{\alpha}{}_{\beta}$ on the parameters s^{ν} of the quaternion of a transformation $\mathbb{S} = s^{\nu}\sigma_{\mu}$.

▷ To conclude, we consider some more definitions and useful identities. We will distinguish the heights of the indices of covariant Pauli matrices. They are raised by the tensor $g^{\mu\nu} = \mathrm{diag}(1,-1,-1,-1)$ by changing the sign of the vector components: $\sigma^{\mu} = g^{\mu\nu}\sigma_{\nu}$. For covariant Pauli matrices the following relation is true:

$$\sigma_{\mu}\, \bar{\sigma}_{\nu} + \sigma_{\nu}\, \bar{\sigma}_{\mu} = \bar{\sigma}_{\mu}\, \sigma_{\nu} + \bar{\sigma}_{\nu}\, \sigma_{\mu} = 2\, g_{\mu\nu}, \tag{11.75}$$

which is verified by direct computation. In addition, define 16 matrices:

$$\sigma_{\mu\nu} = \frac{1}{2}\,(\sigma_{\mu}\bar{\sigma}_{\nu} - \sigma_{\nu}\bar{\sigma}_{\mu}), \qquad \sigma_{\mu}\bar{\sigma}_{\nu} = g_{\mu\nu} + \sigma_{\mu\nu}, \tag{11.76}$$

where the $\mu\nu$ are the numbers of the matrices rather than their components.

▷ A product of quaternions $\mathbb{A}\bar{\mathbb{B}}$ can be written as follows:

$$\mathbb{A}\bar{\mathbb{B}} = a^\mu b^\nu \, \sigma_\mu \bar{\sigma}_\nu = \mathrm{a}\cdot\mathrm{b} + a^\mu b^\nu \, \sigma_{\mu\nu}, \tag{11.77}$$

where $\mathrm{a}\cdot\mathrm{b} = a^0 b^0 - \mathbf{ab}$.

▷ Multiplying the algebra (11.7) by σ_k from the right and again using (11.7), we have:

$$\sigma_i\, \sigma_j\, \sigma_k = \delta_{ij}\, \sigma_k - \delta_{ik}\, \sigma_j + \delta_{jk}\, \sigma_i + \imath\, \varepsilon_{ijk}. \tag{11.78}$$

The trace (the sum of the diagonal elements) of any Pauli matrix is zero, and the trace of the identity matrix is two. Therefore, from (11.7) and (11.78) we have:

$$\mathrm{Tr}\, \sigma_i = 0, \qquad \mathrm{Tr}(\sigma_i\, \sigma_j) = 2\, \delta_{ij}, \qquad \mathrm{Tr}(\sigma_i\, \sigma_j\, \sigma_k) = 2\, \imath\, \varepsilon_{ijk}. \tag{11.79}$$

Multiplying (11.78) by σ_l and taking the trace, we obtain:

$$\mathrm{Tr}(\sigma_i\, \sigma_j\, \sigma_k\, \sigma_l) = 2\, (\delta_{ij}\, \delta_{kl} - \delta_{ik}\, \delta_{jl} + \delta_{il}\, \delta_{jk}). \tag{11.80}$$

The traces of the covariant matrices can now easily be written:

$$\frac{1}{2}\, \mathrm{Tr}\, \sigma_\mu = \delta_{\mu 0}, \qquad \frac{1}{2}\, \mathrm{Tr}(\sigma_\mu \sigma_\nu) = \delta_{\mu\nu}, \qquad \frac{1}{2}\, \mathrm{Tr}(\sigma_\mu \bar{\sigma}_\nu) = g_{\mu\nu}. \tag{11.81}$$

▷ The coefficients in the expansion of the quaternion $\mathbb{Q} = Q^\mu\, \sigma_\mu$ over the Pauli matrices can be found using the trace:

$$Q^\mu = \frac{1}{2}\, \mathrm{Tr}(\mathbb{Q}\, \bar{\sigma}^\mu). \tag{11.82}$$

▷ We now substitute (11.82) into $\mathbb{Q} = Q^\mu \sigma_\mu$ and write the matrix indices explicitly:

$$Q_{ij} = \frac{1}{2}\, Q_{lk}\, (\bar{\sigma}^\mu)_{kl}\, (\sigma_\mu)_{ij}.$$

Take a matrix with all zero elements, except for one equal to unity. Let this element have indices $i = i_0$ and $j = j_0$, where i_0, j_0 are fixed numbers. Then $Q_{ij} = \delta_{ii_0}\delta_{jj_0}$ and, therefore,

$$\delta_{ii_0}\, \delta_{jj_0} = \frac{1}{2}\, (\sigma_\mu)_{ij}\, (\bar{\sigma}^\mu)_{j_0 i_0}. \tag{11.83}$$

This identity is called the ***completeness relation*** (remember the summation over μ!).

▷ By directly multiplying three arbitrary quaternions $\mathbb{A}\bar{\mathbb{B}}\mathbb{C}$ and omitting their coefficients, we ($\lessdot$ H$_{53}$) obtain:

$$\sigma_\alpha\, \bar{\sigma}_\beta\, \sigma_\gamma = g_{\alpha\beta}\, \sigma_\gamma - g_{\alpha\gamma}\, \sigma_\beta + g_{\beta\gamma}\, \sigma_\alpha + \imath\, \varepsilon_{\alpha\beta\gamma\mu}\, \sigma^\mu. \tag{11.84}$$

Multiplying (11.84) by $\bar{\sigma}_\nu$, taking the trace, and using that (11.81), we have:

$$\frac{1}{2}\, \mathrm{Tr}(\sigma_\alpha\, \bar{\sigma}_\beta\, \sigma_\mu\, \bar{\sigma}_\nu) = g_{\alpha\beta}\, g_{\mu\nu} - g_{\alpha\mu}\, g_{\beta\nu} + g_{\alpha\nu}\, g_{\beta\mu} + \imath\, \varepsilon_{\alpha\beta\mu\nu}. \tag{11.85}$$

We will return to considering quaternions in the next volume.

12 Velocity space

If considered as a three-vector, a velocity can be represented as a point in a three-dimensional "velocity space". From general considerations it follows that such a space should be homogeneous and isotropic. There are only three possible geometries in this case. One of them (Euclidean space) corresponds to the Galilean velocity-addition rule and to classical mechanics. Another possibility, hyperbolic geometry (Bolyai-Lobachevsky geometry), is realized by the velocity space in the theory of relativity.

We will first study in detail the geometry of the sphere, and then use it to obtain hyperbolic geometry via a simple change of the fundamental constant. We will then consider the pseudo-Euclidean space whose pseudo-sphere is a "plane" in Lobachevsky (hyperbolic) space, before proceeding to a consideration of the velocity space as such.

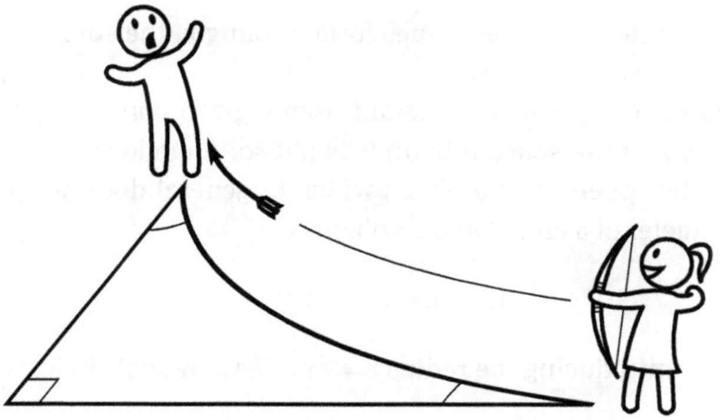

https://doi.org/10.1515/9783110515886-012

12.1 Flatland

Let us imagine that the surface of a sphere of radius λ is inhabited by two-dimensional creatures, flatlanders, unable to perceive the third dimension. We also live on Earth's surface, but we have access to three-dimensional space. The space of the flatlanders is "truly" two-dimensional, and all of its points can be numbered with only two coordinates. For the flatlanders a line segment is the shortest distance between two points on the surface of the sphere. For us it is a segment of a circle whose centre coincides with the centre of the sphere (first figure below):

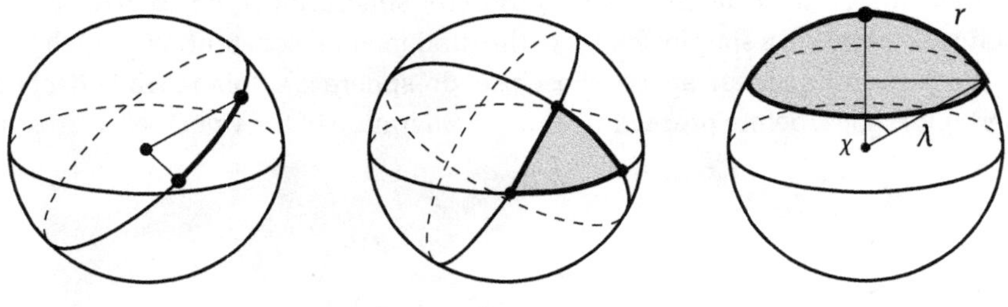

Three points connected by straight lines form a triangle, the sum of whose angles is greater than π (the second figure).

A sphere is a set of points equidistant from a given point. A sphere in Flatland is called a circle, and the space it bounds is a disc. A circle in Flatland is obtained by intersecting the sphere with a plane, which in general does not pass through its centre. The perimeter of a circle on the sphere is

$$L = 2\pi\lambda \sin\chi$$

(the third figure). Introducing the radius $r = \lambda\chi$ of flatlanders' circle, we have

$$L = 2\pi\lambda \sin\left(\frac{r}{\lambda}\right) < 2\pi r. \tag{12.1}$$

The area of the circle is the sum of the areas of the narrow rings $dS = L(r)\,dr$:

$$S = 2\pi\lambda^2 \left[1 - \cos\left(\frac{r}{\lambda}\right)\right] < \pi r^2. \tag{12.2}$$

The volume $4\pi\lambda^3/3$ of Flatland's space (the area of the sphere) is finite, so although this space is without boundary, it is finite. The Euclidean plane, by contrast, is both without boundary and infinite.

A small region of a sphere has approximately Euclidean properties. However, when great distances are considered, the fundamental constant λ of Flatland becomes important. We will henceforth assume $\lambda = 1$. To recover it, all quantities with dimensions of length to some power in all formulas should be divided by λ to the same power; for example, $r \mapsto r/\lambda$, $S \mapsto S/\lambda^2$, etc.

▷ Let us draw three unit vectors **A**, **B**, **C** from the centre of the sphere to the vertices of the triangle ABC with sides a, b, c:

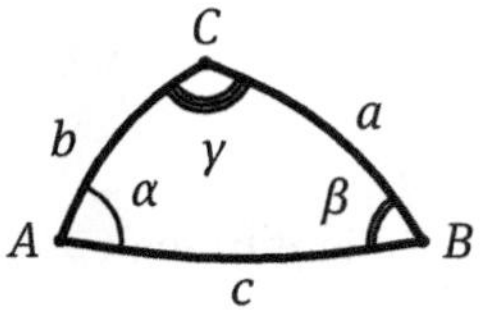

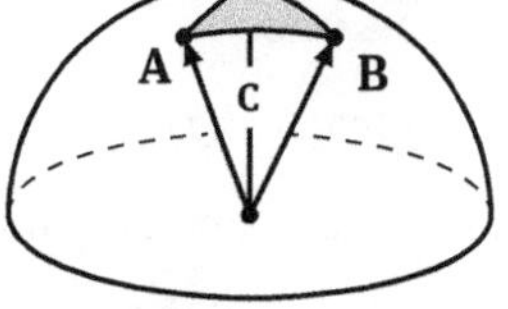

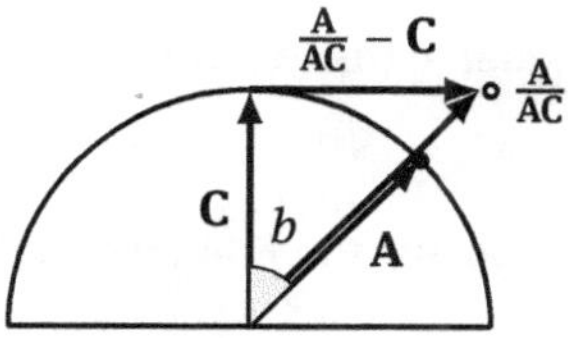

For a unit sphere, the angles between the vectors pointing to the vertices are equal to the lengths of the arcs between them. Therefore, the lengths of the triangle's sides are found from the following relationships:

$$\cos a = \mathbf{BC}, \qquad \cos b = \mathbf{AC}, \qquad \cos c = \mathbf{AB}.$$

The angle γ is equal to the angle between the vectors $\mathbf{A}/(\mathbf{AC}) - \mathbf{C}$ and $\mathbf{B}/(\mathbf{BC}) - \mathbf{C}$, which are tangent vectors to the sphere at the vertex **C** (the third figure above):

$$\cos \gamma = \frac{\mathbf{AB} - (\mathbf{AC})(\mathbf{BC})}{\sqrt{1 - (\mathbf{AC})^2}\,\sqrt{1 - (\mathbf{BC})^2}}. \tag{12.3}$$

Hence, we easily obtain the spherical *law of cosines*:

$$\cos c = \cos a \, \cos b + \sin a \, \sin b \, \cos \gamma. \tag{12.4}$$

▷ For a right triangle ($\gamma = \pi/2$), this gives the spherical *Pythagorean theorem*:

$$\cos c = \cos a \, \cos b. \tag{12.5}$$

If the triangle is small, the "usual" Pythagorean theorem $c^2 \approx a^2 + b^2$ can be obtained from $\cos x \approx 1 - x^2/2$. Writing the law of cosines for the side a opposite to angle α of the right triangle, and given (12.5), we can easily (< H_{56}) find the projections of the hypotenuse onto the legs:

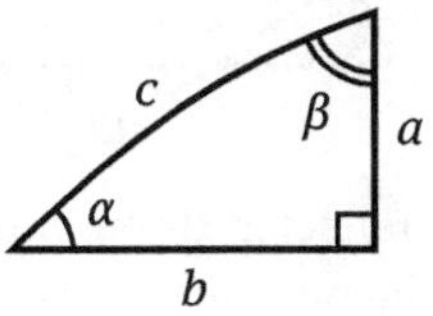

$$\begin{aligned} \operatorname{tg} b &= \operatorname{tg} c \, \cos \alpha, \\ \sin a &= \sin c \, \sin \alpha, \\ \operatorname{tg} a &= \sin b \, \operatorname{tg} \alpha. \end{aligned} \tag{12.6}$$

▷ Finally, by dividing an arbitrary triangle into two right triangles, we obtain (< H_{57}) the spherical law of sines:

$$\frac{\sin a}{\sin \alpha} = \frac{\sin b}{\sin \beta} = \frac{\sin c}{\sin \gamma}. \tag{12.7}$$

Let us cite some more formulas which have no direct parallels in Euclidean geometry.

▷ From (12.6) it follows that a right triangle is fully defined by the two angles α and β ($\lessdot$ H$_{58}$):

$$\cos c = \operatorname{ctg} \alpha \, \operatorname{ctg} \beta.$$

For small c this relationship transforms into

$$1 = \operatorname{ctg} \alpha \, \operatorname{ctg} \beta,$$

which is just the Euclidean relationship between the angles of a right triangle

$$\alpha + \beta = \frac{\pi}{2}.$$

▷ For an arbitrary triangle ABC we can always ($\lessdot$ H$_{59}$) construct a *dual triangle* $A'B'C'$ such that

$$a' = \pi - \alpha, \quad b' = \pi - \beta, \quad c' = \pi - \gamma, \quad \alpha' = \pi - a, \quad \beta' = \pi - b, \quad \gamma' = \pi - c.$$

Its law of cosines (12.4) gives the *dual law of cosines* for the initial triangle:

$$\cos \gamma = -\cos \alpha \, \cos \beta + \sin \alpha \, \sin \beta \, \cos c. \tag{12.8}$$

▷ Finally, there is another remarkable relationship between the area $S_\triangle$ of a triangle and the sum of its angles. The greater the area of the triangle, the more by which this sum exceeds π:

$$S_\triangle = \alpha + \beta + \gamma - \pi. \tag{12.9}$$

To prove this, construct a *digon* formed by two circles intersecting at an angle α at opposite points of the sphere (the shaded area in the first figure). It is composed of two identical triangles, each of them having two right angles. The area of the sphere is 4π (it has unit radius); therefore, the area of the digon is $S = 2\alpha$:

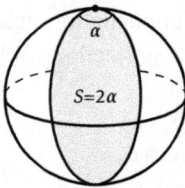

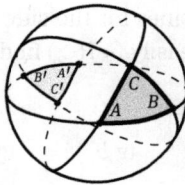

$$S_A = 4\alpha, \quad S_B = 4\beta, \quad S_C = 4\gamma$$

$$S_A + S_B + S_C - 6S_\triangle + 2S_\triangle = 4\pi$$

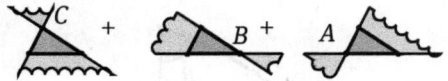

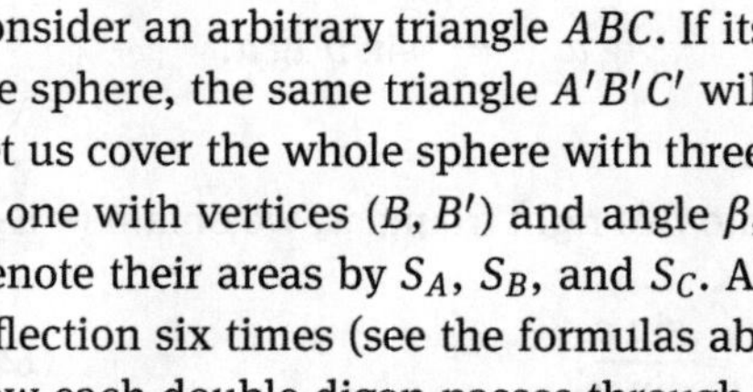

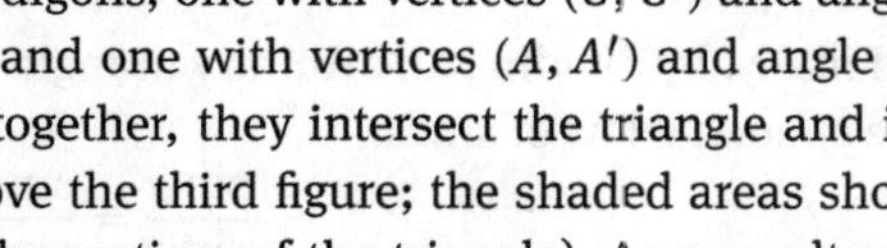

Consider an arbitrary triangle ABC. If its vertices are reflected through the centre of the sphere, the same triangle $A'B'C'$ will appear on the opposite side of the sphere. Let us cover the whole sphere with three digons, one with vertices (C, C') and angle γ, one with vertices (B, B') and angle β, and one with vertices (A, A') and angle α. Denote their areas by S_A, S_B, and S_C. Altogether, they intersect the triangle and its reflection six times (see the formulas above the third figure; the shaded areas show how each double digon passes through the vertices of the triangle). As a result, we obtain equation (12.9).

⊳ Any point of the sphere can be represented by a unit three-dimensional vector $\mathbf{p}$ drawn to it from the centre. Every straight line is also characterized by a unit vector $\mathbf{n}$. This vector is normal to the plane of the circle, whose centre coincides with the centre of the sphere:

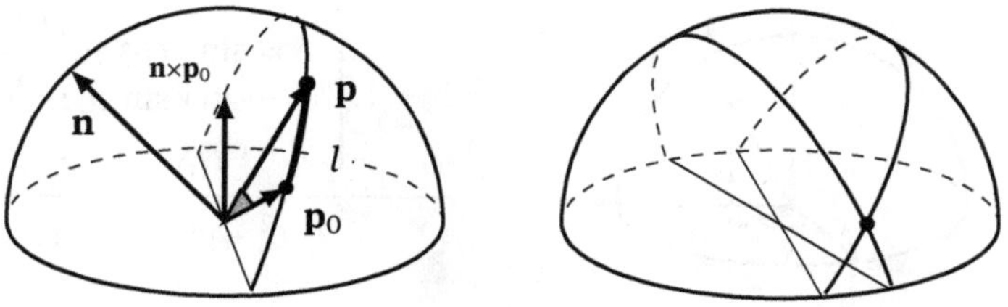

The parametric equation of a straight line passing through the point $\mathbf{p}_0$ has the form

$$\mathbf{p}(l) = \mathbf{p}_0 \cos l + [\mathbf{n} \times \mathbf{p}_0] \sin l. \tag{12.10}$$

The parameter l is the length of the segment from the point $\mathbf{p}_0$ along the straight line (for a sphere of non-unit radius $l \mapsto l/\lambda = \phi$, where ϕ is the angle of rotation of the vector $\mathbf{p}$).

Two lines defined by vectors $\mathbf{n}_1$ and $\mathbf{n}_2$ intersect at the diametrically opposite points

$$\pm[\mathbf{n}_1 \times \mathbf{n}_2]/\sqrt{1 - (\mathbf{n}_1\mathbf{n}_2)^2}$$

(normal to both $\mathbf{n}_1$ and $\mathbf{n}_2$). The angle α between them is found from $\mathbf{n}_1\mathbf{n}_2 = \cos\alpha$.

⊳ The surface of a sphere, like a plane, is a homogeneous and isotropic space. In the plane, there is a three-parameter family transformations (given by compositions of translations and rotations) which preserve the lengths of segments and the angles between them. If a plane is considered as being embedded in three-dimensional Euclidean space, its orientation is given by a unit vector $\mathbf{k}$ normal to it. If the radius vector of a point $\mathbf{p}$ lies in such a plane, which we also assume passes through the coordinate origin, then $\mathbf{kp} = 0$. The general form of a combined rotation and translation in the plane then has the form ($\mathbf{k}^2 = 1$):

$$\mathbf{p}' = \mathbf{a} + \mathbf{p} \cos\phi + [\mathbf{k} \times \mathbf{p}] \sin\phi,$$

where $\mathbf{a}$ is the vector of translation in the plane ($\mathbf{ak} = 0$), and ϕ is the angle of rotation.

The sphere can also be embedded in three-dimensional Euclidean space, where it is realised as the set of points satisfying $\mathbf{p}^2 = 1$. The three-parameter transformation ($\mathbf{n}^2$=1)

$$\mathbf{p}' = \mathbf{p} \cos\phi + \mathbf{n}\,(\mathbf{np})\,(1 - \cos\phi) - [\mathbf{n} \times \mathbf{p}] \sin\phi \tag{12.11}$$

is the analogue of a translation and rotation of the plane on the sphere. Any geometric figure can be rotated about an axis passing through its centre, and then translated along the sphere with the additional help of rotation. Two consecutive rotations are equivalent to a single rotation, whose transformation was written above (see p. 283).

12.2 Coordinates on the sphere

The coordinates used to "number" the points of a sphere are quite arbitrary. Consider as an example spherical coordinates ($r = \lambda = 1$):

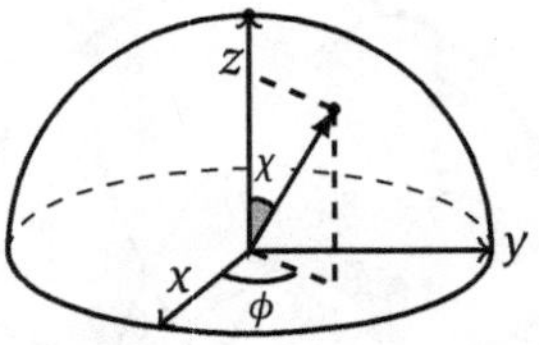

$$\begin{cases} x = \sin\chi\cos\phi \\ y = \sin\chi\sin\phi \\ z = \cos\chi \end{cases} \tag{12.12}$$

$$x^2 + y^2 + z^2 = 1.$$

The *polar angle* χ varies from 0 to π, and the *azimuth angle* ϕ varies from 0 to 2π. These coordinates uniquely characterize every point of a sphere. However, the angle ϕ is an indeterminate quantity at the "north" ($\chi = 0$) and the "south" ($\chi = \pi$) poles (all of its values are equivalent).

The distance between two infinitesimally close points on the surface of a sphere can be calculated with the help of the *Euclidean distance* in three-dimensional space:

$$dl^2 = dx^2 + dy^2 + dz^2. \tag{12.13}$$

Consider the following relationship between the coordinates (12.12):

$$dl^2 = (c_\chi c_\phi\, d\chi - s_\chi s_\phi\, d\phi)^2 + (c_\chi s_\phi\, d\chi + s_\chi c_\phi\, d\phi)^2 + (-s_\chi\, d\chi)^2$$

(for brevity, $c_\chi \equiv \cos\chi$, etc.). After expanding the squares and performing some simple manipulations, we obtain:

$$dl^2 = d\chi^2 + \sin^2\chi\, d\phi^2. \tag{12.14}$$

This element of length can be written using the metric tensor. Let $q^i = \{\chi,\ \phi\}$, and let us sum from 1 to 2 over the indices i, j. Then:

$$dl^2 = g_{ij}\, dq^i\, dq^j, \qquad g_{ij} = \begin{pmatrix} 1 & 0 \\ 0 & \sin^2\chi \end{pmatrix}.$$

The metric tensor g_{ij} is diagonal and its elements are positive. Such a metric is referred to as having positive signature $(+, +)$, where the number of plus signs indicates the dimension of the space. The signature of three-dimensional Euclidean space (12.13) is $(+, +, +)$.

If $\chi = const$, the element of length $dl = \sin\chi\, d\phi$ gives (12.1) when integrated over ϕ from 0 to 2π. The area of any domain on the sphere is obtained by summing up rectangles, the sides of which are obtained by holding one coordinate fixed and taking a small displacement in the other. In spherical coordinates these displacements are orthogonal, and $dS = (d\chi)(\sin\chi d\phi)$. Integrating this expression gives, once again, (12.2) (< H_{60}).

▷ If only the upper hemisphere is considered, the distance dl can be expressed in terms of the coordinates (x, y) of the plane onto which the points of the hemisphere vertically project.

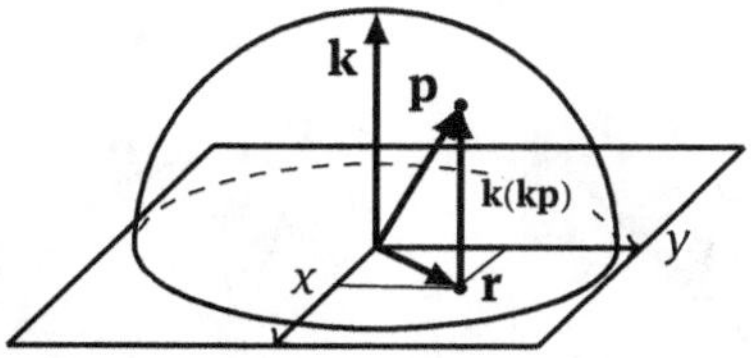

$$dl^2 = dx^2 + dy^2 + \frac{(xdx + ydy)^2}{1 - x^2 - y^2}. \tag{12.15}$$

Indeed, taking the differential of the equation of the sphere, $x^2 + y^2 + z^2 = 1$, we obtain

$$xdx + ydy + zdz = 0.$$

Using this equation to remove dz from the Euclidean three-dimensional distance (12.13), we obtain (12.15). If the radius of the sphere is large, the expression for the distance is approximately Euclidean ($dl^2 \approx dx^2 + dy^2$) for small x and y (in the vicinity of the north pole).

Let us obtain the same result in vector notation. Introduce a unit vector $\mathbf{k} = \{0, 0, 1\}$ pointing along the z-axis. Let us write the relationship between the vector $\mathbf{r}$ directed to the projection of a point on the sphere and the unit vector $\mathbf{p}$ connecting the origin to this point (see the figure):

$$\mathbf{r} = \mathbf{p} - \mathbf{k}\,(\mathbf{kp}). \tag{12.16}$$

We further obtain

$$r^2 = x^2 + y^2 = \mathbf{r}^2 = 1 - (\mathbf{kp})^2.$$

Therefore, the inverse transformation of (12.16) is

$$\mathbf{p} = \mathbf{r} + \mathbf{k}\,\sqrt{1 - r^2}. \tag{12.17}$$

The square of the increment of the radius vector in the plane is $(d\mathbf{r})^2 = dx^2 + dy^2$, and the element of length on the sphere is

$$dl^2 = (d\mathbf{p})^2 = (d\mathbf{r} + \mathbf{k}\,d\sqrt{1 - r^2})^2.$$

Given that $dr^2 = 2\,(xdx + ydy)$ and $\mathbf{k}d\mathbf{r} = 0$, we again obtain (12.15).

Using (12.17) we can easily find the projections of various geometric objects. For example, a straight line for the flatlanders is the intersection of the sphere $\mathbf{p}^2 = 1$ with a plane $\mathbf{np} = 0$ passing through its centre and normal to the unit vector $\mathbf{n}$. Substituting (12.17) into equation $\mathbf{np} = 0$,

$$\mathbf{nr} + \mathbf{nk}\,\sqrt{1 - r^2} = 0.$$

For $\mathbf{nk} = 0$ we obtain the equation of a line $n_x\,x + n_y\,y = 0$, and for $\mathbf{nk} \neq 0$ the equation of an ellipse. For example, if $n_y = 0$, we have $x^2/n_z^2 + y^2 = 1$.

⊳ Another way of projecting a hemisphere onto a plane is to use *Beltrami coordinates*. Let a plane be tangent to the sphere at its north pole N. If we draw a line from the centre of the sphere O, it will intersect a point of the upper hemisphere and a point of the tangent plane. As a result, we obtain the *central projection*:

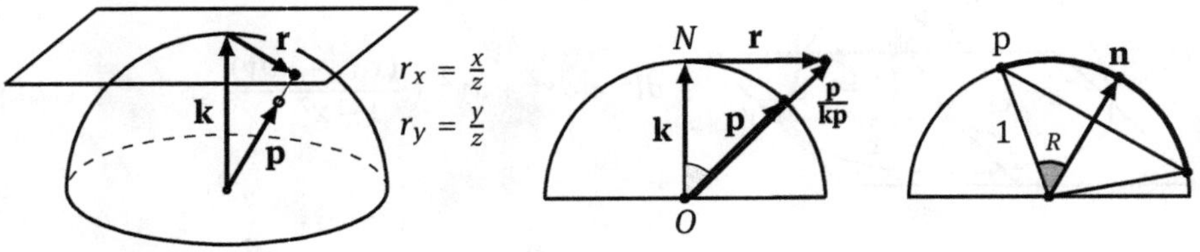

The second figure shows the intersection between the sphere and the plane formed by $\mathbf{k}$ and $\mathbf{p}$. From geometric considerations we can easily find the relationship between the unit vector $\mathbf{p}$ directed to a point on the sphere and the (non-unit) vector $\mathbf{r}$ of its projection onto the tangent plane (the second figure above):

$$\mathbf{r} = \frac{\mathbf{p}}{\mathbf{kp}} - \mathbf{k} = \frac{\mathbf{p} - \mathbf{k}\,(\mathbf{kp})}{\mathbf{kp}}. \tag{12.18}$$

The same relationship can be obtained in another way. The parametric equation of a straight line passing through the centre of the sphere has the form

$$\mathbf{x} = \mathbf{p}\,t.$$

This line intersects the plane described by the equation $\mathbf{kx} = 1$, where $\mathbf{k} = \{0, 0, 1\}$. Therefore, we have $\mathbf{kp}\,t = 1$, or

$$t = \frac{1}{\mathbf{kp}}$$

in the plane. By substituting t into $\mathbf{x} = \mathbf{p}\,t$ and introducing $\mathbf{r} = \mathbf{x} - \mathbf{k}$, we obtain (12.18). From (12.18) it follows that:

$$r^2 = x^2 + y^2 = \mathbf{r}^2 = \frac{1}{(\mathbf{kp})^2} - 1, \qquad \mathbf{p} = \frac{\mathbf{r} + \mathbf{k}}{\sqrt{1 + r^2}}.$$

Using the fact that $(d\mathbf{r})^2 = dx^2 + dy^2$ on the plane, and $(d\mathbf{p})^2 = dl^2$ on the sphere, we obtain the metric of the sphere in Cartesian Beltrami coordinates:

$$dl^2 = \frac{dx^2 + dy^2}{1 + x^2 + y^2} - \frac{(xdx + ydy)^2}{(1 + x^2 + y^2)^2}. \tag{12.19}$$

A straight line on the sphere is a line formed by the intersection of the sphere with a plane passing through its centre. The intersection of this plane with the tangent plane is also a straight line, which projects back onto the original line on the sphere. Spherical triangles get projected onto Euclidean triangles. However, the angles between their sides get distorted (unless they intersect at the tangential point N). A circle of radius R on the sphere is obtained by intersecting it with the plane $\mathbf{np} = \cos R$ (the third figure above), where $\mathbf{n}$ is a unit vector normal to the plane of the circle. The projection of this circle onto the tangent plane is an ellipse (⋖ H_{61}).

▷ Finally, we consider the ***stereographic projection***. Take the plane passing through the equator of the sphere. If we draw a line from the south pole, it will intersect both the plane and the sphere. The interior of the disc of unit radius in the plane is covered by the points of the upper hemisphere, while the points outside the disc are covered by the points of the lower hemisphere. The south pole corresponds to an infinitely distant point:

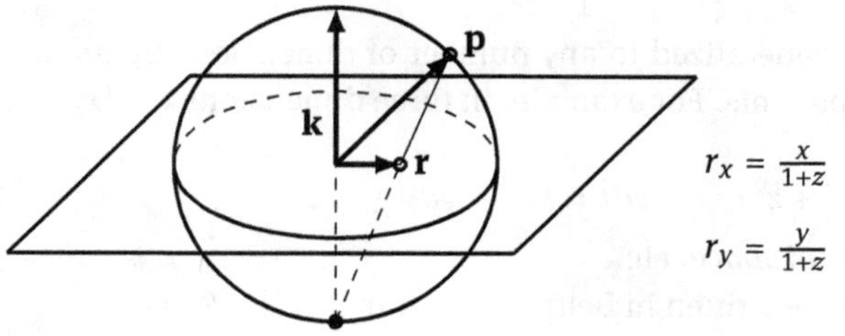

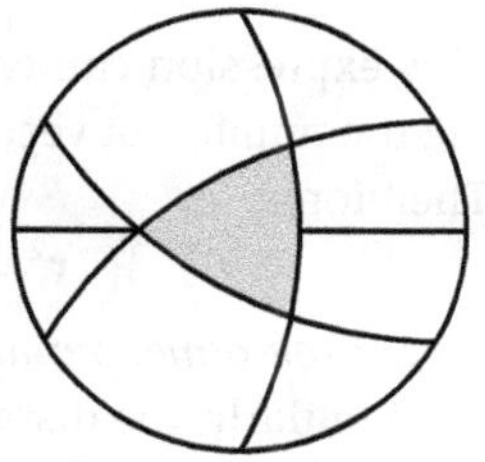

Suppose $\mathbf{k} = \{0, 0, 1\}$. The equation of a line passing through the south pole and the point $\mathbf{p}$ on the sphere has the form $\mathbf{x} = -\mathbf{k} + (\mathbf{p} + \mathbf{k})\,t$. This line intersects the projection plane $\mathbf{kx} = 0$, whence $t = 1/(1 + \mathbf{kp})$. Therefore, for a vector $\mathbf{r} = \mathbf{x}|_{z=0} = \{x, y, 0\}$ in the plane, we have:

$$\mathbf{r} = \frac{\mathbf{p}+\mathbf{k}}{1+\mathbf{kp}} - \mathbf{k} = \frac{\mathbf{p}-\mathbf{k}\,(\mathbf{kp})}{1+\mathbf{kp}}, \qquad \mathbf{kp} = \frac{1-r^2}{1+r^2}. \tag{12.20}$$

Consider a curve $\mathbf{p}(l)$ on the sphere, where l is the length along the curve from the point $\mathbf{p}(0)$. Then we have $\dot{\mathbf{p}}^2 = (d\mathbf{p}/dl)^2 = 1$ and $\mathbf{p}\dot{\mathbf{p}} = 0$ (< H_{62}). The projection of this curve onto the plane is the curve $\mathbf{r}(l)$ such that:

$$\dot{\mathbf{r}} = \frac{d\mathbf{r}}{dl} = \frac{\dot{\mathbf{p}}}{1+\mathbf{kp}} - \frac{(\mathbf{p}+\mathbf{k})\,(\mathbf{k}\dot{\mathbf{p}})}{(1+\mathbf{kp})^2}, \qquad \dot{\mathbf{r}}^2 = \frac{1}{(1+\mathbf{kp})^2}. \tag{12.21}$$

Since $(d\mathbf{r}/dl)^2 = (dx^2 + dy^2)/dl^2$, from the last relationship we immediately obtain the metric on the sphere in the coordinates of its projection onto the plane:

$$dl^2 = 4\,\frac{dx^2+dy^2}{(1+x^2+y^2)^2}. \tag{12.22}$$

Any circle on the sphere transforms into a circle in the plane. For example, for $\mathbf{np} = 0$ the equation of a "spherical line" in the plane has the form:

$$\Big(x - \frac{n_x}{n_z}\Big)^2 + \Big(x - \frac{n_y}{n_z}\Big)^2 = \frac{1}{n_z^2}. \tag{12.23}$$

The angle between two curves $\mathbf{p}_1(l)$ and $\mathbf{p}_2(l)$ at the point of their intersection on the sphere $\mathbf{p}_1(0) = \mathbf{p}_2(0)$ is equal to the angle between their projections in the plane (< H_{63}):

$$\frac{\dot{\mathbf{r}}_1\dot{\mathbf{r}}_2}{|\dot{\mathbf{r}}_1|\,|\dot{\mathbf{r}}_2|} = \dot{\mathbf{p}}_1\,\dot{\mathbf{p}}_2. \tag{12.24}$$

Such angle preserving projections are called *conformal*.

12.3 A three-dimensional spherical world

Until now all of our reasoning has concerned the world of flatlanders. Let us now consider creatures capable of perceiving three dimensions. Consider the distance (12.15) in vector form by introducing the radius vector in the plane with components $\mathbf{r} = \{x, y\}$:

$$dl^2 = (d\mathbf{r})^2 + \frac{(\mathbf{r}\,d\mathbf{r})^2}{1 - \mathbf{r}^2}. \tag{12.25}$$

This expression can easily be generalized to any number of dimensions by increasing the number of vector components. For example, in three dimensions $\mathbf{r} = \{x, y, z\}$. Therefore,

$$\mathbf{r}^2 = x^2 + y^2 + z^2, \qquad \mathbf{r}d\mathbf{r} = xdx + ydy + zdz,$$

for a *three-dimensional spherical space*, etc.

Similarly, the distance can be written in Beltrami coordinates (12.19) as

$$dl^2 = \frac{(d\mathbf{r})^2}{1 + \mathbf{r}^2} - \frac{(\mathbf{r}\,d\mathbf{r})^2}{(1 + \mathbf{r}^2)^2} = \frac{(d\mathbf{r})^2 + [\mathbf{r} \times d\mathbf{r}]^2}{(1 + \mathbf{r}^2)^2}, \tag{12.26}$$

where in the second equality we have made use of the formula for the vector triple product in three-dimensional space. The distance in stereographic coordinates differs from the Euclidean distance $dl^2 = d\mathbf{r}^2$ only by a scaling factor (another definition of conformality):

$$dl^2 = \frac{4\,(d\mathbf{r})^2}{(1 + \mathbf{r}^2)^2}. \tag{12.27}$$

Consider a four-dimensional space with coordinates x, y, z, w. The three-dimensional analogues of the metrics (12.25)–(12.27) are determined by distances "on" the three-dimensional sphere embedded in this four-dimensional space, which is given by

$$x^2 + y^2 + z^2 + w^2 = 1. \tag{12.28}$$

The coordinate axes in four-dimensional space are pairwise orthogonal. For example, the two axes z and w, are normal to the (x, y)-plane and to each other. In three-dimensional space, the coordinate (x, y)-plane (a two-dimensional object) is given by the equation

$$z = const.$$

In four-dimensional space, this equation describes the *hyperplane* with coordinates (x, y, w). We perceive it as a three-dimensional volume. In vector notation, a general hyperplane is given by the equation

$$\mathbf{k}\,(\mathbf{x} - \mathbf{x}_0) = 0,$$

where $\mathbf{k}$ is some vector normal to the hyperplane, and $\mathbf{x}_0$ is some point in it.

In spite of the common notation, the vector $\mathbf{r}$ has different meanings in the above metrics. It lies inside the projection hyperplane. However, the position of this hyperplane, as well as the method by which the points of the spherical space are projected onto it, is different.

▷ Similarly to Euclidean space, spherical coordinates can also be defined in a three-dimensional spherical space:

$$x = r\sin\theta\cos\phi, \quad y = r\sin\theta\sin\phi, \quad z = r\cos\theta,$$

so that

$$\mathbf{r}^2 = r^2, \quad \mathbf{r}d\mathbf{r} = rdr, \quad (d\mathbf{r})^2 = dr^2 + r^2\,d\theta^2 + r^2 s_\theta^2\,d\phi^2.$$

The metrics (12.25), (12.26) and (12.27) are written in these coordinates as follows:

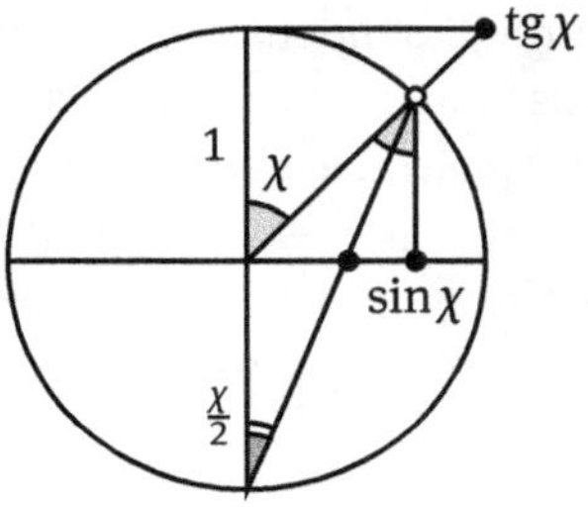

$$dl^2 = d\chi^2 + \sin^2\chi\,(d\theta^2 + \sin^2\theta\,d\phi^2). \qquad (12.29)$$

$r = \sin\chi$	–	vertical projection
$r = \mathrm{tg}\,\chi$	–	Beltrami projection
$r = \mathrm{tg}\,\frac{\chi}{2}$	–	stereographic projection

The coordinate r should be substituted by $r = \sin\chi$ in the first case, by $r = \mathrm{tg}\,\chi$ in the second case, and by $r = \mathrm{tg}(\chi/2)$ in the third case (see the figure above).

The metric (12.29) can also be obtained by changing coordinates in the element of Euclidean length $dl^2 = dx^2 + dy^2 + dz^2 + dw^2$:

$$x = \sin\chi\,\sin\theta\,\cos\phi, \quad y = \sin\chi\,\sin\theta\,\sin\phi, \quad z = \sin\chi\,\cos\theta, \quad w = \cos\chi.$$

This is analogous to taking spherical coordinates on a three-dimensional space with unit radius and projecting onto the (x, y, z)-hyperplane such that the w-coordinate obeys (12.28).

▷ If the coordinates χ and θ are fixed, a small displacement in $d\phi$ corresponds to a segment $dl = s_\chi s_\theta\,d\phi$ in the spherical space. Similarly, displacing χ and θ gives the segments $d\chi$ and $s_\chi d\theta$, respectively. These three segments are orthogonal ($\lessdot$ H_{64}) and determine the volume element $dV = (d\chi)(s_\chi d\theta)(s_\chi s_\theta d\phi)$ In particular, the volume of a two-dimensional sphere of radius $r = \chi$ in a three-dimensional spherical space of unit radius is:

$$V = \int_0^r\int_0^\pi\int_0^{2\pi} \sin^2\chi\sin\theta\,d\chi\,d\theta\,d\phi = \pi\{2r - \sin(2r)\}. \qquad (12.30)$$

Its maximum $V = 2\pi^2$ is reached at $r = \pi$. This is the volume of the entire spherical space. The surface element of the sphere at $\chi = r = const$ is $dS = (s_\chi\,d\theta)(s_\chi s_\theta\,d\phi)$, and the total area is

$$S = 4\pi\sin^2(r). \qquad (12.31)$$

For small r, the Euclidean values $S = 4\pi r^2$ and $V = 4\pi r^3/3$ are obtained.

12.4 Hyperbolic geometry

The geometry of a two-dimensional spherical space is specified by the fundamental constant λ. For us, this constant has the meaning of the radius of the sphere where the flatlanders live. The parameter $1/\lambda^2$ is called the curvature of the sphere. The curvature is the same at all points of a sphere. This is why spherical spaces are homogeneous and isotropic. The larger λ is, the smaller the curvature of the sphere. Our ancestors believed that the Earth was flat because all of the distances available to them were substantially smaller than its radius.

Having considered the geometry of a sphere, we turned to the three-dimensional spherical space. For us (three-dimensional creatures) it is hard to imagine such a space as a sphere embedded in a four-dimensional space. Nevertheless, as regards the mathematics, the increase of dimension is quite trivial.

Let us use another mathematical trick and consider the space of *constant negative curvature* or *Lobachevsky space*. To do this, we make the following substitutions in all previous formulas:

$$\lambda^2 \mapsto -\lambda^2, \qquad \lambda \mapsto \imath\lambda. \tag{12.32}$$

The properties of the space are still determined by a single constant λ. Similarly to the spherical space, we can expect it to be homogeneous and isotropic. Geometric relations established in one part of the space will also be true in any other part.

After making the substitutions (12.32), we see from (12.9) that the sum of the angles of a triangle in Lobachevsky space is smaller than π by the area of the triangle (we first restore λ: $S \mapsto S/\lambda^2$, and then make the substitution $\lambda^2 \mapsto -\lambda^2$):

$$\alpha + \beta + \gamma = \pi - \frac{S}{\lambda^2}.$$

Therefore, triangles in Lobachevsky space are conventionally depicted *in the Euclidean plane* as having concave sides:

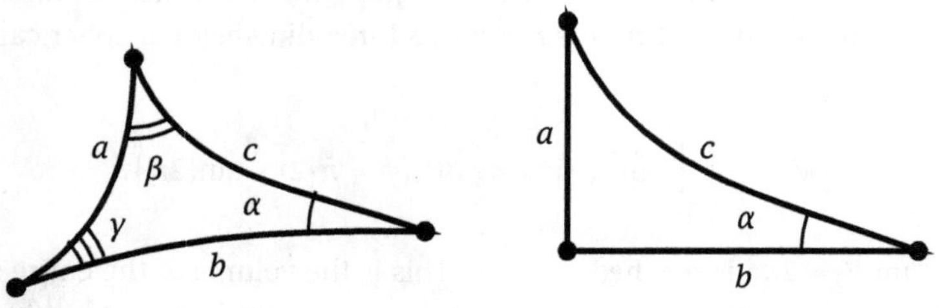

Of course, these sides are really straight lines (in the sense that they cover the shortest distance between two points). Therefore, this convention emphasizes only the fact that the sum of the angles is smaller than 180 degrees.

When discussing spherical trigonometry, we assumed the radius of the sphere to be unity ($\lambda = 1$). To restore λ in all of our formulas, we need to rescale the sides of all triangles, which have dimensions of length, e.g. $a \mapsto a/\lambda$, etc. The angles α, β, γ do not change in this case. When passing to Lobachevsky space ($\lambda \mapsto \imath\lambda$), we make use of the following relationship between hyperbolic and trigonometric functions:

$$\sin(\imath\, x) = \imath\, \mathrm{sh}(x), \qquad \cos(\imath\, x) = \mathrm{ch}(x).$$

As a result, the spherical law of cosines (12.4) for a unit "sphere" with imaginary radius takes the form (< H_{65}):

$$\mathrm{ch}\, c = \mathrm{ch}\, a\; \mathrm{ch}\, b - \mathrm{sh}\, a\; \mathrm{sh}\, b\; \cos\gamma. \tag{12.33}$$

Similarly, for the law of sines (12.7) we have:

$$\frac{\mathrm{sh}\, a}{\sin\alpha} = \frac{\mathrm{sh}\, b}{\sin\beta} = \frac{\mathrm{sh}\, c}{\sin\gamma}, \tag{12.34}$$

and for the dual law of cosines (12.8):

$$\cos\gamma = -\cos\alpha\; \cos\beta + \sin\alpha\; \sin\beta\; \mathrm{ch}\, c. \tag{12.35}$$

For a right triangle ($\gamma = \pi/2$) the "Pythagorean theorem" holds:

$$\mathrm{ch}\, c = \mathrm{ch}\, a\; \mathrm{ch}\, b, \tag{12.36}$$

as do the following rules for projecting the hypotenuse c onto the legs a and b:

$$\mathrm{th}\, b = \mathrm{th}\, c\; \cos\alpha, \qquad \mathrm{sh}\, a = \mathrm{sh}\, c\; \sin\alpha.$$

Additionally, the hypotenuse (and the entire triangle) is completely determined by two angles:

$$\mathrm{ch}\, c = \mathrm{ctg}\,\alpha\; \mathrm{ctg}\,\beta.$$

The perimeter of a circle of radius r and the area of a disc in two-dimensional Lobachevsky space are:

$$L = 2\pi\, \mathrm{sh}\, r, \qquad S = 2\pi(\mathrm{ch}\, r - 1).$$

The surface area and volume of a two-dimensional sphere of radius r in three-dimensional Lobachevsky space have the form:

$$S = 4\pi\, \mathrm{sh}^2 r, \qquad V = \pi(\mathrm{sh}(2r) - 2r).$$

In contrast to the spherical space from above, the volume of Lobachevsky space is infinite.

12.5 Pseudo-Euclidean space

Suppose that each point in an $n+1$-dimensional space is given ("numbered") by $n+1$ coordinates $\{x^0, x^1, \ldots, x^n\}$. Such a space is said to be pseudo-Euclidean of type $(1, n)$ if the distance between two infinitely close points x and x + dx is given by

$$ds^2 = (dx^0)^2 - (dx^1)^2 - \ldots - (dx^n)^2. \tag{12.37}$$

In the theory of relativity, space-time has type $(1, 3)$ and signature $(+, -, -, -)$. We will also consider the three-dimensional space $(1, 2)$ and the two-dimensional space $(1, 1)$. In this case, the coordinates of a point, in the space $(1, 2)$, for example, are often written as $\{t, x, y\}$, although t is not necessarily the time. In general, t is simply a coordinate of the pseudo-Euclidean space whose signature differs from that of the coordinates x and y.

The vectors of an $n+1$-dimensional space have $n+1$ components:

$$\mathrm{a} = \{a^0, a^1, \ldots, a^n\} = \{a^0, \mathbf{a}\}.$$

For the time component, the position of the index has no importance: $a^0 = a_0$, while spatial components will often be written as Euclidean components: $a_x = a^1$, $a_y = a^2$, etc. By definition, the scalar product of two vectors is:

$$\mathrm{a} \cdot \mathrm{b} \equiv \mathrm{a}\,\mathrm{b} = a^0 b^0 - a^1 b^1 - \ldots - a^n b^n = a_0 b_0 - \mathbf{ab}.$$

The square of a vector $\mathrm{a}^2 = \mathrm{aa}$ can be positive $\mathrm{a}^2 > 0$ (for *time-like* vectors), negative $\mathrm{a}^2 < 0$ (for *space-like* vectors) or zero $\mathrm{a}^2 = 0$ (for *light-like* vectors). If the scalar product of two vectors is zero, $\mathrm{ab} = 0$, they are referred to as *orthogonal*.

The *line* passing through the point x_0 in the direction of the vector u is the set of points

$$\mathrm{x} = \mathrm{x}_0 + \mathrm{u}\, s,$$

where s is a real parameter (for $s = 0$, we have $\mathrm{x} = \mathrm{x}_0$). If $\mathrm{u}^2 = 1$, then s is the length (12.37) along the line counted from the point x_0.

A *hyperplane* is a space whose dimension is one unit smaller than the dimension of the initial space. Fix a point x_0, and let n be a vector. Then the set points of satisfying

$$\mathrm{n} \cdot (\mathrm{x} - \mathrm{x}_0) = 0$$

form a hyperplane. If a line passes through the point x_0 and lies in the hyperplane, then

$$\mathrm{n} \cdot \mathrm{u} = 0.$$

Therefore, the vector n is said to be normal to the hyperplane. A hyperplane in the space $(1, 2)$ is a two-dimensional plane, and in the space $(1, 1)$ it is a straight line.

There are three different types of pseudosphere (points equidistant from a given point; hereafter, simply "spheres") in a pseudo-Euclidean space. The sphere with unit radius and centre at the coordinate origin in the space $(1,1)$ has the equation $t^2 - x^2 = 1$. Similarly, $t^2 - x^2 = -1$ is a sphere with radius -1, and the lines $t^2 = x^2$ constitute the sphere of radius zero (the first figure below). In the space $(1,n)$ with $n > 1$, the sphere of positive radius is still two-connected (the second figure), and a sphere of negative radius is simply connected (the third figure):

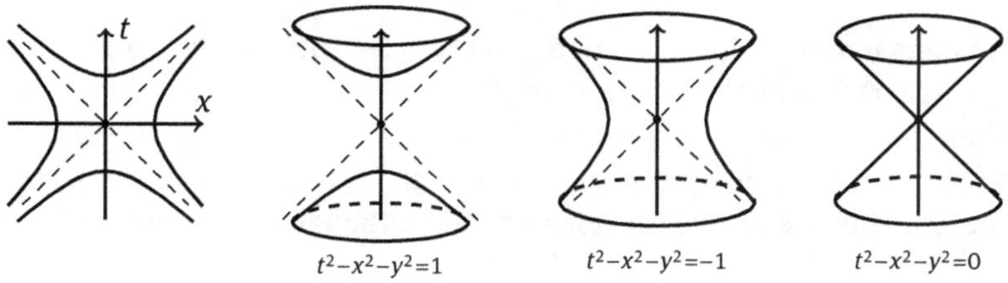

Pseudo-Euclidean spaces are represented on paper using coordinates. A point in the (t,x)-plane is represented as a point with the Cartesian coordinates (t,x). Similarly, vector components are plotted along the axes. Therefore, vectors can be added using the usual parallelogram rule:

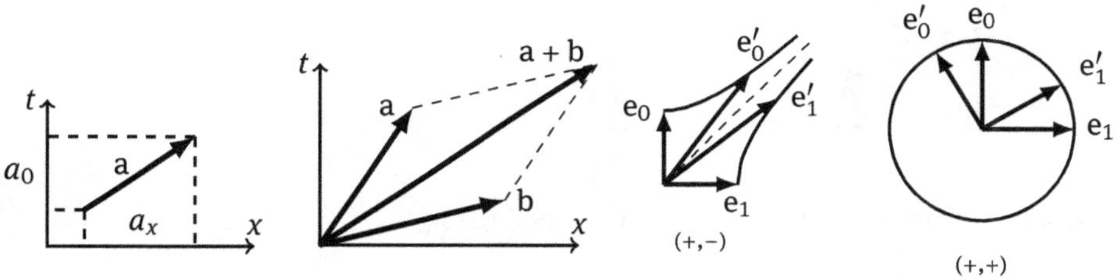

However, the metric properties of the pseudo-Euclidean space are not preserved by this mapping onto the "Euclidean paper". Introduce two orthogonal *basis vectors* in the space $(1,1)$:

$$\mathrm{e}_0^2 = 1, \qquad \mathrm{e}_1^2 = -1, \qquad \mathrm{e}_0\mathrm{e}_1 = 0, \tag{12.38}$$

so that one of them is a time-like vector (e_0) and the other is a space-like vector (e_1). Any vector can be decomposed in terms of the basis: $\mathrm{a} = a^0\mathrm{e}_0 + a^1\mathrm{e}_1$. In this case, $\mathrm{a}^2 = (a^0)^2 - (a^1)^2$ and $\mathrm{ab} = a^0b^0 - a^1b^1$.

The vectors e_0' and e_1' in the third figure also satisfy the condition (12.38), since they are tangential to the spheres. However, they appear longer than the vectors e_0 and e_1, and are also not orthogonal. As is easily verified, $(\mathrm{e}_0 + \mathrm{e}_1)^2 = 0$ is true for (12.38). This means that the vector of the sum of the orthogonal unit vectors lies on the light cone (the dashed line in the third figure). The last figure shows the Euclidean analogue of the third figure (two pairs of rotated orthogonal vectors).

▷ The modulus of the scalar product of two unit vectors can exceed unity in pseudo-Euclidean space. Therefore, the notion of the angle between two vectors is quite specific. Let two unit vectors, a time-like vector "a" and a space-like vector "b", belong to the space (1, 1). Decompose them in terms of the basis (12.38). Using the identity $\mathrm{ch}^2 x - \mathrm{sh}^2 x = 1$, we have:

$$\mathrm{a} = \mathrm{e}_0 \,\mathrm{ch}\,\alpha + \mathrm{e}_1 \,\mathrm{sh}\,\alpha, \qquad \mathrm{a}^2 = +1,$$
$$\mathrm{b} = \mathrm{e}_0 \,\mathrm{sh}\,\beta + \mathrm{e}_1 \,\mathrm{ch}\,\beta, \qquad \mathrm{b}^2 = -1.$$

The parameters α and β will be referred to as "angles" (the quotation marks will be omitted hereafter). The angle α is measured from the basis vector e_0, so that $\mathrm{a} = \mathrm{e}_0$ for $\alpha = 0$. Similarly, the angle β is measured from e_1. When $\alpha, \beta \to \infty$, the unit vectors approach the light cone (dashed line) from above ("a") and from below ("b"). Their lengths become infinite on the Euclidean paper, although we still have $\mathrm{a}^2 = 1$ and $\mathrm{b}^2 = -1$.

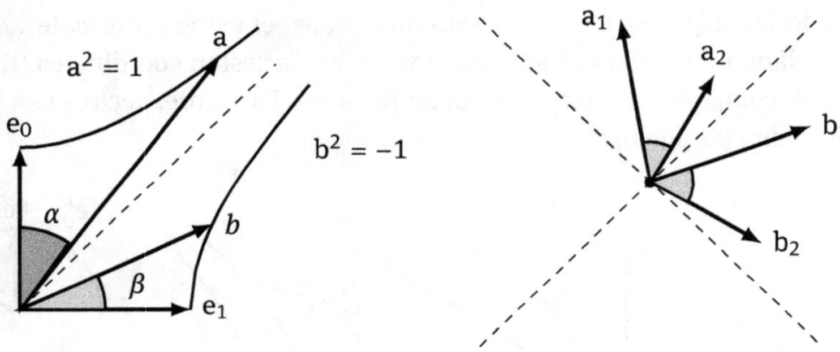

Using hyperbolic identities, it is easily verified that

$$\mathrm{a}_1 \cdot \mathrm{a}_2 = \mathrm{ch}(\alpha_2 - \alpha_1), \quad \mathrm{b}_1 \cdot \mathrm{b}_2 = -\,\mathrm{ch}(\beta_2 - \beta_1), \quad \mathrm{a}_1 \cdot \mathrm{b}_2 = \mathrm{sh}(\beta_2 - \alpha_1),$$

where the a_i are time-like vectors, and the b_i are space-like vectors.

A vector of arbitrary length can be obtained by multiplying the components of a unit vector by a positive number. Therefore, in general, the angles between vectors which are both time-like or both space-like are defined by the following relationships:

$$\frac{\mathrm{a}_1 \cdot \mathrm{a}_2}{\sqrt{\mathrm{a}_1^2\,\mathrm{a}_2^2}} = \mathrm{ch}(\alpha), \qquad \frac{\mathrm{b}_1 \cdot \mathrm{b}_2}{\sqrt{\mathrm{b}_1^2\,\mathrm{b}_2^2}} = -\,\mathrm{ch}(\beta), \tag{12.39}$$

where we still have $\mathrm{a}_i^2 = 1$, $\mathrm{b}_i^2 = -1$, and $a_i^0 > 0$ (the vectors a_i lie inside the "upper" light cone). If two vectors have squares of opposite sign, their scalar product can take any value:

$$\frac{\mathrm{a} \cdot \mathrm{b}}{\sqrt{-\mathrm{a}^2\,\mathrm{b}^2}} = \mathrm{sh}(\gamma). \tag{12.40}$$

If $\gamma > 0$, the vector "b" is closer to the light cone than "a", and vice versa for $\gamma < 0$. The value $\gamma = 0$ corresponds to perpendicular vectors.

▷ Let us introduce three basis vectors in the three-dimensional space (1, 2):

$$\mathrm{e}_0^2 = 1, \quad \mathrm{e}_1^2 = \mathrm{e}_2^2 = -1, \quad \mathrm{e}_0\mathrm{e}_1 = \mathrm{e}_0\mathrm{e}_2 = \mathrm{e}_1\mathrm{e}_2 = 0.$$

We can then parametrize the components of a time-like vector $\mathrm{a}^2 > 0$ and a space-like vector $\mathrm{b}^2 < 0$ as follows:

$$\mathrm{a} = a\,\{\mathrm{ch}\,\alpha,\ \mathbf{n}_a\ \mathrm{sh}\,\alpha\}, \qquad \mathrm{b} = b\,\{\mathrm{sh}\,\beta,\ \mathbf{n}_b\ \mathrm{ch}\,\beta\},$$

where $\mathbf{n}_a$ and $\mathbf{n}_b$ are Euclidean unit two-vectors, $\mathbf{n}_a^2 = \mathbf{n}_b^2 = 1$, and a and b are the absolute values of the magnitudes of the pseudo-euclidean vectors: $\mathrm{a}^2 = a^2$, $\mathrm{b}^2 = -b^2$. Similarly to the two-dimensional case, the angle between two time-like vectors inside one light cone can be determined from the formula:

$$\frac{\mathrm{a}_1 \cdot \mathrm{a}_2}{\sqrt{\mathrm{a}_1^2\,\mathrm{a}_2^2}} = \mathrm{ch}(\alpha).$$

Let the beginnings of the vectors a_1 and a_2 coincide at a point, and consider a unit sphere centred at this point. Two lines directed along these vectors intersect the sphere at some points. The angle α has the meaning of the distance between these points on the sphere (p. 321).

▷ There are ***space-like planes*** in the space (1, 2). All vectors lying in such a plane have negative squares. The plane of the coordinates (x, y) orthogonal to the t-axis is one example. If $\mathrm{b}_1 = \{0, \mathbf{b}_1\}$ and $\mathrm{b}_2 = \{0, \mathbf{b}_2\}$, then we have b_1^2, b_2^2, $(\mathrm{b}_1 + \mathrm{b}_2)^2 < 0$. In general, we have $\mathrm{b}^2 = -(\alpha^2 + \beta^2) < 0$ for any vector

$$\mathrm{b} = \alpha_1\mathrm{e}_1 + \beta\mathrm{e}_2.$$

Any two vectors b_1 and b_2 lying in a space-like plane obey the Cauchy–Schwarz inequality. Let s be a real number. Then the vector $\mathrm{b}_1 - s\,\mathrm{b}_2$ also lies in this plane and for it, we have:

$$(\mathrm{b}_1 - s\,\mathrm{b}_2)^2 = \mathrm{b}_1^2 - 2\,(\mathrm{b}_1\mathrm{b}_2)\,s + \mathrm{b}_2^2\,s^2 < 0. \tag{12.41}$$

Since $\mathrm{b}_2^2 < 0$, this defines a parabola (as a function of s) with downward-pointing opening. The maximum of the parabola occurs at $s = \mathrm{b}_1\mathrm{b}_2/\mathrm{b}_2^2$. Substituting this into the inequality (12.41), we obtain

$$\mathrm{b}_1^2 - \frac{(\mathrm{b}_1\mathrm{b}_2)^2}{\mathrm{b}_2^2} < 0 \quad \Rightarrow \quad \frac{(\mathrm{b}_1\mathrm{b}_2)^2}{\mathrm{b}_1^2\,\mathrm{b}_2^2} < 1$$

(when divided by $\mathrm{b}_1^2 < 0$, the sign "less than" gets reversed). Thus, for vectors lying in a space-like plane (in the space $(1, n)$, hyperplane) the "usual" angle can be defined using the cosine function:

$$\frac{\mathrm{b}_1 \cdot \mathrm{b}_2}{\sqrt{\mathrm{b}_1^2\,\mathrm{b}_2^2}} = -\cos\alpha. \tag{12.42}$$

The minus sign is chosen to make the angle α coincide with the Euclidean angle between the spatial two-vectors $\mathbf{b}_1$ and $\mathbf{b}_2$ in the (x, y)-plane.

12.6 Pseudospheres in Minkowski space

There is a striking relation between hyperbolic geometry and the Minkowski pseudo-Euclidean geometry. We consider the three-dimensional pseudo-Euclidean space $(1, 2)$, which carries the metric

$$ds^2 = dt^2 - dx^2 - dy^2. \tag{12.43}$$

Let us draw the upper pseudo-hemisphere (hereafter, "sphere") of unit radius (the first figure below):

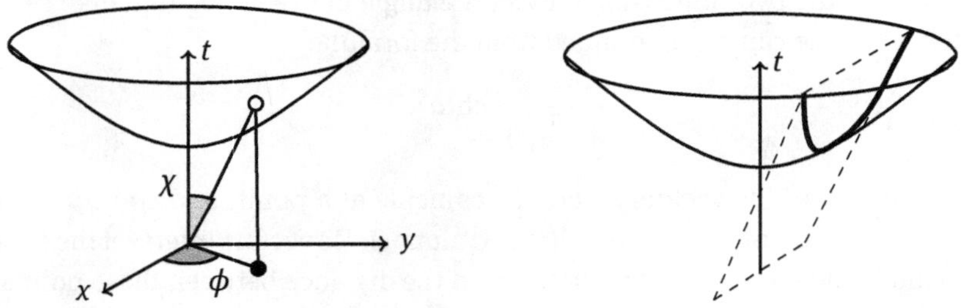

and define the analogue of spherical coordinates:

$$\begin{cases} x &= \operatorname{sh}\chi \cos\phi, \\ y &= \operatorname{sh}\chi \sin\phi, \\ t &= \operatorname{ch}\chi. \end{cases} \tag{12.44}$$

The coordinates (χ, ϕ) are chosen in such a way as to make the equation of the unit sphere,

$$\mathrm{x}^2 = t^2 - x^2 - y^2 = 1,$$

an identity. Substituting (12.44) into the distance (12.43), we obtain a two-dimensional metric:

$$dl^2 = -ds^2 = d\chi^2 + \operatorname{sh}^2\chi\, d\phi^2. \tag{12.45}$$

The normal to any point of the sphere lies inside the light cone (it is a ***space-like surface***), and the distance (12.43) is always negative on it. Therefore, it is convenient to introduce a positive squared length dl^2 with the signature of the surface $(+, +)$. As is easily seen, this distance is obtained from the distance on the sphere (12.14) by making the substitution $\lambda \mapsto \imath\lambda$ (after making the substitutions $\chi \mapsto \chi/\lambda$, $l \mapsto l/\lambda$). Therefore, a sphere embedded in three-dimensional pseudo-Euclidean space is a "plane" with hyperbolic geometry.

Let a time-like ***secant plane*** $\mathrm{nx} = 0$ pass through the coordinate origin (its normal has negative square: $\mathrm{n}^2 = -1$). The intersection of this plane with the upper hemisphere is a straight line in the hyperbolic plane (the second figure above). In contrast to lines on the Euclidean sphere, hyperbolic lines on the sphere of a pseudo-Euclidean space are open, and an infinite number of parallel lines can be drawn through a given point to a given line, if the point does not belong to this line.

▷ Let's find the equation of a straight line in the hyperbolic plane. Let orthogonal basis vectors e_0 and e_1 lie in a secant plane, and the third basis vector e_2 be normal to this plane (as before, $e_0^2 = 1$ and $e_1^2 = e_2^2 = -1$). A unit time-like vector

$$p = e_0 \operatorname{ch} l + e_1 \operatorname{sh} l, \qquad p^2 = 1, \tag{12.46}$$

connects the origin to a point on the hyperbolic line. Its derivative with respect to l is also a unit (space-like) vector, normal to p (and tangential to the sphere):

$$\dot{p} = \frac{dp}{dl} = e_0 \operatorname{sh} l + e_1 \operatorname{ch} l, \qquad \dot{p}^2 = -1, \qquad p\dot{p} = 0.$$

Thus, we have $(dp)^2 = -dl^2 = ds^2$ along the line. Therefore, the angle l between the vectors p and e_0 is equal to the length of the line (in terms of the metric (12.45)) between the points on the sphere which correspond to p and e_0.

▷ Let a segment of a hyperbolic line be given by points A and B, where $A^2 = B^2 = 1$. Its equation can easily be obtained from (12.46). Direct the vector e_0 along A and decompose the vector B in terms of the basis:

$$B = A \operatorname{ch} l_0 + e_1 \operatorname{sh} l_0, \qquad \operatorname{ch} l_0 = A \cdot B,$$

where l_0 is the length of the segment. Rearranging for e_1 and substituting into (12.46), we obtain ($l = [0 \dots l_0]$, $p(0) = A$ and $p(l_0) = B$):

$$p_{AB}(l) = A\,\frac{\operatorname{sh}(l_0 - l)}{\operatorname{sh} l_0} + B\,\frac{\operatorname{sh}(l)}{\operatorname{sh} l_0}. \tag{12.47}$$

▷ Consider a triangle ABC on the pseudosphere. Draw two tangential vectors at the vertex C in the directions of the sides A and B:

$$\dot{p}_{CA}(0) = \frac{A - C \operatorname{ch} b}{\operatorname{sh} b}, \qquad \dot{p}_{CB}(0) = \frac{B - C \operatorname{ch} a}{\operatorname{sh} a},$$

where the lengths of the triangle's sides are defined to be

$$AB = \operatorname{ch} c, \qquad AC = \operatorname{ch} b, \qquad BC = \operatorname{ch} a.$$

The angle γ between these vectors is the angle between the sides CA and CB of the triangle. Therefore, we have

$$\cos\gamma = -\dot{p}_{CA}(0)\dot{p}_{CB}(0) = -\frac{\operatorname{ch} c - \operatorname{ch} a \operatorname{ch} b}{\operatorname{sh} a \operatorname{sh} b},$$

which coincides with the law of cosines (12.33). Other trigonometric identities of hyperbolic geometry can be obtained similarly.

12.7 Coordinates in Lobachevsky space

We now consider an analogue of stereographic projection which takes the points of the upper hemisphere onto the (x, y)-plane. Let $\mathrm{k} = \{1, 0, 0\}$ be a unit time-like vector directed from the origin to the pole of the upper hemisphere (the upper hyperboloid in the Euclidean picture). Let another time-like vector p be directed to an arbitrary point on it:

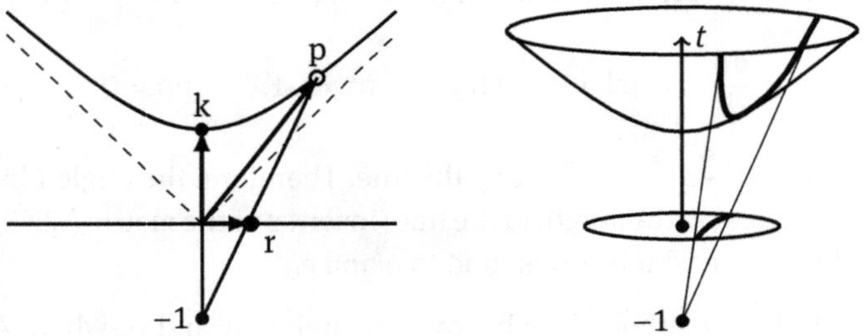

Draw a line through the lower pole $\{-1, 0, 0\}$ and the point p of the upper hemisphere. This line intersects the hemisphere and the (x, y)-plane, giving the stereographic projection of p onto the plane. The equation of the line is

$$\mathrm{x} = -\mathrm{k} + (\mathrm{p} + \mathrm{k})\,\tau.$$

The value $\tau = 0$ corresponds to the pole, and $\tau = 1$ corresponds to the point p. Since $\mathrm{kx} = 0$ in the plane, we have

$$\tau = \frac{1}{1 + \mathrm{kp}}.$$

As a result, the vector $\mathrm{r} = \mathrm{x}|_{t=0} = \{0, x, y\}$ of the projection of the point p satisfies:

$$\mathrm{r} = \frac{\mathrm{p} + \mathrm{k}}{1 + \mathrm{kp}} - \mathrm{k}, \qquad 1 + \mathrm{r}^2 = \frac{2}{1 + \mathrm{kp}}, \tag{12.48}$$

where the second relationship is obtained by squaring the first relationship. Note that $\mathrm{r}^2 = -x^2 - y^2 = -\mathbf{r}^2$, where $\mathbf{r}$ is a Euclidean vector.

Consider an arbitrary curve $\mathrm{p}(l)$ on the surface of the hemisphere, where l is the positive length coming from the metric (12.45). A vector tangential to the curve (and to the hemisphere) $\dot{\mathrm{p}} = d\mathrm{p}/dl$ is a unit space-like vector, $\dot{\mathrm{p}}^2 = -1$ (since $(d\mathrm{p})^2 = ds^2 = -dl^2$). Besides this, it is orthogonal to p: $\mathrm{p}\dot{\mathrm{p}} = 0$ (since $\mathrm{p}^2 = 1$). Thus, from (12.48) we obtain:

$$\dot{\mathrm{r}} = \frac{\dot{\mathrm{p}}}{1 + \mathrm{kp}} - \frac{(\mathrm{p} + \mathrm{k})\,(\mathrm{k}\dot{\mathrm{p}})}{(1 + \mathrm{kp})^2}, \qquad \dot{\mathrm{r}}^2 = -\frac{1}{(1 + \mathrm{kp})^2}. \tag{12.49}$$

This immediately gives us the element of length:

$$dl^2 = \frac{4\,(d\mathbf{r})^2}{(1 - \mathbf{r}^2)^2}. \tag{12.50}$$

This can also be obtained from (12.27), after the fundamental constant has been restored, $l \mapsto l/\lambda$, $\mathbf{r} \mapsto \mathbf{r}/\lambda$, and the substitution $\lambda^2 \mapsto -\lambda^2$ has been made.

▷ Let a space-like vector n be normal to a plane passing through the origin, which is then given by the equation np $= 0$, $\mathrm{n}^2 = -1$. The intersection of this plane and the upper hemisphere is a line in the hyperbolic plane. Let us find the curve obtained as the stereographic projection of this line. Multiplying (12.48) by n, we obtain

$$2(\mathrm{nr}) = (\mathrm{nk})\,(\mathrm{r}^2 - 1).$$

If nk $= 0$ (that is, if the time axis lies in the plane), then this is the equation of a line, $n_x x + n_y y = 0$, passing through the coordinate origin. Otherwise, we obtain the equation of a circle:

$$\left(x - \frac{n_x}{n_0}\right)^2 + \left(y - \frac{n_y}{n_0}\right)^2 = \frac{1}{n_0^2}, \tag{12.51}$$

where we have used the fact that $\mathrm{r} = \{0, x, y\}$, nk $= n_0$ and $n_0^2 - n_x^2 - n_y^2 = -1$. The points of the hemisphere are mapped onto the inside of the unit "boundary circle" (the bold line in the first figure below). Hyperbolic lines get mapped onto segments of circles (12.51), which are always orthogonal to the boundary circle (the Pythagorean theorem in the first figure):

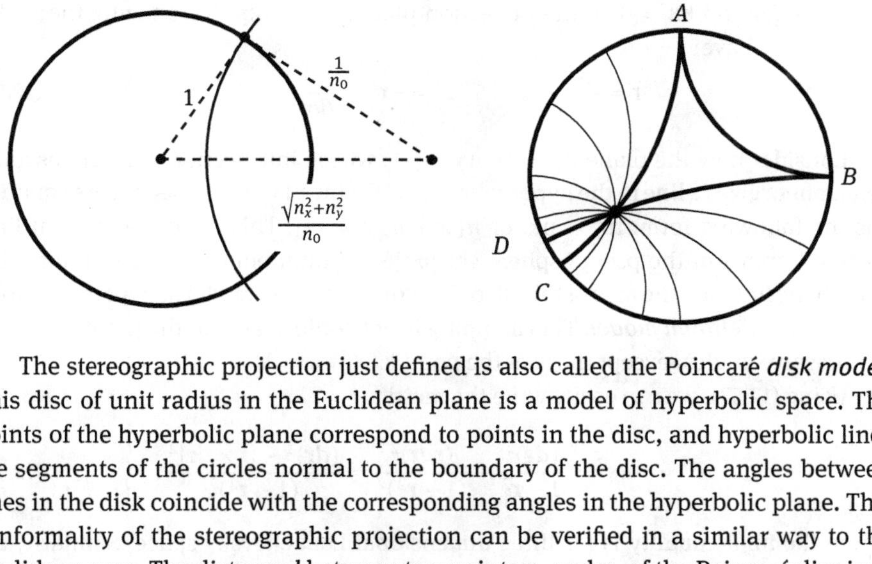

The stereographic projection just defined is also called the Poincaré *disk model.* This disc of unit radius in the Euclidean plane is a model of hyperbolic space. The points of the hyperbolic plane correspond to points in the disc, and hyperbolic lines are segments of the circles normal to the boundary of the disc. The angles between lines in the disk coincide with the corresponding angles in the hyperbolic plane. This conformality of the stereographic projection can be verified in a similar way to the Euclidean case. The distance l between two points $\mathbf{r}_1$ and $\mathbf{r}_2$ of the Poincaré disc is:

$$\operatorname{sh}\frac{l}{2} = \frac{|\mathbf{r}_1 - \mathbf{r}_2|}{\sqrt{(1 - \mathbf{r}_1^2)(1 - \mathbf{r}_2^2)}}. \tag{12.52}$$

This relationship is easily obtained from the relationship $\mathrm{p}_1\mathrm{p}_2 = \operatorname{ch} l$ and the mapping (12.48) between the points p_1 and p_2 in the hyperbolic plane and their projections $\mathbf{r}_1$ and $\mathbf{r}_2$ in the Poincaré disc.

The second figure above shows a number of lines passing through some point and parallel to the line AB ($\lessdot$ H_{66}). Their extremes are the lines AC and BD.

▷ Let us consider also hyperbolic Beltrami coordinates. Consider the plane tangential to the pole of a hemisphere in the pseudo-Euclidean space (1, 2). This the plane onto which we will project. Suppose that a line passes through the origin. It intersects the plane and the upper hemisphere (the first figure below shows the section):

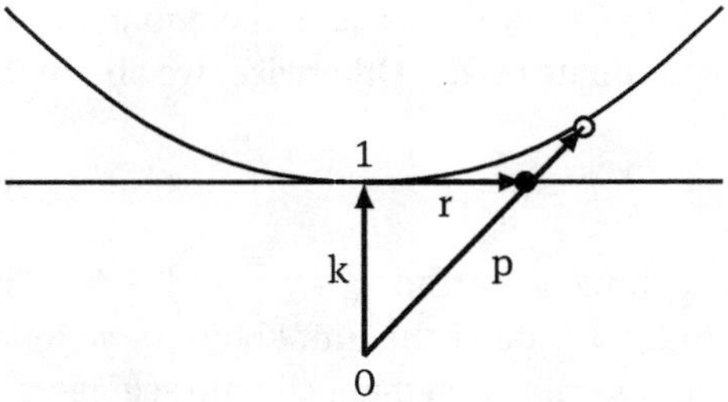

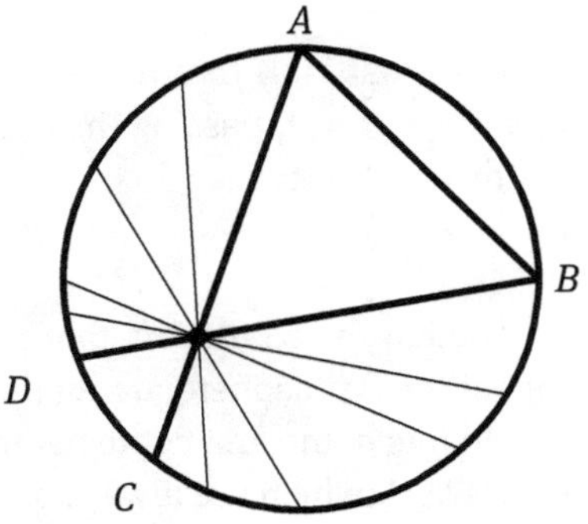

The reasoning is similar to that above. Substitute the equation of the line $x = p\tau$ into the equation $kx = 1$ of the projection plane: $\tau = 1/kp$. Introducing the vector $r = x - k$, we have:

$$r = \frac{p}{kp} - k, \qquad r^2 = -\mathbf{r}^2 = \frac{1}{(kp)^2} - 1. \tag{12.53}$$

Consider now the plane $np = 0$, $n^2 = -1$, whose intersection with the pseudo-hemisphere gives a line in the hyperbolic plane. On the plane of projection its equation has the following form: $nr = nk$, or $n_x x + n_y y = n_0$. This defines a straight line. Since all points of the pseudosphere get projected onto points in the unit disc, the hyperbolic lines are the arcs of this disc. The corresponding model of hyperbolic space is called the ***Beltrami model***. This mapping is not conformal, and the projection does not preserve angles, except those at the point k.

Using (12.53), we can easily write the metric:

$$dl^2 = -ds^2 = \frac{(d\mathbf{r})^2}{1 - \mathbf{r}^2} + \frac{(\mathbf{r}d\mathbf{r})^2}{(1 - \mathbf{r}^2)^2} = \frac{(d\mathbf{r})^2 - [\mathbf{r} \times d\mathbf{r}]^2}{(1 - \mathbf{r}^2)^2}, \tag{12.54}$$

where the final equality is for three-dimensional Lobachevsky space. Similarly, the metric can be represented in spherical coordinates:

$$dl^2 = d\chi^2 + \operatorname{sh}^2 \chi \, (d\theta^2 + \sin^2 \theta \, d\phi^2). \tag{12.55}$$

In this case, χ is the angle coming from the scalar product $kp = \operatorname{ch}\chi$, and θ and ϕ are the azimuth and the polar angles of an ordinary three-dimensional spherical coordinate system. This can be expressed in the other coordinate systems (12.50) and (12.54) by using the substitutions $r = \operatorname{th}(\chi/2)$ (stereographic projection) and $r = \operatorname{th}(\chi)$ (Beltrami projection).

•* Spaces of constant negative curvature can also be partially represented on a Euclidean surface. Suppose that a curve in the (ρ, z)-plane is constructed so that the following property is true: the length of a segment of tangent line AB between the tangency point and the horizontal z-axis is constant and is equal to unity for any point A on the curve (the first picture below).

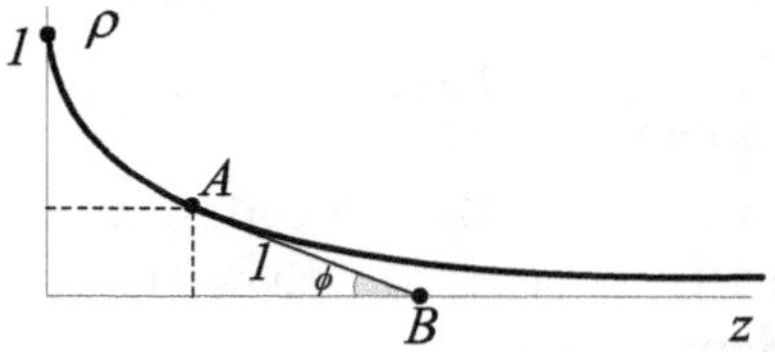

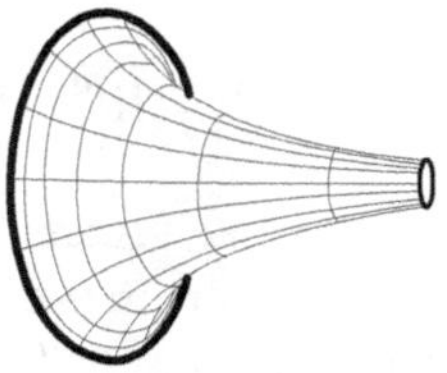

Such a curve is called a *tractrix*. Let us find its equation. The slope of the curve is equal to the derivative $\rho'(z) = -\operatorname{tg}\phi$ (with the opposite sign above). On the other hand, since $\operatorname{tg}\phi = \rho/\sqrt{1-\rho^2}$, we have:

$$z'(\rho) = -\frac{\sqrt{1-\rho^2}}{\rho}.$$

We now rotate this curve around the z-axis to form a ***surface of revolution***, depicted in the second figure. Using cylindrical coordinates (ρ, ϕ, z), we find the element of length:

$$dl^2 = d\rho^2 + \rho^2\,d\phi^2 + dz^2 = [1 + z'^2(\rho)]\,d\rho^2 + \rho^2\,d\phi^2 = \frac{d\rho^2}{\rho^2} + \rho^2\,d\phi^2 = \frac{du^2 + dv^2}{v^2},$$

where to obtain the final equality we have made the substitutions $v = 1/\rho$ and $u = \phi$. This metric can also be written in stereographic coordinates. In this case, the computations can be shortened if complex numbers are introduced via the linear fractional substitutions:

$$z = x + \imath y, \qquad w = u + iv, \qquad z = \frac{1+\imath w}{1-\imath w}, \qquad dz = \frac{2\imath\,dw}{(1-\imath w)^2}.$$

As a result:

$$dl^2 = \frac{dx^2 + dy^2}{(1 - x^2 - y^2)^2/4} = \frac{4dz\,dz^*}{(1 - zz^*)^2} = \frac{-4\,dw\,dw^*}{(w^* - w)^2} = \frac{du^2 + dv^2}{v^2}.$$

The constructed surface covers only a part of Lobachevsky space contained between two parallel lines. These lines run along the axial section of the surface (in the figure, any "horizontal" line with $\phi = const$). In particular, the surface has an edge $(z = 0)$, and any triangle will sooner or later cease to fit on the surface as it moves in the direction of the taper.

12.8 Velocity space

So far we have considered an abstract pseudo-Euclidean space and a surface (hypersurface) $p^2 = 1$ in it which possesses a hyperbolic geometry. Let us now return to relativity theory.

Recall (p. 178) that a four-vector of velocity $U = dx/ds$ with components

$$U^\alpha = \{U^0, \mathbf{U}\} = \left\{\frac{1}{\sqrt{1-\mathbf{u}^2}}, \frac{\mathbf{u}}{\sqrt{1-\mathbf{u}^2}}\right\} = \left\{\mathrm{ch}\,\alpha, \frac{\mathbf{u}}{|\mathbf{u}|}\,\mathrm{sh}\,\alpha\right\} \tag{12.56}$$

is a unit time-like vector: $U^2 = 1$. If drawn from the origin, it corresponds to some point on the sphere in pseudo-Euclidean space. The set of all such points is called the *velocity space*. From the above considerations it follows that this space possesses a hyperbolic geometry.

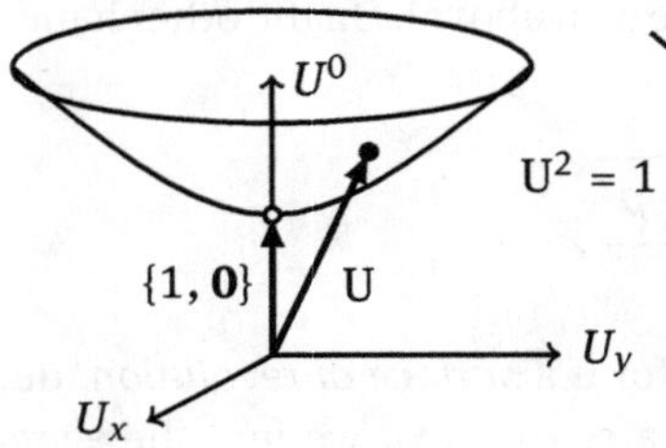

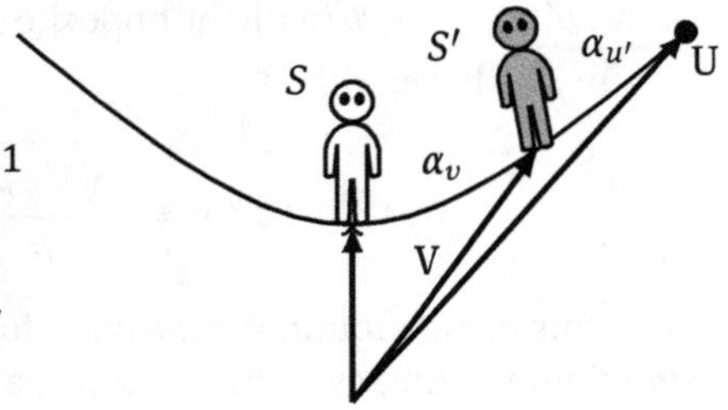

▷ Let an observer in a frame S consider themself at rest, and let them measure the velocities of other objects. Introduce a vector $K = \{1, \mathbf{0}\}$ of four-velocity with zero three-speed. Let there be a moving observer (attached to the frame S') at some other point of the velocity space defined by the unit vector V. Finally, let some object move with four-velocity U.

The parameter α in (12.56) is the angle between four-vectors K and U. It is also equal to the distance between the points on the sphere corresponding to these four-vectors. From (12.56) it follows that $|\mathbf{u}| = \mathrm{th}\,\alpha$. If the four-vectors K, V and U are in the same plane, then $\alpha_u = \alpha_v + \alpha'_u$, where

$$KU = \mathrm{ch}\,\alpha_u, \qquad KV = \mathrm{ch}\,\alpha_v, \qquad VU = \mathrm{ch}\,\alpha_{u'}.$$

A standard hyperbolic identity,

$$\mathrm{th}(\alpha_u) = \mathrm{th}(\alpha_v + \alpha_{u'}) = \frac{\mathrm{th}(\alpha_v) + \mathrm{th}(\alpha_{u'})}{1 + \mathrm{th}(\alpha_v)\,\mathrm{th}(\alpha_{u'})} = \frac{v + u'}{1 + vu'}$$

gives the relativistic velocity-addition law.

▷ Let us show how the law of cosines in hyperbolic geometry can be derived from the vector velocity-addition law. Write the expression for squared velocities (1.23), p. 20:

$$\gamma_{u'} = \gamma_u \gamma_v\,(1 - \mathbf{vu}) \quad \Leftrightarrow \quad 1 - \mathbf{u}'^2 = \frac{(1-\mathbf{u}^2)\,(1-\mathbf{v}^2)}{(1-\mathbf{vu})^2} \tag{12.57}$$

and use the following notation (the first figure below):

$$v = \operatorname{th} a, \qquad u = \operatorname{th} b, \qquad u' = \operatorname{th} c, \qquad \mathbf{vu} = \operatorname{th} a \operatorname{th} b \cos\gamma.$$

Given that $1 - \operatorname{th}^2 x = 1/\operatorname{ch}^2 x$, we have

$$\operatorname{ch}^2 c = \operatorname{ch}^2 a \operatorname{ch}^2 b\,(1 - \operatorname{th} a \operatorname{th} b \cos\gamma)^2.$$

After extracting the root, we obtain the law of cosines:

$$\operatorname{ch} c = \operatorname{ch} a \operatorname{ch} b - \operatorname{sh} a \operatorname{sh} b \cos\gamma.$$

The angle γ between the vectors $\mathbf{v}$ and $\mathbf{u}$ is constant on the surface of the sphere and on the plane tangential to the sphere *at the point of tangency* (the second figure):

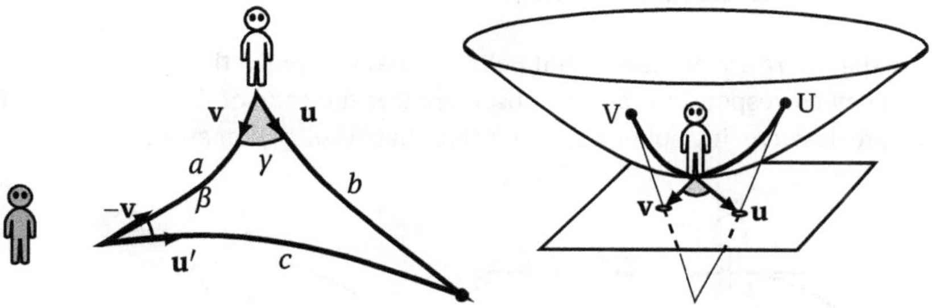

▷ Consider also the angle β, where $-\mathbf{vu}' = \operatorname{th} a \operatorname{th} c \cos\beta$, and the special case of a right triangle ($\gamma = \pi/2$ or $\mathbf{uv} = 0$). Then, from the velocity-addition law (1.24), p. 20, it follows (after multiplication by $\mathbf{v}$) that:

$$\mathbf{u}' = \frac{\mathbf{u}}{\gamma} - \mathbf{v} \qquad \Rightarrow \qquad \mathbf{v}^2 = -\mathbf{vu}' = u' v \cos\beta.$$

In hyperbolic notation this gives the projection of the hypotenuse c onto the leg a:

$$\operatorname{th} a = \operatorname{th} c \cos\beta.$$

Note that the angle γ which touches the tangency point is the only angle which remains the same under projection onto the tangent plane. Other angles are distorted as a result of the mapping. So is the hypotenuse c, which is equal to $\operatorname{ath} u'$ only after projection onto the plane tangential to S'.

▷ Let us show by an explicit calculation that the usual three-velocity $\mathbf{u}$ is a vector of a three-dimensional hypersurface in *Beltrami coordinates*. Rewrite the transformation for the squared velocity (12.57) as follows:

$$\mathbf{u}'^2 = \frac{(\mathbf{u} - \mathbf{v})^2 - [\mathbf{u} \times \mathbf{v}]^2}{(1 - \mathbf{uv})^2}. \tag{12.58}$$

Assuming $\mathbf{u} = \mathbf{v} + d\mathbf{v}$, we have that the numerator is proportional to $d\mathbf{v}$. Since this velocity is small, we can neglect $d\mathbf{v}$ in the denominator, up to the first order of smallness, and assume that $1 - \mathbf{uv} \approx 1 - \mathbf{v}^2$. As a result:

$$dl^2 = \mathbf{u}'^2 = \frac{d\mathbf{v}^2 - [\mathbf{v} \times d\mathbf{v}]^2}{(1 - \mathbf{v}^2)^2}.$$

The squared velocity $(\mathbf{u}')^2 = dl^2$ coincides with the distance in Lobachevsky space.

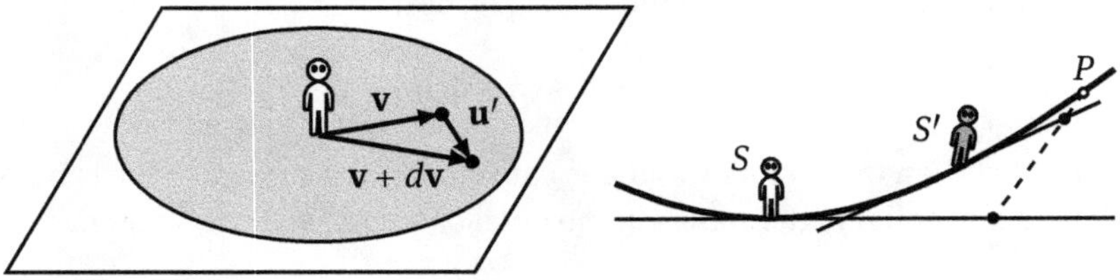

⊳ Note that there are *two* tangential Beltrami planes used in the velocity-addition formula. They correspond to the two observers (for the sake of illustration, the figures below are drawn using spherical rather than hyperbolic geometry):

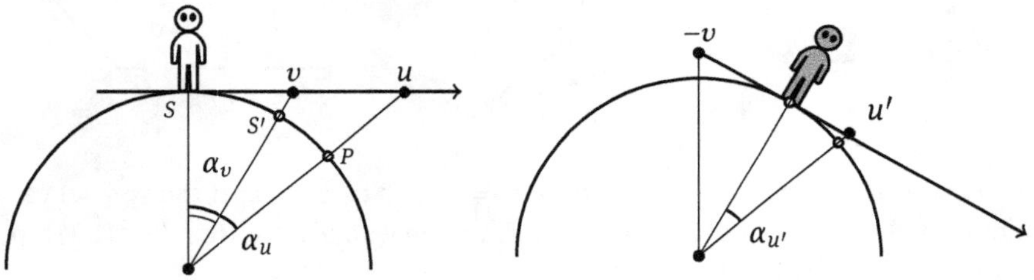

The first figure shows the tangential plane of the frame S, and the second shows the same for the frame S'. The projection of an arc $S'P$ onto the plane of the resting observer S is not equal to þ$\alpha_{u'}$. The required relationship between the distances on the "sphere" $(\alpha_v, \alpha_u, \alpha_{u'})$ and the coordinate values

$$v = \operatorname{th} \alpha_v, \qquad u = \operatorname{th} \alpha_u \quad \text{and} \quad u' = \operatorname{th} \alpha_{u'}$$

(which are physical velocities) is true *only* with respect to the point connected with the given frame.

⊳ Velocity space provides another method of axiomatically justifying the theory of relativity. The relativity principle asserts the equality of all inertial frames of reference. Geometrically, this means that the space of velocities should be homogeneous and isotropic (having no preferential directions or points). According to differential geometry, there can be three, and only three, such spaces: 1) Euclidean space (zero curvature); 2) a space with positive curvature (spherical geometry); 3) a space with negative curvature (hyperbolic geometry). Qualitatively, this is clear. If the curvature of the space changes (for example, if a sphere has dents on it), a rigid figure (say, a triangle) cannot move in this space without being subject to distortions.

Euclidean space is a special case of the spaces with non-zero curvature, obtained as the radius $\lambda \to \infty$. Topologically, it is closer to Lobachevsky space, since it has infinite volume (is not closed). This is the principle of parametric incompleteness (p. 31) in action, as Euclidean geometry requires more axioms than, e.g., hyperbolic geometry, where Euclid's fifth postulate is rejected. As a result of having less axiomatic information, we obtain a new fundamental constant with dimensions of velocity and the theory of relativity itself.

M. Mathematical appendix

This appendix is not a mathematics textbook. However, the material may be useful for those who want to get acquainted with the notation used in this book, or to recall what was once learned and forgotten. The reader is advised to refer to the book [1] and to solve the problems in it to study these questions in more detail.

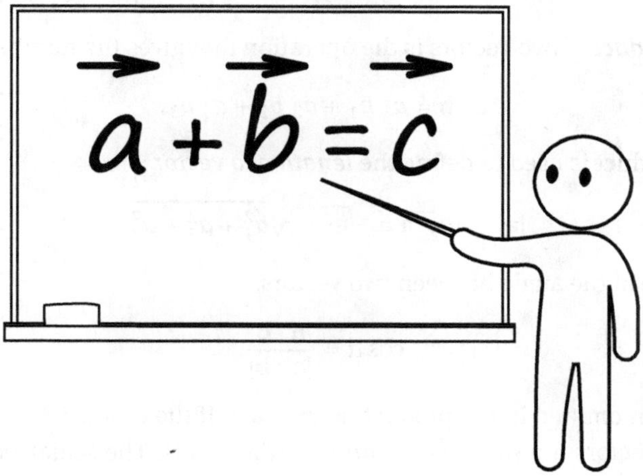

https://doi.org/10.1515/9783110515886-013

M.1 Vectors

A *vector* in three-dimensional space is an array of three real numbers

$$\mathbf{a} = \{a_1, a_2, a_3\},$$

which are called the components of the vector. For example, the velocity of a particle is a vector.

Vectors can be added (in a component-wise manner) and multiplied by a number to obtain a new vector:

$$\mathbf{a} + \mathbf{b} = \{a_1 + b_1,\ a_2 + b_2,\ a_3 + b_3\}, \qquad \mu\,\mathbf{a} = \{\mu\, a_1,\ \mu\, a_2,\ \mu\, a_3\}.$$

The *scalar product* of two vectors is the operation that gives the number:

$$\mathbf{a} \cdot \mathbf{b} = a_1\, b_1 + a_2\, b_2 + a_3\, b_3. \tag{1}$$

The scalar product is used to define the *length of a vector*:

$$|\mathbf{a}| = \sqrt{\mathbf{a} \cdot \mathbf{a}} \equiv \sqrt{\mathbf{a}^2} = \sqrt{a_1^2 + a_2^2 + a_3^2}, \tag{2}$$

and the cosine of the angle between two vectors:

$$\cos\alpha = \frac{\mathbf{a} \cdot \mathbf{b}}{|\mathbf{a}|\,|\mathbf{b}|}. \tag{3}$$

The dot is often omitted in the product: $\mathbf{a} \cdot \mathbf{b} \equiv \mathbf{a}\,\mathbf{b}$. If the cosine is zero (for non-zero vectors), the vectors are said to be *normal* (orthogonal). The scalar product of such vectors is zero: $\mathbf{a} \cdot \mathbf{b} = 0$.

▷ Geometrically, a vector **a** has the meaning of a directed line segment with projections $\{a_x,\ a_y,\ a_z\}$ on the x-, y- and z-axes of the *Cartesian coordinate system*:

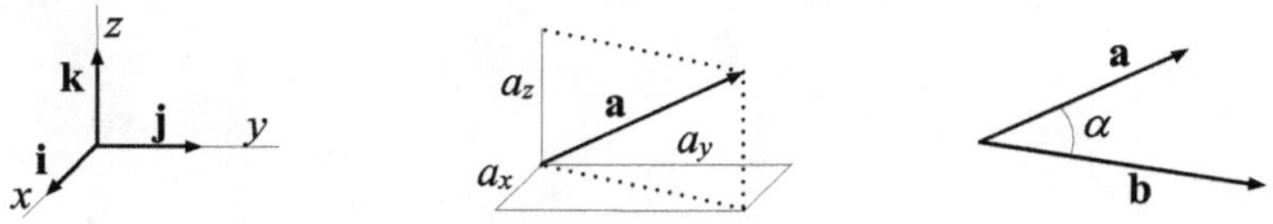

It is convenient to associate with the axes three mutually *normal unit* vectors **i**, **j**, **k**, or the *basis*:

$$\mathbf{i}^2 = \mathbf{j}^2 = \mathbf{k}^2 = 1, \qquad \mathbf{i} \cdot \mathbf{j} = \mathbf{i} \cdot \mathbf{k} = \mathbf{j} \cdot \mathbf{k} = 0. \tag{4}$$

Any vector can be decomposed with respect to this basis and expressed as a linear combination:

$$\mathbf{a} = a_x\,\mathbf{i} + a_y\,\mathbf{j} + a_z\,\mathbf{k}, \tag{5}$$

where $a_x = \mathbf{a} \cdot \mathbf{i}$, $a_y = \mathbf{a} \cdot \mathbf{j}$, $a_z = \mathbf{a} \cdot \mathbf{k}$ are the projections of **a** onto the x-, y- and z- axes. By writing the vectors **a** and **b** in this form, we obtain: $\mathbf{a} \cdot \mathbf{b} = a_x b_x + a_y b_y + a_z b_z$.

▷ Geometrically, the addition of vectors is described by the *parallelogram law*, and multiplication by a number μ makes a vector μ times longer:

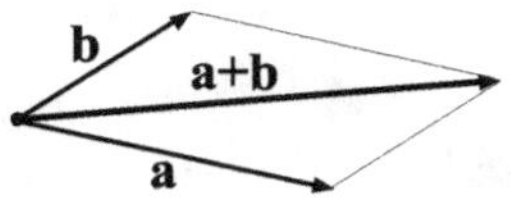

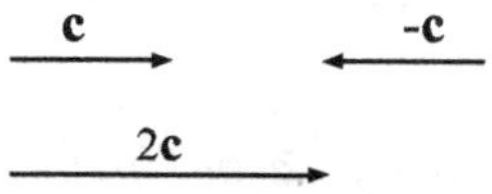

The parallelogram law has a simple physical interpretation. If two mice drag a piece of cheese in different directions with different forces, **a** and **b**, the resulting effort will be directed diagonally as $\mathbf{a} + \mathbf{b}$.

▷ Algebraically, vectors are objects which carry the operations of addition and multiplication by a number, both of which return another vector. Vectors can also be multiplied by each other to obtain a number. The properties of these *operations* are given by axioms. For example, scalar multiplication obeys the axioms of symmetry

$$\mathbf{a} \cdot \mathbf{b} = \mathbf{b} \cdot \mathbf{a}$$

and distributivity

$$\mathbf{a} \cdot (\mathbf{b} + \mathbf{c}) = (\mathbf{a} \cdot \mathbf{b}) + (\mathbf{a} \cdot \mathbf{c}).$$

The parentheses emphasize the order of operations. The order of multiplication is important if there are several scalar products, since in the general case

$$(\mathbf{a} \cdot \mathbf{b})\,\mathbf{c} \neq \mathbf{a}\,(\mathbf{b} \cdot \mathbf{c}).$$

In fact, the vector **c** is multiplied by the *number* $(\mathbf{a} \cdot \mathbf{b})$ in the first case, while *another vector* **a** is multiplied by the number $(\mathbf{b} \cdot \mathbf{c})$ in the second case. If **c** and **a** have different directions, they cannot be made equal by multiplying by numbers.

▷ Similarly, a vector **a** in a space of arbitrary dimension n is defined as an array of n numbers $\{a_1, a_2, \ldots, a_n\}$ with the same properties as in three-dimensional space. If the array of n vectors

$$\mathbf{e}_1\,, \ldots, \,\mathbf{e}_n$$

is a *complete* system of *linearly independent* vectors, it is called a *basis*. Linear independence means that none of the basis vectors can be represented as a linear combination of the others. Completeness ensures that any arbitrary vector can be decomposed in terms of the basis:

$$\mathbf{a} = a_1\,\mathbf{e}_1 + \ldots + a_n\,\mathbf{e}_n = \sum_{i=1}^{n} a_i\,\mathbf{e}_i.$$

The *dimension of the space* n is the number of basis vectors. In general, basis vectors are not necessarily normal or of unit length. If they are, the coordinate system defined by the basis is called *Cartesian*.

▷ A *vector product* can be defined for any two vectors in a three-dimensional space. In contrast to the scalar product (which is a number), the vector product is a vector, with the following components:

$$\mathbf{a} \times \mathbf{b} = \{\, a_2 b_3 - a_3 b_2,\; a_3 b_1 - a_1 b_3,\; a_1 b_2 - a_2 b_1 \,\} = \det \begin{pmatrix} \mathbf{i} & \mathbf{j} & \mathbf{k} \\ a_1 & a_2 & a_3 \\ b_1 & b_2 & b_3 \end{pmatrix},$$

where **i, j** and **k** are basis vectors, and det is the determinant of the matrix (p. 339). The vector product is *associative*:

$$\mathbf{a} \times (\mathbf{b} + \mathbf{c}) = [\mathbf{a} \times \mathbf{b}] + [\mathbf{a} \times \mathbf{c}], \tag{6}$$

however, in contrast to the scalar product, it is *antisymmetric*:

$$[\mathbf{a} \times \mathbf{b}] = -[\mathbf{b} \times \mathbf{a}]. \tag{7}$$

In particular, the vector product of a vector with itself is the *zero vector*

$$\mathbf{0} = \{0, 0, 0\}$$

(with zero length):

$$\mathbf{a} \times \mathbf{a} = \mathbf{0}. \tag{8}$$

▷ For vectors **a** and **b** directed along parallel lines there exists a number μ such that

$$\mathbf{b} = \mu\, \mathbf{a}.$$

Their vector product is zero: $\mathbf{a} \times \mathbf{b} = \mathbf{0}$. Otherwise, the vector $\mathbf{a} \times \mathbf{b}$ is normal to the plane containing the vectors **a** and **b**:

$$\mathbf{a} \cdot [\mathbf{a} \times \mathbf{b}] = a_1(a_2 b_3 - a_3 b_2) + a_2(a_3 b_1 - a_1 b_3) + a_3(a_1 b_2 - a_2 b_1) = 0.$$

▷ Similarly, the following property of the *mixed product* (combination of the scalar and vector products) can be verified component-wise:

$$\mathbf{a} \cdot [\mathbf{b} \times \mathbf{c}] = [\mathbf{a} \times \mathbf{b}] \cdot \mathbf{c}. \tag{9}$$

This property can be called the *"push" rule*: the scalar product "pushes" one vector from the vector product and all vectors are shifted to the right. The rule works both from left to right and from right to left (if read in the opposite direction).

For the vector triple product the following "*BAC - CAB*" rule is true:

$$\mathbf{a} \times [\mathbf{b} \times \mathbf{c}] = \mathbf{b}\,(\mathbf{ac}) - \mathbf{c}\,(\mathbf{ab}). \tag{10}$$

It can also be verified by computing component-wise.

▷ Another important identity concerns the scalar product of two vector products:

$$[\mathbf{a} \times \mathbf{b}] \cdot [\mathbf{c} \times \mathbf{d}] = (\mathbf{ac})(\mathbf{bd}) - (\mathbf{ad})(\mathbf{bc}). \tag{11}$$

The first vectors inside the vector products on the right-hand side are multiplied by each other, and then the same is done with the second vectors. Then a similar "crosswise" product (the first vectors multiplied by second vectors) is subtracted. This identity is easily obtained from the previous ones by sequential use of the "push" rule, the antisymmetry rule, and the "BAC - CAB" rule. The square of a vector product is a special case of this identity:

$$[\mathbf{a} \times \mathbf{b}]^2 = \mathbf{a}^2\,\mathbf{b}^2 - (\mathbf{ab})^2 = \mathbf{a}^2\,\mathbf{b}^2 \sin^2 \alpha. \tag{12}$$

The latter equality is written using the angle α between the vectors, defined in terms of the scalar product:

$$\mathbf{ab} = |\mathbf{a}|\,|\mathbf{b}| \cos \alpha.$$

Thus, the magnitude of a vector product is the area of the parallelogram formed by the vectors **a** and **b** (the height and base of the parallelogram in the first figure are $|\mathbf{b}| \sin \alpha$ and $|\mathbf{a}|$, respectively):

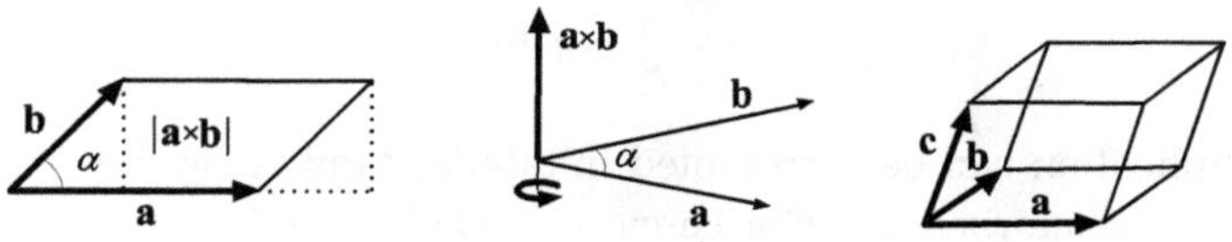

The vector product **a**×**b** is normal to the vectors **a** and **b**, and its direction is determined by the *"right-hand screw" rule* (or simply the *"right-hand" rule*). If a screw is twisted from **a** towards **b**, it will move in the direction $\mathbf{a} \times \mathbf{b}$. For example, the basis vectors in the usual Cartesian axes (x, y, z) constitute a *right-handed triple*, so that:

$$\mathbf{k} = \mathbf{i} \times \mathbf{j}, \qquad \mathbf{j} = \mathbf{k} \times \mathbf{i}, \qquad \mathbf{i} = \mathbf{j} \times \mathbf{k}. \tag{13}$$

The mixed product

$$V = [\mathbf{a} \times \mathbf{b}] \cdot \mathbf{c}$$

is the volume V of the parallelepiped built from the vectors **a, b, c.** This volume is positive if (**a, b, c**) is a right-handed triple as shown in the third figure above. Interchanging any pair of vectors changes the sign of the mixed product.

M.2 Matrices

A *matrix* is an array of numbers arranged in the form of a rectangular table. While the components of a vector have only one index a_i, matrix elements have two indices: a_{ij}.

Consider a *square matrix* **A** of size $n \times n$:

$$\mathbf{A} = \begin{pmatrix} a_{11} & a_{12} & \dots & a_{1n} \\ a_{21} & a_{22} & \dots & a_{2n} \\ \vdots & \vdots & \ddots & \vdots \\ a_{n1} & a_{n2} & \dots & a_{nn} \end{pmatrix}.$$

The first index of a_{ij} grows when passing to the next line, and the second does the same when passing to the next column. The numbers a_{ij} are called *matrix elements*. The indices of *diagonal elements* coincide: a_{11}, a_{22}, ... The number of rows (or columns) is called the dimension of the matrix. A *zero matrix* is a matrix, all of whose elements are zero.

▷ Matrices can be added element-wise and multiplied by numbers. The result of these operations is:

$$\mathbf{C} = \alpha\,\mathbf{A} + \beta\,\mathbf{B} \quad \Rightarrow \quad c_{ij} = \alpha\,a_{ij} + \beta\,b_{ij}.$$

Besides this, there is a way of multiplying matrices by one another to get another *matrix*. The product of the matrices **A** and **B** with elements a_{ij} and b_{ij}, respectively, is the matrix **C** with elements

$$c_{ij} = \sum_{k=1}^{n} a_{ik}\,b_{kj}. \tag{14}$$

Matrix multiplication can be represented in tabular form using the *"crowbar" rule*. The i-th row of the matrix **A** (crowbar) breaks a "hole" in the j-th column of the matrix **B** (wall) to obtain c_{ij}:

$$\mathbf{C} = \mathbf{A}\cdot\mathbf{B} = \begin{pmatrix} a_{11} & a_{12} & \dots & a_{1n} \\ \mathbf{a_{21}} & \mathbf{a_{22}} & \dots & \mathbf{a_{2n}} \\ \vdots & \vdots & \ddots & \vdots \\ a_{n1} & a_{n2} & \dots & a_{nn} \end{pmatrix} \cdot \begin{pmatrix} b_{11} & \mathbf{b_{12}} & \dots & b_{1n} \\ b_{21} & \mathbf{b_{22}} & \dots & b_{2n} \\ \vdots & \vdots & \ddots & \vdots \\ b_{n1} & \mathbf{b_{n2}} & \dots & b_{nn} \end{pmatrix}.$$

After that, the crowbar and the wall elements are multiplied component-wise by each other and the sum of these products is written in the hole of the matrix **C**:

$$c_{ij} = a_{i1}b_{1j} + a_{i2}b_{2j} + \ldots + a_{in}b_{nj}.$$

This operation is repeated for each element of the matrix **C**.

▷ As follows from definition (14), matrix multiplication is *associative*:

$$(\mathbf{A}\cdot\mathbf{B})\cdot\mathbf{C} = \mathbf{A}\cdot(\mathbf{B}\cdot\mathbf{C}), \tag{15}$$

but is *not commutative* (the order of the matrices cannot be changed):

$$\mathbf{A}\cdot\mathbf{B} \neq \mathbf{B}\cdot\mathbf{A}. \tag{16}$$

In other words, the situation is exactly opposite to the scalar product of vectors! Two matrices are said to *commute* if their product does not depend on the order of multiplication.

▷ The *transpose* $\mathbf{A}^T$ of a matrix $\mathbf{A}$ is a matrix obtained from $\mathbf{A}$ by switching its columns and rows:

$$\mathbf{A}^T = \begin{pmatrix} a_{11} & a_{21} & \dots & a_{n1} \\ a_{12} & a_{22} & \dots & a_{n2} \\ \vdots & \vdots & \ddots & \vdots \\ a_{1n} & a_{2n} & \dots & a_{nn} \end{pmatrix}.$$

Thus, if the elements of $\mathbf{A}$ are a_{ij}, its transpose $\mathbf{A}^T$ is obtained by the permutation of indices:

$$a_{ij}^T = a_{ji}.$$

A matrix is called *symmetric* if it is equal to its transpose:

$$a_{ij} = a_{ji}.$$

Sometimes, the matrix coefficients are related by

$$a_{ij} = -a_{ji}.$$

Such matrices are called *skew-symmetric.*

The contraction (sum of the products of the elements) of a symmetric and a skew-symmetric matrix over two indices is zero:

$$\sum_{i,j} s_{ij}\, a_{ij} = \sum_{i,j} s_{ji}\, a_{ji} = -\sum_{i,j} s_{ij}\, a_{ij} = 0, \tag{17}$$

where the indices are renamed in the first equality ($i \mapsto j$, $j \mapsto i$) and switched due to symmetry in the second equality: $s_{ij} = s_{ji}$, $a_{ij} = -a_{ji}$. A quantity equal to itself with opposite sign can only be zero.

▷ Of special importance is the *identity matrix*, with ones on the main diagonal and zeros elsewhere. For example:

$$\mathbf{1} = \begin{pmatrix} 1 & 0 & 0 \\ 0 & 1 & 0 \\ 0 & 0 & 1 \end{pmatrix}.$$

No matrix $\mathbf{A}$ is changed when multiplied by the identity matrix:

$$\mathbf{1} \cdot \mathbf{A} = \mathbf{A} \cdot \mathbf{1} = \mathbf{A}. \tag{18}$$

The identity commutes with every other matrix.

▷ The matrix $\mathbf{A}^{-1}$ is the *inverse* of the matrix $\mathbf{A}$ if the following equation is satisfied:

$$\mathbf{A}^{-1} \cdot \mathbf{A} = \mathbf{A} \cdot \mathbf{A}^{-1} = \mathbf{1}. \tag{19}$$

The matrix $\mathbf{A}$ and its inverse $\mathbf{A}^{-1}$ can be interchanged (commuted) when multiplied. Indeed, assume that $\mathbf{A}\cdot\mathbf{A}^{-1} = \mathbf{1}$ and prove that $\mathbf{A}^{-1}\cdot\mathbf{A} = \mathbf{1}$. Multiplying this equality from the left by $\mathbf{A}$, we have:

$$\mathbf{A}\cdot(\mathbf{A}^{-1}\cdot\mathbf{A}) = \mathbf{A}\cdot\mathbf{1} = \mathbf{A} \quad\Rightarrow\quad (\mathbf{A}\cdot\mathbf{A}^{-1})\cdot\mathbf{A} = \mathbf{1}\cdot\mathbf{A} = \mathbf{A}.$$

Another identity,

$$(\mathbf{A}\cdot\mathbf{B})^{-1} = \mathbf{B}^{-1}\cdot\mathbf{A}^{-1}, \tag{20}$$

can be verified by multiplying both sides by $\mathbf{A}\cdot\mathbf{B}$ and using (19), (15).

▷ The *trace* of a matrix is the sum of its diagonal elements:

$$\mathrm{Tr}\,\mathbf{A} = a_{11} + a_{22} + \ldots + a_{nn}. \tag{21}$$

The trace of the identity matrix is equal to its dimension. It is worth verifying that the order of matrices can be switched under the trace sign: $\mathrm{Tr}(\mathbf{A}\cdot\mathbf{B}) = \mathrm{Tr}(\mathbf{B}\cdot\mathbf{A})$.

▷ Matrix functions are understood in in terms of power series. For example, the exponent is an abbreviated record of the following infinite series:

$$e^{\mathbf{A}} = \mathbf{1} + \frac{\mathbf{A}}{1!} + \frac{\mathbf{A}^2}{2!} + \frac{\mathbf{A}^3}{3!} + \ldots,$$

where $\mathbf{A}^2 = \mathbf{AA}$, etc. Another decomposition:

$$(1 - \mathbf{A})^{-1} = \mathbf{1} + \mathbf{A} + \mathbf{A}^2 + \ldots$$

can be verified by multiplying the right-hand and the left-hand side by $(1 - \mathbf{A})$.

▷ The elements of the identity matrix have their own special notation suggested by Kronecker:

$$\delta_{ij} = \begin{cases} 1 & \text{if } i = j \\ 0 & \text{if } i \neq j. \end{cases} \tag{22}$$

The symbol δ_{ij} "eats up" the summation and "dies", to replace the summation index by its second index:

$$\sum_{j=1}^{n} \delta_{ij}\, a_j = a_i. \tag{23}$$

All terms of the sum over j will be zero, except those with $j = i$. Obviously, the Kronecker symbol is symmetric: $\delta_{ij} = \delta_{ji}$.

▷ Matrices are a special case of *index* symbols. In general there can be more indices, for example a_{ijk}. In three-dimensional space, each index ranges over the values 1,2,3. Vectors are also indexed objects. Their components $a_i = \{a_1, a_2, a_3\}$ can be represented in tabular form as a row or a column. Indexed expressions are also called ***tensors***. However, besides having indices, tensors should transform in a certain way under coordinate changes (p. 349).

Einstein's notation, which implies summation over repeated indices without using the summation sign, is a convenient way to make formulas shorter when working with index expressions. In this notation, the multiplication of a matrix by a vector can be written in three equivalent ways:

$$\sum_{j=1}^{3} a_{ij}\, b_j = a_{i1} b_1 + a_{i2} b_2 + a_{i3} b_3 = a_{ij}\, b_j .$$

The summation is not written but is implied, in the last equation, since the index j repeats.

The scalar product of two vectors **a** and **b** with components

$$a_i = \{a_1, a_2, a_3\} \quad \text{and} \quad b_i = \{b_1, b_2, b_3\}$$

can be written in index form as follows:

$$\mathbf{a} \cdot \mathbf{b} = a_i\, b_i .$$

The summation rule for the Kronecker symbol and some a_{ijk} is: $a_{ijk}\, \delta_{kq} = a_{ijq}$.

The summation index, just like an integration variable, can be denoted by any letter which is not used as an index in the expression. Because of this interchangeability, such indices are called *dummy indices*. For example:

$$a_{ij} b_j = a_{ik} b_k$$

(k is substituted for j), but

$$a_{ij} b_j \neq a_{ii} b_i$$

(this is impossible, since the index i is already used in the expression).

▷ When differentiating index expressions, the components of a radius vector are conveniently written as $\mathbf{r} = \{x_1, x_2, x_3\}$. Then

$$\frac{\partial x_j}{\partial x_i} \equiv \partial_i x_j = \delta_{ij}, \qquad \partial_i \equiv \frac{\partial}{\partial x_i} . \tag{24}$$

For indices which coincide the derivative is unity, for example, $\partial x_1/\partial x_1 = 1$. Otherwise the derivative is zero. Suppose as an example that $a_i = const$. Then,

$$\partial_i (x_j a_j) = \delta_{ij}\, a_j = a_i .$$

In vector notation, $\nabla(\mathbf{ra}) = \mathbf{a}$, where the del operator (see Volume II) has components ∂_i, and the a_i are the components of vector **a**.

M.3 The determinant

For a square 2×2 matrix, a number called the *determinant of the matrix* is defined as:

$$\det \mathbf{A} \equiv \det \begin{pmatrix} a_{11} & a_{12} \\ a_{21} & a_{22} \end{pmatrix} \equiv \begin{vmatrix} a_{11} & a_{12} \\ a_{21} & a_{22} \end{vmatrix} = a_{11} a_{22} - a_{12} a_{21} . \tag{25}$$

The determinant is calculated by taking the "cross-wise" products of the matrix elements and subtracting them from each other. The determinant of the product of two matrices is the product of their determinants:

$$\det(\mathbf{A} \cdot \mathbf{B}) = \det \mathbf{A} \det \mathbf{B}. \tag{26}$$

This is a very interesting property. Multiplication of matrices is a specific procedure (the "crowbar" rule). Therefore, it is quite unexpected that each matrix can be related to a number such that the usual arithmetic rule of multiplication (26) is satisfied.

A determinant having the same property can also be defined for 3×3 matrices:

$$\det \mathbf{A} = a_{11} \det \begin{pmatrix} a_{22} & a_{23} \\ a_{32} & a_{33} \end{pmatrix} - a_{12} \det \begin{pmatrix} a_{21} & a_{23} \\ a_{31} & a_{33} \end{pmatrix} + a_{13} \det \begin{pmatrix} a_{21} & a_{22} \\ a_{31} & a_{32} \end{pmatrix}.$$

It is calculated by taking each matrix element of the first line and deleting the row and column which contain this element. Then, the determinant of the resulting 2×2 matrix should be multiplied by this element. All such products are then summed, with plus signs if the sum of the line number and the column number is even, and with minus signs otherwise.

This definition can be extended to matrices of arbitrary size N using recursive calculations to reduce their determinants to the sum (or the difference) of the determinants of matrices of size $N - 1$. The determinant of a 2×2 matrix also obeys this rule, if the determinant of a 1×1 "matrix" is assumed equal to its element.

The determinant of a *diagonal matrix* is the product of its diagonal elements:

$$\det \begin{pmatrix} a_1 & 0 & 0 \\ 0 & a_2 & 0 \\ 0 & 0 & \ddots \end{pmatrix} = a_1 \cdot a_2 \cdot \ldots$$

The product of two diagonal matrices is again a diagonal matrix, and, obviously, the rule (26) is also satisfied.

▷ A number of useful properties can be used to calculate determinants:

- Transposition does not change the determinant, $\det \mathbf{A}^T = \det \mathbf{A}$.
- Interchanging any pair of rows or columns changes the sign of the determinant.
- Element-wise addition of a row, multiplied by an arbitrary number λ, to another row, does not change the determinant. The same is true for the columns.

▷ Consider a *system of linear equations* with respect to n unknowns:

$$\begin{pmatrix} a_{11} & a_{12} & \ldots & a_{1n} \\ a_{21} & a_{22} & \ldots & a_{2n} \\ \vdots & \vdots & \ddots & \vdots \\ a_{n1} & a_{n2} & \ldots & a_{nn} \end{pmatrix} \cdot \begin{pmatrix} x_1 \\ x_2 \\ \vdots \\ x_n \end{pmatrix} = \begin{pmatrix} b_1 \\ b_2 \\ \vdots \\ b_n \end{pmatrix}.$$

It can be written in matrix notation, and the solution can be expressed using the matrix inverse (by multiplying from the left by $\mathbf{A}^{-1}$):

$$\mathbf{A}\cdot\mathbf{x} = \mathbf{b} \quad \Rightarrow \quad \mathbf{x} = \mathbf{A}^{-1}\cdot\mathbf{b}. \tag{27}$$

The solutions of the system can also be obtained from *Cramer's rule*:

$$x_i = \frac{\Delta_i}{\Delta}, \tag{28}$$

where

$$\Delta = \det\mathbf{A}$$

is the determinant of the matrix $\mathbf{A}$, and the Δ_i are the determinants of the matrices obtained from $\mathbf{A}$ by substituting $\mathbf{b}$ for the i-th column.

If the right-hand side of the system is zero,

$$\mathbf{A}\cdot\mathbf{x} = 0,$$

a non-zero solution exists only if

$$\det\mathbf{A} = 0.$$

Such a system of equations is called *homogeneous*.

▷ The *algebraic adjunct* to the element a_{ij} is the determinant of the matrix obtained by deleting the i-th row and the j-th column and then multiplying by $(-1)^{i+j}$. To find the inverse $\mathbf{A}^{-1}$ of $\mathbf{A}$, all of the algebraic adjuncts of $\mathbf{A}$ are calculated and divided by $\det\mathbf{A}$. A matrix is constructed from the numbers obtained, and its transpose is taken. As a result, we obtain $\mathbf{A}^{-1}$. For example:

$$\mathbf{A} = \begin{pmatrix} 2 & 1 & 0 \\ 6 & 4 & \mathbf{2} \\ \mathbf{1} & 0 & 0 \end{pmatrix}, \qquad \mathbf{A}^{-1} = \frac{1}{2}\begin{pmatrix} 0 & 2 & -4 \\ 0 & 0 & 1 \\ 2 & -4 & 2 \end{pmatrix}^T = \begin{pmatrix} 0 & 0 & 1 \\ 1 & 0 & -2 \\ -2 & 1/2 & 1 \end{pmatrix}.$$

The algebraic adjunct is $(-1)^{2+3}\cdot(2\cdot 0 - 1\cdot 1) = 1$ for the element a_{23} and $(-1)^{3+1}(1\cdot 2 - 0\cdot 4) = 2$ for the element a_{31}. The determinant is $\det\mathbf{A} = 2$.

M.4 Levi-Civita symbol

The antisymmetric Levi-Civita symbol ε_{ijk} with three indices is defined for three-dimensional space as:

$$\varepsilon_{123} = 1, \qquad \varepsilon_{ijk} = -\varepsilon_{jik} = -\varepsilon_{kji} = -\varepsilon_{ikj}. \tag{29}$$

Interchanging *any* two indices in this symbol changes its sign. Consequently, a symbol having equal indices is zero. For example:

$$\varepsilon_{213} = -1, \qquad \varepsilon_{231} = 1,$$

$$\varepsilon_{112} = 0, \qquad \varepsilon_{323} = 0.$$

▷ The Levi-Civita symbol can be used to write the vector product

$$\mathbf{c} = \mathbf{a} \times \mathbf{b}$$

in component form:

$$c_i = \varepsilon_{ijk}\, a_j\, b_k. \tag{30}$$

Remember that summation is performed over repeated indices (i.e., j and k above). For example:

$$c_3 = \varepsilon_{312}\, a_1\, b_2 + \varepsilon_{321}\, a_2\, b_1 + \ldots = a_1\, b_2 - a_2\, b_1.$$

The terms denoted by the ellipsis are zero, since components of the Levi-Civita tensor with equal indices are zero.

▷ Contracting the product of two Levi-Civita symbols over one index (i below) gives the following identity:

$$\varepsilon_{ijk}\, \varepsilon_{ipq} = \delta_{jp}\, \delta_{kq} - \delta_{jq}\, \delta_{kp}. \tag{31}$$

It is equivalent to the vector triple product (10). The order of the indices can be memorized with the help of the "direct minus cross-wise" rule (p. 335). The scalar product of two vector products can be computed using this identity. Let us write it in component form:

$$[\mathbf{a} \times \mathbf{b}][\mathbf{c} \times \mathbf{d}] = \varepsilon_{ijk}\, a_j b_k\, \varepsilon_{ipq}\, c_p d_q = (\delta_{jp}\, \delta_{kq} - \delta_{jq}\, \delta_{kp})\, a_j b_k c_p d_q.$$

Summing over j and k with the Kronecker symbols gives:

$$[\mathbf{a} \times \mathbf{b}][\mathbf{c} \times \mathbf{d}] = a_p\, b_q\, c_p\, d_q - a_q\, b_p\, c_p\, d_q = (\mathbf{ac})(\mathbf{bd}) - (\mathbf{ad})(\mathbf{bc}).$$

We have interchanged components to obtain the second equality, e.g., $a_p\, c_p = \mathbf{ac}$. This is an advantage of index notation. After the scalar and the vector product have been expressed in component form, there is no need to pay attention to the order of matrices, non-associativity of vector multiplication, etc.

▷ The height of an index has no importance in Cartesian coordinates in three-dimensional Euclidean space, and the Levi-Civita symbol takes the values $\varepsilon_{123} = \varepsilon^{123} = 1$. Let us prove that

$$\varepsilon^{abc}\varepsilon_{ijk} = \begin{vmatrix} \delta^a_i & \delta^b_i & \delta^c_i \\ \delta^a_j & \delta^b_j & \delta^c_j \\ \delta^a_k & \delta^b_k & \delta^c_k \end{vmatrix} = \binom{abc}{ijk} - \binom{abc}{jik} + \binom{abc}{kij} - \binom{abc}{ikj} + \binom{abc}{jki} - \binom{abc}{kji},$$

where

$$\binom{abc}{ijk} = \delta^a_i \delta^b_j \delta^c_k.$$

Indeed, if the indices in the triples abc and ijk are different, the product is:

$$\varepsilon^{123}\, \varepsilon_{123} = \det \mathbf{1} = 1.$$

If any two indices are equal, two columns or two rows of the matrix are equal. For example, for $a = b$ the first two columns are equal. The determinant is zero in this case. Interchanging any pair of indices results in a switching of the corresponding columns for the first symbol ε^{abc}, and the same for the rows, for the second symbol ε_{ijk}. In this case, the determinant changes its sign. Expanding the determinant and contracting it over one, two, and three indices, we have:

$$\varepsilon^{abc}\,\varepsilon_{ijc} = \delta^a_i\,\delta^b_j - \delta^a_j\,\delta^b_i, \qquad \varepsilon^{abc}\,\varepsilon_{ibc} = 2\,\delta^a_i, \qquad \varepsilon^{abc}\varepsilon_{abc} = 6.$$

▷ The Levi-Civita symbol on Minkowski four-space (p. 169) takes different values for upper and lower indices, $\varepsilon_{0123} = 1$, $\varepsilon^{0123} = -1$ (the tensor with upper indices is obtained by contracting with the metric tensor $g_{\mu\nu} = \mathrm{diag}(1, -1, -1, -1)$):

$$\varepsilon^{\alpha\beta\gamma\lambda}\,\varepsilon_{\mu\nu\sigma\tau} = -\begin{vmatrix} \delta^\alpha_\mu & \delta^\beta_\mu & \delta^\gamma_\mu & \delta^\lambda_\mu \\ \delta^\alpha_\nu & \delta^\beta_\nu & \delta^\gamma_\nu & \delta^\lambda_\nu \\ \delta^\alpha_\sigma & \delta^\beta_\sigma & \delta^\gamma_\sigma & \delta^\lambda_\sigma \\ \delta^\alpha_\tau & \delta^\beta_\tau & \delta^\gamma_\tau & \delta^\lambda_\tau \end{vmatrix},$$

which is why the minus sign appears before the determinant. This relation is proven similarly as in the three-dimensional Euclidean case. Expanding the determinant, for example, over the bottom row, contracting over $\lambda = \tau$, and using the fact that $\delta^\lambda_\lambda = 4$, we have:

$$\varepsilon^{\alpha\beta\gamma\lambda}\varepsilon_{\mu\nu\sigma\lambda} = -\left[\binom{\alpha\beta\gamma}{\mu\nu\sigma} - \binom{\alpha\beta\gamma}{\nu\mu\sigma} + \binom{\alpha\beta\gamma}{\sigma\mu\nu} - \binom{\alpha\beta\gamma}{\mu\sigma\nu} + \binom{\alpha\beta\gamma}{\nu\sigma\mu} - \binom{\alpha\beta\gamma}{\sigma\nu\mu}\right].$$

A further contraction of indices gives:

$$\varepsilon^{\alpha\beta\gamma\lambda}\varepsilon_{\mu\nu\gamma\lambda} = -2\left(\delta^\alpha_\mu\delta^\beta_\nu - \delta^\alpha_\nu\delta^\beta_\mu\right), \quad \varepsilon^{\alpha\beta\gamma\lambda}\varepsilon_{\mu\beta\gamma\lambda} = -6\,\delta^\alpha_\mu, \quad \varepsilon^{\alpha\beta\gamma\lambda}\varepsilon_{\alpha\beta\gamma\lambda} = -24.$$

Similar identities can also be obtained for n-dimensional space [1].

M.5 Curvilinear coordinates

Many physical systems are symmetric. For example, the intensity of the field created by a fixed point charge is spherically symmetric. In such situations it is often convenient to use special coordinates which reflect the symmetry.

▷ Consider first two-dimensional ***polar coordinates*** (r, ϕ), where r is the distance from the coordinate origin, and ϕ is the angle between the radius vector and the x-axis (the first figure):

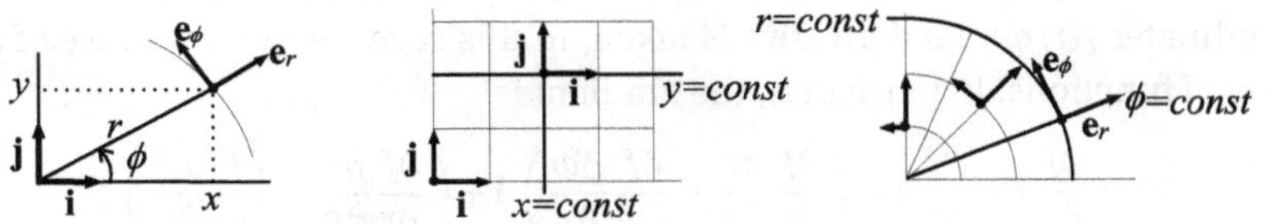

The relationship between Cartesian and polar coordinates is easily found from geometric considerations (by projecting **r** onto the x- and y-axes):

$$\begin{cases} x &= r\cos\phi &= rc_\phi \\ y &= r\sin\phi &= rs_\phi \end{cases}, \qquad \begin{cases} r &= \sqrt{x^2+y^2} \\ \mathrm{tg}\,\phi &= y/x \end{cases} \tag{32}$$

(to make the formulas shorter, we use the symbol c_ϕ for the cosine, and s_ϕ for the sine).

The radius vector can be written in Cartesian coordinates using orthogonal basis vectors ($\mathbf{ij} = 0$, $\mathbf{i}^2 = \mathbf{j}^2 = 1$):

$$\mathbf{r} = x\,\mathbf{i} + y\,\mathbf{j} = r\,(c_\phi\,\mathbf{i} + s_\phi\,\mathbf{j}).$$

Let us define two vectors representing small displacements of the radius vector when one coordinate of the polar coordinate system varies and the other is fixed:

$$\mathbf{e}_r = \frac{\partial \mathbf{r}}{\partial r} = c_\phi\,\mathbf{i} + s_\phi\,\mathbf{j}, \quad \mathbf{e}_\phi = \frac{\partial \mathbf{r}}{\partial \phi} = r\,(-s_\phi\,\mathbf{i} + c_\phi\,\mathbf{j}). \tag{33}$$

Direct multiplication verifies that these vectors are orthogonal, but $\mathbf{e}_\phi$ is not a unit vector:

$$\mathbf{e}_r\mathbf{e}_\phi = 0, \qquad \mathbf{e}_r^2 = 1, \qquad \mathbf{e}_\phi^2 = r^2. \tag{34}$$

The pair of vectors $(\mathbf{e}_r, \mathbf{e}_\phi)$ is called the ***basis of the polar*** coordinate system. In contrast to the basis $(\mathbf{i}, \mathbf{j})$ of Cartesian coordinates, $(\mathbf{e}_r, \mathbf{e}_\phi)$ has different directions at different points of space (the right-hand figure above). The vector $\mathbf{e}_r$ is always directed along the radius vector, and $\mathbf{e}_\phi$ is normal to it and is directed towards the direction of increase of the angle ϕ. Note that $\mathbf{r} = r\,\mathbf{e}_r$.

▷ The decomposition of a vector **a** with respect to the Cartesian and polar bases is defined to be:

$$\mathbf{a} = a^x\,\mathbf{i} + a^y\,\mathbf{j} = a^r\,\mathbf{e}_r + a^\phi\,\mathbf{e}_\phi.$$

Projections onto basis vectors are denoted using upper indices, so a^x, a^r, etc. are signify projections rather than ***exponents***.

Using the relationships (34), the scalar product of two vectors can be written as follows:

$$\mathbf{a}\cdot\mathbf{b} = a^x b^x + a^y b^y = a^r b^r + r^2 a^\phi b^\phi. \tag{35}$$

The first equality is the scalar product in Cartesian coordinates. The second equality is the product of the same vectors in polar coordinates.

▷* By reversing the definition of the polar basis (33), we have:

$$\mathbf{i} = c_\phi\,\mathbf{e}_r - \frac{s_\phi}{r}\,\mathbf{e}_\phi, \qquad \mathbf{j} = s_\phi\,\mathbf{e}_r + \frac{c_\phi}{r}\,\mathbf{e}_\phi. \tag{36}$$

A scalar function of Cartesian coordinates $f(x, y)$ can also be expressed in terms of polar coordinates $f(r, \phi)$. Its derivative is taken, in this case, as the derivative of a composition of functions. In particular, the gradient is:

$$\nabla f = \frac{\partial f}{\partial x}\mathbf{i} + \frac{\partial f}{\partial y}\mathbf{j} = \left(\frac{\partial f}{\partial r}\frac{\partial r}{\partial x} + \frac{\partial f}{\partial \phi}\frac{\partial \phi}{\partial x}\right)\mathbf{i} + \left(\frac{\partial f}{\partial r}\frac{\partial r}{\partial y} + \frac{\partial f}{\partial \phi}\frac{\partial \phi}{\partial y}\right)\mathbf{j}.$$

The derivatives of the radius $r = \sqrt{x^2 + y^2}$ with respect to x and y (32) are:

$$\frac{\partial r}{\partial x} = \frac{x}{r} = c_\phi, \qquad \frac{\partial r}{\partial y} = \frac{y}{r} = s_\phi.$$

Similarly, given that $(\mathrm{tg}\,\phi)' = 1/c_\phi^2$, the derivatives of the polar angle are:

$$\frac{\partial \phi}{\partial x} = -\frac{y}{x^2}\,c_\phi^2 = -\frac{s_\phi}{r}, \qquad \frac{\partial \phi}{\partial y} = \frac{c_\phi^2}{x} = \frac{c_\phi}{r}.$$

Substituting them into the gradient and expressing the basis $(\mathbf{i}, \mathbf{j})$ in terms of $(\mathbf{e}_r, \mathbf{e}_\phi)$, we obtain from (36):

$$\nabla f = \frac{\partial f}{\partial x}\mathbf{i} + \frac{\partial f}{\partial y}\mathbf{j} = \frac{\partial f}{\partial r}\mathbf{e}_r + \frac{1}{r^2}\frac{\partial f}{\partial \phi}\mathbf{e}_\phi. \tag{37}$$

Therefore, ∇f can be decomposed, like any other vector, in terms of the Cartesian and the polar bases.

▷ We now generalize the idea of a curvilinear basis to an arbitrary coordinate system q^i. For example, $q^i = \{x, y\}$ in two-dimensional Cartesian coordinates, $q^i = \{r, \phi\}$ in polar coordinates, $q^i = \{q^1, q^2, q^3\}$ in a three-dimensional space, etc. Remember that upper indices here are coordinate numbers rather than exponents. Suppose that a triple of independent vectors $(\mathbf{e}_1, \mathbf{e}_2, \mathbf{e}_3)$ is defined at each point of a three-dimensional space, where lower indices refer to the number of the vector rather than different components. Independence means that none of the vectors can be decomposed as a linear combination of the other two. They do not belong to a common plane:

$$\mathbf{e}_3 \neq \alpha_1 \mathbf{e}_1 + \alpha_2 \mathbf{e}_2.$$

Let us define basis vectors in the coordinates q^i by:

$$\mathbf{e}_i = \frac{\partial \mathbf{r}}{\partial q^i} \equiv \partial_i \mathbf{r}, \tag{38}$$

where

$$\mathbf{r} = x\,\mathbf{i} + y\,\mathbf{j} + z\,\mathbf{k}$$

is the radius vector in Cartesian coordinates, which depends on the curvilinear coordinates as:

$$x = x(q^1, q^2, q^3), \quad y = y(q^1, q^2, q^3), \quad z = z(q^1, q^2, q^3).$$

▷ An arbitrary vector at a *given point* of space can be decomposed in terms of this basis:

$$\mathbf{a} = a^i\,\mathbf{e}_i = a^1\,\mathbf{e}_1 + a^2\,\mathbf{e}_2 + a^3\,\mathbf{e}_3.$$

As agreed earlier, we sum over repeated indices (the index i above). Furthermore, we will place summation indices at different levels, one at the top and the other at the bottom.

▷ Consider the scalar product of two vectors:

$$\mathbf{a} \cdot \mathbf{b} = (a^i \mathbf{e}_i) \cdot (b^j \mathbf{e}_j) = (\mathbf{e}_i \cdot \mathbf{e}_j)\, a^i\, b^j = g_{ij}\, a^i\, b^j,$$

where the scalar product of the basis vectors is written using special symbols called the coefficients of the *metric tensor*:

$$g_{ij} = \mathbf{e}_i \cdot \mathbf{e}_j = \frac{\partial \mathbf{r}}{\partial q^i} \frac{\partial \mathbf{r}}{\partial q^j}. \tag{39}$$

Note that the calculation of the scalar product required that different indices i and j be used in the expansions of vectors **a** and **b**. These are two independent sums, so the indices should be different. By definition, the metric tensor is symmetric, i.e., it does not change when indices are interchanged: $g_{ij} = g_{ji}$.

▷ The meaning of the term "metric tensor" is as follows. Consider a vector of infinitesimal displacement from a given point in space. By definition of the basis vectors (38), the expansion coefficients of $d\mathbf{r}$ in terms of the basis are the differentials of the coordinates:

$$d\mathbf{r} = \mathbf{e}_i\, dq^i, \qquad ds^2 = (d\mathbf{r})^2 = g_{ij}\, dq^i\, dq^j. \tag{40}$$

The squared infinitesimal displacement $(d\mathbf{r})^2$ is the distance between two infinitesimally close points. It is usually denoted by ds^2.

The metric tensor in polar coordinates is [see (34)]:

$$g_{ij} = \begin{pmatrix} 1 & 0 \\ 0 & r^2 \end{pmatrix}, \tag{41}$$

and it is a diagonal matrix with unit diagonal elements in Cartesian coordinates. Therefore, the distance in Cartesian and in polar coordinates has the form:

$$ds^2 = (dx)^2 + (dy)^2 = (dr)^2 + r^2\,(d\phi)^2. \tag{42}$$

The first equality is the well-known Pythagorean theorem (the figure on the left below). The distance in polar coordinates also has a simple geometric interpretation. The length of the arc with radius $r = const$ and the angle increment $d\phi$ is $r\, d\phi$. Therefore, from the Pythagorean theorem we again obtain the distance $(dr)^2 + (r\, d\phi)^2$ (the figure on the right):

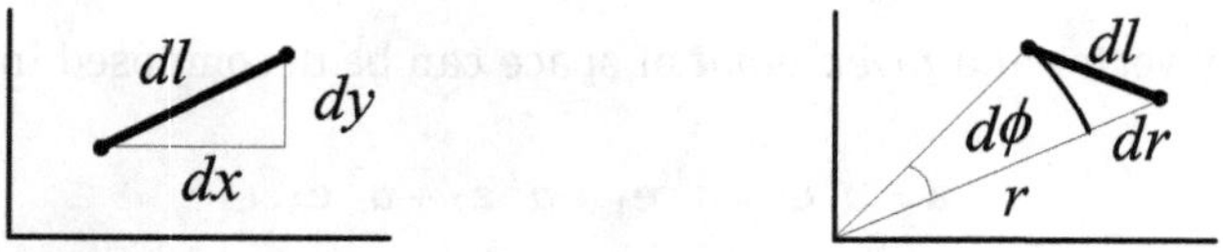

The metric coefficients g_{ij} determine the distance between two points in a given coordinate system, so the g_{ij} are said to define the metric.

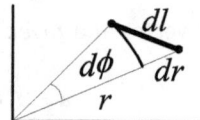

▷ Let us note one more point. The value $d\mathbf{r} = \mathbf{e}_i\, dq^i$ is called the vector of infinitesimal displacement. In Cartesian coordinates, the radius vector

$$\mathbf{r} = \mathbf{i}\,x + \mathbf{j}\,y$$

is also a vector. It can be obtained by integrating

$$d\mathbf{r} = \mathbf{i}\,dx + \mathbf{j}\,dy$$

(the vectors **i** and **j** are constant). The situation is more complex in curvilinear coordinates, where the basis vectors depend on the point in space. Thus, in curved spaces (e.g. those which describe the geometry of some surface), vectors are generally "attached" to a given point of space. Therefore, in the general case, a vector can be decomposed only at a given point of space in terms of the basis at this point. However, in Euclidean space a radius vector can be defined in curvilinear coordinates. For example, $\mathbf{r} = r\,\mathbf{e}_r$ in polar coordinates.

M.6 The reciprocal basis

The basis vectors $(\mathbf{e}_r, \mathbf{e}_\phi)$ are orthogonal to each other in a polar coordinate system. This is not true for an arbitrary coordinate system with the basis $\mathbf{e}_i = \partial_i \mathbf{r}$, and the metric tensor is not diagonal in general. In addition to the basis

$$\mathbf{e}_i = (\mathbf{e}_1, \mathbf{e}_2, \mathbf{e}_3),$$

it is convenient to introduce the so-called *reciprocal* basis

$$\mathbf{e}^i = (\mathbf{e}^1, \mathbf{e}^2, \mathbf{e}^3),$$

which is *orthonormal* to the original basis:

$$\mathbf{e}^i \cdot \mathbf{e}_j = \delta^i_j, \tag{43}$$

where δ^i_j is the Kronecker symbol, which is zero for different indices and unity for indices which coincide. The left-hand figure below shows a curvilinear basis, and the right-hand figure shows its reciprocal basis (the thin lines denote the vectors of the original basis):

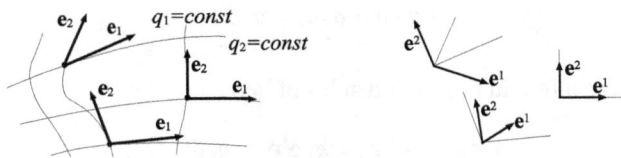

▷ In three-dimensional space, the vectors of the reciprocal basis can be expressed using vector products. As an example, the vector $\mathbf{e}^1$ is normal to $\mathbf{e}_2$ and $\mathbf{e}_3$ (by definition of the Kronecker symbol), and is therefore proportional to $\mathbf{e}_2 \times \mathbf{e}_3$. Similarly, for the rest of basis vectors we have:

$$\mathbf{e}^1 = \frac{\mathbf{e}_2 \times \mathbf{e}_3}{V}, \quad \mathbf{e}^2 = \frac{\mathbf{e}_3 \times \mathbf{e}_1}{V}, \quad \mathbf{e}^3 = \frac{\mathbf{e}_1 \times \mathbf{e}_2}{V}. \tag{44}$$

The denominator contains a normalization factor,

$$V = \mathbf{e}_1 \, [\mathbf{e}_2 \times \mathbf{e}_3],$$

which can be interpreted as the volume built from the vectors of the original basis. It is used here so that the definition (43) is satisfied.

▷ If $V > 0$, the basis $(\mathbf{e}_1, \mathbf{e}_2, \mathbf{e}_3)$ is said to form a right-handed triple:

Respectively, the order of the basis vectors $\mathbf{e}^i$ in the vector product is chosen in such a way that $(\mathbf{e}^1, \mathbf{e}^2, \mathbf{e}^3)$ is a right-handed basis.

▷ Any vector at a *given point* of space can be decomposed both over the original and the reciprocal basis. The expansion coefficients will be different, in general:

$$\mathbf{a} = a^i \, \mathbf{e}_i = a_i \, \mathbf{e}^i. \tag{45}$$

The projections onto the vectors of the original basis denoted by upper indices a^i are called the *contravariant* components of the vector, and the projections with lower indices a_i are called *covariant* components.

▷ The reciprocal basis can be used to simplify the appearance of the scalar product of two vectors. Namely, if one of the vectors is decomposed in terms of the initial basis, and the other in terms of the the reciprocal basis:

$$\mathbf{a} \cdot \mathbf{b} = (a^i \, \mathbf{e}_i) \cdot (b_j \, \mathbf{e}^j) = a^i \, b_j \, \delta^i_j = a^i \, b_i = a^1 \, b_1 + a^2 \, b_2 + a^3 \, b_3.$$

Thus, the obtained expression looks similar to the scalar product in Cartesian coordinates; however, there are different vector components in the sum. Obviously,

$$\mathbf{a} \cdot \mathbf{b} = a^i \, b_i = a_i \, b^i.$$

▷ Multiplying the left- and right-hand sides of (45) by $\mathbf{e}_j$, we obtain:

$$a^i \, \mathbf{e}_i \mathbf{e}_j = a^i \, g_{ij} = a_i \, \mathbf{e}^i \mathbf{e}_j = a_i \, \delta^i_j = a_j.$$

The last equation implies a contraction with the Kronecker symbol. Therefore, the metric tensor can be used to lower indices, thus transforming contravariant components into covariant ones:

$$a_i = g_{ij}\, a^j. \tag{46}$$

In the above equation we renamed the indices ($i \mapsto j$, $j \mapsto i$) and used the symmetry of the metric tensor:

$$g_{ij} = g_{ji}.$$

▷ The metric tensor with upper indices can be defined as a scalar product of the vectors of the reciprocal basis:

$$g^{ij} = \mathbf{e}^i \cdot \mathbf{e}^j. \tag{47}$$

It is used to raise indices:

$$a^i = g^{ij}\, a_j. \tag{48}$$

The metric tensor with upper indices is the inverse of the metric tensor with lower indices:

$$g^{ik}\, g_{kj} = \delta^i_j, \tag{49}$$

as is easily verified by substituting $a_k = g_{kj}\, a^j$ into $a^i = g^{ik}\, a_k$.

M.7 Coordinate transforms

Let us recall the most important curvilinear coordinate systems.

▷ The *cylindrical coordinate system* (r, ϕ, z) adds a third dimension to the polar system (p. 344). A third coordinate z is added to the distance r from the origin of the system and the angle ϕ in the (x, y)-plane:

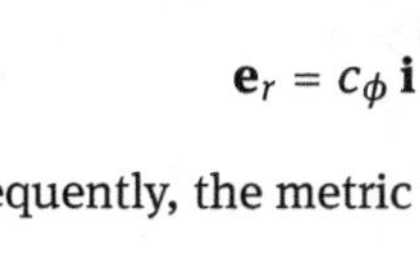

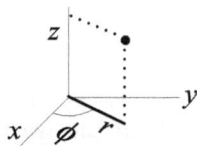

$$\begin{cases} x = r\cos\phi, \\ y = r\sin\phi, \\ z = z. \end{cases}$$

The basis vectors of the cylindrical coordinate system decomposed in terms of the orthonormal basis **i**, **j**, **k** of the Cartesian system have the form:

$$\mathbf{e}_r = c_\phi\,\mathbf{i} + s_\phi\,\mathbf{j}, \qquad \mathbf{e}_\phi = r\,(-s_\phi\,\mathbf{i} + c_\phi\,\mathbf{j}), \qquad \mathbf{e}_z = \mathbf{k}.$$

Consequently, the metric tensor (39) for $\mathbf{e}_i = (\mathbf{e}_r, \mathbf{e}_\phi, \mathbf{e}_z)$ is:

$$g_{ij} = \mathbf{e}_i\,\mathbf{e}_j = \begin{pmatrix} 1 & 0 & 0 \\ 0 & r^2 & 0 \\ 0 & 0 & 1 \end{pmatrix}. \tag{50}$$

Cylindrical coordinates vary over the following ranges to cover all points of space:

$$0 \leqslant r < \infty, \quad 0 \leqslant \phi < 2\pi, \quad -\infty < z < \infty.$$

▷ The *spherical coordinate system*: $(r,\ \theta,\ \phi)$, where r is the distance to the origin, θ is the angle between the radius vector and the z-axis, and ϕ is the angle between the projection of the radius vector onto the (x, y)-plane and the x-axis:

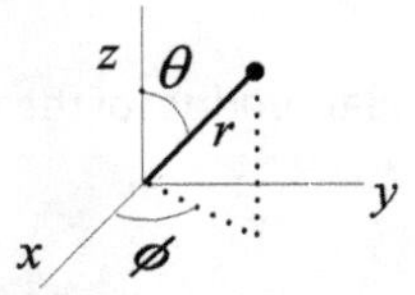

$$\begin{cases} x = r\sin\theta\,\cos\phi, \\ y = r\sin\theta\,\sin\phi, \\ z = r\cos\theta. \end{cases}$$

The basis vectors $\mathbf{e}_i = (\mathbf{e}_r, \mathbf{e}_\theta, \mathbf{e}_\phi)$:

$$\mathbf{e}_r = s_\theta c_\phi\,\mathbf{i} + s_\theta s_\phi\,\mathbf{j} + c_\theta\,\mathbf{k}, \quad \mathbf{e}_\theta = r\,(c_\theta c_\phi\,\mathbf{i} + c_\theta s_\phi\,\mathbf{j} - s_\theta\,\mathbf{k}), \quad \mathbf{e}_\phi = r\,(-s_\theta s_\phi\,\mathbf{i} + s_\theta c_\phi\,\mathbf{j}),$$

give the following metric tensor:

$$g_{ij} = \mathbf{e}_i\mathbf{e}_j = \begin{pmatrix} 1 & 0 & 0 \\ 0 & r^2 & 0 \\ 0 & 0 & r^2\,s_\theta^2 \end{pmatrix}. \tag{51}$$

The coordinates vary over the ranges: $0 \leqslant r < \infty$, $0 \leqslant \theta < \pi$, $0 \leqslant \phi < 2\pi$.

▷ An infinitesimal displacement in space $d\mathbf{r}$ is a vector. A curvilinear basis is a set of linearly independent vectors $\mathbf{e}_1, \mathbf{e}_2, \mathbf{e}_3$, and the expansion components of $d\mathbf{r}$ over these vectors are the differentials of the coordinates of the curvilinear coordinate system (q^1, q^2, q^3):

$$d\mathbf{r} = \mathbf{e}_i\,dq^i. \tag{52}$$

Each coordinate system has its own basis. But a small displacement $d\mathbf{r}$ is a geometric object connecting two infinitely close points and does not depend on the choice of coordinates.

Consider two coordinate systems $(\tilde{q}^1, \tilde{q}^2, \tilde{q}^3)$ and (q^1, q^2, q^3) related by differentiable single-valued functions $q^i(\tilde{q}^1, \tilde{q}^2, \tilde{q}^3)$. According to the chain rule, we have:

$$\frac{\partial \tilde{q}^i}{\partial \tilde{q}^j} = \delta^i_j = \frac{\partial \tilde{q}^i}{\partial q^k}\frac{\partial q^k}{\partial \tilde{q}^j}, \qquad \frac{\partial q^i}{\partial q^j} = \delta^i_j = \frac{\partial q^i}{\partial \tilde{q}^k}\frac{\partial \tilde{q}^k}{\partial q^j}, \tag{53}$$

where the δ^i_j are Kronecker symbols, and summation is performed over k. Therefore, the matrices

$$\frac{\partial q^i}{\partial \tilde{q}^j} \quad \text{and} \quad \frac{\partial \tilde{q}^i}{\partial q^j}$$

are inverse to each other. These equations are easily remembered if read in reverse order: the contraction (summation) of the differentials with equal coordinates "reduces", and the remaining derivative gives the Kronecker symbol.

▷ Decompose the vector of infinitesimal displacement in terms of the basis of each of the coordinate systems:

$$d\mathbf{r} = \tilde{\mathbf{e}}_i\, d\tilde{q}^{\,i} = \mathbf{e}_j\, dq^j = \mathbf{e}_j \frac{\partial q^j}{\partial \tilde{q}^{\,i}}\, d\tilde{q}^{\,i},$$

where basis vectors in the coordinate system $(\tilde{q}^1, \tilde{q}^2, \tilde{q}^3)$ are marked with a tilde, and the expression for the differential of the functions $q^j(\tilde{q}^{\,i})$ has been substituted into the last equation. Comparing the first and the last equation, we obtain the relationship between the basis vectors of the two coordinate systems (the first relation below):

$$\tilde{\mathbf{e}}_i = \frac{\partial q^j}{\partial \tilde{q}^{\,i}}\, \mathbf{e}_j, \qquad \mathbf{e}_i = \frac{\partial \tilde{q}^j}{\partial q^i}\, \tilde{\mathbf{e}}_j. \tag{54}$$

The inverse transform (the second relation) is obtained in a similar way, or with the help of the matrix properties (53). For the reciprocal basis, the corresponding transformations have the form:

$$\tilde{\mathbf{e}}^i = \frac{\partial \tilde{q}^{\,i}}{\partial q^j}\, \mathbf{e}^j, \qquad \mathbf{e}^i = \frac{\partial q^i}{\partial \tilde{q}^j}\, \tilde{\mathbf{e}}^j. \tag{55}$$

They are obtained from the definition $\mathbf{e}^i \mathbf{e}_j = \delta^i_j$ and the conditions (53).

▷ An arbitrary vector $\mathbf{a}$ is a geometric object that can be decomposed with respect to the basis of any curvilinear system:

$$\mathbf{a} = a^i\, \mathbf{e}_i = \tilde{a}^{\,i}\, \tilde{\mathbf{e}}_i.$$

Substituting in the transformations of the basis vectors, we obtain vector components with upper indices (contravariant components):

$$\tilde{a}^i = \frac{\partial \tilde{q}^{\,i}}{\partial q^j}\, a^j, \qquad a^i = \frac{\partial q^i}{\partial \tilde{q}^j}\, \tilde{a}^j. \tag{56}$$

The transformation laws for components with lower indices (covariant components) are found in a similar way. We can also do this using the fact that the scalar product of two vectors **a** and **b** does not depend on the coordinate system. The decomposition of the product has the form:

$$\mathbf{a} \cdot \mathbf{b} = a_i\, b^i = \tilde{a}_i\, \tilde{b}^i.$$

Substituting the transformations for the components with upper indices for b^i and $\tilde{b}^i$ and given that arbitrary vectors are used, we obtain:

$$\tilde{a}_i = \frac{\partial q^j}{\partial \tilde{q}^{\,i}}\, a_j, \qquad a_i = \frac{\partial \tilde{q}^j}{\partial q^i}\, \tilde{a}_j. \tag{57}$$

As examples of transformations of components with upper and lower indices, consider the differentials dq^i and partial derivatives of a scalar function. From the general rules of calculus, we have:

$$d\tilde{q}^i = \frac{\partial \tilde{q}^{\,i}}{\partial q^j}\, dq^j, \qquad \frac{\partial f}{\partial \tilde{q}^{\,i}} = \frac{\partial q^j}{\partial \tilde{q}^{\,i}} \frac{\partial f}{\partial q^j}.$$

Their contraction gives the *invariant* differential, which does not depend on the choice of coordinate system:

$$df = \frac{\partial f}{\partial q^i}\, dq^i = \frac{\partial f}{\partial \tilde{q}^{\,i}}\, d\tilde{q}^{\,i}. \tag{58}$$

Thus, vector components with upper indices are transformed *like differentials* of the coordinates dq^i, and those with lower indices *like the partial derivatives* of a scalar function

$$\partial_i f = \frac{\partial f}{\partial q^i}.$$

Keeping this in mind, the form of the transformation matrices can easily be remembered.

When the coordinates change, the functional form of the scalar function changes as well, $f(q^i) = f(q^i(\tilde{q}^j)) = \tilde{f}(\tilde{q}^j)$; however, its value at the *given* point of space is constant, so we the tilde in $\tilde{f}$ is dropped.

▷ A vector **a** can be expressed in a given coordinate system in terms of its projections with upper a^i or lower a_i indices. There are four possible combinations of these projections:

$$a^i b^j, \qquad a^i b_j, \qquad a_i b^j, \qquad a_i b_j.$$

Each of them contains n^2 values, where n is the dimension of the space. Since the components are transformed independently, the product of two vector components will transform under a coordinate change as follows:

$$\tilde{a}^i \tilde{b}^j = \frac{\partial \tilde{q}^{\,i}}{\partial q^\alpha} \frac{\partial \tilde{q}^j}{\partial q^\beta}\, a^\alpha b^\beta, \qquad \tilde{a}^i \tilde{b}_j = \frac{\partial \tilde{q}^{\,i}}{\partial q^\alpha} \frac{\partial q^\beta}{\partial \tilde{q}^j}\, a^\alpha b_\beta.$$

In general, a tensor of order (μ, ν) is a quantity with $\mu + \nu$ indices, μ on top, and ν on the bottom. The order of a tensor determines its transformation law under a change of the coordinate system. By *definition*, this is the product of μ transformation matrices for the vector components with upper indices, and ν matrices for those with lower indices. For example, for a tensor of order (1, 2) we have:

$$\tilde{a}^i_{jk} = \frac{\partial \tilde{q}^{\,i}}{\partial q^\alpha} \frac{\partial q^\beta}{\partial \tilde{q}^j} \frac{\partial q^\gamma}{\partial \tilde{q}^k}\, a^\alpha_{\beta\gamma}. \tag{59}$$

▷ Any contraction (summation) over upper and lower indices returns another tensor:

$$f = A^i_i, \qquad a_i = B^j_{ik}\, C^k\, D_j, \qquad F_{ij} = g_{i\alpha}\, g_{j\beta}\, F^{\alpha\beta}.$$

In the first case, a scalar (number, index-free value) is obtained. In the second case, a tensor of order (0,1), i.e., vector (covariant) components with lower indices. In the third case, two metric tensors are used to lower the indices of a tensor of order (2,0), transforming it into a (0,2)-tensor, which is traditionally denoted by the same symbol.

▷ The metric tensors g_{ij} and g^{ij} of order (0,2) and (2,0), respectively, obey the tensor transformation law. This is a direct consequence of the fact that the distance between two infinitely close points

$$ds^2 = g_{ij}\, dq^i\, dq^j$$

does not depend on the choice of coordinate system, and that dq^i is a vector.

Using the relation (53), we can verify that the Kronecker symbol δ^i_j is a tensor. Also, it has the same components in every coordinate system. The tensorial nature of the Kronecker symbols also follows from the relation

$$g^{ik} g_{kj} = \delta^i_j$$

(a contraction of two tensors is again a tensor).

M.8 Complex numbers

We see no negative numbers in the surrounding world. Minus one apples is a man-made abstraction. The equation $x + 2 = 3$ has the solution $x = 1$, but the equation $x + 3 = 2$ has no solution amongst the natural numbers. However, if zero and the *negative numbers* $-1, -2, -3, \ldots$ are added to the natural numbers, such equations will have solutions, albeit amongst the "unreal" invented negative numbers.

Similarly, not all quadratic equations have solutions. Therefore, the number $\imath$ was invented to give a solution to the equation $x^2 = -1$. It was called the *imaginary unit*:

$$\imath^2 = -1. \tag{60}$$

The term "imaginary" should not be misleading, and $\imath$ is no more "imaginary" than the number -1. Any complex number z can be written in terms of two real numbers and the imaginary unit:

$$z = x + \imath\, y. \tag{61}$$

The number x is the *real* part, and y is the *imaginary* part of the complex number z. In fact, a complex number is an ordered pair of two real numbers $z = (x, y)$. For such a pair, the operations of addition, multiplication and division are defined, with the same properties as in the case of real numbers. The symbol $\imath$ is used to easily perform various calculations involving complex numbers according to the usual algebraic rules. For example, for

$$z_1 = 1 + 2\imath \quad \text{and} \quad z_2 = 3 + 4\imath,$$

we have:

$$z_1 \cdot z_2 = (1 + 2\imath) \cdot (3 + 4\imath) = 1 \cdot 3 + 1 \cdot 4\imath + 2 \cdot 3\imath + 2 \cdot 4\imath^2 = -5 + 10\imath,$$

$$\frac{z_1}{z_2} = \frac{1 + 2\imath}{3 + 4\imath} = \frac{(1 + 2\imath)(3 - 4\imath)}{(3 + 4\imath)(3 - 4\imath)} = \frac{11 + 2\imath}{3^2 - (4\imath)^2} = \frac{11 + 2\imath}{9 + 16} = \frac{11}{25} + \frac{2}{25}\imath.$$

When calculating the quotient of two complex numbers, the numerator and the denominator are multiplied by z_2, with the difference that a minus sign is placed before its imaginary part. This number is called the *complex conjugate* of a number z and is denoted by an asterisk:

$$z^* = x - \imath y. \tag{62}$$

The product of a number z and its complex conjugate z^* is always a non-negative real number:

$$|z|^2 = z \cdot z^* = (x + \imath y)(x - \imath y) = x^2 + y^2, \tag{63}$$

whose root

$$|z| = \sqrt{x^2 + y^2}$$

is called the *modulus of the complex number.*

▷ Leonhard Euler derived a remarkable formula in 1748:

$$e^{\imath\phi} = \cos\phi + \imath\sin\phi, \tag{64}$$

which replaces all of trigonometry by providing an easy way to derive any relationship involving sin and cos. For example, when raised to a power n, Euler's formula gives *de Moivre's formula*:

$$\cos(n\phi) + \imath\sin(n\phi) = (\cos\phi + \imath\sin\phi)^n, \tag{65}$$

which can be used to find the values of trigonometric functions at integer multiples of angles. The formula (64) can be verified, for example, by expanding its left- and right-hand sides in Taylor series.

▷ An arbitrary complex number

$$z = x + \imath y = |z|\, e^{\imath\phi} = |z|\,(\cos\phi + \imath\, \sin\phi)$$

is defined by its real x and imaginary y parts *or* by the modulus

$$|z| = \sqrt{x^2 + y^2}$$

and the *argument* (angle)

$$\phi = \operatorname{arctg}(y/x).$$

Consider the Cartesian plane. Let the real parts of the complex numbers correspond to the x-axis, and the imaginary parts to the y-axis. Then any complex number is a point in this plane, and its real and imaginary parts are its projections onto the coordinate axes:

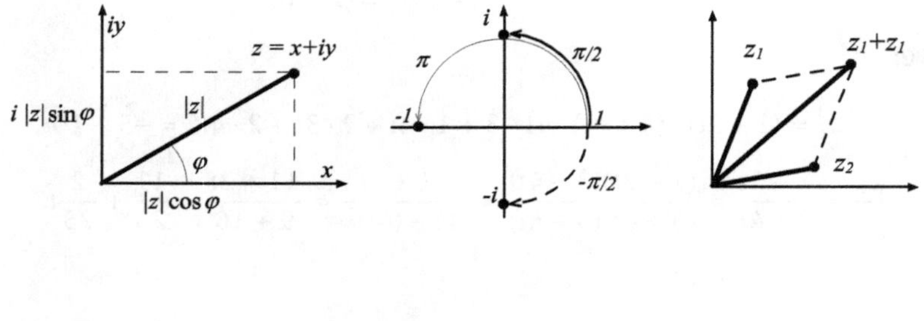

The imaginary unit has unit modulus and the argument $\pi/2$, and the argument of "-1" is π. The addition of two complex numbers is geometrically equivalent to the addition of two vectors (the third figure above).

▷ The argument of a complex number is defined up to an integer number of complete revolutions. This means that $|z|\, e^{\imath\phi}$ and $|z|\, e^{\imath\,(\phi+2\pi k)}$ are the same number ($k = 1, 2, \ldots$). Therefore, when extracting the n-th root:

$$\sqrt[n]{z} = |z|^{1/n} \exp\left\{\imath\, \frac{\phi + 2\pi k}{n}\right\}, \quad k = 0, 1, .., n-1, \tag{66}$$

we obtain different complex numbers for $k = 0, 1, .., n-1$. *Any* algebraic equation of n-th degree always has n solutions, some of which may be complex. In the particular case of a quadratic equation with $n = 2$, there are two solutions.

M.9 Hyperbolic functions

The basic hyperbolic functions are the hyperbolic sine and cosine:

$$\mathrm{sh}(x) = \frac{e^x - e^{-x}}{2}, \qquad \mathrm{ch}(x) = \frac{e^x + e^{-x}}{2}. \tag{67}$$

Hyperbolic sine is an odd function: $\mathrm{sh}(-x) = -\,\mathrm{sh}(x)$, and hyperbolic cosine is an even function: $\mathrm{ch}(-x) = \mathrm{ch}(x)$.

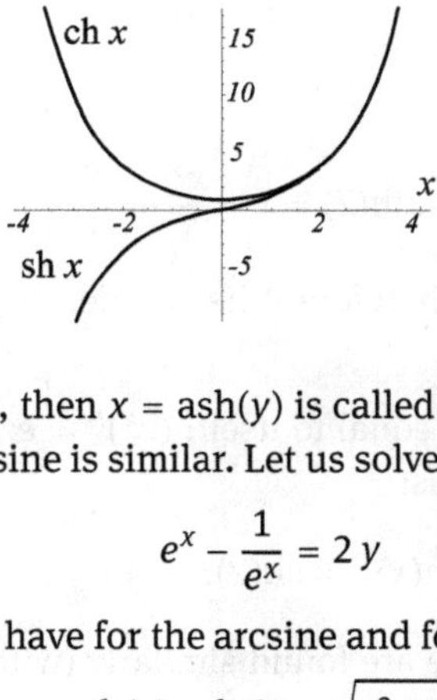

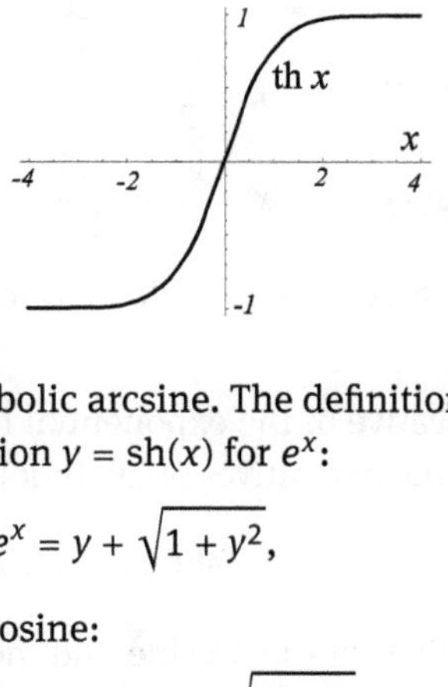

If $y = \mathrm{sh}(x)$, then $x = \mathrm{ash}(y)$ is called the hyperbolic arcsine. The definition of hyperbolic arccosine is similar. Let us solve the equation $y = \mathrm{sh}(x)$ for e^x:

$$e^x - \frac{1}{e^x} = 2\,y \quad \Rightarrow \quad e^x = y + \sqrt{1 + y^2},$$

whence we have for the arcsine and for the arccosine:

$$\mathrm{ash}(y) = \ln(y + \sqrt{y^2 + 1}), \qquad \mathrm{ach}(y) = \ln(y \pm \sqrt{y^2 - 1}). \tag{68}$$

Hyperbolic cosine is always positive for real x; therefore, the domain of the arccosine is $|y| \geqslant 1$. Besides this, $\mathrm{ch}(x)$ is an even function, so its inverse is a two-valued function.

▷ By definition, the hyperbolic tangent is:

$$y = \mathrm{th}(x) = \frac{\mathrm{sh}(x)}{\mathrm{ch}(x)} = \frac{e^x - e^{-x}}{e^x + e^{-x}} \quad \Rightarrow \quad x = \mathrm{ath}(y) = \frac{1}{2} \ln \frac{1 + y}{1 - y}.$$

The arctangent is obtained by solving this equation for e^x, as was done for the arcsine.

From the definition of the hyperbolic functions in terms of the exponential function we immediately obtain:

$$\mathrm{ch}^2(x) - \mathrm{sh}^2(x) = 1, \qquad \mathrm{th}(x+y) = \frac{\mathrm{th}(x) + \mathrm{th}(y)}{1 + \mathrm{th}(x)\,\mathrm{th}(y)},$$

$$\mathrm{ch}^2(x) + \mathrm{sh}^2(x) = \mathrm{ch}(2x), \qquad 2\,\mathrm{sh}(x)\,\mathrm{ch}(x) = \mathrm{sh}(2x),$$

as well as many other useful identities.

▷ With the help of Euler's formula (64):

$$e^{\imath x} = \cos(x) + \imath\,\sin(x),$$

hyperbolic functions can be related to trigonometric functions using the imaginary unit $\imath$:

$$\mathrm{sh}(\imath\,x) = \imath\,\sin(x), \qquad \mathrm{ch}(\imath\,x) = \cos(x). \tag{69}$$

Therefore, all trigonometric identities can easily be rewritten for hyperbolic functions, and vice versa.

▷ Using the known series expansions for the exponential function and the logarithm, we can write power series for the hyperbolic functions:

$$\mathrm{sh}(x) = x + \frac{x^3}{6} + \frac{x^5}{120} + \dots \qquad \mathrm{ash}(x) = x - \frac{x^3}{6} + \frac{3x^5}{40} + \dots$$

$$\mathrm{ch}(x) = 1 + \frac{x^2}{2} + \frac{x^4}{24} + \dots$$

$$\mathrm{th}(x) = x - \frac{x^3}{3} + \frac{2x^5}{15} + \dots \qquad \mathrm{ath}(x) = x + \frac{x^3}{3} + \frac{x^5}{5} + \dots$$

Hyperbolic arccosine is undefined for $x = 0$; therefore it has no decomposition for small x.

▷ The derivative of the exponential function is equal to itself: $(e^x)' = e^x$, so we can easily find the derivatives of hyperbolic functions:

$$(\mathrm{sh}(x))' = \mathrm{ch}(x), \qquad (\mathrm{ch}(x))' = \mathrm{sh}(x). \tag{70}$$

The derivatives of the arcsine and the arccosine are found similarly (with the upper sign, plus, in (68)):

$$(\mathrm{ash}(x))' = \frac{1}{\sqrt{1+x^2}}, \qquad (\mathrm{ach}(x))' = \frac{1}{\sqrt{x^2-1}}. \tag{71}$$

Therefore, we have the following integrals:

$$\int \frac{dx}{\sqrt{1+x^2}} = \mathrm{ash}(x) + const, \qquad \int \frac{dx}{\sqrt{x^2-1}} = \mathrm{ach}(x) + const.$$

A useful integral is also obtained by differentiating the hyperbolic arctangent:

$$(\mathrm{ath}(x))' = \frac{1}{1-x^2}, \qquad \int \frac{dx}{1-x^2} = \mathrm{ath}(x) + const,$$

which follows from the definition $\mathrm{ath}(x) = \ln[(1+x)/(1-x)]/2$.

C. Comments

In the text, the symbol (< C_i) refers to the comment under the number *i*. These comments *only* need to be read if something is not clear, or causes disagreement with the author, at the point where the symbol appears. The corresponding comment may give you some clarification, or more or less convincing additional arguments. Of course, even the comments cannot provide answers to ALL questions....

• **C_1** *Is there a preferred frame?* Yes, there is. While our universe is conceived of as empty space, there are no reasons to consider any inertial frame to be "better" than any other frame. However, such a frame can be identified as soon as the space is filled with matter. Indeed, imagine a "gas" composed of particles moving with different velocities. We can always find a frame such that the *average velocity* of all these particles is zero with respect to this frame. Or, for systems of separate particles, it is more appropriate to refer to the zero *average momentum*. As an example, stars can be such particles. The photons of the cosmic microwave background (CMB) are even more suitable objects, since they fill space in a highly uniform and isotropic manner. The temperature (average energy of the photon gas) of the CMB is 2.725 K. By measuring its temperature change in the direction of the Earth's motion, we can find its "absolute" velocity.

Even though this is a preferred frame, there is no reason to believe that it has some special properties compared to any other inertial frame. The temperature of the photon gas can be measured by relatively local experiments. Technically, an observer can detect the motion of a Galilean ship from inside its hold. However, if the hold is shielded not only from the wind at sea but also from the CMB, then absolute motion will probably be undetectable.

• **C_2** *Expansion into a series* (p. 16, 61, 368).
The Taylor formula is a universal tool for expanding an arbitrary function $f(x)$ in a series in x, provided that it has no singularities at $x = 0$:

$$f(x) = f(0) + f'(0)\,\frac{x}{1!} + f''(0)\,\frac{x^2}{2!} + \dots,$$

where the primes of the functions denote the derivative of the corresponding order at $x = 0$. However, expansions can often be found without using the Taylor formula. Let us write as an example:

$$\sqrt{1+x} = 1 + \alpha\,x + \dots,$$

or

$$1 + x = 1 + 2\alpha\,x + \dots,$$

https://doi.org/10.1515/9783110515886-014

where the right- and left-hand sides of the equality are squared, and the term x^2 is omitted as being substantially smaller than $2\alpha\, x$ for $x \to 0$. Equating the coefficients of x, we obtain $\alpha = 1/2$. Similarly:

$$\frac{1}{1+x} = 1 + \alpha\, x + \ldots,$$

or

$$1 = (1+x)(1+\alpha x) = 1 + (1+\alpha)\, x + \ldots,$$

whence we have $\alpha = -1$.

• C_3 *Orientation of coordinate axes* (p. 17).
Let two frames move relative to each other with arbitrary directed velocity $\mathbf{v}$. Since $\mathbf{v}$ is a vector, we have $\mathbf{v} = \{v_x, v_y, v_z\}$ (and the same but with the opposite sign for S'). Constant velocity components mean that some specific *orientation* is chosen for the coordinate axes in each frame. Generally speaking, the velocity components will not change if the coordinate basis rotates about the vector $\mathbf{v}$. Therefore, one more direction, in addition to the velocity vector, is necessary to unambiguously specify the orientation of the axes. For example, the observers can assume two rulers to be parallel if they are normal to the relative speed (similarly, the y- and z-axes are assumed parallel when moving along the x-axis).

• C_4 *The transitivity of mass* (p. 92).
Let us illustrate the transitivity of mass with an experiment involving a collision between two particles C and B, and then a collision between a third particle A and C:

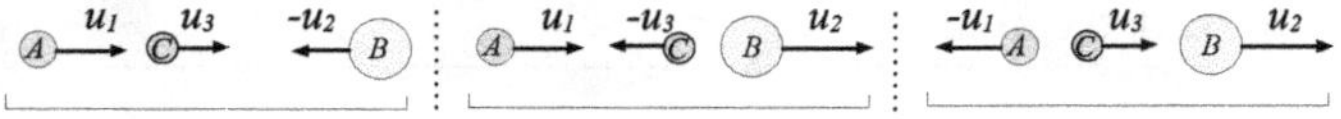

Assume that A and B are moving with the same velocities as they would in the absence of C, and that the velocity u_3 is chosen to ensure the symmetry of the positions of C and B after the collision: $m_3/m_2 = F(u_2, u_3)$. *If* the velocities of particles A and C are reversed after the collision, particle C will return have returned to its initial state and will be moving to the right with velocity u_3. This final situation is equivalent to that where C does not participate in the collision at all, and A and B collide according to the ratio of their masses.

• C_5 *Tangent vectors to the sphere* (p. 79).
When the angles change, the radius vector moves along the sphere, maintaining a constant length. The displacement vector is tangential to the sphere and normal to $\mathbf{n}$. The derivatives (increments) of the vector $\mathbf{n}$ with respect to the angles are proportional to tangential vectors to the sphere:

$$\mathbf{n}_\phi = \frac{\partial \mathbf{n}}{\partial \phi} = (-s_\theta s_\phi,\ s_\theta c_\phi, 0), \qquad \mathbf{n}_\theta = \frac{\partial \mathbf{n}}{\partial \theta} = (c_\theta c_\phi,\ c_\theta s_\phi, -s_\theta).$$

To normalize the vectors, we divide them by their lengths:

$$\mathbf{n}_\phi^2 = s_\theta^2 s_\phi^2 + s_\theta^2 c_\phi^2 = s_\theta^2, \qquad \mathbf{n}_\theta^2 = c_\theta^2 c_\phi^2 + c_\theta^2 s_\phi^2 + s_\theta^2 = 1.$$

• **C**$_6$*Projections of tangent vectors to the sphere* (p. 80).
We have to turn to the coordinates of the Earth (the celestial sphere of the terrestrial observer). However, the projections $\mathbf{ne}_\phi = -\mathbf{Pe}_\phi$ and $\mathbf{ne}_\theta = -\mathbf{Pe}_\theta$ will be equal for both spheres, up to the first order of smallness with respect to the parallax.

• **C**$_7$ *Coordinates on the celestial sphere* (p. 81).
All celestial bodies are perceived by the observer as being located on the "*celestial sphere*". In the *altazimuth system* the fundamental plane (x, y) coincides with the circle of the visible horizon. The point directly overhead (on the z-axis) is called the *zenith*:

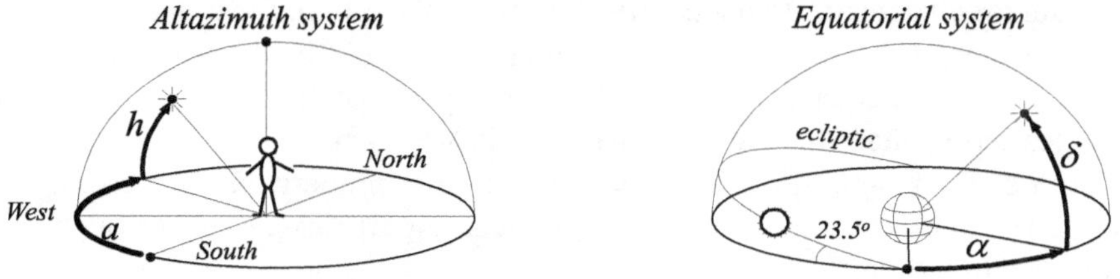

The position of an object is specified by the angle h (*altitude*), which is the angular distance from the horizon to the object, measured along the large circle passing through the zenith, and a (*azimuth*) , which is the angle of the point on the horizon below the object, measured from the south and increasing towards the west.

Since the Earth rotates about its axis, the celestial sphere will also rotate from east to west in the altazimuth system. In the northern hemisphere, the axis of this rotation passes through the *north celestial pole*, located in the vicinity of the Pole Star.

The *equatorial* coordinate system is used to specify the coordinates of "fixed stars". It does not depend on the rotation of Earth or the observer's position on it. Its centre is the centre of the Earth, and the fundamental plane is a circle called the *celestial equator* which surrounds the Earth's equator. The two angular coordinates of the system are called the declination δ and the right ascension α.

The *declination* δ is an angle similar to the altitude; it is counted from the line of the celestial equator towards the point on the sphere along a large circle passing through the north and the south poles of the sphere. The declination is measured in degrees, from -90 to +90. Positive and negative angles indicate the directions toward the north and the south poles respectively.

The *right ascension* α is an angle counted eastward from a fixed point on the equator (vernal equinox) to the projection of the star onto the celestial equator; it is measured in hours and minutes, so that 24 hours correspond to 360 degrees. The *vernal equinox* is the point where the trajectory of the Sun along the celestial sphere intersects the equator *in the course of a year*. This trajectory is called the *ecliptic* and is inclined to the celestial equator at an angle of 23.5 degrees, since the Earth's axis is at an incline to the plane of the orbit.

The *ecliptic* coordinate system is also used. It is attached to the plane of the Earth's orbit or, in the equatorial system, to the ecliptic lines. The ecliptic latitude β is measured "up" or "down" from the ecliptic, and the longitude λ is measured eastward of the vernal equinox.

• **C$_8$** *Why do we need fixed-target accelerators* (p. 104)?
A question may arise as to why not all accelerators are colliders. The answer is simple. High energies are important for the discovery of new heavy particles which simply cannot be produced at lower energies, due to the law of conservation of energy and momentum. However, the comprehensive study of their properties requires many statistics, i.e., a great number of events associated with their occurrence. Besides fundamental reasons related to the nature of particle interaction, the probability of such reactions depends on the density of the colliding beams. The density of particles in a beam is much smaller than the possible density of a stationary target. Therefore, target scattering often substantially increases the accuracy.

Besides the kinetic energy, *luminosity* is another property characterizing accelerators. The luminosity is defined as the number of particles passing through a unit surface per unit time:

$$L = \frac{N}{\Delta t \cdot S} = \frac{N}{\Delta x \cdot S}\frac{\Delta x}{\Delta t} = \rho\, u,$$

where ρ is the density of the beam and u is its speed. Luminosity is measured in cm^{-2} s^{-1}. The higher the luminosity, the higher the probability of reactions occurring during the collisions. For example, the luminosity of the Large Hadron Collider is of order 10^{34} cm^{-2} s^{-1}. The protons used there are not continuous beams but *packets*, each separated by a distance of several metres. Each packet contains about 10^{10} protons. Geometrically, a packet is a "needle" , a few dozen centimetres long and less than a millimetre thick.

• **C$_9$** *Clocks in a non-inertial frame* (p. 139).
Obviously, some clock types will not work in a non-inertial frame, and vice versa. For example, a pendulum clock with suspension on the x-axis will "tick" *uniformly* under the influence of *constant* inertial forces, but will break down in the inertial frame due to "weightlessness". Nevertheless, we assume that there is a wide class of synchronized clocks in the vicinity of a given point in space in a uniformly accelerated frame.

• **C$_{10}$** *The interval for a light signal is* $ds = 0$ (p. 161).
This statement follows from the method by which the interval ds is obtained in the NIRF. Its expression is obtained by substituting transformations into $ds^2 = dT^2 - dX^2 - dY^2 - dZ^2$. Obviously, if $ds = 0$ in the inertial system, it must also be so in the NIRF between the same events.

H. Help

This chapter contains solutions to the problems and cumbersome derivations marked with symbol ($<$ H_i) in the main text.

Logical background

• H_1 *Projective invariance of the equation of a line* (p. 12).
Substituting the transform (1.6) in equation $x' = x'_0 + u'\,t'$, we obtain

$$\frac{D\,x + E\,t}{1 + a\,t + b\,x} = x'_0 + u'\,\frac{A\,t + B\,x}{1 + a\,t + b\,x}.$$

Multiplying the right and the left sides by the denominator then gives

$$x = x_0 + u \cdot t,$$

where $x_0 = x'_0/(D - x'_0\,b - u'\,B)$, $u = (x'_0\,a + u'\,A - E)/(D - x'_0\,b - u'\,B)$. Thus, we again obtain a linear time dependence for the coordinate with new constants x_0 and u.

• H_2 *Linearity of transforms* (p. 12).
From the previous problem we have $u = (x'_0\,a + u'\,A - E)/(D - x'_0\,b - u'\,B)$. The velocity will be independent of the initial condition x'_0 only if $a = b = 0$ (otherwise, two particles with equal velocities but different initial positions x'_0 in S' will have different velocities in S).

• H_3 *Deriving linear fractional transformations* (p. 12)*.
Let the coordinate and time transformations $t' = f(t, x)$, $x' = g(t, x)$, satisfy the definition of an inertial frame:

$$\frac{du}{dt} = 0 \qquad \Rightarrow \qquad \frac{du'}{dt'} = 0.$$

The velocities in the frames are $u = dx/dt$ and $u' = dx'/dt'$, therefore ($<$ H_4):

$$u' = \frac{g_x\,u + g_t}{f_x\,u + f_t},$$

where $f_x = \partial f(x,t)/\partial x$, etc. Differentiating u' with respect to $dt' = (f_x u + f_t)\,dt$ and setting to zero the coefficients of those terms with powers of u (due to its arbitrariness), we obtain ($<$ H_5) the following system of equations:

$$f_{xx}\,g_x = g_{xx}\,f_x, \tag{1}$$

$$f_{tt}\,g_t = g_{tt}\,f_t, \tag{2}$$

$$f_{xx}\,g_t + 2\,f_{xt}\,g_x = g_{xx}\,f_t + 2\,g_{xt}\,f_x, \tag{3}$$

$$f_{tt}\,g_x + 2\,f_{xt}\,g_t = g_{tt}\,f_x + 2\,g_{xt}\,f_t. \tag{4}$$

https://doi.org/10.1515/9783110515886-015

Multiplying (3) by f_t, and (4) by f_x, subtracting one from the other and using (1) and (2), we obtain a differential equation for f only:

$$2 f_{xt} = f_{xx} \frac{f_t}{f_x} + f_{tt} \frac{f_x}{f_t} \tag{5}$$

(and similarly for g, due to the symmetry of $f \mapsto g$ and $g \mapsto f$).

Let us differentiate the Jacobian of the transformation $D = f_x g_t - f_t g_x \neq 0$ with respect to x and t and rewrite the mixed derivatives with the help of (1)-(4):

$$2 \frac{D_x}{D} = 3 \frac{f_{xx}}{f_x}, \qquad 2 \frac{D_t}{D} = 3 \frac{f_{tt}}{f_t}.$$

These equations are easily integrated:

$$\frac{f_x}{A(t)} = \frac{f_t}{B(x)} = \frac{g_x}{\bar{A}(t)} = \frac{g_t}{\bar{B}(x)} = D^{2/3}, \tag{6}$$

where $A(t)$, $B(x)$, $\bar{A}(t)$ and $\bar{B}(x)$ are some functions ("constants" of integration) and the equalities in g_x and g_t follow from symmetry. We next find f_{xx} and f_{tt} from $f_x = f_t\, A(t)/B(x)$ and substitute them into (5):

$$\dot{A}(t) = -B'(x) = \alpha.$$

Since t and x are independent variables, α is an arbitrary constant. For symmetry reasons, similar equations also hold for $\bar{A}(t)$, $\bar{B}(x)$:

$$A(t) = \alpha\, t + \beta, \quad B(x) = -\alpha\, x + \gamma, \quad \bar{A}(t) = \bar{\alpha}\, t + \bar{\beta}, \quad \bar{B}(x) = -\bar{\alpha}\, x + \bar{\gamma},$$

where α, β, γ, $\bar{\alpha}$ and $\bar{\beta}$, $\bar{\gamma}$ are constants. Integrating $f_x\, \bar{A}(t) = g_x\, A(t)$ with respect to x and, similarly, $f_t\, \bar{B}(x) = g_t\, B(x)$ with respect to t, we have:

$$f(x,t)\, \bar{A}(t) = g(x,t)\, A(t) + M(t), \qquad f(x,t)\, \bar{B}(x) = g(x,t)\, B(x) + N(x).$$

We differentiate the first relation with respect to t and the second relation with respect to x. Rewriting first derivatives with the help of (6), we obtain:

$$\dot{M}(t) = -N'(x) = \sigma,$$

where σ is also a constant, due to the independence of t and x. Therefore:

$$M(t) = \sigma\, t + \lambda, \qquad N(x) = -\sigma\, x + \mu,$$

which give linear fractional transformations.

This result is true for any number of dimensions. In fact, assuming that the motion takes place along the x-axis only, we conclude that the coefficients of the linear fractional transformation are also functions of y. However, for motion along y, the whole derivation is exactly the same, and we again obtain linear fractional transformations with respect to t and y, with coefficients depending on x. Since x and y are equivalent, the linear fractional transformation should be valid also with respect to t, x and y. In this case, both the numerator and denominator should contain xy terms, which are linear with respect to both x and y. Substituting the trajectory $x_i = u_i t + x_{0i}$ into these transformations, we can see that linearity is preserved only if the coefficients of the xy terms are zero.

• **H**$_4$ *Deriving projective transformations 1* (p. 363).
From the definition of the differentials we have:

$$dt' = \frac{\partial f}{\partial x}\,dx + \frac{\partial f}{\partial t}\,dt = f_x\,dx + f_t\,dt, \qquad dx' = \frac{\partial g}{\partial x}\,dx + \frac{\partial g}{\partial t}\,dt = g_x\,dx + g_t\,dt.$$

Their ratio gives the velocity transformation:

$$\frac{dx'}{dt'} = \frac{g_x\,dx + g_t\,dt}{f_x\,dx + f_t\,dt} = \frac{g_x\,u + g_t}{f_x\,u + f_t},$$

where the numerator and the denominator have been divided by dt in the final equality.

• **H**$_5$ *Deriving projective transformations 2* (p. 363).

$$du' = d\Big(\frac{g_x\,u + g_t}{f_x\,u + f_t}\Big) = \frac{(g_{xx}\,dx + g_{xt}\,dt)\,u + g_{tx}\,dx + g_{tt}\,dt}{f_x\,u + f_t}$$
$$- \frac{g_x\,u + g_t}{(f_x\,u + f_t)^2}\,((f_{xx}\,dx + f_{xt}\,dt)\,u + f_{tx}\,dx + f_{tt}\,dt).$$

Dividing by dt, multiplying by $(f_x\,u + f_t)^2$, and equating to zero:

$$(f_x\,u + f_t)\,(g_{xx}\,u^2 + 2g_{xt}\,u + g_{tt}) - (g_x\,u + g_t)\,(f_{xx}\,u^2 + 2f_{xt}\,u + f_{tt}) = 0.$$

• **H**$_6$ *Transformation of the squared velocity* (p. 20).
Let us write down the transformations for the velocity components (1.20):

$$\mathbf{u}'^2 = u_x'^2 + u_y'^2 = \frac{(u_x - v)^2 + (1 - v^2)\,u_y^2}{(1 - u_x v)^2},$$

whence:

$$1 - \mathbf{u}'^2 = (1 + \mathbf{u}^2\,v^2 - \mathbf{u}^2 - v^2)\,(1 - u_x v)^{-2} = \frac{(1 - \mathbf{u}^2)(1 - v^2)}{(1 - u_x v)^2}.$$

• **H**$_7$ *Identity (1.26)* (p. 21).

$$u_x = \frac{u_x' + v}{1 + u_x' v} \qquad \Rightarrow \qquad 1 \pm u_x = \frac{1 + u_x' v \pm (u_x' + v)}{1 + u_x' v} = \frac{(1 \pm u_x')(1 \pm v)}{1 + u_x' v}.$$

• H_8 *Invariance of the interval* (p. 22).
Using transformations for the increments (1.19), p. 18, we have:

$$(\Delta s)^2 = (\Delta t')^2 - (\Delta x')^2 - (\Delta y')^2 = \frac{(\Delta t - v\Delta x)^2 - (\Delta x - v\Delta t)^2}{1 - v^2} - (\Delta y)^2.$$

Expanding the squares, we obtain the invariance of the interval:

$$(\Delta s)^2 = (\Delta t)^2 - (\Delta x)^2 - (\Delta y)^2.$$

A little philosophy and history

• H_9 *Anisotropic Lorentz transformations* (p. 37).
Define a new function of the velocity in (1.11), p. 14:

$$\gamma(-v)\gamma(v) = \frac{1}{1 - \alpha\, v^2}, \qquad \gamma(v) = \frac{f(v)}{\sqrt{1 - \alpha\, v^2}}.$$

Whence $f(-v)\,f(v) = 1$. Take the first equation in the system of successive transformations (p. 13) and substitute $\sigma(v) = \alpha\, v$ into it:

$$x_3 = \gamma_2\gamma_1\,[(1 + \alpha\, v_1 v_2)\, x_1 - (v_1 + v_2)\, t_1\,] = \gamma_3\,[x_1 - v_3\, t_1].$$

Let us equate the coefficients of x_1: $\gamma_3 = (1+\alpha v_1 v_2)\,\gamma_1\gamma_2$, and write this equation using the function $f(v)$:

$$\frac{f_3}{\sqrt{1 - \alpha\, v_3^2}} = (1 + \alpha\, v_1 v_2)\,\frac{f_1 f_2}{\sqrt{1 - \alpha\, v_1^2}\,\sqrt{1 - \alpha\, v_2^2}}.$$

Substituting v_3 (1.21), (1.23) into the left-hand side, we have $f_3 = f_1 f_2$. As a result:

$$f\left(\frac{v_1 + v_2}{1 + \alpha\, v_1 v_2}\right) = f(v_1) f(v_2).$$

Since the velocities v_1 and v_2 are independent, differentiate with respect to v_2 gives

$$f'\left(\frac{v_1 + v_2}{1 + \alpha\, v_1 v_2}\right)\frac{1 - \alpha v_1^2}{(1 + \alpha\, v_1 v_2)^2} = f(v_1) f'(v_2).$$

Assuming now that $v_2 = 0$:

$$f'(x)\,(1 - \alpha x^2) = a\, f(x),$$

where $x = v_1$, and the constant $a = f'(0)$. This equation is not difficult to solve. Suppose $\alpha = 1/c^2 > 0$, so that

$$\int \frac{df}{f} = \int \frac{c^2\, a\, dx}{c^2 - x^2} = \frac{a\, c}{2} \int \left[\frac{1}{c - x} + \frac{1}{c + x}\right] dx = \frac{a\, c}{2}\ln\frac{c + x}{c - x} = \ln\frac{f}{f_0}.$$

Introducing a constant $\mu = a\, c/2$ and given that $\gamma(0) = 1$ (the frames S and S' coincide), we obtain the transformations

$$x' = \left(\frac{c + v}{c - v}\right)^{\mu} \frac{x - vt}{\sqrt{1 - v^2/c^2}}, \qquad t' = \left(\frac{c + v}{c - v}\right)^{\mu} \frac{t - vx/c^2}{\sqrt{1 - v^2/c^2}}.$$

Kinematics

• $\mathbf{H}_{10}$ *Recovering "c" in the formulas for the Doppler effect* (p. 61).

$$\nu = \nu_0 \sqrt{\frac{c-v}{c+v}}, \qquad \nu = \nu_0 \sqrt{\frac{c+v}{c-v}}, \qquad \nu = \nu_0 \sqrt{1 - v^2/c^2}.$$

• $\mathbf{H}_{11}$ *Motion of a cube in two reference frames* (p. 70).

Let some fixed point in S' move along the trajectory $x' = x'_0$, $y' = y'_0 + u't'$, where x'_0, y'_0, and u' are constants. Substitute this trajectory into the inverse Lorentz transformations: $t = \gamma(t' + vx'_0)$, $x = \gamma(x'_0 + vt')$, $y = y'_0 + u't'$. Rearranging to remove the time t', we obtain the trajectory of the point in S:

$$x = \frac{x'_0}{\gamma} + vt, \qquad y = (y'_0 - u'v x'_0) + u\, t,$$

where $u = u'/\gamma$ is the velocity of the point along the y-axis in S, and $\gamma = 1/\sqrt{1-v^2}$. Suppose that the bottom left corner of the square (point D in the figure on p. 69) coincides with the origin $x'_0 = y'_0 = 0$ at $t = t' = 0$, so that the bottom right corner (point C) has coordinates $y'_0 = 0$, $x'_0 = L_0$. If the square's sides have length L_0 in the proper frame, then it looks shrunken when moving along the axis y' in S', and the coordinates of points A and B are $\{0,\ L_0/\gamma'\}$ and $\{L_0,\ L_0/\gamma'\}$, respectively, where $\gamma' = 1/\sqrt{1-u'^2}$. As a result, the coordinates $\{x, y\}$ of the corners in S are ($t = 0$):

$$\Big\{0,\ 0\Big\}_D \qquad \Big\{\frac{L_0}{\gamma},\ -u'vL_0\Big\}_C \qquad \Big\{0,\ \frac{L_0}{\gamma'}\Big\}_A \qquad \Big\{\frac{L_0}{\gamma},\ \frac{L_0}{\gamma'} - u'v\,L_0\Big\}_B.$$

The proportions in the figures on p. 69 correspond to the velocities $v = 4/5 = 0.8$ ($\gamma = 5/3$) and $u' = 3/5 = 0.6$ ($\gamma' = 5/4$).

• $\mathbf{H}_{12}$ *The total number of stars in the sky* (p. 82).

Integrate the star density distribution over the solid angle:

$$\frac{N}{4\pi} \int \frac{1-v^2}{(1 - v\cos\theta')^2} \sin\theta' d\theta' d\phi' = \frac{N}{2} \int_{-1}^{1} \frac{1-v^2}{(1-vz)^2}\, dz,$$

where the integral over $d\phi'$ is 2π, and we have made the substitution $z = \cos\theta'$. Since $\theta' = [0 \ldots \pi]$, we have $z = [1 \ldots -1]$. By interchanging the limits of integration we remove the minus sign, $\sin\theta'\, d\theta' = -d(\cos\theta')$, and evaluate the integral over z:

$$\frac{N}{2} \int_{-1}^{1} \frac{1-v^2}{(1-vz)^2}\, dz = \frac{N}{2}\frac{1}{v}\frac{1-v^2}{1-vz}\Big|_{-1}^{+1} = \frac{N}{2v}\left[(1+v) - (1-v)\right] = N.$$

• $\mathbf{H}_{13}$ *The trajectory of accelerated motion* (p. 85).
Integrate the velocity of the body:

$$x(t) = x(0) + \int_0^t u(t)\,dt = x_0 + \int_0^t \frac{\pi_0 + at}{\sqrt{1 + (\pi_0 + at)^2}}\,dt.$$

Substituting in $\tau = \pi_0 + at$ and $d\tau = a\,dt$, we obtain:

$$x(t) - x_0 = \frac{1}{a}\int_{\pi_0}^{\pi_0+at} \frac{\tau}{\sqrt{1+\tau^2}}\,d\tau = \frac{1}{2a}\int_{\pi_0}^{\pi_0+at} \frac{d(\tau^2)}{\sqrt{1+\tau^2}} = \frac{1}{a}\sqrt{1+\tau^2}\Big|_{\pi_0}^{\pi_0+at}.$$

• $\mathbf{H}_{14}$ *Recovering "c" for uniformly accelerated motion* (p. 86).
The dimensions of π_0 are those of velocity; therefore $\pi_0 \mapsto \pi_0/c$. For the time and the acceleration we have $t \mapsto ct$, $a \mapsto a/c^2$ and

$$x(t) = x_0 + \frac{c^2}{a}\left(\sqrt{1 + (\pi_0 + a\,t)^2/c^2} - \sqrt{1 + \pi_0^2/c^2}\right).$$

• $\mathbf{H}_{15}$ *Expanding the travel time in a series* (p. 87).
Expand the hyperbolic arcsine with the following formula:

$$\operatorname{ash} y \approx y - \frac{y^3}{6} + \dots$$

This can be derived from the general Taylor formula or from the exponential expansion:

$$e^x \approx 1 + x + \frac{x^2}{2} + \frac{x^3}{6} + \dots \quad => \quad \operatorname{sh} x = \frac{e^x - e^{-x}}{2} \approx x + \frac{x^3}{6} + \dots$$

Since $y = \operatorname{sh} x$, we have $y^3 = \operatorname{sh}^3 x \approx x^3$ (up to the x^3 order of smallness). Therefore,

$$\operatorname{ash} y = x = \operatorname{sh} x - \frac{x^3}{6} = y - \frac{y^3}{6}.$$

To expand the second term, the following expansions are used: ($\lessdot C_2$):

$$\sqrt{1+x} \approx 1 + \frac{x}{2}, \qquad \frac{1}{1+x} \approx 1 - x.$$

Relativistic dynamics

• $\mathbf{H}_{16}$ *Conservation of energy in the one-dimensional case* (p. 99).
Consider a collision of two identical particles with masses m moving with velocities u towards each other. After the collision we have a single resting particle with mass M:

Suppose that $E = m\,f(u)$, and f is an even function such that $f(-u) = f(u)$ and $f(0) = 1$. Consider another frame moving to the left with speed v. The law of conservation of energy in this frame will have the form:

$$mf\left(\frac{v+u}{1+uv}\right) + mf\left(\frac{v-u}{1-uv}\right) = Mf\,(v)\,.$$

Assume $v = 0$. Then, given the symmetry of f, we have $M = 2mf(u)$ for the final particle. Substituting this mass into the conservation law for an arbitrary speed v, we obtain the functional equation:

$$f\left(\frac{v+u}{1+uv}\right) + f\left(\frac{v-u}{1-uv}\right) = 2f(u)\,f(v).$$

To find its solution, we use the hyperbolic tangents $v = \mathrm{th}(x)$ and $u = \mathrm{th}(y)$, and the function $g(x) = f(\mathrm{th}(x))$:

$$g(x+y) + g(x-y) = 2g(x)g(y),$$

Differentiating the right- and left-hand sides with respect to y,

$$g'\,(x+y) - g'\,(x-y) = 2g(x)\,g'\,(y).$$

For $y = 0$, we have $g'(0) = 0$. The second derivative gives ($g''(0) = a$):

$$g''(x+y) + g''(x+y) = 2g(x)\,g''(y).$$

Assuming again $y = 0$, we have:

$$g''(x) = a\,g(x) \qquad => \qquad g(x) = A\,e^{ax} + B\,e^{-ax}.$$

Using the conditions $g(0) = 1$ and $g'(0) = 0$, we obtain $A+B = 1$, $A-B = 0$. Therefore, $A = B = 1/2$, and we finally have:

$$g(x) = \mathrm{ch}(ax) \qquad => \qquad E = \frac{m}{2}\,\frac{(1+u)^a + (1-u)^a}{(1-u^2)^{a/2}}.$$

The value $a = 1$ gives the usual relativistic expression.

• **H**$_{17}$ *Derivation of the momentum expression* (p. 101).
Assuming that $\mathbf{p} = m\mathbf{u}f(\mathbf{u}^2)$ and that the momentum is conserved, we obtain the function $f(\mathbf{u}^2)$ from the relativistic velocity-addition formula. Consider two identical particles which collide and are stuck together thereafter. The final particle has zero velocity in the center-of-mass frame. If the observer moves to the *left* with speed $v = -u$, he will see that the second particle is at rest. Due to the relation

$$u'_x = (u_x - v)/(1 - u_x v),$$

the first particle hits the second from the left with speed $u'_x = w = 2u/(1 + u^2)$:

The law of conservation of momentum has the following form for the observer:

$$m\,w\,f\left(w^2\right) = Mu\,f(u^2). \tag{7}$$

Let us turn the picture so that the particle with velocity $w = 2u/(1 + u^2)$ hits the identical resting particle up from below. The resulting particle will also go up with speed u. Write the law of conservation of momentum for the observer in the frame S' moving *to the left* along the x-axis with speed v relative to S:

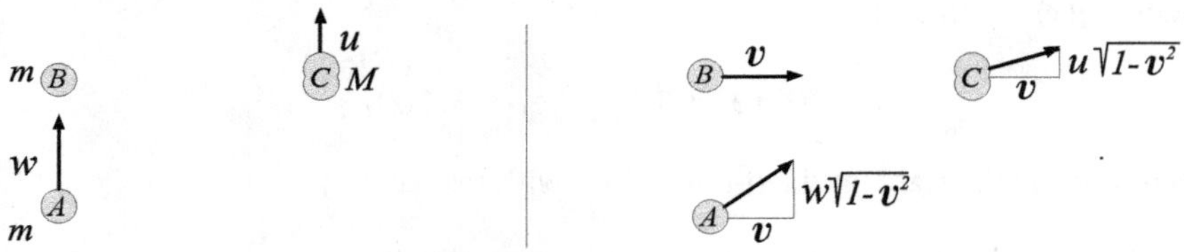

The horizontal velocity component is v for all particles in S'. Since it was zero in S, we have from the velocity-addition formula (1.20), p. 19:

$$u'_y = u_y\sqrt{1 - v^2}.$$

Let us write down the law of conservation for the projection of the momentum onto the x-axis, given that $\mathbf{u}^2 = u_x^2 + u_y^2$:

$$\underbrace{mvf\left(v^2 + w^2\left(1 - v^2\right)\right)}_{A} + \underbrace{mvf(v^2)}_{B} = \underbrace{Mvf(v^2 + u^2(1 - v^2))}_{C}.$$

Reduce v, then assume it to be zero $v = 0$ and divide the expression by (7): $\frac{f(0)}{f(w^2)} = \frac{w}{u} - 1$. Expressing u in terms of $w = 2u/(1 + u^2)$, we have

$$u = (1 - \sqrt{1 - w^2})/w,$$

and finally obtain $f(w^2) = f(0)/\sqrt{1 - w^2}$.

• $\mathbf{H}_{18}$ *Maximum scattering angle* (p. 108).
Denote $E_1' = x$ and find the maximum of the expression:

$$F = \frac{(E_1 + m_2)x - (m_1^2 + E_1 m_2)}{p_1\sqrt{x^2 - m_1^2}}.$$

Differentiating and setting the derivative to zero:

$$F' = \frac{E_1 + m_2}{p_1\sqrt{x^2 - m_1^2}} - \frac{(E_1 + m_2)x - (m_1^2 + E_1 m_2)}{p_1(x^2 - m_1^2)^{3/2}}\,x = 0,$$

whence:

$$x = \frac{m_1^2(E_1 + m_2)}{m_1^2 + E_1 m_2},$$

or

$$\sqrt{x^2 - m_1^2} = \frac{p_1 m_1\sqrt{m_1^2 - m_2^2}}{m_1^2 + E_1 m_2},$$

where we used that $E_1^2 - m_1^2 = p_1^2$. Substituting x and $\sqrt{x^2 - m_1^2}$ into F, we obtain:

$$F_{max} = \frac{\sqrt{m_1^2 - m_2^2}}{m_1} = (\cos\theta_1)_{max}.$$

Given that $\sin^2\theta_1 = 1 - \cos^2\theta_1$, we obtain the maximum value of the sine.

• $\mathbf{H}_{19}$ *Solution of the equation of jet motion* (p. 113).
By integration, we obtain $\ln(f/f_0)$ on the left-hand side, where f_0 is the constant of integration. We make the substitutions $v^2 = z$ and $2v\,dv = dz$ for the first term on the right-hand side to obtain the logarithm. For the second and third terms the logarithms are obtained immediately:

$$\ln\frac{f}{f_0} = -\frac{1}{2}\ln(1 - v^2) - \frac{1}{2u_0}\left[-\ln(1 - v) + \ln(1 + v)\right].$$

• $\mathbf{H}_{20}$ *Non-relativistic limit of the Tsiolkovsky formula* (p. 113).
Let us find the non-relativistic limit of the relation (4.27), p. 113, where the speed of light has been restored:

$$\frac{M}{M_0} = \left(\frac{1 - v/c}{1 + v/c}\right)^{c/(2u_0)} \approx \left(1 - \frac{2v}{c}\right)^{c/(2u_0)}.$$

The approximate equality is written for small velocities, so that $(1 + v/c)^{-1} \approx 1 - v/c$, and only the first order with respect to v/c is considered when multiplying out the parentheses. We use the limit used to define Euler's number and the exponential function: $e^x = (1 + \alpha x)^{1/\alpha}$ for $\alpha \to 0$. Write $\alpha = 2u_0/c$ and $x = -v/u_0$. The non-relativistic limit corresponds to $c \to \infty$ or $\alpha \to 0$, whence $M/M_0 = e^{-v/u_0}$.

Force and equations of motion

• $\mathbf{H}_{21}$ *Completing the square* (p. 122).
After some elementary algebraic calculations, we have:

$$2\mathbf{w}_0\mathbf{a}\,t + \mathbf{a}^2\,t^2 = \mathbf{a}^2\left(t^2 + 2\frac{\mathbf{w}_0\mathbf{a}}{\mathbf{a}^2}\,t\right) = \mathbf{a}^2\left(t + \frac{\mathbf{w}_0\mathbf{a}}{\mathbf{a}^2}\right)^2 - \frac{(\mathbf{w}_0\mathbf{a})^2}{\mathbf{a}^2}.$$

In addition, $[\mathbf{w}_0 \times \mathbf{a}]^2 = \mathbf{w}_0^2\,\mathbf{a}^2 - (\mathbf{w}_0\mathbf{a})^2$.

• $\mathbf{H}_{22}$ *Equations in component form* (p. 124).
Assuming that $\mathbf{b} = \{0, 0, 1\}$, we obtain from (5.17) and (5.18) the following equations for $\mathbf{u} = \{u_x,\ u_y,\ u_z\}$:

$$\frac{du_x}{dt} = \omega\,u_y, \qquad \frac{du_y}{dt} = -\omega\,u_x, \qquad \frac{du_z}{dt} = 0.$$

Differentiating the first equation and substituting the second equation into the first one, we obtain two oscillator equations with the same frequency:

$$\frac{d^2u_x}{dt^2} + \omega^2\,u_x = 0, \qquad \frac{d^2u_y}{dt^2} + \omega^2\,u_y = 0.$$

• $\mathbf{H}_{23}$ *The time derivative* (p. 125).
The time derivative of the radius vector can be found like so:

$$\frac{dr}{dt} = \frac{d\sqrt{x^2+y^2+z^2}}{dt} = \frac{1}{2\sqrt{x^2+y^2+z^2}}\left(2x\frac{dx}{dt} + 2y\frac{dy}{dt} + 2z\frac{dz}{dt}\right) = \frac{\mathbf{ru}}{r}.$$

• $\mathbf{H}_{24}$ *Multiplicative energy* (p. 126).

$$\frac{d\mathcal{E}}{dt} = \left(\mathbf{uF} + E\,V'\,\frac{\mathbf{ru}}{r}\right)e^{V(r)} = 0.$$

Substituting in (5.26), we obtain zero, as expected.

• $\mathbf{H}_{25}$ *The relation for* f_2 (p. 127).
Since

$$\mathbf{r}\times[\mathbf{u}\times[\mathbf{r}\times\mathbf{u}]] = \mathbf{r}\times(\mathbf{r}\,\mathbf{u}^2 - \mathbf{u}\,(\mathbf{ru})) = -[\mathbf{r}\times\mathbf{u}]\,(\mathbf{ru}),$$

substituting in the expression (5.26) for the force, with $f_3 = 0$, gives

$$\frac{d\mathbf{L}}{dt} = -E\,g'(E)\,\frac{V'(r)}{r}\,(\mathbf{ru})\,[\mathbf{r}\times\mathbf{p}] + g(E)\,f_2\,[\mathbf{r}\times\mathbf{u}]\,(\mathbf{ru}) = 0,$$

and, given that $\mathbf{p} = \mathbf{u}\,E$, we obtain f_2.

Non-inertial framesNon-inertial frames

• $\mathbf{H}_{26}$ *Trajectory of the second ship* (p. 141).

Use the identities following from the definition of hyperbolic functions (p. 355):

$$\mathrm{ch}(\alpha) + \mathrm{sh}(\alpha) = e^{\alpha}, \qquad \mathrm{ch}(\alpha) - \mathrm{sh}(\alpha) = e^{-\alpha}.$$

• $\mathbf{H}_{27}$ *Sending and receiving times for the signal* (p. 142).

Since $l_0 = \tau_0/2 = \ln(1 + ax_0)/a$, we have

$$e^{at_2} = aT + \sqrt{e^{a\tau_0} + (aT)^2} = (1 + ax_0)\left[\mathrm{sh}\left(a\, e^{-al_0}\, t'\right) + \mathrm{ch}\left(a\, e^{-al_0}\, t'\right)\right],$$

where we have substituted in the time T from the second relation in (6.12). Given the identities of the previous problem, we obtain the time t_2 at which the signal is received. The time of its sending is $t_1 = t_2 - \tau_0 = t_2 - 2l_0$.

• $\mathbf{H}_{28}$ *Velocity of the point* $X = 0$ (p. 147).

Using the transformations (6.21), write the expression for the trajectory of the point $X = 0$ in S_0 relative to the observer in $x = 0$:

$$x = -\frac{1}{a}\ln[\mathrm{ch}(at)] \quad => \quad u = -\frac{dx}{dt} = -\,\mathrm{th}(at) = -\frac{aT}{\sqrt{1 + (aT)^2}}.$$

In the final equality, we have substituted the time t with the time of the clock at $x = 0$, equal to $\mathrm{sh}(at) = aT$ for observers in S_0. As follows from this relation, the velocity of the observer at $X = 0$ relative to S is the velocity (with the opposite sign) of the observer at $x = 0$ relative to S_0.

• $\mathbf{H}_{29}$ *Decomposing non-inertial transformations with respect to* $1/c^2$ (p. 150).

Write the transformations (6.29) on p. 150 in the initial noncompact form and make the substitutions $T \mapsto cT$, $t \mapsto ct$, $a \mapsto a/c^2$, and $U_0 \mapsto U_0/c$:

$$\frac{aX}{c^2} = \gamma_0\left[\mathrm{ch}\left(\frac{at}{c}\right) + \frac{U_0}{c}\,\mathrm{sh}\left(\frac{at}{c}\right)\right] e^{ax/c^2} - \gamma_0,$$

$$\frac{aT}{c} = \gamma_0\left[\mathrm{sh}\left(\frac{at}{c}\right) + \frac{U_0}{c}\,\mathrm{ch}\left(\frac{at}{c}\right)\right] e^{ax/c^2} - \gamma_0\,\frac{U_0}{c}.$$

When expanding in a series with respect to $1/c$, we have used that:

$$e^x \approx 1 + x + \ldots, \qquad \mathrm{ch}(x) \approx 1 + \frac{x^2}{2} + \ldots, \qquad \mathrm{sh}(x) \approx x + \ldots$$

Substituting in these expansions, we obtain:

$$X \approx x + U_0 t + \frac{1}{2}at^2, \qquad T \approx t.$$

Covariant formalism

• $\mathbf{H}_{30}$ *Lorentz transformations for a 2nd-order tensor* (p. 186).

Let us find the explicit vector form for the transformation laws of a 2nd-order tensor $T^{\alpha\beta}$. Since it is transformed as the product of the components of two four-vectors, the particular case $T^{\alpha\beta} = A^{\alpha}B^{\beta}$ can be considered. Write the Lorentz transformations (7.4), p. 174, for the four-vectors A^{α} and B^{β} in component form:

$$A'^{0} = \gamma \cdot (A^{0} - v_k A^{k}), \qquad A'^{i} = A^{i} - \gamma v_i A^{0} + \Gamma\, v_i v_j A^{j},$$

$$B'^{0} = \gamma \cdot (B^{0} - v_k B^{k}), \qquad B'^{j} = B^{j} - \gamma v_j B^{0} + \Gamma\, v_j v_k B^{k},$$

where Latin indices vary from 1 to 3, and upper and lower indices are not distinguished for the three-dimensional velocity $\mathbf{v}$, assuming that $v_i = v^{i}$. Multiplying these transformations, we obtain:

$$T'^{00} = \gamma^2 T^{00} - \gamma^2 v_k (T^{0k} + T^{k0}) + \gamma^2 v_n v_m T^{nm},$$

$$T'^{0i} = \gamma(T^{0i} - v_k T^{ki}) - \gamma^2 v_i (T^{00} - v_k T^{k0}) + \gamma\Gamma\, v_i (v_k\, T^{0k} - v_n v_m\, T^{nm}),$$

$$T'^{i0} = \gamma(T^{i0} - v_k T^{ik}) - \gamma^2 v_i (T^{00} - v_k T^{0k}) + \gamma\Gamma\, v_i (v_k\, T^{k0} - v_n v_m\, T^{nm}),$$

$$T'^{ij} = T^{ij} + \gamma^2 v_i v_j T^{00} - \gamma (v_i T^{0j} + v_j T^{i0}) + \Gamma (v_j v_k T^{ik} + v_i v_k T^{kj})$$
$$-\gamma\Gamma\, v_i v_j\, (v_k T^{0k} + v_k T^{k0}) + \Gamma^2 v_i v_j v_n v_m T^{nm}.$$

These relations can be simplified for an antisymmetric tensor $T^{\alpha\beta} = -T^{\alpha\beta}$. Let us use the fact that $v_n v_m T^{nm} = 0$ in this case. Indeed, by renaming the summation indices, we can write

$$v_n v_m T^{nm} = v_m v_n T^{mn} = -v_n v_m T^{nm}.$$

The antisymmetry property was used in the second equality. An expression equal to itself with the opposite sign can only be zero. As a result, we have for the zero components $T'^{00} = \gamma^2 T^{00} = 0$, i.e., if the component T^{00} of the antisymmetric tensor is zero in one coordinate system, it will also be zero in any other system. For the remaining components, we have:

$$T'^{0i} = \gamma(T^{0i} - v_k T^{ki}) - \Gamma\, v_i v_k\, T^{0k},$$

$$T'^{ij} = T^{ij} - \gamma(v^{i} T^{0j} - v^{j} T^{0i}) + \Gamma(v^{j} T^{ik} - v^{i} T^{jk}) v_k.$$

An antisymmetric 2nd-order four-tensor has six non-zero components. They can be expressed in terms of the components of two vectors in a three-dimensional space. In Chapter 5 we will use these two vectors to make these transformations even more compact.

Dynamics in covariant notation

• **H**$_{31}$ *Scattering angle in the center-of-mass frame* (p. 196).
Let us write down

$$\cos\chi = \frac{t - m_1^2 - m_3^2 + 2\,E_1E_3}{2\,|\mathbf{p}||\mathbf{q}|}$$

and substitute in E_1, E_3, $|\mathbf{p}|$ and $|\mathbf{q}|$:

$$\cos\chi = \frac{2\,s\,(t - m_1^2 - m_3^2) + (s + m_1^2 - m_2^2)\,(s + m_3^2 - m_4^2)}{\sqrt{\lambda(s,\ m_1^2,\ m_2^2)\,\lambda(s,\ m_3^2,\ m_4^2)}}.$$

Multiplying the parentheses in the numerator and replacing $m_1^2 + m_2^2 + m_3^2 + m_4^2$ by $s + t + u$, we obtain (8.18).

• **H**$_{32}$ *Scattering angle in the laboratory frame* (p. 197).
Note that the triangle function $\lambda(x,\ y,\ z)$ is symmetric and $\lambda(u,\ m_3^2,\ m_2^2) = \lambda(u,\ m_2^2,\ m_3^2)$.

• **H**$_{33}$ *Interchanging indices in the Levi-Chevita symbol* (p. 205).

$$\varepsilon_{\alpha\beta\mu\nu} = -\varepsilon_{\alpha\mu\beta\nu} = \varepsilon_{\alpha\mu\nu\beta} = -\varepsilon_{\mu\alpha\nu\beta} = \varepsilon_{\mu\nu\alpha\beta}.$$

• **H**$_{34}$ *Invariants of antisymmetric tensors* (p. 207).
Substitute the Lorentz transformation for the vectors **a** and **b** (the same as those for the momentum tensor $L^{\alpha\beta} = (\mathbf{G}, \mathbf{L})$) into the expressions $\mathbf{a}'^2 - \mathbf{b}'^2$. Given that

$$[\mathbf{v}\times\mathbf{a}]^2 = \mathbf{v}^2\mathbf{a}^2 - (\mathbf{va})^2,$$

and similarly for the vector **b**, we have:

$$\gamma^2\,(\mathbf{a}^2 + \mathbf{v}^2\,\mathbf{b}^2 - (\mathbf{vb})^2 + 2[\mathbf{v}\times\mathbf{b}]\,\mathbf{a}) + (\Gamma^2\,\mathbf{v}^2 - 2\gamma\Gamma)\,(\mathbf{va})^2,$$

$$-\gamma^2\,(\mathbf{b}^2 + \mathbf{v}^2\,\mathbf{a}^2 - (\mathbf{va})^2 - 2[\mathbf{v}\times\mathbf{a}]\,\mathbf{b}) + (\Gamma^2\,\mathbf{v}^2 - 2\gamma\Gamma)\,(\mathbf{vb})^2.$$

The vector products simplify, since by the push rule:

$$[\mathbf{v}\times\mathbf{b}]\,\mathbf{a} = \mathbf{v}\,[\mathbf{b}\times\mathbf{a}] = -\mathbf{v}\,[\mathbf{a}\times\mathbf{b}], \qquad [\mathbf{v}\times\mathbf{a}]\,\mathbf{b} = \mathbf{v}\,[\mathbf{a}\times\mathbf{b}].$$

Therefore, the expression $\mathbf{a}'^2 - \mathbf{b}'^2$ equals:

$$\gamma^2(1 - v^2)\,(\mathbf{a}^2 - \mathbf{b}^2) + ((\mathbf{va})^2 - (\mathbf{vb})^2)\,(\gamma^2 - 2\,\gamma\Gamma + \Gamma^2 v^2) = \mathbf{a}^2 - \mathbf{b}^2,$$

where we have substituted in the expression for Γ and used the following transformations:

$$\gamma^2 - 2\gamma\Gamma + \Gamma^2 v^2 = \frac{\gamma^2 - 1 - 2\,\gamma\,(\gamma - 1) + (\gamma - 1)^2}{v^2} = 0.$$

Curvilinear coordinates

• $\mathbf{H}_{35}$ *The Lorentz-Møller-Nelson interval* (p. 219) is

$$ds^2 = \left\{(1 + \mathbf{Wr})^2 - [\Omega \times \mathbf{r}]^2\right\} dt^2 - 2[\Omega \times \mathbf{r}]\, d\mathbf{r}dt - d\mathbf{r}^2,$$

where

$$\mathbf{W} = \gamma \dot{\mathbf{v}} + \frac{\gamma(\gamma - 1)}{v^2} (\dot{\mathbf{v}}\mathbf{v})\mathbf{v}, \qquad \Omega = \frac{\gamma - 1}{v^2} [\mathbf{v} \times \dot{\mathbf{v}}].$$

• $\mathbf{H}_{36}$ *Transformation for* γ_{ij} (p. 230).

The components of the metric tensor are transformed as follows:

$$g_{00} = (\partial_0 t')^2 g'_{00}, \qquad g_{0i} = (\partial_0 t')(\partial_i t' g'_{00} + \partial_i x'^p g'_{0p})$$

$$g_{ij} = \partial_i t' \partial_j t' g'_{00} + (\partial_i x'^p \partial_j t' + \partial_j x'^p \partial_i t') g'_{p0} + \partial_i x'^p \partial_j x'^q g'_{pq}.$$

Furthermore,

$$\frac{g_{0i} g_{0j}}{g_{00}} = \frac{1}{g'_{00}} (\partial_i t' g'_{00} + \partial_i x'^p g'_{0p})(\partial_j t' g'_{00} + \partial_j x'^q g'_{0q})$$

$$= \partial_i t' \partial_j t' g'_{00} + \partial_i t' \partial_j x'^q g'_{0q} + \partial_i x'^p \partial_j t' g'_{0p} + \partial_i x'^p \partial_j x'^q \frac{g'_{0p} g'_{0q}}{g'_{00}}.$$

By subtracting these, we obtain the required relation.

• $\mathbf{H}_{37}$ *Trajectory of a particle in a rotating frame* (p. 241).

The system of equations

$$\begin{cases} r\cos(\phi + \omega t) = r_0 \cos\phi_0 + V_0 t \cos\alpha \\ r\sin(\phi + \omega t) = r_0 \sin\phi_0 + V_0 t \sin\alpha \end{cases}$$

can be conveniently rewritten in complex form using Euler's formula (the second equation is multiplied by the imaginary unit and added to the first equation):

$$r\, e^{\imath(\phi + \omega t)} = r_0\, e^{\imath \phi_0} + V_0 t\, e^{\imath \alpha}.$$

The modulus of this expression gives the equation (9.71), p. 241, for r^2. Multiplying both sides by $e^{-\imath \omega t}$ and again expanding the exponent in terms of the cosine and sine, we obtain the final solution (9.70). The time derivative

$$\left[\dot{r} + \imath r(\dot{\phi} + \omega)\right] e^{\imath(\phi + \omega t)} = V_0\, e^{\imath \alpha},$$

together with its complex modulus, gives the first relation (9.72). The second relation (9.72) is most easily obtained in the laboratory frame $R^2 \dot{\Phi} = V_0 (X \sin\alpha - Y\cos\alpha) = V_0 R_0 \sin(\alpha - \Phi_0)$ by differentiating the trajectory of motion, written in polar coordinates with respect to time.

Motion of rods and gyroscopes

• $\mathbf{H}_{38}$ *Recovering "c" in the equation for the rod* (p. 255).

$$\frac{d\mathbf{s}}{dt} = -\frac{\gamma^2}{c^2}(\mathbf{vs})\,\mathbf{a}.$$

• $\mathbf{H}_{39}$ *Rotation and the relativity of simultaneity* (p. 256).

Let the observers in S' start lifting the rod upwards with velocity $\mathbf{u}'$. Since $\mathbf{u}'\mathbf{v} = 0$, it follows from the velocity-addition law (1.24), p. 20, that in the rest frame S this velocity is:

$$\mathbf{u} = \mathbf{v} + \mathbf{u}'/\gamma.$$

Denote by $d\mathbf{v} = \mathbf{u}'/\gamma$ the vertical component of the velocity.

The left and the right ends of the horizontal rod start moving upwards simultaneously ($\Delta t' = 0$) in S'. Since $\Delta t' = 0$, the right end will start moving $\Delta t = v\Delta x$ later for the observers in S, and the left end will start moving earlier than the right end ($\Delta x = x_2 - x_1 > 0$, therefore $\Delta t = t_2 - t_1 > 0$).

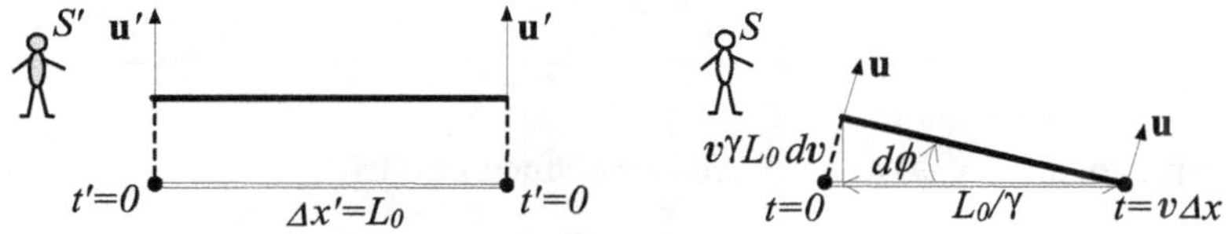

Let the left end of the rod coincide with the origins of the frames $x = x' = 0$ at $t' = t = 0$. At the same time, according to the synchronized clock in S', the right end also lies on the x'-axis (the rod is horizontal). Due to the Lorentz transformation, $x = \gamma\,(x' + vt')$, and for simultaneous events in S' ($\Delta t' = 0$) we have $\Delta x = \gamma\Delta x'$. In this case $\Delta x' = L_0$ is the proper length of the rod in S'. Therefore, the coordinate of the right event in S is γL_0. For the time $\Delta t = v\Delta x = v\gamma L_0$ the left end of the rod will move $(v\gamma L_0)\cdot dv$ upwards and $(v\gamma L_0)\cdot v$ to the right. Therefore, the projection of the rod onto the x-axis is:

$$\gamma L_0 - v^2\gamma L_0 = L_0/\gamma.$$

The tangent of the angle of rotation for a small speed dv is approximately equal to the angle:

$$\mathrm{tg}(d\phi) \approx d\phi = \frac{v\gamma L_0\, dv}{L_0/\gamma} = \gamma^2 v dv.$$

Thus, the relativity of simultaneity cause the rod to rotate through an angle $\gamma^2 v dv$ for the observers in S. The same result can be obtained from equation (10.8), p. 255.

• $\mathbf{H}_{40}$ *The length of a rod undergoing circular motion* (p. 259).
Summation of the squared equations gives the squared length of the rod, $L^2 = x^2 + y^2$. If the angle it makes with the x-axis is ϕ_0 at $t = 0$ and $x_0 = \bar{L}_0 c_0$, $y_0 = \bar{L}_0 s_0$, where $s_0 = \sin\phi_0$, $c_0 = \cos\phi_0$, then

$$\frac{L^2}{\bar{L}_0^2} = 1 + \gamma v^2 s_0 c_0 \sin(2\omega\gamma t) + \frac{v^2}{2}(\gamma^2 s_0^2 - c_0^2)(1 - \cos(2\omega\gamma t)).$$

Using the relation (3.8), p. 68, we can find the relation between the initial length of the rod $\bar{L}_0$ in the resting frame and its proper length L_0:

$$\bar{L}_0 = L_0 / \sqrt{1 + (\gamma^2 - 1)\sin^2\phi_0}.$$

The length recovers ($L = \bar{L}_0$) at times $\omega t = \pi k/\gamma$, where $k = 1, 2, \ldots$. If $\operatorname{tg}\phi_0 = y_0/x_0$ is the initial orientation of the rod, then after a time $\omega t = \pi/\gamma$ its angle ϕ will be such that $\operatorname{tg}(\phi - \pi/\gamma) = \operatorname{tg}\phi_0$, or $\phi - \phi_0 = -\pi k + \pi/\gamma$. Since the rod rotates clockwise, the angle increment is negative, and $k = 1$ is to be chosen.

• $\mathbf{H}_{41}$ *Small angle of rotation* (p. 265).
The vector product (10.31) contains a small $d\mathbf{v}$; therefore, only values of zeroth with respect to $d\mathbf{v}$ contribute to the multiplier next to it: $\gamma_1 = \gamma_2 = \gamma = 1/\sqrt{1 - \mathbf{v}^2}$, $\gamma_w = \gamma^2(1 - \mathbf{v}^2) = 1$.

• $\mathbf{H}_{42}$ *Averaging coordinates* (p. 272).
Let the functions $x = x(\phi)$ and $y = y(\phi)$ be defined implicitly as:

$$x = \cos(\phi - \sigma x), \qquad y = \sin(\phi - \sigma x), \tag{8}$$

where $\sigma = const$. Obviously, $x^2 + y^2 = 1$. The functions $x(\phi)$ and $y(\phi)$ are periodic, with period 2π. The averages (10.51) of these functions satisfy the relations: $\langle f' \rangle = 0$ and $\langle f g' \rangle = -\langle g f' \rangle$, where the prime is the derivative with respect to ϕ. Differentiate (8) with respect to ϕ:

$$x' = -y + \sigma y x', \qquad y' = x - \sigma x x'. \tag{9}$$

Since $xx' = (x^2)'/2$, the average of the second equation gives $\langle x \rangle = 0$. Multiplying the second equation of (9) by x and calculating again its average value: $\langle x^2 \rangle = \langle x y' \rangle$. Similarly, by averaging the first equation (9), we have:

$$\langle y \rangle = \sigma \langle y x' \rangle = -\sigma \langle x y' \rangle = -\sigma \langle x^2 \rangle.$$

We now multiply the first equation (9) by x and y:

$$\langle xy \rangle = \sigma \langle y x x' \rangle = \frac{\sigma}{2} \langle y (x^2)' \rangle = -\frac{\sigma}{2} \langle y (y^2)' \rangle = 0.$$

$$\langle y^2 \rangle = \sigma \langle y^2 x' \rangle - \langle y x' \rangle = -\sigma \langle x (y^2)' \rangle - \langle y x' \rangle = \sigma \langle x (x^2)' \rangle - \langle y x' \rangle = \langle x y' \rangle,$$

whence $\langle y^2 \rangle = \langle x^2 \rangle = 1/2$, $\langle y \rangle = -\sigma/2$.

Quaternions

• $\mathbf{H}_{43}$ $\mathbb{SAS}^{+}$ *transformation* (p. 276).
Hermitian conjugation interchanges the order of matrices, $(\mathbb{AB})^{+} = \mathbb{B}^{+}\mathbb{A}^{+}$, and $(\mathbb{A}^{+})^{+} = \mathbb{A}$. Therefore, we have $(\mathbb{SAS}^{+})^{+} = (\mathbb{S}^{+})^{+}\mathbb{A}^{+}\mathbb{S}^{+} = \mathbb{SAS}^{+}$.

• $\mathbf{H}_{44}$ *The algebra of Pauli matrices* (p. 277).

$$\sigma_1\sigma_2 = \begin{pmatrix} 0 & 1 \\ 1 & 0 \end{pmatrix}\begin{pmatrix} 0 & -\imath \\ \imath & 0 \end{pmatrix} = \begin{pmatrix} \imath & 0 \\ 0 & -\imath \end{pmatrix} = \imath\sigma_3.$$

Also, $\varepsilon_{12k}\sigma_k = \varepsilon_{123}\sigma_3 = \sigma_3$, since the antisymmetric Levi-Civita tensor differs from zero only for different indices, and $\varepsilon_{123} = 1$.

• $\mathbf{H}_{45}$ *Identity involving Pauli matrices* (p. 277).
The required identity is obtained by contracting (11.7) with a_i and b_j and using the definition of the vector product $[\mathbf{a}\times\mathbf{b}]_k = \varepsilon_{ijk}a_i b_j = \varepsilon_{kij}a_i b_j$.

• $\mathbf{H}_{46}$ *The conjugate property* $\overline{\mathbb{AB}} = \bar{\mathbb{B}}\,\bar{\mathbb{A}}$ (p. 279).
Since $\bar{\mathbb{B}} = \{b_0,\ -\mathbf{b}\}$, $\bar{\mathbb{A}} = \{a_0,\ -\mathbf{a}\}$, we have:

$$\bar{\mathbb{B}}\,\bar{\mathbb{A}} = \{a_0 b_0 + \mathbf{ab}, -a_0\mathbf{b} - b_0\mathbf{a} + \imath\mathbf{b}\times\mathbf{a}\}.$$

Interchanging the multipliers in $\mathbf{b}\times\mathbf{a}$, we have: $\bar{\mathbb{B}}\,\bar{\mathbb{A}} = \overline{\mathbb{AB}}$, since its vector part is of opposite sign to $\mathbb{AB}$.

• $\mathbf{H}_{47}$ *A boost resulting from the product of two boosts* (p. 290).
From $\mathbf{nm} = 0$ and the relations

$$c_\alpha c_\phi = c_1 c_2 + (\mathbf{m}_1\mathbf{m}_2)\, s_1 s_2, \qquad \mathbf{n}\, c_\alpha s_\phi = -[\mathbf{m}_1\times\mathbf{m}_2]\, s_1 s_2,$$

$$\mathbf{m} s_\alpha c_\phi + [\mathbf{m}\times\mathbf{n}]\, s_\alpha s_\phi = \mathbf{m}_1\, s_1 c_2 + \mathbf{m}_2\, c_1 s_2$$

α and $\mathbf{m}$ can be found. Since $\mathbf{n}$ and $\mathbf{m}$ are perpendicular, $\mathbf{m}\times\mathbf{n}$ is a unit vector. Squaring the third ratio, we have

$$s_\alpha^2 c_\phi^2 + s_\alpha^2 s_\phi^2 = s_\alpha^2 = s_1^2 c_2^2 + c_1^2 s_2^2 + 2\mathbf{m}_1\mathbf{m}_2\, s_1 c_1 s_2 c_2.$$

Since $s_\alpha = \operatorname{sh}(\alpha/2)$ is a hyperbolic sine, we obtain $\operatorname{ch}\alpha = 1 + 2s_\alpha^2$:

$$\operatorname{ch}\alpha = 1 + 2s_1^2 c_2^2 + 2c_1^2 s_2^2 + 4(\mathbf{m}_1\mathbf{m}_2)s_1 c_1 s_2 c_2 = \operatorname{ch}\alpha_1 \operatorname{ch}\alpha_2 + (\mathbf{m}_1\mathbf{m}_2)\operatorname{sh}\alpha_1 \operatorname{sh}\alpha_2,$$

where we made use of the relation $1 + 2s_1^2 c_2^2 + 2c_1^2 s_2^2 = (c_1^2 + s_1^2)(c_2^2 + s_2^2)$. Taking the vector product of the third relation with $\mathbf{n}$, using that $\mathbf{n}\times[\mathbf{m}\times\mathbf{n}] = \mathbf{m}$ for $\mathbf{nm} = 0$, and removing $\mathbf{m}\times\mathbf{n}$ (using again the third relation), we find that:

$$\mathbf{m}\, s_\alpha = \mathbf{m}_1 s_1 c_2 c_\phi + \mathbf{m}_2 c_1 s_2 c_\phi + [\mathbf{n}\times\mathbf{m}_1] s_1 c_2 s_\phi + [\mathbf{n}\times\mathbf{m}_2] c_1 s_2 s_\phi.$$

Substituting in the above expressions for c_ϕ and $\mathbf{n}s_\phi$, we obtain

$$\mathbf{m}\operatorname{sh}\alpha = \mathbf{m}_1 \operatorname{sh}\alpha_1 \operatorname{ch}\alpha_2 + \mathbf{m}_2 \operatorname{sh}\alpha_2 + \mathbf{m}_1(\mathbf{m}_1\mathbf{m}_2)\operatorname{sh}\alpha_2(\operatorname{ch}\alpha_1 - 1).$$

• $\mathbf{H}_{48}$ *Angle and axis of rotation of the product of two boosts* (p. 290).

$$c_\alpha c_\phi = c_1 c_2 + (\mathbf{m}_1\mathbf{m}_2)\, s_1 s_2, \qquad \mathbf{n}\, c_\alpha s_\phi = -[\mathbf{m}_1 \times \mathbf{m}_2]\, s_1 s_2.$$

From the second relation it follows that $\mathbf{n} \sim -\mathbf{m}_1 \times \mathbf{m}_2$, and since $\mathbf{n}$ is a unit vector, we have for $s_\phi > 0$ (c_α, s_1, s_2 is always greater than zero):

$$\mathbf{n} = -[\mathbf{m}_1 \times \mathbf{m}_2]/|\mathbf{m}_1 \times \mathbf{m}_2|.$$

Multiplying the first and the second relations, we obtain

$$\mathbf{n}\, c_\alpha^2 c_\phi s_\phi = -[\mathbf{m}_1 \times \mathbf{m}_2]\,(s_1 c_1 s_2 c_2 + \mathbf{m}_1\mathbf{m}_2\, s_1^2 s_2^2).$$

Multiplying this expression by four, and given that the identities $2c_\alpha^2 = \operatorname{ch}\alpha + 1$, $2s_\alpha^2 = \operatorname{ch}\alpha - 1$ are true for $c_\alpha = \operatorname{ch}(\alpha/2)$, we find:

$$\mathbf{n}\, \sin\phi\,(\operatorname{ch}\alpha + 1) = -[\mathbf{m}_1 \times \mathbf{m}_2]\,(\operatorname{sh}\alpha_1 \operatorname{sh}\alpha_2 + \mathbf{m}_1\mathbf{m}_1\,(\operatorname{ch}\alpha_1 - 1)(\operatorname{ch}\alpha_2 - 1)).$$

Substituting in the velocities $\mathbf{m}\operatorname{sh}\alpha = \mathbf{v}\gamma$ and $\operatorname{ch}\alpha = \gamma$, and removing $\mathbf{v}_1\mathbf{v}_2$ using the formula $\gamma = \gamma_1\gamma_2(1 + \mathbf{v}_1\mathbf{v}_2)$, we obtain the required result.

• $\mathbf{H}_{49}$ *The inverse product* $\mathbb{L}\mathbb{R} = (c_\alpha - s_\alpha\, \mathbf{m}\boldsymbol{\sigma})\,(c_\phi + \imath s_\phi\, \mathbf{n}\boldsymbol{\sigma})$ (p. 291).
Multiplying the quaternions, we have

$$\mathbb{L}\,\mathbb{R} = c_\alpha c_\phi - \imath\, \mathbf{m}\mathbf{n}\, s_\alpha s_\phi, +(\imath\mathbf{n}\, c_\alpha s_\phi - \mathbf{m} s_\alpha c_\phi + [\mathbf{m} \times \mathbf{n}]\, s_\alpha s_\phi)\boldsymbol{\sigma}.$$

By comparing this with $\mathbb{L}_2\mathbb{L}_1$ (p. 290), we see that the resulting velocity and the rotation axis are *perpendicular* ($\mathbf{n}\mathbf{m} = 0$), and the following equations are true:

$$c_\alpha c_\phi = c_1 c_2 + (\mathbf{m}_1\mathbf{m}_2)\, s_1 s_2, \qquad \mathbf{n}\, c_\alpha s_\phi = -[\mathbf{m}_1 \times \mathbf{m}_2]\, s_1 s_2,$$

$$\mathbf{m} s_\alpha c_\phi - [\mathbf{m} \times \mathbf{n}]\, s_\alpha s_\phi = \mathbf{m}_1\, s_1 c_2 + \mathbf{m}_2\, c_1 s_2.$$

They coincide with the relations of the problem ($\lessdot \mathrm{H}_{47}$) after the vector $\mathbf{n}$ has been substituted by its negative $\mathbf{n} \mapsto -\mathbf{n}$, and the indices 1 and 2 of the values $\alpha_i$and $\mathbf{m}_i$ have been interchanged. As a result, the expressions for the axis $\mathbf{n}$, angle of rotation ϕ, and rapidity α (or γ) will not change. However, the velocities are to be interchanged in the velocity expression of the final boost $\mathbf{v}$ (or the vector $\mathbf{m}$).

• $\mathbf{H}_{50}$ *The condition* $\mathbb{L}\bar{\mathbb{L}} = \mathbb{I}$ (p. 291).
Calculate the product of the numerators, given that $\mathbb{S}\bar{\mathbb{S}} = \bar{\mathbb{S}}\mathbb{S} = \mathbb{I}$, and similarly for the Hermitian conjugate:

$$(\mathbb{S} + \mathbb{S}^+)(\bar{\mathbb{S}} + \bar{\mathbb{S}}^+) = 2 \cdot \mathbb{I} + \mathbb{S}^+\bar{\mathbb{S}} + \mathbb{S}\bar{\mathbb{S}}^+.$$

On the other hand, we have for the denominator:

$$(\mathbb{I} + \mathbb{S}\bar{\mathbb{S}}^+)\,\overline{(\mathbb{I} + \mathbb{S}\bar{\mathbb{S}}^+)} = (\mathbb{I} + \mathbb{S}\bar{\mathbb{S}}^+)\,(\mathbb{I} + \mathbb{S}^+\bar{\mathbb{S}}) = 2\mathbb{I} + \mathbb{S}^+\bar{\mathbb{S}} + \mathbb{S}\bar{\mathbb{S}}^+ = |\mathbb{I} + \mathbb{S}^+\bar{\mathbb{S}}|^2\,\mathbb{I},$$

which was to be proved.

• **H**$_{51}$ *The quaternion* $\mathbb{R} = \mathbb{S}\bar{\mathbb{L}}$ (p. 291).

$$\mathbb{R} = \mathbb{S}\bar{\mathbb{L}} = \frac{\mathbb{S}(\bar{\mathbb{S}} + \bar{\mathbb{S}}^+)}{|\mathbb{I} + \mathbb{S}\bar{\mathbb{S}}^+|} = \frac{\mathbb{I} + \mathbb{S}\bar{\mathbb{S}}^+}{|\mathbb{I} + \mathbb{S}\bar{\mathbb{S}}^+|}.$$

The fact that this quaternion has no pure imaginary vector part can be verified by direct multiplication:

$$\mathbb{S}\bar{\mathbb{S}}^+ = (S_0 + \mathbf{S}\boldsymbol{\sigma})(S_0^* - \mathbf{S}^*\boldsymbol{\sigma}) = |S_0|^2 - \mathbf{S}\mathbf{S}^* + (\mathbf{S}S_0^* - \mathbf{S}^* S_0)\boldsymbol{\sigma} - \imath[\mathbf{S} \times \mathbf{S}^*]\boldsymbol{\sigma}$$

(the last term in this expression is real). Besides this, the denominator in $\mathbb{R}$ is obviously real (see < H$_{50}$).

• **H**$_{52}$ *Expressing* $\Lambda^{\alpha}{}_{\beta}$ *in terms of the quaternion parameters* s^{ν} (p. 300).
From the conjugate (11.74) and given that matrices σ_μ are Hermitian, we have:

$$\Lambda_{\alpha}^{\ \mu}\, \bar{\sigma}_\mu = \mathbb{S}^+\, \bar{\sigma}_\alpha\, \mathbb{S} = s^{*\mu} s^{\nu}\, \sigma_\mu \bar{\sigma}_\alpha \sigma_\nu.$$

Multiplying this from right by $\bar{\sigma}_\beta$ and taking the trace on both sides:

$$\Lambda_{\alpha}^{\ \mu}\ \mathrm{Tr}(\bar{\sigma}_\mu \bar{\sigma}_\beta) = s^{*\mu} s^{\nu}\ \mathrm{Tr}(\sigma_\mu \bar{\sigma}_\alpha \sigma_\nu \bar{\sigma}_\beta).$$

Using the traces (11.81) and (11.85), we finally obtain:

$$\Lambda_{\alpha}^{\ \beta} = s_\alpha^* s_\beta + s_\beta^* s_\alpha - g_{\alpha\beta}\, s_\mu^* s^\mu + \imath\, \varepsilon_{\alpha\beta\mu\nu}\, s^\mu s^{*\nu}. \tag{10}$$

Since the trace of the matrix $\bar{\sigma}_\mu \bar{\sigma}_\beta$ is proportional to $\delta_{\mu\beta}$ rather than to $g_{\mu\beta}$, this relation is not completely covariant. However, its values are correct for specific indices. We suggest (< H$_{55}$) writing down the coefficients $\Lambda^{\beta}{}_{\alpha}$ and finding the conditions under which they are symmetric: $\Lambda^{\beta}{}_{\alpha} = \Lambda^{\alpha}{}_{\beta}$.

• **H**$_{53}$ *The* $\mathbb{A}\bar{\mathbb{B}}\mathbb{C}$ *product* (p. 301).
Consider three quaternions

$$\mathbb{A} = a^\mu \sigma_\mu = a^0 + \mathbf{a}\boldsymbol{\sigma}, \qquad \bar{\mathbb{B}} = b^\mu \bar{\sigma}_\mu = b^0 - \mathbf{b}\boldsymbol{\sigma}, \qquad \mathbb{C} = c^\mu \sigma_\mu = c^0 + \mathbf{c}\boldsymbol{\sigma},$$

and write down their product:

$$\mathbb{A}\bar{\mathbb{B}}\mathbb{C} = \{\mathrm{a} \cdot \mathrm{b} + (b^0\mathbf{a} - a^0\mathbf{b})\boldsymbol{\sigma} - \imath[\mathbf{a} \times \mathbf{b}]\boldsymbol{\sigma}\}(c^0 + \mathbf{c}\boldsymbol{\sigma}),$$

where $\mathrm{a} \cdot \mathrm{b} = a^0 b^0 - \mathbf{ab}$. Multiplying out the parentheses, we have:

$$\begin{aligned} \mathbb{A}\bar{\mathbb{B}}\mathbb{C} &= (\mathrm{a} \cdot \mathrm{b})\,\mathbb{C} + c^0 b^0 \mathbf{a}\boldsymbol{\sigma} - c^0 a^0 \mathbf{b}\boldsymbol{\sigma} + b^0(\mathbf{ac}) - a^0(\mathbf{bc}) + [[\mathbf{a} \times \mathbf{b}] \times \mathbf{c}]\boldsymbol{\sigma} \\ &- \imath c^0[\mathbf{a} \times \mathbf{b}]\boldsymbol{\sigma} + \imath b^0[\mathbf{a} \times \mathbf{c}]\boldsymbol{\sigma} - \imath a^0[\mathbf{b} \times \mathbf{c}]\boldsymbol{\sigma} - \imath[\mathbf{a} \times \mathbf{b}]\mathbf{c}. \end{aligned}$$

Expanding the vector triple product $[\mathbf{a} \times \mathbf{b}] \times \mathbf{c} = \mathbf{b}(\mathbf{ac}) - \mathbf{a}(\mathbf{bc})$ and adding $a^0 b^0 c^0 - a^0 b^0 c^0$, we obtain:

$$\begin{aligned} \mathbb{A}\bar{\mathbb{B}}\mathbb{C} &= (\mathrm{a} \cdot \mathrm{b})\,\mathbb{C} - (\mathrm{a} \cdot \mathrm{c})\,\mathbb{B} + (\mathrm{b} \cdot \mathrm{c})\,\mathbb{A} \\ &- \imath c^0[\mathbf{a} \times \mathbf{b}]\boldsymbol{\sigma} + \imath b^0[\mathbf{a} \times \mathbf{c}]\boldsymbol{\sigma} - \imath a^0[\mathbf{b} \times \mathbf{c}]\boldsymbol{\sigma} - \imath[\mathbf{a} \times \mathbf{b}]\mathbf{c}. \end{aligned}$$

The last four terms can be contracted (< H$_{54}$) into a single expression using the Levi-Civita symbol.

• **H**$_{54}$ *The* $\mathbb{A}\bar{\mathbb{B}}\mathbb{C}$ *product* (p. 381).
Write down the sum over the first index in the following contraction:

$$\varepsilon_{\mu\nu\sigma\tau}a^\mu b^\nu c^\sigma d^\tau = \varepsilon_{0\nu\sigma\tau}a^0 b^\nu c^\sigma d^\tau + \varepsilon_{i\nu\sigma\tau}a^i b^\nu c^\sigma d^\tau.$$

Since all indices must be different in the Levi-Civita four-symbol, the indices ν, σ and τ in the first term are not zero and can be replaced by Latin indices ranging over 1,2,3. Similarly, consider setting each of the four-indices of the second term to zero:

$$\varepsilon_{\mu\nu\sigma\tau}a^\mu b^\nu c^\sigma d^\tau = \varepsilon_{0ijk}a^0 b^i c^j d^k + \varepsilon_{i0jk}a^i b^0 c^j d^k + \varepsilon_{ij0k}a^i b^j c^0 d^k + \varepsilon_{ijk0}a^i b^j c^k d^0.$$

Since $\varepsilon_{0123} = 1$, and this symbol is antisymmetric for all indices, the following relationship between the four- and three-dimensional Levi-Civita symbols is obviously true : $\varepsilon_{0ijk} = \varepsilon_{ijk}$. Therefore:

$$\varepsilon_{\mu\nu\sigma\tau}a^\mu b^\nu c^\sigma d^\tau = a^0\,[\mathbf{b}\times\mathbf{c}]\mathbf{d} - b^0\,[\mathbf{a}\times\mathbf{c}]\mathbf{d} + c^0\,[\mathbf{a}\times\mathbf{b}]\mathbf{d} - d^0\,[\mathbf{a}\times\mathbf{b}]\mathbf{c}.$$

Replacing d^μ by $\sigma^\mu = \{1, -\boldsymbol{\sigma}\}$ and comparing this expression with the last four terms in the product $\mathbb{A}\bar{\mathbb{B}}\mathbb{C}$, we finally obtain:

$$\mathbb{A}\bar{\mathbb{B}}\mathbb{C} = (\mathrm{a}\cdot\mathrm{b})\,\mathbb{C} - (\mathrm{a}\cdot\mathrm{c})\,\mathbb{B} + (\mathrm{b}\cdot\mathrm{c})\,\mathbb{A} + \imath\varepsilon_{\alpha\beta\gamma\mu}a^\alpha b^\beta c^\gamma \sigma^\mu.$$

We note that this covariant form is due to the presence of the conjugate of the quaternion $\mathbb{B}$ in the product of the three quaternions on the left-hand side, while on the right-hand side $\mathbb{B}$ is not conjugated.

• **H**$_{55}$ *Elements of the matrix of Lorentz transformations* (p. 381).

$$\Lambda^0{}_0 = |s^0|^2 + |\mathbf{s}|^2,$$

$$\Lambda^i{}_0 = (s^{*0}\mathbf{s} + s^0\mathbf{s}^* + \imath\,\mathbf{s}\times\mathbf{s}^*)^i, \qquad \Lambda^0{}_i = (s^{*0}\mathbf{s} + s^0\mathbf{s}^* - \imath\,\mathbf{s}\times\mathbf{s}^*)^i,$$

$$\Lambda^i{}_j = s^{*i}s^j + s^{*j}s^i + \delta^i_j(|s^0|^2 - |\mathbf{s}|^2) + \imath\,\varepsilon_{ijk}\,(s^0 s^{*k} - s^{*0}s^k),$$

where the indices are interchanged: $\Lambda_0{}^0 = \Lambda^0{}_0$ and $\Lambda_0{}^i = -\Lambda^0{}_i$, $\Lambda_i{}^j = \Lambda^i{}_j$. The latter relation can be rewritten in a more compact form:

$$\Lambda^i{}_j = 2\Re(s^{*i}s^j) + \delta^i_j \mathrm{s}\cdot\mathrm{s}^* + 2\imath\,\varepsilon_{ijk}\,\Im(s^0 s^{*k}).$$

It is worth writing down the matrix $\Lambda_\alpha{}^\beta$ for the vector $s^\mu = \{\mathrm{ch}(\alpha/2), -\mathbf{m}\,\mathrm{sh}(\alpha/2)\}$, where $\mathrm{ch}(\alpha) = \gamma$, $\mathbf{m}\,\mathrm{sh}(\alpha) = \mathbf{v}\gamma$, and comparing it with the matrix on p. 180.

The elements of the matrix $\Lambda^0{}_i$ will be symmetric $\Lambda^0{}_i = \Lambda^i{}_0$, if the vector $\mathbf{s}$ is either real, or purely imaginary, or if $\Im\mathbf{s}$ is parallel to $\Re\mathbf{s}$. The elements $\Lambda^i{}_j$ are symmetric if $\Im(s^0 s^{*k}) = 0$.

Velocity space

• H_{56} *Projections in a spherical right triangle* (p. 305).
Renaming the sides and the angles in the law of cosines, we have:

$$\cos a = \cos c \cos b + \sin c \sin b \cos \alpha.$$

Substituting $\cos a = \cos c / \cos b$ from the "Pythagorean theorem", we obtain:

$$\sin c \sin b \cos \alpha = \frac{\cos c(1 - \cos^2 b)}{\cos b} = \frac{\cos c \sin^2 b}{\cos b},$$

whence $\operatorname{tg} b = \operatorname{tg} c \cos \alpha$. Similarly for the second relation.

• H_{57} *The law of sines* (p. 305).
Draw a line CD normal to the side AB of an arbitrary triangle and use the relations for two right triangles CDA and CDB:

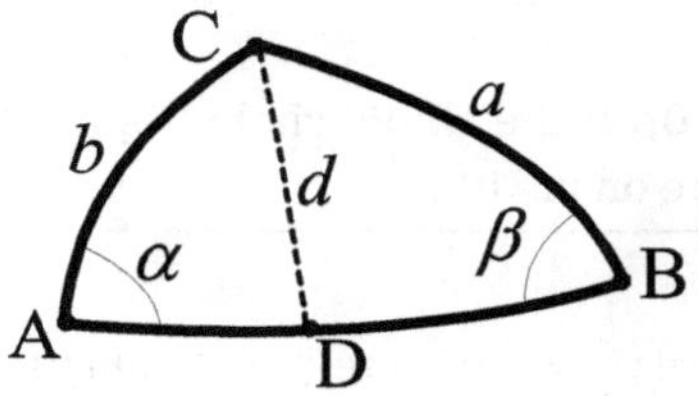

$$\sin d = \sin b \sin \alpha,$$
$$\sin d = \sin a \sin \beta.$$

$$\Rightarrow \quad \frac{\sin a}{\sin \alpha} = \frac{\sin b}{\sin \beta}.$$

• H_{58} *Hypotenuse of a spherical triangle in terms of two angles* (p. 306).
For a right spherical triangle we have:

$$\operatorname{tg} \alpha = \frac{\sin \alpha}{\cos \alpha} = \frac{\sin a / \sin c}{\operatorname{tg} a / \operatorname{tg} c} = \frac{\cos a}{\cos c}.$$

Therefore:

$$\operatorname{tg} \alpha \operatorname{tg} \beta = \frac{\cos a \cos b}{\cos^2 c} = \frac{\cos c}{\cos^2 c} = \frac{1}{\cos c},$$

by the spherical Pythagorean theorem.

• H_{59} *Dual triangle* (p. 306).
For a given line on a sphere (the intersection of a sphere and a large disk passing through its center) a *pole* is a point at a distance of $\pi/2$ from the points of the line (special cases are the north and south poles with respect to the equator). For a given spherical triangle we can always construct a *polar triangle* whose vertices are the poles of large circles passing through the sides of the triangle (in the direction of the third vertex). The angles and sides of a spherical triangle are complemented by the sides and angles of the corresponding polar triangle with respect to the angle π.

• $\mathbf{H}_{60}$ *Area of a disk on a sphere* (p. 308).
Let the center of a disk be at the north pole of a sphere. Then its area is

$$S=\int_0^r\int_0^{2\pi} s_\chi\, d\chi d\phi=-2\pi c_\chi\Big|_0^r=2\pi(1-\cos r)\,.$$

The limits of integration with respect to χ correspond to moving from 0 to $\chi = r$ along the radius of the circle.

• $\mathbf{H}_{61}$ *Circles under Beltrami projection* (p. 310).
Direct the axes so that $\mathbf{n}=\{n_x,0,n_z\}$. From the equation $\mathbf{nr}+\mathbf{nk}=\sqrt{1+r^2}\cos R$ we have

$$\frac{(c^2-n_x^2)^2}{c^2s^2}\Big(x-\frac{n_xn_z}{c^2-n_x^2}\Big)^2+\frac{c^2-n_x^2}{s^2}\,y^2=1,$$

where $c=\cos R$, $s=\sin R$ and $\mathbf{r}=\{x,y,0\}$. If the whole circle lies on the upper hemisphere, then $c^2>n_x^2$ (see the third figure on p. 310).

• $\mathbf{H}_{62}$ *Orthogonality* (p. 311).
Let us differentiate the equation of the sphere $\mathbf{p}^2=1$ with respect to length l along a line on it: $2\,\mathbf{p}\,d\mathbf{p}/dl=0$. Therefore, we have that the vectors $\mathbf{p}$ and $\dot{\mathbf{p}}$ are orthogonal, i.e. that $\dot{\mathbf{p}}$ is tangent to the sphere. Furthermore, by definition we have $dl^2=(d\mathbf{p})^2$, so $\dot{\mathbf{p}}$ is a unit vector.

• $\mathbf{H}_{63}$ *Invariance of angles* (p. 311).

$$\dot{\mathbf{r}}_1\dot{\mathbf{r}}_2=\Big(\frac{\dot{\mathbf{p}}_1}{1+\mathbf{kp}}-\frac{(\mathbf{p}+\mathbf{k})\,(\mathbf{k}\dot{\mathbf{p}}_1)}{(1+\mathbf{kp})^2}\Big)\Big(\frac{\dot{\mathbf{p}}_2}{1+\mathbf{kp}}-\frac{(\mathbf{p}+\mathbf{k})\,(\mathbf{k}\dot{\mathbf{p}}_2)}{(1+\mathbf{kp})^2}\Big),$$

where $\mathbf{p}=\mathbf{p}_1(0)=\mathbf{p}_2(0)$ is the point of intersection of the curves, and $\dot{\mathbf{r}}_i=\dot{\mathbf{r}}_i(0)$, $\dot{\mathbf{p}}_i=\dot{\mathbf{p}}_i(0)$. Multiply out the parentheses:

$$\dot{\mathbf{r}}_1\dot{\mathbf{r}}_2=\frac{\dot{\mathbf{p}}_1\dot{\mathbf{p}}_2}{(1+\mathbf{kp})^2}-2\,\frac{(\mathbf{k}\dot{\mathbf{p}}_1)(\mathbf{k}\dot{\mathbf{p}}_2)}{(1+\mathbf{kp})^3}+\frac{(\mathbf{p}+\mathbf{k})^2}{(1+\mathbf{kp})^4}\,(\mathbf{k}\dot{\mathbf{p}}_1)\,(\mathbf{k}\dot{\mathbf{p}}_2).$$

Given that $\mathbf{p}\dot{\mathbf{p}}_i=0$, $\mathbf{p}^2=\mathbf{k}^2=1$, we obtain:

$$\dot{\mathbf{r}}_1\dot{\mathbf{r}}_2=\frac{\dot{\mathbf{p}}_1\dot{\mathbf{p}}_2}{(1+\mathbf{kp})^2},$$

It only remains to use the second relation (12.21), p. 311, for the magnitudes of the vectors $|\dot{\mathbf{r}}_i|=1/(1+\mathbf{kp})$.

• $\mathbf{H}_{64}$ *Orthogonality of displacements* (p. 313).
Let us write the relation between four-dimensional Cartesian coordinates x, y, z, w and spherical coordinates χ, θ, ϕ, for a point that stays on the surface of a three-sphere:

$$x = s_\chi\, s_\theta\, c_\phi, \quad y = s_\chi\, s_\theta\, s_\phi, \quad z = s_\chi\, c_\theta, \quad w = c_\chi.$$

The components of the radius vector are $\mathbf{r} = \{x, y, z, w\}$. Its partial derivatives with respect to the coordinates are displacements tangent to the sphere:

$$\begin{aligned}
\mathbf{e}_\chi &= \frac{\partial \mathbf{r}}{\partial \chi} = \{\ c_\chi\, s_\theta\, c_\phi,\quad c_\chi\, s_\theta\, s_\phi,\quad c_\chi\, c_\theta,\quad -s_\chi\ \},\\
\mathbf{e}_\theta &= \frac{\partial \mathbf{r}}{\partial \theta} = \{\ s_\chi\, c_\theta\, c_\phi,\quad s_\chi\, c_\theta\, s_\phi,\quad -s_\chi\, s_\theta,\quad 0\ \},\\
\mathbf{e}_\phi &= \frac{\partial \mathbf{r}}{\partial \phi} = \{-s_\chi\, s_\theta\, s_\phi,\quad s_\chi\, s_\theta\, c_\phi,\quad -s_\chi\, s_\theta,\quad 0\ \}.
\end{aligned}$$

Direct multiplication of these vectors verifies that:

$$\mathbf{e}_\chi \mathbf{e}_\theta = \mathbf{e}_\chi \mathbf{e}_\phi = \mathbf{e}_\theta \mathbf{e}_\phi = 0, \qquad \mathbf{e}_\phi^2 = 1, \qquad \mathbf{e}_\phi^2 = s_\chi^2, \quad \mathbf{e}_\phi^2 = s_\chi^2 s_\theta^2.$$

The scalar products of the tangent vectors $\mathbf{e}_k$ (k is the number rather than the component of the vector) are the coefficients of the metric tensor:

$$d\mathbf{r} = \frac{\partial \mathbf{r}}{\partial q^k}\, dq^k = \mathbf{e}_k q^k, \qquad \Rightarrow \qquad dl^2 = d\mathbf{r}^2 = \mathbf{e}_i \mathbf{e}_j dq^i dq^j = g_{ij} dq^i dq^j,$$

where the $q^k = \{q^1, q^2, q^3\}$ are coordinates (e.g., χ, θ, ϕ), and summation is performed with respect to k, i, j. If the metric tensor is diagonal, the tangent vectors are orthogonal, and vice versa.

• $\mathbf{H}_{65}$ *The law of cosines in the Lobachevsky geometry* (p. 315).
Restore the constant λ in the law of cosines (12.4), p. 305:

$$\cos(c/\lambda) = \cos(a/\lambda)\, \cos(b/\lambda) + \sin(a/\lambda)\, \sin(b/\lambda)\, \cos\gamma.$$

The angle γ is a dimensionless quantity, therefore, no substitution is used in the latter cosine. Substituting $\lambda \mapsto \imath\lambda$, we obtain

$$\mathrm{ch}(c/\lambda) = \mathrm{ch}(a/\lambda)\, \mathrm{ch}(b/\lambda) + \imath\, \mathrm{sh}(a/\lambda)\, \imath\, \mathrm{sh}(b/\lambda)\, \cos\gamma.$$

Using that $\imath^2 = -1$, and assuming $\lambda = 1$, we obtain the final result.

• $\mathbf{H}_{66}$ *A line passing through two points* (p. 323).
To draw such a picture, we need to have the parameters of a circle (a line under stereographic projection) passing through two points $\mathbf{r}_1 = \{x_1, y_1\}$ and $\mathbf{r}_2 = \{x_2, y_2\}$ in the projection plane. Using equation (12.51), p. 323, we obtain:

$$\frac{n_x}{n_0} = \frac{1}{2}\, \frac{(r_1^2 + 1)\, y_2 - (r_2^2 + 1)\, y_1}{x_1 y_2 - x_2 y_1}, \qquad \frac{n_y}{n_0} = -\frac{1}{2}\, \frac{(r_1^2 + 1)\, x_2 - (r_2^2 + 1)\, x_1}{x_1 y_2 - x_2 y_1}.$$

Therefore, we have: $R^2 = (n_x/n_0)^2 + (n_y/n_0)^2$ and $n_0 = 1/\sqrt{R^2 - 1}$.

Bibliography

[1] Stepanov S.S. – "*Vectors, tensors and forms. Instructions for use*" (2010)

[2] Poincaré A. – "*The Value of Science*", New York (1958)

[3] Galilei, Galileo – "*Dialogue Concerning the Two Chief World Systems*", Translated by Drake, Stillman. Berkeley, CA: University of California Press. (1953)

[4] Newton, Isaac – "*The Principia: Mathematical Principles of Natural Philosophy*", University of California Press, (1999)

[5] Einstein, Albert – "*Zur Elektrodynamik bewegter Körper*", Annalen der Physik. 17 (10) 891-921 (1905)

[6] Einstein, Albert – "*The meaning of relativity*", Princeton Univ. Press. Princeton, N. Y., 1921.

[7] Born M. – *Die Theorie des starren Elektrons in der Kinematik des Relativitätsprinzips*, Annalen der Physik 335 (11), 1-56 (1909)

[8] Ginzburg V. "*How and who created the special theory of relativity*", Nauka (1976)

[9] "*The Principle of Relativity*", Atomizdat (1973), (ru)

[10] W. A. von Ignatowsky – "*Einige allgemeine Bemerkungen zum Relativitätsprinzip*" Berichte der Deutschen Physikalischen Gesellschaft, p. 788 ff. (1910), Archiv der Mathematik und Physik, 17. p. 1 ff. (1910).

[11] W. A. von Ignatowsky – "*Einige allgemeine Bemerkungen über das Relativitätsprinzip*", Physikalische Zeitschrift (Phys.Z.) 11, (1910), S. 972-976,

[12] P. Frank and H. Rothe – "*Öber die Transformation der Raumzeitkoordinaten von ruhenden auf bewegte Systeme*", Ann. Phys. 34, 825-853 (1911).

[13] Bailey J. et al. – "*Measurements of relativistic time dilatation for positive and negative muons in circular orbit*", Nature, v.268, p.301-305 (1977)

[14] Manida S.N. – "*Fock-Lorentz transformation and time-varying speed of light.*" arXiv:: gr-qc/9905046 (1999)

[15] Stepanov S.S. – "*Fundamental physical constants and the principle of parametric incompleteness*" arXiv: physics/9909009, (1999)

[16] Stepanov S.S. – "*A time-space varying speed of light and the Hubble Law in static Universe*", Phys. Rev. D 62 (2000) 023507, arXiv: astro-ph/9909311 (1999).

[17] Pauli W. – "*Theory of relativity*", Dover Publications (1981)

[18] Möller C. – "*The Theory of Relativity*" (3rd ed.), Oxford University Press (1952)

[19] Logunov A.A. – "*Lectures on the theory of relativity and gravitation: A modern analysis of the problem*" Nauka (1987) (ru)

[20] Fock V. – "*The Theory of Space, Time and Gravitation*", Pergamon Press (1964)

[21] Jackson J. D. – "*Classical Electrodynamics*", John Wiley & Sons (1962)

[22] Thomas L.H. ""*Motion of the spinning electron*", Nature, **117**, 514 (1926)

[23] Stapp H.P. – "*Relativistic Theory of Polarization Phenomena*", Phys.Rev. **103**, 2, pp.425-434, (1956)

[24] Wigner E.P. – Rev.Mod.Phys., **29**, p.255, (1957)

[25] Stepanov S.S. – "*Thomas precession for spin and for a rod*" Phys.Part.Nuclei (2012) 43: 128.

https://doi.org/10.1515/9783110515886-016

Index

https://doi.org/10.1515/9783110515886-017

www.ingramcontent.com/pod-product-compliance
Lightning Source LLC
Chambersburg PA
CBHW082104220326
41598CB00066BA/5247